Developments in Theoretical and Applied Mechanics

VOLUME 1

Proceedings of the First Southeastern Conference on Theoretical and Applied Mechanics

DEVELOPMENTS IN THEORETICAL AND APPLIED MECHANICS

VOLUME 1

Proceedings of the First Southeastern Conference on
Theoretical and Applied Mechanics
held at
Gatlinburg, Tennessee
May 3–4, 1962

SPRINGER SCIENCE+BUSINESS MEDIA, LLC 1963

ISBN 978-1-4899-5698-9 ISBN 978-1-4899-5696-5 (eBook)
DOI 10.1007/978-1-4899-5696-5
Softcover reprint of the hardcover 1st edition 1963

Foreword

It is well known that a scientific community wherein ideas are exchanged on both formal and informal bases provides stimuli to the members and attracts nonmembers. A need for such a community in engineering mechanics was recognized by many persons in the southeastern area of the United States. This need was manifested to industry by an undersupply of personnel with the required training, to colleges and universities by the lack of both students and faculty members, and to all by the desire for an exchange of ideas between workers in the field. Thus a means for establishing the community was sought, with the result that the Southeastern Conference on Theoretical and Applied Mechanics was formed.

The purpose of the Conference is to promote dissemination of the results of scientific and engineering research of both an applied and a theoretical nature and to foster informal exchanges of ideas through biennial meetings. In particular, younger faculty members, beginning industrial research workers, and graduate students are encouraged to attend and join in the discussions. The area served includes Louisiana, Arkansas, Kentucky, Virginia, and all states to the east and south thereof. Persons from outside this region are also invited to attend and participate.

The first session, called the First Southeastern Conference on Theoretical and Applied Mechanics, was held in Gatlinburg, Tennessee, on May 3–4, 1962. A total of 32 technical papers, a special lecture, and an invited paper were presented. There were 123 attendees, representing 8 industrial organizations, 10 government laboratories, 30 colleges and universities throughout the United States, and 1 organization in western Europe. Although 22 states and the District of Columbia were represented, the majority of attendees were from the southeastern area.

It is planned that future Conference sessions will perpetuate the spirit and objectives of the first meeting. These are exemplified by emphasis on applied research and by attempting to serve all persons working in the field of engineering mechanics, rather than a relatively small, select group.

The organizers of the Conference are indebted to the individuals and organizations that participated in and contributed to the Conference.

W. L. Greenstreet
Oak Ridge National Laboratory
Executive Chairman

C. E. Stoneking
Georgia Institute of Technology
Executive Vice-Chairman

R. L. Maxwell
University of Tennessee
Secretary

W. A. Shaw
Auburn University
Organizational Committee Chairman

Program

The First Southeastern Conference on Theoretical and Applied Mechanics was held in Gatlinburg, Tennessee, May 3–4, 1962, under the sponsorship of the Oak Ridge National Laboratory (operated by Union Carbide Corporation for the U.S. Atomic Energy Commission). The conference was organized into eight regular sessions plus a special lecture and an invited paper; a condensed program follows:

THURSDAY, MAY 3, 1962

9:00 am – **Welcoming Remarks,** H. G. MACPHERSON, Associate Director, Oak Ridge National Laboratory

9:30 am – **Solid Mechanics I,** Chairman: W. A. SHAW, Auburn University, Auburn, Ala.

Fluid Mechanics I, Chairman: C. H. PARR, Rohm & Haas Co., Huntsville, Ala.

1:30 pm – **Special Lecture,** J. P. DEN HARTOG

2:30 pm – **Dynamics and Vibrations I,** Chairman: C. E. STONEKING, Georgia Institute of Technology, Atlanta, Ga.

Experimental and Applied I, Chairman: R. L. MAXWELL, University of Tennessee, Knoxville, Tenn.

6:00 pm – **Reception,** Mountain View Hotel

7:00 pm – **Banquet,** Mountain View Hotel, Principal Speaker: A. M. WEINBERG, Director, Oak Ridge National Laboratory

Topic: Fashions in Science

FRIDAY, MAY 4, 1962

9:00 am – **Invited Paper,** R. M. EVAN-IWANOWSKI

10:00 am – **Dynamics and Vibrations II,** Chairman: DANIEL FREDERICK, Virginia Polytechnic Institute, Blacksburg, Va.

Experimental and Applied II, Chairman: W. B. STILES, University of Arkansas, Little Rock, Ark.

2:00 pm – **Solid Mechanics II,** Chairman: C. C. STEYER, University of Alabama, University, Ala.

Fluid Mechanics II, Chairman: H. F. HRUBECKY, Tulane University, New Orleans, La.

4:45 pm – **Policy Committee Meeting,** Huff House Auditorium, Lower Level

The Second Southeastern Conference on Theoretical and Applied Mechanics will be held at the Georgia Institute of Technology, Atlanta, Georgia, in 1964.

H. W. Hoffman
Oak Ridge National Laboratory
Program Committee Chairman

W. F. Swinson
Auburn University
Publicity Committee Chairman

Preface

The response and interest in the First Southeastern Conference on Theoretical and Applied Mechanics, as typified by the number of papers submitted, have been gratifying to the organizers. Because of the response it was possible to form a program balanced between the theoretical and the applied aspects of mechanics.

The Proceedings consist of the 32 technical papers and the invited paper. Copies of the welcoming remarks and the banquet speech are included on the pages immediately following the Preface. The technical papers are grouped by subject matter into the four categories listed in the Program (p vii). The individual papers are arranged so that the authors' names appear in alphabetical order within each category, with the exception of the invited paper appearing first in the solid mechanics category.

Thanks are extended to the authors who submitted papers to the Conference, for it is to them that the success of the Conference must be attributed. Secondly, a particular expression of thanks is made to our reviewers for their diligence and the time spent on what must remain an anonymous effort.

Grateful acknowledgment is also made of the contribution of the organizations for preparing the manuscripts and producing these Proceedings. The manuscript was edited, VariTyped, and made up by the Technical Publications Department, Technical Information Division, Oak Ridge National Laboratory, and the printing and final production of the book were done by Plenum Press, Inc.

T. C. Huang
University of Florida
Publications Committee Chairman

C. H. Parr
Rohm and Haas Company
Editorial Committee Chairman

Welcoming Remarks

H. G. MacPherson

Scientists and engineers in the age groups represented here should pause now and then to reflect on how very fortunate they are to be spending their allotted term on earth at the middle of this century. Progress in science, engineering, and invention, after lagging with little or no progress throughout most of man's history, has built up to an explosive pace in this century. We who now inhabit the earth are privileged to see the development of science and engineering during its most exciting and most rapidly changing period.

For example, people my age (I realize that some of you are much younger) can remember the infancy of the automobile industry, which has now reached full maturity. The very start of commercial aviation is within most of our memories, and we now travel routinely in fourth-generation planes that no one would have seriously dreamed of forty years ago. We can remember when the radio was a novel toy, and now our habits are strongly affected by TV. One could go on indefinitely listing the many items, blessings or not, that have become common through engineering achievements in the last forty years. Thinking of these things, one can wonder what the next century can hold that will compare with the recent past.

I won't try to answer this question, but I would like to point out a very significant change in the character of our engineering that has been developing with our expanding technology, and which I believe holds the key to our engineering and industrial future in this country.

We have been entering an era of *precision engineering*. In the distant past we built many machines without much in the way of engineering calculation. If a part failed, there was usually no great disaster – one just replaced it with a heavier member. When calculations were made, one could, in most cases, allow a factor of safety of 2 to 4 in the calculation and another factor of safety of 2 in estimating the properties of the materials. In this era of precision engineering, however, such laxity is not possible. Such an unlikely looking candidate for flight as an Electra plane would certainly not get off the ground without some rather precise engineering. And, as we well know, it isn't safe for 90 passengers unless that engineering is sophisticated enough to foresee all contingencies. As another example, I recently heard a popular newscaster say that the Saturn booster flight would not be a success if there was a malfunction in any one of 250,000 parts. If true, then this is really precision engineering.

Some years ago the Oak Ridge National Laboratory also became engaged in precision engineering tasks. We were trying to build two reactor systems, both using fluid fuels, one for civilian power and the other for possible aircraft use. Particularly for the aircraft use, where size and weight must be kept low and temperature high, the engineering tasks bordered on the just barely possible, and we found it necessary to call on the analytical talents and experimental know-how of our best scientists and engineers to solve the problems. These skills and knowledge were later applied to other reactor projects at the Laboratory. In addition to permanent personnel, we obtained the temporary help of many college and university staff members during the summers. The idea for this conference was born in these summer research participation groups, and the many people who contributed toward getting it started are to be congratulated on their foresight and perseverance.

The founders of this conference have worked carefully to organize it in such a way that it can survive and benefit everyone in the Southeast. Every interested company or school in the Southeast is invited to participate. The governing body is a policy committee made up of one designated representative from each such institution. The 1964 meeting will be sponsored by the Georgia Institute of Technology, and the 1966 sponsor will be chosen this week by the policy committee.

The Oak Ridge National Laboratory is especially proud to sponsor this First Southeastern Conference on Theoretical and Applied Mechanics. The Laboratory stands to benefit greatly from the existence of this group and from the interest that it will stimulate in the field of mechanics. Furthermore, we are convinced that these same benefits will prove increasingly important to all industry in the area as the progress of technology and competition from the rest of the world force us to take greater advantage of this era of precision engineering.

It is with great pleasure that we welcome you to this meeting, and we hope that you will find it profitable.

Fashions in Science*

Alvin M. Weinberg

I shall talk this evening about the natural history of scientific fields – that is, about the birth and death of certain pieces of man's intellectual universe. All of us know that science does not stand still – that a field of endeavor which stirred our teachers may strike our generation as being unbearably dull. I propose to examine this process of scientific obsolescence and to argue that all too often the process goes ahead in a capricious way that has little to do with the overall well-being of science: that the enforced obsolescence of certain fields and the tendency, especially in education, to jump onto new scientific bandwagons are harmful to the society that is supporting the science.

We are all familiar with scientific fashions. At the time of Newton, the burning scientific questions lay in what we now call classical mechanics; despite this meeting of people interested in theoretical mechanics, most modern physicists would judge classical mechanics to be a rather closed, if not dull, subject. In the late nineteenth century, at the time of Maxwell and Hertz, electromagnetic theory was at the forefront of physical science; yet, at least during the decades before World War II, classical electromagnetic theory commanded relatively few major thinkers. Many other examples of such presumably "closed" fields can be given.

One can discern several reasons for a shift in fashion from one scientific field to another. The most obvious reason is that, in some sense, a field of science may become so clear that it no longer fully challenges people's intellects. But "clarity" is surely a vague criterion. A field is "cracked" open by a major discovery or insight, only in some order of approximation. There always exists a higher order of approximation, a closer approximation to reality, which remains to be investigated. Though "classical" mechanics is closed, the elucidation of the mechanical behavior of real objects remains interesting and fruitful, as this conference attests.

Perhaps a more important reason for shifts in fashion is that people get tired of unsolved problems: scientific problems very often are not solved; they are forgotten. Thus, a tremendous amount of work went into meson field theory immediately after the war, especially as a result of the invention of Tomanaga, Dyson, Schwinger, and Feynman of the methods of renormalization. The work was unsuccessful in that nuclear forces remain an unsolved mystery, and most people have given up batting their heads against the difficulties of nonconvergent meson field theories. Or again, take superconductivity before the

---

*After-dinner talk, First Southeastern Conference on Theoretical and Applied Mechanics, Gatlinburg, Tennessee, May 3, 1962.

discovery of the isotope effect – that is, the observation that the transition temperature went as $M^{-1/2}$, which observation suggested that superconductivity must be connected with lattice vibration, and led finally to the Bardeen-Cooper-Schrieffer theory of the phenomenon. Until these brilliant successes, superconductivity was a "closed" field because no one had a very good idea about how to open it; no less an authority than E. P. Wigner said in 1950: "... the phenomena connected with superconductivity are apparently being left by the wayside, most physicists not feeling an acute sense of unhappiness about it."[1]

Another reason for shifts in scientific fashion is that, should a field become of practical importance, it is often ruled, by definition, no longer to be a part of science but rather of engineering or of medicine. I suppose that the field represented at this meeting – theoretical mechanics – is now generally considered part of engineering, not of physics, and that this happy fate is inevitable for any part of science which is found to be useful: like microwaves, solid state, superconductors (now that high-field superconducting magnets are successful), and even antibiotics.

But there are many serious dangers in allowing a scientific field to become prematurely closed – either because some people get tired of its problems, or because the field is declared, by fiat, no longer to be part of science but part of engineering. In the first place, science is a unified structure; what comes later depends on what comes earlier, and if certain parts of science are forgotten by the scientists, the whole structure is in danger of crumbling. Fields of science that were considered closed have a bad habit of coming to life with a big bang as the result of some crucial discovery. I have already mentioned superconductivity; at the moment, the field of optics, which ten years ago we thought was pretty dead, has suddenly become enormously exciting as the result of the invention of the laser, and the discovery of nonlinearities in optical absorption and emission.

Again, the premature passage of science into engineering seems to me to make for a kind of intellectual snobbishness among scientists which detracts from both science and engineering. I recently chaired a panel in Pittsburgh on the question, "Is the Present Day Training of Physicists Suited to the Needs of Industry and Government?" The general view, held both by the university and industrial physicists, was that it is not; and at least one physicist ardently, and perhaps properly, defended the university physicists with the comment that physics was only what interested physicists at the time! If all that physicists were interested in was elementary particles, then physicists could not rightly claim their name: they would have to be called "elementary particlists," and the title "physicist" would have to be bestowed only on physicists whose departments didn't own big accelerators, or on engineering physics types.

This snobbishness of purism is manifested in many ways, to my mind all bad. Thus, according to my physics department friends, the best physics students go into, and tend to remain in the field of, elementary particle physics: other branches get less than their fair share of really brilliant people. (Incidentally, a similar trend seems to be at work in medicine: the number of

[1]E. P. Wigner, "The Limits of Science," *Proc. Am. Phil. Soc.* **94**, 422–27 (1950).

medical-school applicants, which was 24,000 in 1949, fell to only 15,000 in 1960. The number of "A" students among applicants dropped from forty percent to fifteen percent and failures rose from four percent to about eight percent in this period. It is my impression, though this is hard to prove, that many of the "A" students who formerly went into medicine are now attracted to the purer, possibly more abstract, branches of biology.)

THE FASHION FOR ABSTRACTION

But equally striking, and perhaps more dangerous, is the rise of the cult of abstraction in parts of our science, notably mathematics; this I believe deserves very serious attention. It is a curious fact that the United States, which by tradition has been the land of application – of big bridges, of skyscrapers, of a generally materialist outlook – is the home of such art movements as abstract expressionism. Now abstract expressionism in art never hurt anyone – one either buys the pictures, if one likes them, or one ignores them (in which case the abstract expressionist dies of starvation). But it seems to me that a kind of abstractionism has become the main stream of American mathematical thinking, and that, since this trend affects many more than the relatively few mathematicians who are concerned with research in mathematics, notice must be taken of it.

The dangers of too strong an inbreeding of abstract mathematics, of a separation between the applied and pure streams, have been commented upon by some of America's foremost mathematicians – J. von Neumann, H. Kac, R. Courant. To quote from von Neumann: "As a ... discipline travels far from its empirical source [as it loses its contact with the sister sciences] it is beset with very grave dangers. It becomes more and more pure aestheticizing, more and more purely l'art pour l'art. This need not be bad if the field is surrounded by correlated subjects which still have closer empirical connections or if the discipline is under the influence of men with an exceptionally well-developed taste. But there is a grave danger that the subject will develop along the line of least resistance, that the stream, so far from its source, will separate into a multitude of insignificant branches, and that the discipline will become a disorganized mass of details and complexities. In other words, at a great distance from its empirical source, or after much 'abstract' inbreeding, a mathematical subject is in danger of degeneration. At the inception the style is usually classical; when it shows signs of becoming baroque, then the danger signal is up."[2] Or from Mark Kac: "The extreme purists in our midst urge division and separation. Like Plato, centuries before them, they see in applications of mathematics 'mere corruption and annihilation of the one good in mathematics' and they want mathematics to return to 'the unembodied objects of pure intelligence.' In their horror of one kind of professionalism they are creating another – professional purism.

---

[2] R. B. Heywood (ed.), "The Mathematician," p 196 in *The Works of the Mind,* University of Chicago Press, 1947.

"Our extreme pragmatists preach division and separation on different grounds.

"I would view separation as a tragedy, and I consider its prevention to be a major challenge of the years to come. The two great streams of mathematical creativity are a tribute to the universality of the human genius. Each carries its ... own passions. Apart both may die – one in a kind of unembodied sterility of medieval scholasticism and the other as a part of military art."[3]

The non-mathematician is always at a great disadvantage in criticizing a trend within mathematics since his impressions are necessarily secondhand. That I can find great mathematicians who agree with my view – that the current fashion for extreme abstraction in mathematics is dangerous – merely means, I am told, that I quote the wrong great mathematicians. In support of my concern is a recent very excellent study of Russian mathematics by a special mathematics panel of RIAS, chaired by J. P. LaSalle. The conclusions of the RIAS study are so interesting that I shall read most of them.

"1. In mathematics the Soviet Union and the United States lead the world and are about at the same level. Other than algebraic geometry there is no broad area of mathematics in which Soviet mathematicians are not working and in which they are not making significant contributions.

"2. The Soviet Union will move at a faster rate than the United States in the practical application of mathematical theories. Further, there is a distinct possibility that the Soviet Union will gradually surpass us in certain areas more or less directly connected with applications, for example, control theory, numerical analysis, ordinary and partial differential equations.

"Our main reason for believing this is that in Russia the application of mathematics by mathematicians is encouraged in many ways. In this country, however, there is a growing gap between pure mathematicians, and those interested in applications. Consider the following:

> "a. Leading Soviet mathematicians have an interest in and contribute to both pure and applied mathematics. In the United States a pure mathematician rarely has an interest in applications.
>
> "b. Soviet scientists and engineers have a high regard for mathematics. There are a large number of Soviet mathematical-engineers who keep up with the latest mathematical developments and who exploit the deepest kind of mathematics. These mathematical-engineers have no counterpart in this country.
>
> "c. Soviet mathematicians make serious and successful efforts to communicate the latest theoretical advances to engineers and scientists. Long expository articles are encouraged and many fine books are written by Soviet mathematicians for engineers and others. In the United States the space in mathematical journals is inadequate and conciseness in presentation is often enforced to the point of unintelligibility. Our specialists write for each other.

---

[3]Mark Kac, "Mathematics: Its Trends and Its Tensions," p 8, Occasional Paper Number Ten, The Rockefeller Institute Press, New York, 1961.

"d. The above points indicate that applied mathematics is quite respectable in the USSR. In addition, certain areas (control theory, numerical analysis, differential equations) are being emphasized in Russia [see (3) and (4) below], and this tends to attract good young people to these fields. In the United States, among pure mathematicians, there is an increasing tendency towards greater and greater abstraction, with a lessening regard for applications, which tends to discourage good young people from ever approaching the applications. From this it is obvious that the quality of applied mathematics in Russia will tend to improve while in this country the quality of applied mathematics will tend to decrease.

"3. An increasingly large number of Soviet mathematicians are working on control theory and numerical analysis. In quality they match the United States. In quantity of output and in numbers of people they lead the world.

"4. Their effort in control theory is part of an all-out Soviet scientific effort to advance their knowledge and technical competence in the field of automatic control. Applications are to missiles and space vehicles and to the automation of industrial processes and industries."[4]

The three greatest mathematicians, Archimedes, Newton, and Gauss, made little distinction between applied and pure mathematics; they were *not* professional purists in Kac's sense. The greatest mathematician of the twentieth century, Henri Poincaré, worked in this broad tradition, as did the greatest contemporary mathematician, J. von Neumann. I therefore believe there is a good probability that the fashion for abstraction which we find in modern American mathematics – and to some extent in parts of American physics – will prove to be a passing phase; that, if experience is any basis for predicting the future, the most fruitful lines of mathematical thought will continue to be those which are fertilized by contact with sister sciences.

But in the meantime harm can be, and is being, done. That the professional purists divert mathematics from what appears to me to be its proper course, which is parallel to and in intimate contact with empirical science, is clearly none of my business, and I have been so informed by distinguished and knowledgeable mathematicians. But what I believe is the business of those of us who are interested in the applied branches of mathematics is the effect that this trend in higher mathematics has on the *teaching* of school mathematics. The highly abstract fashion in mathematics has found its way into many of the newer mathematics curricula at the high school level. Now the teaching of school mathematics is too important a matter to be left entirely to the mathematicians. We are justified, for example, in asking whether our new curricula spoil the taste of our new generation for mathematics in its broadest, unified tradition – the tradition which has created the world's greatest mathematicians, the tradition which keeps mathematics and empirical science together. That universal experimentation in elementary and secondary education can have horrifying results is too well documented to need belaboring – phonic reading

[4]"Summary Report of the Panel on Soviet Mathematics," Research Institute for Advanced Study, Baltimore, Maryland, McGraw-Hill, New York (in press).

has created a generation of poor spellers. Or, to paraphrase the British mathematician, Hardy: "The Mathematical Tripos was the greatest pedagogical tool mathematics has ever known. It created in England a race of problem solvers of unparalleled competence, but it ruined English mathematics for two generations!"

Complaining about new curricula is an old business which every self-respecting scientist is expected to indulge in. Before I sound off too vehemently I should say exactly what I am complaining about. It seems to me that the new curricula – the PSSC in physics as well as the various modifications in mathematics – have two quite separate aspects. On the one hand the curricula are tougher than were the old curricula; they demand much more of the student in the way of hard work, insight, and discipline. The physics curriculum also is wonderfully imaginative in its use of technical props, audio-visual aids, collateral reading. With these reforms I am completely in agreement. All of us are delighted to see that even American children can work hard when they have to, and that school systems are willing to make such demands on their brighter students.

But I disagree with the second aspect of the new curricula – to some extent in physics, much more in mathematics. It seems to me that, with some notable exceptions, most of the new mathematics curricula tend to represent the current fashion in American mathematics – its professional purism, its scholasticism, its isolation from the remainder of science. I have examined four of the new mathematics curricula, and I find that three of them are highly objectionable in this regard; on the other hand, the intermediate SMSG curriculum strikes me as admirable in content as well as in method. To a lesser extent, the same criticism can be leveled at the new physics curriculum: its fundamental viewpoint is professional "modern" physics, and it pays relatively little attention to those aspects of physics which underlie conventional technology.

I believe that enough very extreme curricula have been circulating for a long enough time to cause one to worry. As you may know, the issue which I raise has been argued at length among mathematicians and to a lesser extent among physicists. There has been, for example, a heated public argument between Professor Morris Kline of New York University and Professor E. G. Begle of the SMSG which has caused many wry smiles among innocent bystanders. The argument has led to the issuance of a manifesto by an ad hoc committee of mathematicians urging that in the revamping of the curriculum one try to maintain the tried and true Poincaré tradition – that "... the new curricula should reflect more the connection between mathematics and science. . . ."[5]

But such manifestos are insufficient. Those of us who believe that the current trend is dangerous have no business complaining about the situation unless we are prepared to do something about it. The mathematicians very wisely have produced not one curriculum but several, since they realize that in diversity lies insurance against serious error. (Oh that John Dewey were five different people!) I should think therefore that a mathematics curriculum created

[5]L. V. Ahlfors *et al.*, "Mathematics Curriculum of the High School," *Am. Math. Monthly* **69**, 189–93 (1962).

by mathematicians who work in the Poincaré tradition – who are sensitive always to the connections between mathematics and empirical science – would be welcomed by the mathematics community. To be accepted, the curriculum will have to be good; it will have to be demanding, pedagogically sound; in short it must have all the *pedagogic* attributes of the successful new mathematics curricula now available; but instead of reflecting the current fashion in American mathematical thinking, it should reflect the much sounder Poincaréan tradition which denies the separation of mathematics into the applied and basic streams. Similarly, I believe there is room for a high school physics curriculum which has in it the beautifully painstaking work and charming elegance of PSSC, but whose orientation is as much toward physics for engineering as the PSSC is toward physics for modern physics.

To carry out these programs the work must be supplied by people who are willing to devote themselves to the task as wholeheartedly as did the pioneers in high school science curriculum reform, notably Professors J. R. Zacharias and F. L. Friedman. I therefore call on people, such as you who are here this evening, to give serious consideration to forming curriculum study groups in physics and in mathematics which will represent the unified tradition both in mathematics and in physics. I believe that the NSF has ample money to support efforts of this sort and that if the proposal is put forth seriously, money will be forthcoming.

THE PROBLEM OF SCIENTIFIC CRITICISM

I close my remarks on a rather different but related subject, a subject which has been raised by Mortimer Taube in the Introduction to his "Computers and Common Sense: The Myth of Thinking Machines." When I comment on fashions in science, and especially when I claim that certain trends in mathematics may have deleterious effects both on mathematics and on the rest of science, I am indulging in a kind of scientific criticism which is close to scientific lèse majesté. Criticism in the arts has long since been accepted as a valid intellectual activity though perhaps grudgingly by the artists who are criticized. Criticism in detail of separate pieces of science is of course the essence of science. Science progresses only by ruthless exposure of error.

But the kind of criticism I am making here is of a different, less tangible, sort. What I am criticizing is not scientific validity or accuracy but scientific taste. Since taste in all things, including science, is highly individual, there has been no particular reason to raise the question previously. However, as Taube points out, with science becoming such a large public activity, and with the effect of science being felt so broadly, we must accept scientific criticism, in the sense I use the term, as a necessary evil. Whether scientific criticism can ever be made in a truly constructive manner, whether self-appointed scientific critics will ever be listened to, whether, for that matter, they have anything worth saying – this I do not know. One general principle does seem clear to me – that the most useful criticism of scientific taste must come not directly from the practitioners of the science under criticism but from scientists in neighboring fields who see with broader perspective what their cousins do. Can scientists in neighboring fields really know enough to be able to criticize their colleagues – does the expert in theoretical mechanics know enough to criticize the trend toward abstraction in pure mathematics? Certainly not in detail, but he has a perfect right, in fact a duty, to criticize those parts of the neighboring field which bear strongly on his business. Moreover, where distortion in taste indirectly affects his work – as in the curricular reforms – the specialist in theoretical mechanics has a duty not only to criticize but to roll up his sleeves and get to work creating a curriculum that more closely embodies his mathematical taste. I hope therefore that the pure mathematicians in the audience will be lenient with me for criticizing what they do. My criticism and my intentions are constructive – at least, I shall consider them so if they imbue some of our more practically oriented mathematicians to participate in the arduous task of school mathematics reform.

Contents

SOLID MECHANICS

FLUID MECHANICS

EXPERIMENTAL AND APPLIED MECHANICS

Dynamics and Vibrations

APPLICATION OF REISSNER'S VARIATIONAL PRINCIPLE TO THE VIBRATION ANALYSIS OF SQUARE FLAT PLATES WITH VARIOUS ROOT SUPPORT CONDITIONS[1]

R. N. Austin, D. A. Caughfield, and H. J. Plass, Jr.
The University of Texas

A variational principle due to E. Reissner has been rewritten in a form which is applicable to small deflection problems for thin plates with various root support conditions. The modified principle is used to obtain approximations to vibration problems of square plates of uniform thickness. Theoretical results are obtained for several choices of shear boundary conditions to be initially satisfied in the problem formulation and are compared with results obtained experimentally.

INTRODUCTION

The static and dynamic analysis of thin cantilever plates having various plan forms and root support conditions has become increasingly important with the knowledge that the approximate behavior of idealized missile fins can be obtained from a study of these plate configurations. Although no known exact solutions exist for problems of this type, various investigators have obtained approximate solutions by means of the Ritz method [*1, 2*].

Plass [*3*] modified a variational theorem of Reissner [*4, 5*] for application to the deflection and vibration problems of a cantilever plate with a partially supported root. The main advantage of the modified Reissner theorem over the other approximate methods is that independent variations can be made on the deflections and moments. In applying this modified theorem to square flat plates with complete root support and root hub and pin support, separate expressions were found for the deflection and the moment distributions, and the unknown parameters in each of the expressions were selected to make the modified Reissner integral an extremum. This same technique has been applied to square and skew cantilever plates of uniform thickness, and the results are presented in ref [*6*].

There are difficulties encountered in formulating the problem in this manner for use in Reissner's principle. First, although the proper selection of an adequate deflection expression can be made by simply considering the geometry of

[1]Supported by the U.S. Naval Bureau of Weapons (contract No. NOrd-16498).

NOTATION

x, y, z	Rectangular coordinates
w	Deflection of plate in z direction
M_x, M_y	Plate bending moments
M_{xy}	Plate twisting moment
D	Flexural rigidity = $Eh^3/12(1-\nu^2)$
E	Young's modulus
h	Plate thickness
ν	Poisson's ratio
ρ	Mass per unit area
ω	Angular frequency
a	Plate dimension
ξ	x/a
η	y/a
$\bar{w}$	w/a
M_ξ	$M_x a/D$
M_η	$M_y a/D$
$M_{\xi\eta}$	$M_{xy} a/D$
Ω	$(\omega^2 \rho a^4/D)^{1/2}$
$[T]$	Matrix containing coefficients for trapezoidal rule for double integration
$\bar{a}$	Distance between mesh points

the particular configuration being studied, the choice of realistic moment expressions is quite difficult due to a limited knowledge of what is taking place in the interior of the plate. Secondly, when these deflections and moments are represented by a series of analytical expressions, the integrals in the theorem become more tedious to evaluate as the number of terms in the various expressions increases.

In ref [7] these conditions are circumvented by having the deflections and moments both appear as unknowns at discrete points and using the techniques of finite differences, numerical integration, and matrix algebra to formulate the problem. The disadvantage of this completely numerical approach is that as the plate is divided into a greater number of parts, the resulting matrices become prohibitively large and the usual processes of matrix analysis (e.g., inversion) become unwieldy. A compromise approach was therefore suggested in which the best features of each of the two previous techniques are incorporated into the analysis in order to obtain an accurate solution with a reasonable amount of computational effort. This alternate method consists of expressing the plate deflections in terms of a series of analytical expressions with undetermined coefficients, and allowing the moments to appear as discrete unknowns at a finite number of equally spaced mesh points.

In this paper the modified Reissner principle is used in conjunction with this alternate technique to study the vibration characteristics of two square flat plates, one with the root completely clamped, and the other with the root supported by a bending and shear-resisting hub at the midchord point and a shear-resisting pin at the quarter-chord point. Results are presented for the two plate configurations for four different choices of moment and shear boundary conditions, and are compared with results obtained experimentally.

REISSNER'S VARIATIONAL THEOREM APPLIED TO THIN PLATES

The details of the form of Reissner's principle suitable for application to plate vibration problems have been discussed at some length in recent publications [3, 6, 7]. For this reason, the principle is simply stated in this paper without development from Reissner's original form, or proof of its validity:

$$\delta\left(\iint_A \left\{-M_x \frac{\partial^2 w}{\partial x^2} - M_y \frac{\partial^2 w}{\partial y^2} + 2M_{xy} \frac{\partial^2 w}{\partial x\, \partial y} - \frac{1}{2D(1-\nu^2)}\left[M_x^2 + M_y^2 - 2\nu M_x M_y + 2(1+\nu)M_{xy}^2\right]\right\} dx\, dy - \frac{\omega^2 \rho}{2}\iint_A w^2\, dx\, dy\right) = 0\,, \qquad (1)$$

where the only restriction is that w be required to satisfy the geometrical boundary conditions, that is, zero deflection or zero slope at the appropriate support points.

For convenience in analysis the following dimensionless quantities are introduced:

$$\xi = x/a\,, \qquad \eta = y/a\,, \qquad \overline{w} = w/a\,,$$

$$M_\xi = M_x a/D\,, \qquad M_\eta = M_y a/D\,, \qquad M_{\xi\eta} = M_{xy} a/D\,,$$

$$\Omega^2 = \omega^2 \rho a^4/D\,,$$

where a is a plate dimension (e.g., length of one edge). Substituting these quantities into Eq. (1), the following relation for the plate vibrating in a natural mode is obtained:

$$\delta\Bigg(\iint_A \Bigg\{-M_\xi \frac{\partial^2\overline{w}}{\partial\xi^2} - M_\eta \frac{\partial^2\overline{w}}{\partial\eta^2} + 2M_\eta \frac{\partial^2\overline{w}}{\partial\xi\,\partial\eta} - \frac{1}{2(1-\nu^2)}\Big[M_\xi^2 + M_\eta^2 - 2\nu M_\xi M_\eta + 2(1+\nu)M_{\xi\eta}^2\Big]\Bigg\}\, d\xi\, d\eta - \frac{\Omega^2}{2}\iint_A \overline{w}^2\, d\xi\, d\eta\Bigg) = 0\,. \qquad (2)$$

METHOD OF ANALYSIS

Discussion of the Technique of the Problem Solution

The deflection expressions to be selected for both the square flat cantilever plate and the square flat hub-pin-supported plate will be of the general form

$$\begin{aligned}\overline{w}(\xi,\eta) &= A_1 f_1(\xi) + A_2 f_2(\xi) + \dots \\ &+ B_1 g_1(\eta) + B_2 g_2(\eta) + \dots \\ &+ C_1 h_1(\xi,\eta) + C_2 h_2(\xi,\eta) + \dots\,, \end{aligned} \qquad (3)$$

where $A_1, A_2, \dots, B_1, B_2, \dots, C_1, C_2, \dots$ are unknown coefficients, and the quantities $f_1(\xi)$, $f_2(\xi)$, ..., $g_1(\eta)$, $g_2(\eta)$, ..., $h_1(\xi,\eta)$, $h_2(\xi,\eta)$, ... are simple polynomials or beam functions, or combinations of these which may be functions of ξ or η or a combination of ξ and η as indicated. The moments appearing in the variational expression will be treated as discrete unknowns, the location of each governed by the "size" of the finite-difference net. For the two particular plate configurations under investigation, the net size indicated in Fig. 1 was adopted for formulating the governing matrices. This choice of the plate divided into 16 parts was made both from a consideration of the geometry of the root supports and of the maximum number of points that can be conveniently handled by using numerical methods. Initially, 75 unknown moments are introduced, 25 for each of the three moments corresponding to the 25 mesh points. Application of the boundary moment and shear conditions on the free edges of the plates will reduce the number of unknowns considerably. Expressions involving derivatives of moments in the free-edge conditions are represented in terms of finite differences of moments at boundary and interior points only. For all the numerical procedures, central difference operators are used whenever possible, but for points on the boundary it becomes necessary to use forward or backward difference operators in order not to use points off the plate. These forward and backward difference operators have the same order of error as that of the central difference operators.

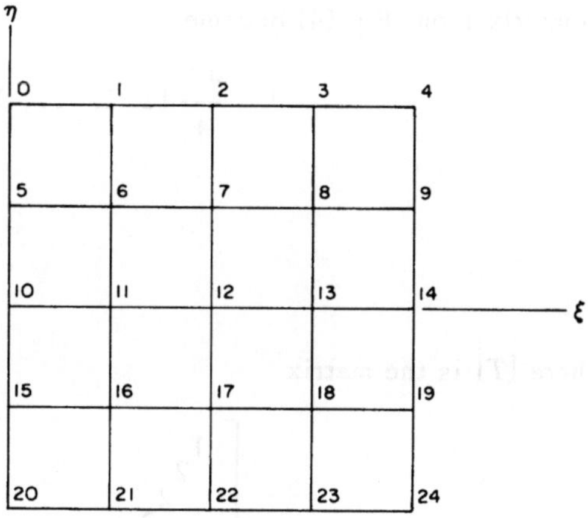

Fig. 1. Finite-Difference Net Showing Plate Divided into 16 Parts and Numbering of Mesh Points.

The numerical integration is performed by using the trapezoidal rule, which for double integration is illustrated in Fig. 2 for the plate divided into 16 parts, having 25 mesh points. With matrix formulation, the trapezoidal rule for integration, and the notation of Fig. 1, the double integration of a function $f(\xi,\eta)$ can be approximated in the following manner:

$$I = \iint f(\xi,\eta)\, d\xi\, d\eta$$

$$= \frac{\bar{a}^2}{4}\, [1f_0 + 2f_1 + 2f_2 + 2f_3 + 1f_4 + 2f_5 + 4f_6 + 4f_7 + 4f_8 + 2f_9$$

$$+ 2f_{10} + 4f_{11} + 4f_{12} + \ldots + 1f_{24}] \; . \qquad (4)$$

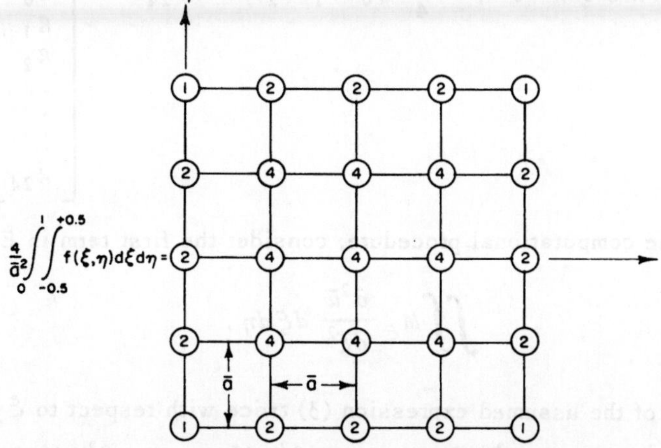

Fig. 2. Trapezoidal Rule for Double Integration.

In matrix form, Eq. (4) becomes

$$I = \frac{\bar{a}^2}{4} [1, 1, \ldots, 1] [T] \begin{bmatrix} f_0 \\ f_1 \\ f_2 \\ \cdot \\ \cdot \\ \cdot \\ f_{24} \end{bmatrix}, \tag{5}$$

where $[T]$ is the matrix

$$[T] = \begin{bmatrix}
1 \\
& 2 \\
&& 2 \\
&&& 2 \\
&&&& 1 \\
&&&&& 2 \\
&&&&&& 4 \\
&&&&&&& 4 \\
&&&&&&&& 4 \\
&&&&&&&&& 2 \\
&&&&&&&&&& 2 \\
&&&&&&&&&&& 4 \\
&&&&&&&&&&&& 4 \\
&&&&&&&&&&&&& 4 \\
&&&&&&&&&&&&&& 2 \\
&&&&&&&&&&&&&&& 2 \\
&&&&&&&&&&&&&&&& 4 \\
&&&&&&&&&&&&&&&&& 4 \\
&&&&&&&&&&&&&&&&&& 4 \\
&&&&&&&&&&&&&&&&&&& 2 \\
&&&&&&&&&&&&&&&&&&&& 1 \\
&&&&&&&&&&&&&&&&&&&&& 2 \\
&&&&&&&&&&&&&&&&&&&&&& 2 \\
&&&&&&&&&&&&&&&&&&&&&&& 2 \\
&&&&&&&&&&&&&&&&&&&&&&&& 1
\end{bmatrix}. \tag{6}$$

Similarly, the integral of the product of two functions can be written thus:

$$\iint f(\xi,\eta)\, g(\xi,\eta)\, d\xi\, d\eta = \frac{\bar{a}^2}{4} [f_0, f_1, f_2, \ldots, f_{24}] [T] \begin{bmatrix} g_0 \\ g_1 \\ g_2 \\ \cdot \\ \cdot \\ \cdot \\ g_{24} \end{bmatrix}. \tag{7}$$

To illustrate the computational procedure, consider the first term in Eq. (2):

$$\iint M_\xi \frac{\partial^2 \bar{w}}{\partial \xi^2}\, d\xi\, d\eta \,.$$

Differentiation of the assumed expression (3) twice with respect to ξ yields

$$\frac{\partial^2 \bar{w}(\xi,\eta)}{\partial \xi^2} = A_1 \frac{d^2 f_1(\xi)}{d\xi^2} + A_2 \frac{d^2 f_2(\xi)}{d\xi^2} + \ldots + C_1 \frac{\partial^2 h_1(\xi,\eta)}{\partial \xi^2} + C_2 \frac{\partial^2 h_2(\xi,\eta)}{\partial \xi^2} + \ldots . \tag{8}$$

In matrix form, Eq. (8) can be written as

$$\frac{\partial^2 \overline{w}(\xi,\eta)}{\partial \xi^2} = [A_1, A_2, \ldots, C_1, C_2, \ldots] \begin{bmatrix} \frac{d^2 f_1(\xi)}{d\xi^2} \\ \frac{d^2 f_2(\xi)}{d\xi^2} \\ \vdots \\ \frac{\partial^2 b_1(\xi,\eta)}{\partial \xi^2} \\ \frac{\partial^2 b_2(\xi,\eta)}{\partial \xi^2} \\ \vdots \end{bmatrix} . \tag{9}$$

Using Eq. (9) the integral under consideration may be written as

$$\iint M_\xi \frac{\partial^2 \overline{w}}{\partial \xi^2} d\xi\, d\eta = \iint \frac{\partial^2 \overline{w}}{\partial \xi^2} M_\xi\, d\xi\, d\eta$$

$$= [A_1, A_2, \ldots, C_1, C_2, \ldots] \begin{bmatrix} \iint \frac{d^2 f_1(\xi)}{d\xi^2} M_\xi\, d\xi\, d\eta \\ \iint \frac{d^2 f_2(\xi)}{d\xi^2} M_\xi\, d\xi\, d\eta \\ \vdots \\ \iint \frac{\partial^2 b_1(\xi,\eta)}{\partial \xi^2} M_\xi\, d\xi\, d\eta \\ \iint \frac{\partial^2 b_2(\xi,\eta)}{\partial \xi^2} M_\xi.\, d\xi\, d\eta \\ \vdots \end{bmatrix} . \tag{10}$$

On using (7), one obtains

$$\iint M_\xi \frac{\partial^2 \bar{w}}{\partial \xi^2}\, d\xi\, d\eta$$

$$= \frac{\bar{a}^2}{4}[A_1, A_2, \ldots, C_1, C_2, \ldots]\begin{bmatrix} \left(\frac{d^2 f_1}{d\xi^2}\right)_0 & \left(\frac{d^2 f_1}{d\xi^2}\right)_1 & \cdots & \left(\frac{d^2 f_1}{d\xi^2}\right)_{24} \\ \left(\frac{d^2 f_2}{d\xi^2}\right)_0 & \left(\frac{d^2 f_2}{d\xi^2}\right)_1 & \cdots & \left(\frac{d^2 f_2}{d\xi^2}\right)_{24} \\ \left(\frac{\partial^2 b_1}{\partial\xi^2}\right)_0 & \left(\frac{\partial^2 b_2}{\partial\xi^2}\right)_1 & \cdots & \left(\frac{\partial^2 b_1}{\partial\xi^2}\right)_{24} \\ \left(\frac{\partial^2 b_2}{\partial\xi^2}\right)_0 & \left(\frac{\partial^2 b_1}{\partial\xi^2}\right)_1 & \cdots & \left(\frac{\partial^2 b_2}{\partial\xi^2}\right)_{24} \end{bmatrix}[T]\begin{bmatrix}(M_\xi)_0 \\ (M_\xi)_1 \\ \vdots \\ (M_\xi)_{24}\end{bmatrix}. \quad (11)$$

Equation (11) can be represented symbolically as

$$\iint M_\xi \frac{\partial^2 \bar{w}}{\partial \xi^2}\, d\xi\, d\eta = \frac{\bar{a}^2}{4}\tilde{\bar{w}} U T M_\xi\,, \quad (12)$$

where $\tilde{\bar{w}}$ is the transpose of $\bar{w}$. Equation (12) is now rewritten as follows:

$$\iint M_\xi \frac{\partial^2 \bar{w}}{\partial \xi^2}\, d\xi\, d\eta = \frac{\bar{a}^2}{4}\tilde{\bar{w}} U T G \bar{M}_\xi\,, \quad (13)$$

where $\bar{M}_\xi$ is a reduced matrix, consisting of the collection of numerical values of M_ξ at the points where M_ξ is not prescribed, and the matrix G is such that when it operates on the reduced matrix $\bar{M}_\xi$, zero values result at the appropriate boundary points. It can be similarly shown that the remaining terms in Eq. (2) can be expressed in the following general forms:

$$\iint M_\eta \frac{\partial^2 \bar{w}}{\partial \eta^2}\, d\xi\, d\eta = \frac{\bar{a}^2}{4}\tilde{\bar{w}} V T H \bar{M}_\eta\,, \quad (14)$$

$$\iint M_{\xi\eta} \frac{\partial^2 \bar{w}}{\partial \xi\, \partial \eta}\, d\xi\, d\eta = \frac{\bar{a}^2}{4}[\tilde{\bar{w}} W T R \bar{M}_\xi + \tilde{\bar{w}} W T S \bar{M}_\eta + \tilde{\bar{w}} W T J \bar{M}_{\xi\eta}]\,, \quad (15)$$

$$\iint M_\xi^2\, d\xi\, d\eta = \frac{\bar{a}^2}{4}\tilde{\bar{M}}_\xi \tilde{G} T G \bar{M}_\xi\,, \quad (16)$$

$$\iint M_\eta^2\, d\xi\, d\eta = \frac{\bar{a}^2}{4}\tilde{\bar{M}}_\eta \tilde{H} T H \bar{M}_\eta\,, \quad (17)$$

$$\iint M_\xi M_\eta\, d\xi\, d\eta = \frac{\bar{a}^2}{4}\tilde{\bar{M}}_\xi \tilde{G} T H \bar{M}_\eta\,, \quad (18)$$

$$\iint M_{\xi\eta}^2 \, d\xi \, d\eta = \frac{\bar{a}^2}{4} [\tilde{\bar{M}}_\xi \tilde{R} T R \bar{M}_\xi + 2\tilde{\bar{M}}_\xi \tilde{R} T S \bar{M}_\eta + 2\tilde{\bar{M}}_\xi \tilde{R} T J \bar{M}_{\xi\eta}$$

$$+ \tilde{\bar{M}}_\eta \tilde{S} T S \bar{M}_\eta + 2\tilde{\bar{M}}_\eta \tilde{S} T J \bar{M}_{\xi\eta} + \tilde{\bar{M}}_{\xi\eta} \tilde{J} T J \bar{M}_{\xi\eta}] , \tag{19}$$

$$\iint \bar{w}^2 \, d\xi \, d\eta = \tilde{\bar{w}} \tilde{\Phi} T \Phi \bar{w}$$

$$= \frac{\bar{a}^2}{4} \tilde{\bar{w}} N \bar{w} , \tag{20}$$

where the barred quantities $\bar{M}_\xi$, $\bar{M}_\eta$, and $\bar{M}_{\xi\eta}$ represent the reduced moment matrices obtained by satisfying the boundary conditions on the free edges. Substituting these equations into Eq. (2), one has

$$\delta \left\{ - \frac{\bar{a}^2}{4} \tilde{\bar{w}} U T G \bar{M}_\xi - \frac{\bar{a}^2}{4} \tilde{\bar{w}} V T H \bar{M}_\eta + 2 \frac{\bar{a}^2}{4} [\tilde{\bar{w}} W T R \bar{M}_\xi + \tilde{\bar{w}} W T S \bar{M}_\eta + \tilde{\bar{w}} W T J \bar{M}_{\xi\eta}] \right.$$

$$- \frac{1}{2(1 - \nu^2)} \frac{\bar{a}^2}{4} [\tilde{\bar{M}}_\xi \tilde{G} T G \bar{M}_\xi + \tilde{\bar{M}}_\eta \tilde{H} T H \bar{M}_\eta - 2\nu \tilde{\bar{M}}_\xi \tilde{G} T H \bar{M}_\eta$$

$$+ 2(1 + \nu) (\tilde{\bar{M}}_\xi \tilde{R} T R \bar{M}_\xi + 2\bar{M}_\xi \tilde{R} T S \bar{M}_\eta + 2\tilde{\bar{M}}_\xi \tilde{R} T J \bar{M}_{\xi\eta} + \tilde{\bar{M}}_\eta \tilde{S} T S \bar{M}_\eta$$

$$\left. + 2\tilde{\bar{M}}_\eta \tilde{S} T J \bar{M}_{\xi\eta} + \tilde{\bar{M}}_{\xi\eta} \tilde{J} T J \bar{M}_{\xi\eta})] - \frac{\bar{a}^2}{4} \frac{\Omega^2}{2} \tilde{\bar{w}} N \bar{w} \right\} = 0 . \tag{21}$$

Performing the indicated variations and grouping terms, the resulting equation is of the form

$$\delta \tilde{\bar{w}} [U T G \bar{M}_\xi + V T H \bar{M}_\eta - 2 W T R \bar{M}_\xi - 2 W T S \bar{M}_\eta - 2 W T J \bar{M}_{\xi\eta} + \Omega^2 N \bar{w}]$$

$$+ \delta \tilde{\bar{M}}_\xi \left[\tilde{G} T \tilde{U} \bar{w} - 2\tilde{R} T \tilde{W} \bar{w} + \frac{1}{1 - \nu^2} \tilde{G} T G \bar{M}_\xi - \frac{\nu}{1 - \nu^2} \tilde{G} T H \bar{M}_\eta + \frac{2}{1 - \nu} \tilde{R} T R \bar{M}_\xi \right.$$

$$\left. + \frac{2}{1 - \nu} \tilde{R} T S \bar{M}_\eta + \frac{2}{1 - \nu} \tilde{R} T J \bar{M}_{\xi\eta} \right] + \delta \tilde{\bar{M}}_\eta \left[\tilde{H} T \tilde{V} \bar{w} - 2\tilde{S} T \tilde{W} \bar{w} + \frac{1}{1 - \nu} \tilde{H} T H \tilde{M}_\eta \right.$$

$$\left. - \frac{\nu}{1 - \nu^2} \tilde{H} T G \bar{M}_\xi + \frac{2}{1 - \nu} \tilde{S} T R \bar{M}_\xi + \frac{2}{1 - \nu} \tilde{S} T S \bar{M}_\eta + \frac{2}{1 - \nu} \tilde{S} T J \bar{M}_{\xi\eta} \right]$$

$$+ \delta \tilde{\bar{M}}_{\xi\eta} \left[- 2\tilde{J} T \tilde{W} \bar{w} + \frac{2}{1 - \nu} \tilde{J} T R \bar{M}_\xi + \frac{2}{1 - \nu} \tilde{J} T S \bar{M}_\eta + \frac{2}{1 - \nu} \tilde{J} T J \bar{M}_{\xi\eta} \right] = 0 . \tag{22}$$

Since Eq. (22) is true for all values of $\delta\bar{w}$, $\delta\bar{M}_{\xi}$, $\delta\bar{M}_{\eta}$, and $\delta\bar{M}_{\xi\eta}$, each of the bracketed terms must be independently zero. These equations may be expressed in the following matrix form:

$$\begin{bmatrix} 0 & UTG - 2WTR & VTH - 2WTS & -2WTJ \\ \tilde{G}T\tilde{U} - 2\tilde{R}T\tilde{W} & \frac{1}{1-\nu^2}\tilde{G}TG + \frac{2}{1-\nu}\tilde{R}TR & -\frac{\nu}{1-\nu^2}\tilde{G}TH + \frac{2}{1-\nu}\tilde{R}TS & \frac{2}{1-\nu}\tilde{R}TJ \\ \tilde{H}T\tilde{V} - 2\tilde{S}T\tilde{W} & -\frac{\nu}{1-\nu^2}\tilde{H}TG + \frac{2}{1-\nu}\tilde{S}TR & \frac{1}{1-\nu^2}\tilde{H}TH + \frac{2}{1-\nu}\tilde{S}TS & \frac{2}{1-\nu}\tilde{S}TJ \\ -2\tilde{J}T\tilde{W} & \frac{2}{1-\nu}\tilde{J}TR & \frac{2}{1-\nu}\tilde{J}TS & \frac{2}{1-\nu}\tilde{J}TJ \end{bmatrix} \begin{bmatrix} \bar{w} \\ \bar{M}_{\xi} \\ \bar{M}_{\eta} \\ \bar{M}_{\xi\eta} \end{bmatrix} = -\Omega^2 \begin{bmatrix} N & 0 & 0 & 0 \\ 0 & 0 & 0 & 0 \\ 0 & 0 & 0 & 0 \\ 0 & 0 & 0 & 0 \end{bmatrix} \begin{bmatrix} \bar{w} \\ \bar{M}_{\xi} \\ \bar{M}_{\eta} \\ \bar{M}_{\xi\eta} \end{bmatrix} . \qquad (23)$$

Equation (23) may be written as follows:

$$\begin{bmatrix} 0 & B \\ \tilde{B} & C \end{bmatrix} \begin{bmatrix} \bar{w} \\ \bar{M} \end{bmatrix} = -\Omega^2 \begin{bmatrix} N & 0 \\ 0 & 0 \end{bmatrix} \begin{bmatrix} \bar{w} \\ \bar{M} \end{bmatrix} , \qquad (24)$$

where $\tilde{B}$, B, C, and N are the submatrices separated by the heavy lines in Eq. (23). Two equations are then obtained,

$$B\bar{M} = -\Omega^2 N\bar{w} , \qquad (25a)$$

$$\tilde{B}\bar{w} + C\bar{M} = 0 . \qquad (25b)$$

Solving these two equations for $\bar{w}$, one obtains the eigenvalue problem

$$\Omega^2\bar{w} = N^{-1}(BC^{-1}\tilde{B})\bar{w} . \qquad (26)$$

The values of Ω obtained from this eigenvalue problem are the nondimensional natural frequencies for the vibrating plate, and the values of $A_1, A_2, \ldots, C_1, C_2, \ldots$ (the eigenvectors) used in Eq. (3) will yield the corresponding mode shapes.

Statement of Boundary Conditions Investigated

The conditions for zero effective shear force on the boundaries of the plate are given below for the case of a square plate, Fig. 3, for $0 \leqq x \leqq a$, and $-a/2 \leqq y \leqq a/2$:

$$\frac{\partial M_x}{\partial x} - 2\frac{\partial M_{xy}}{\partial y} = 0 \text{ , along free edge where } x \text{ is a constant,} \qquad (27a)$$

$$\frac{\partial M_y}{\partial y} - 2\,\frac{\partial M_{xy}}{\partial x} = 0\ ,\ \text{along}\ y = \pm\frac{a}{2}\ . \tag{27b}$$

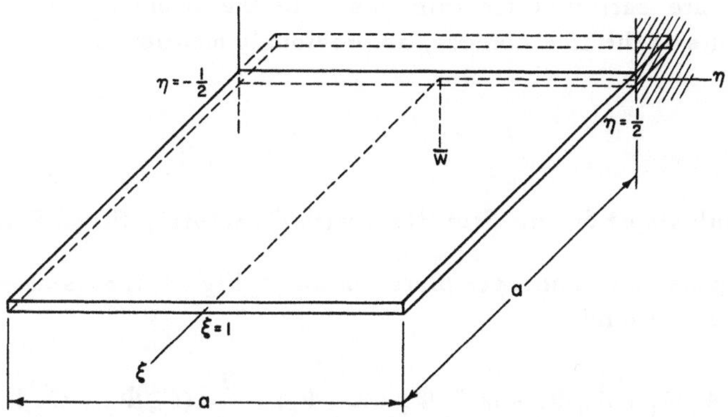

Fig. 3. Square Cantilever Plate.

In nondimensional form, the above conditions are

$$\frac{\partial M_\xi}{\partial \xi} - 2\,\frac{\partial M_{\xi\eta}}{\partial \eta} = 0\ ,\ \text{along free edge where}\ \xi\ \text{is a constant,} \tag{28a}$$

$$\frac{\partial M_\eta}{\partial \eta} - 2\,\frac{\partial M_{\xi\eta}}{\partial \xi} = 0\ ,\ \text{along}\ \eta = \pm\frac{1}{2}\ . \tag{28b}$$

In expressing $\partial M_\xi/\partial\xi$, for example, by finite differences, it is necessary to use two interior points as well as the free-edge value of M_ξ in the backward difference equation. As the finite-difference net size gets larger, the number of interior points decreases, and for a relatively large net size it would be possible to have nearly all the interior moments specified in satisfying the boundary conditions before the variational principle is applied. To determine the effect that a particular choice of boundary conditions, initially satisfied in the matrix formulation, has on the natural frequencies and mode shapes obtained from Reissner's principle, four different sets of boundary conditions are investigated. Since for each set the boundary moment equation is satisfied, the results obtained will actually indicate the effect of the manner in which the shear boundary equation is satisfied. The twisting moments are taken to be zero at the free corners for all cases.

Case 1. Shear Boundary Conditions Partially Satisfied. – For this case $\partial M_{\xi\eta}/\partial\eta = 0$ along the free edge where ξ is a constant, and $\partial M_{\xi\eta}/\partial\xi = 0$ along $\eta = \pm 0.5$.

Case 2. Shear Boundary Conditions Completely Satisfied. – These conditions require $\partial M_\xi/\partial\xi - 2\,\partial M_{\xi\eta}/\partial\eta = 0$ along the free edge where ξ is a constant, and $\partial M_\eta/\partial\eta - 2\,\partial M_{\xi\eta}/\partial\xi = 0$ along $\eta = \pm 0.5$.

Case 3. Shear Boundary Conditions Overspecified. – In the matrix formulation the following conditions are satisfied: $\partial M_{\xi}/\partial\xi = 0$, $\partial M_{\xi\eta}/\partial\eta = 0$ along the free edge where ξ is a constant, and $\partial M_{\eta}/\partial\eta = 0$, $\partial M_{\xi\eta}/\partial\xi = 0$ along $\eta = \pm 0.5$.

Case 4. Shear Boundary Conditions Not Satisfied. – The only boundary conditions which are satisfied for this case are the boundary moments and the twisting moments at the free corners, as previously mentioned.

Analysis of Square Flat Plate with Completely Fixed Root

For the square flat cantilever plate shown in Fig. 3, the assumed deflection expression is of the form

$$\bar{w} = A_1\Phi_1 + A_2\Phi_2 + \eta(C_1\Phi_1 + C_2\Phi_2) + \frac{\eta^2}{2}(C_3\Phi_1 + C_4\Phi_2) , \tag{29}$$

where Φ_1 is the first clamped-free beam function tabulated in ref [8], and Φ_2 is the second clamped-free beam function tabulated in ref [8]. These functions describe the shape of a uniform beam vibrating in a natural mode. The quantities Φ_1 and Φ_2 are functions of ξ. Consequently,

$$\frac{\partial^2\bar{w}}{\partial\xi^2} = A_1\frac{d^2\Phi_1}{d\xi^2} + A_2\frac{d^2\Phi_2}{d\xi^2} + \eta\left(C_1\frac{d^2\Phi_1}{d\xi^2} + C_2\frac{d^2\Phi_2}{d\xi^2}\right) + \frac{\eta^2}{2}\left(C_3\frac{d^2\Phi_1}{d\xi^2} + C_4\frac{d^2\Phi_2}{d\xi^2}\right) , \tag{30}$$

$$\frac{\partial^2\bar{w}}{\partial\eta^2} = C_3\Phi_1 + C_4\Phi_2 , \tag{31}$$

$$\frac{\partial^2\bar{w}}{\partial\xi\,\partial\eta} = C_1\frac{d\Phi_1}{d\xi} + C_2\frac{d\Phi_2}{d\xi} + \eta\left(C_3\frac{d\Phi_1}{d\xi} + C_4\frac{d\Phi_2}{d\xi}\right) . \tag{32}$$

The quantities $d\Phi_1/d\xi$, $d\Phi_2/d\xi$, $d^2\Phi_1/d\xi^2$, and $d^2\Phi_2/d\xi^2$ are also tabulated in ref [8].

Since a similar technique is used for handling each of the four shear boundary conditions, only case 1 will be considered here. Again considering Fig. 3, the moment boundary conditions require the following:

$$M_{\xi} = 0 \text{ at points 4, 9, 14, 19, 24} , \tag{33}$$

$$M_{\eta} = 0 \text{ at points 1, 2, 3, 4, 21, 22, 23, 24} . \tag{34}$$

The conditions on the free-corner twisting moments imply

$$M_{\xi\eta} = 0 \text{ at points 4 and 24} . \tag{35}$$

The boundary shear conditions require that along $\xi = 1.0$, $\partial M_{\xi\eta}/\partial\eta = 0$, and along $\eta = \pm 0.5$, $\partial M_{\xi\eta}/\partial\xi = 0$. Considering point 4 along $\xi = 1.0$, the backward difference equation for satisfying $\partial M_{\xi\eta}/\partial\eta = 0$ is of the form

$$\frac{3(M_{\xi\eta})_4 - 4(M_{\xi\eta})_9 + (M_{\xi\eta})_{14}}{2\bar{a}} = 0 .$$

Since $(M_{\xi\eta})_4 = 0$, one obtains

$$4(M_{\xi\eta})_9 = (M_{\xi\eta})_{14} .$$

Similar operations along the remaining free-edge points will show that

$$M_{\xi\eta} = 0 \text{ at points 1, 2, 3, 9, 14, 19, 21, 22, 23} . \tag{36}$$

Equations (33)–(36) allow the construction of the G, H, and J matrices appearing in Eqs. (13)–(19). These three "boundary condition matrices" and the reduced moment matrices are shown on pp 14 and 15 to indicate the manner in which these conditions are imposed in the problem formulation.

The remaining submatrices corresponding to Eqs. (13)–(19) are formed in a similar manner, and the composite matrix of Eq. (23) may be formed and the eigenvalues and eigenvectors calculated.

Analysis of Square Flat Plate with Hub-Pin Support

For the square flat plate with the hub-pin support as shown in Fig. 4, the assumed deflection expression is the following:

$$\bar{w} = A_1\left(\xi^2 - \frac{1}{3}\xi^3\right) + A_2\left(\xi^2 - \frac{91}{75}\xi^3\right) + A_3\left(\xi^2 - \frac{14}{5}\xi^3 + \frac{28}{15}\xi^4\right) + B_1\eta$$
$$+ B_2\xi\eta + B_3\xi^2\eta + C_1\left[\frac{1}{2}\left(\eta^2 - \frac{2}{3}\eta^4\right)\right] + C_2\xi\left[\frac{1}{2}\left(\eta^2 - \frac{2}{3}\eta^4\right)\right.$$
$$\left. + C_3\xi^2\left[\frac{1}{2}\left(\eta^2 - \frac{2}{3}\eta^4\right)\right] = 0 , \tag{37}$$

where A_1, A_2, and A_3 are orthogonal to each other. Since the deflection is zero at the pin support, the coefficient $C_1 = -(192/23)B_1$. The second partial derivatives are

$$\frac{\partial^2\bar{w}}{\partial\xi^2} = 2\left\{A_1(1 - \xi) + A_2\left(1 - \frac{91}{25}\xi\right) + A_3\left(1 - \frac{42}{5}\xi + \frac{56}{5}\xi^2\right)\right.$$
$$\left. + B_3\eta + C_3\left[\frac{1}{2}\left(\eta^2 - \frac{2}{3}\eta^4\right)\right]\right\}, \tag{38}$$

$$\frac{\partial^2\bar{w}}{\partial\eta^2} = (1 - 4\eta^2)\left(-\frac{192}{23}B_1 + C_2\xi + C_3\xi^2\right), \tag{39}$$

$$M_\xi = \begin{bmatrix} (M_\xi)_0 \\ (M_\xi)_1 \\ (M_\xi)_2 \\ (M_\xi)_3 \\ (M_\xi)_4 \\ (M_\xi)_5 \\ (M_\xi)_6 \\ (M_\xi)_7 \\ (M_\xi)_8 \\ (M_\xi)_9 \\ (M_\xi)_{10} \\ (M_\xi)_{11} \\ (M_\xi)_{12} \\ (M_\xi)_{13} \\ (M_\xi)_{14} \\ (M_\xi)_{15} \\ (M_\xi)_{16} \\ (M_\xi)_{17} \\ (M_\xi)_{18} \\ (M_\xi)_{19} \\ (M_\xi)_{20} \\ (M_\xi)_{21} \\ (M_\xi)_{22} \\ (M_\xi)_{23} \\ (M_\xi)_{24} \end{bmatrix} = \begin{bmatrix} 10000000000000000000 \\ 01000000000000000000 \\ 00100000000000000000 \\ 00010000000000000000 \\ 00000000000000000000 \\ 00001000000000000000 \\ 00000100000000000000 \\ 00000010000000000000 \\ 00000001000000000000 \\ 00000000000000000000 \\ 00000000100000000000 \\ 00000000010000000000 \\ 00000000001000000000 \\ 00000000000100000000 \\ 00000000000000000000 \\ 00000000000010000000 \\ 00000000000001000000 \\ 00000000000000100000 \\ 00000000000000010000 \\ 00000000000000000000 \\ 00000000000000001000 \\ 00000000000000000100 \\ 00000000000000000010 \\ 00000000000000000001 \\ 00000000000000000000 \end{bmatrix} \begin{bmatrix} (M_\xi)_0 \\ (M_\xi)_1 \\ (M_\xi)_2 \\ (M_\xi)_3 \\ (M_\xi)_5 \\ (M_\xi)_6 \\ (M_\xi)_7 \\ (M_\xi)_8 \\ (M_\xi)_{10} \\ (M_\xi)_{11} \\ (M_\xi)_{12} \\ (M_\xi)_{13} \\ (M_\xi)_{15} \\ (M_\xi)_{16} \\ (M_\xi)_{17} \\ (M_\xi)_{18} \\ (M_\xi)_{20} \\ (M_\xi)_{21} \\ (M_\xi)_{22} \\ (M_\xi)_{23} \end{bmatrix} = G\bar{M}_\xi$$

$$M_\eta = \begin{bmatrix} (M_\eta)_0 \\ (M_\eta)_1 \\ (M_\eta)_2 \\ (M_\eta)_3 \\ (M_\eta)_4 \\ (M_\eta)_5 \\ (M_\eta)_6 \\ (M_\eta)_7 \\ (M_\eta)_8 \\ (M_\eta)_9 \\ (M_\eta)_{10} \\ (M_\eta)_{11} \\ (M_\eta)_{12} \\ (M_\eta)_{13} \\ (M_\eta)_{14} \\ (M_\eta)_{15} \\ (M_\eta)_{16} \\ (M_\eta)_{17} \\ (M_\eta)_{18} \\ (M_\eta)_{19} \\ (M_\eta)_{20} \\ (M_\eta)_{21} \\ (M_\eta)_{22} \\ (M_\eta)_{23} \\ (M_\eta)_{24} \end{bmatrix} = \begin{bmatrix} 10000000000000000 \\ 00000000000000000 \\ 00000000000000000 \\ 00000000000000000 \\ 00000000000000000 \\ 01000000000000000 \\ 00100000000000000 \\ 00010000000000000 \\ 00001000000000000 \\ 00000100000000000 \\ 00000010000000000 \\ 00000001000000000 \\ 00000000100000000 \\ 00000000010000000 \\ 00000000001000000 \\ 00000000000100000 \\ 00000000000010000 \\ 00000000000001000 \\ 00000000000000100 \\ 00000000000000010 \\ 00000000000000001 \\ 00000000000000000 \\ 00000000000000000 \\ 00000000000000000 \\ 00000000000000000 \end{bmatrix} \begin{bmatrix} (M_\eta)_0 \\ (M_\eta)_5 \\ (M_\eta)_6 \\ (M_\eta)_7 \\ (M_\eta)_8 \\ (M_\eta)_9 \\ (M_\eta)_{10} \\ (M_\eta)_{11} \\ (M_\eta)_{12} \\ (M_\eta)_{13} \\ (M_\eta)_{14} \\ (M_\eta)_{15} \\ (M_\eta)_{16} \\ (M_\eta)_{17} \\ (M_\eta)_{18} \\ (M_\eta)_{19} \\ (M_\eta)_{20} \end{bmatrix} = H\bar{M}_\eta$$

The U matrix of Eq. 13 is formed from Eq. 29 and the table in Ref. [8], and is given by

$$U = \begin{bmatrix} 7.02304 & 4.62543 & 2.38755 & .68492 & 0.00000 & 7.02304 & 4.62543 & 2.38755 & .68492 & 0.00000 \\ 44.06898 & -5.92838 & -31.45049 & -18.38161 & 0.00000 & 44.06898 & -5.92838 & -31.45049 & -18.38161 & 0.00000 \\ 3.51602 & 2.31271 & 1.19377 & .34246 & 0.00000 & 1.75801 & 1.15635 & .59688 & .17123 & 0.00000 \\ 22.03449 & -2.96419 & -15.72524 & -9.19080 & 0.00000 & 11.01724 & -1.48209 & -7.86262 & -4.59540 & 0.00000 \\ .87900 & .57818 & .29844 & .08561 & 0.00000 & .21975 & .14454 & .07461 & .02140 & 0.00000 \\ 5.50862 & -.74105 & -3.93131 & -2.29770 & 0.00000 & 1.37716 & -.18526 & -.98283 & -.57442 & 0.00000 \end{bmatrix}$$

Using the previously defined U, T, and G matrices

$$\iint M_\xi \frac{\partial^2 \bar{w}}{\partial \xi^2} \, d\xi d\eta = \frac{\bar{a}^2}{4} \bar{w} \, UTG\bar{M}_\xi$$

$$= \frac{\bar{a}^2}{4} \begin{bmatrix} A_1, A_2, C_1, C_2, C_3, C_4 \end{bmatrix} \begin{bmatrix} 7.02304 & 9.25086 & 4.77510 & 1.36984 & 14.04608 & 18.50172 & 9.55020 & 2.73968 & 14.04608 & 18.50172 \\ 44.06898 & -11.85676 & -62.90098 & -36.76322 & 88.13796 & -23.71352 & -125.80196 & -73.52644 & 88.13796 & -23.71352 \\ 3.51602 & 4.62542 & 2.38754 & 0.68492 & 3.51602 & 4.62542 & 2.38754 & 0.68492 & 0.00000 & 0.00000 \\ 22.03449 & -5.92838 & -31.45048 & -18.38160 & 22.03449 & -5.92838 & -31.45048 & -18.38160 & 0.00000 & 0.00000 \\ .87900 & 1.15636 & .59688 & .17122 & .43950 & .57816 & .29844 & .08560 & 0.00000 & 0.00000 \\ 5.50862 & -1.48210 & -7.86262 & -4.59540 & 2.75432 & -0.74104 & -3.93132 & -2.29768 & 0.00000 & 0.00000 \end{bmatrix}$$

$$M_{\xi\eta} = \begin{bmatrix} (M_{\xi\eta})_0 \\ (M_{\xi\eta})_1 \\ (M_{\xi\eta})_2 \\ (M_{\xi\eta})_3 \\ (M_{\xi\eta})_4 \\ (M_{\xi\eta})_5 \\ (M_{\xi\eta})_6 \\ (M_{\xi\eta})_7 \\ (M_{\xi\eta})_8 \\ (M_{\xi\eta})_9 \\ (M_{\xi\eta})_{10} \\ (M_{\xi\eta})_{11} \\ (M_{\xi\eta})_{12} \\ (M_{\xi\eta})_{13} \\ (M_{\xi\eta})_{14} \\ (M_{\xi\eta})_{15} \\ (M_{\xi\eta})_{16} \\ (M_{\xi\eta})_{17} \\ (M_{\xi\eta})_{18} \\ (M_{\xi\eta})_{19} \\ (M_{\xi\eta})_{20} \\ (M_{\xi\eta})_{21} \\ (M_{\xi\eta})_{22} \\ (M_{\xi\eta})_{23} \\ (M_{\xi\eta})_{24} \end{bmatrix} \begin{bmatrix} 10000000000000 \\ 00000000000000 \\ 00000000000000 \\ 00000000000000 \\ 00000000000000 \\ 01000000000000 \\ 00100000000000 \\ 00010000000000 \\ 00001000000000 \\ 00000000000000 \\ 00000100000000 \\ 00000010000000 \\ 00000001000000 \\ 00000000100000 \\ 00000000000000 \\ 00000000010000 \\ 00000000001000 \\ 00000000000100 \\ 00000000000010 \\ 00000000000000 \\ 00000000000001 \\ 00000000000000 \\ 00000000000000 \\ 00000000000000 \\ 00000000000000 \end{bmatrix} \begin{bmatrix} (M_{\xi\eta})_0 \\ (M_{\xi\eta})_5 \\ (M_{\xi\eta})_6 \\ (M_{\xi\eta})_7 \\ (M_{\xi\eta})_8 \\ (M_{\xi\eta})_{10} \\ (M_{\xi\eta})_{11} \\ (M_{\xi\eta})_{12} \\ (M_{\xi\eta})_{13} \\ (M_{\xi\eta})_{15} \\ (M_{\xi\eta})_{16} \\ (M_{\xi\eta})_{17} \\ (M_{\xi\eta})_{18} \\ (M_{\xi\eta})_{20} \end{bmatrix} = J\bar{M}_{\xi\eta}$$

7.02304	4.62543	2.38755	.68492	0.00000	7.02304	4.62543	2.38755	.68492	0.00000	7.02304	4.62543	2.38755	.68492	0.00000
44.06898	-5.92838	-31.45049	-18.38161	0.00000	44.06898	-5.92838	-31.45049	-18.38161	0.00000	44.6898	-5.92838	-31.45049	-18.38161	0.00000
0.00000	0.00000	0.00000	0.00000	0.00000	-1.75801	-1.15635	-.59688	-.17123	0.00000	-3.51602	-2.31271	-1.19377	-.34246	0.00000
0.00000	0.00000	0.00000	0.00000	0.00000	-11.01724	1.48209	7.86262	4.59540	0.00000	-22.03449	2.96519	15.72524	9.19080	0.00000
0.00000	0.00000	0.00000	0.00000	0.00000	.21975	.14454	.07461	.02140	0.00000	.87900	.57818	.29844	.08561	0.00000
0.00000	0.00000	0.00000	0.00000	0.00000	1.37716	-.18526	-.98283	-.57442	0.00000	5.50862	-.74105	-3.93131	-2.29770	0.00000

9.55020	2.73968	14.04608	18.50172	9.55020	2.73968	7.02304	9.25086	4.77510	1.36984
-125.80196	-73.52644	88.13796	-23.71352	-125.80196	-73.52644	44.06898	-11.85676	-62.90098	-36.76322
0.00000	0.00000	-3.51602	-4.62542	-2.38754	-0.68492	-3.51602	-4.62542	-2.38754	-0.68492
0.00000	0.00000	-22.03449	5.92838	31.45048	18.38160	-22.03449	5.92838	31.45048	18.38160
0.00000	0.00000	.43950	.57816	.29844	.08560	.87900	1.15636	.59688	.17122
0.00000	0.00000	2.75432	-.74104	-3.93132	-2.29768	5.05862	-1.48210	-7.86262	-4.59540

$$\begin{bmatrix} (M_{\xi})_0 \\ (M_{\xi})_1 \\ (M_{\xi})_2 \\ (M_{\xi})_3 \\ (M_{\xi})_5 \\ (M_{\xi})_6 \\ (M_{\xi})_7 \\ (M_{\xi})_8 \\ (M_{\xi})_{10} \\ (M_{\xi})_{11} \\ (M_{\xi})_{12} \\ (M_{\xi})_{13} \\ (M_{\xi})_{15} \\ (M_{\xi})_{16} \\ (M_{\xi})_{17} \\ (M_{\xi})_{18} \\ (M_{\xi})_{20} \\ (M_{\xi})_{21} \\ (M_{\xi})_{22} \\ (M_{\xi})_{23} \end{bmatrix}$$

$$\frac{\partial^2 \bar{w}}{\partial \xi \, \partial \eta} = B_2 + 2B_3 \xi + \left(\eta - \frac{4}{3} \eta^3 \right) (C_2 + 2C_3 \xi) \, . \tag{40}$$

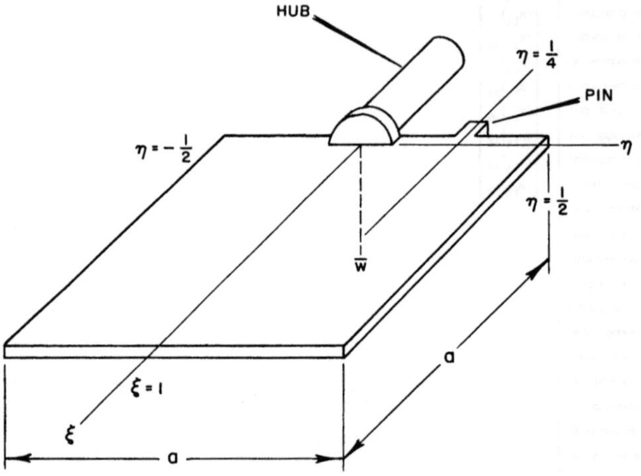

Fig. 4. Square Plate with Hub-Pin Root Support.

Considering case 1 and referring to Fig. 4, the moment boundary conditions require the following:

$$M_\xi = 0 \text{ at points } 0,\ 4,\ 5,\ 9,\ 14,\ 15,\ 19,\ 20,\ 24 \ , \tag{41}$$

$$M_\eta = 0 \text{ at points } 0,\ 1,\ 2,\ 3,\ 4,\ 20,\ 21,\ 22,\ 23,\ 24 \ . \tag{42}$$

The conditions on the free-corner twisting moment imply

$$M_{\xi\eta} = 0 \text{ at points } 0,\ 4,\ 20,\ 24 \ .$$

The boundary shear conditions require that

$$\frac{\partial M_{\xi\eta}}{\partial \eta} = 0 \text{ along the free edge where } \xi \text{ is constant} \ ,$$

$$\frac{\partial M_{\xi\eta}}{\partial \xi} = 0 \text{ along } \eta = \pm 0.5 \ .$$

By using the same analysis as explained in the preceding section, it is found that

$$M_{\xi\eta} = 0 \text{ at points } 1,\ 2,\ 3,\ 9,\ 14,\ 15,\ 19,\ 21,\ 22,\ 23 \ . \tag{43}$$

Equations (41)–(43) allow the construction of the G, H, and J matrices that appear in Eqs. (13)–(19). These three ''boundary condition matrices'' are shown on pp 18 and 19.

The remaining submatrices corresponding to Eqs. (13)–(19) are formed in a similar manner, and the composite matrix of Eq. (23) may be formed and the eigenvalues and eigenvectors calculated.

Complete Results of Boundary Conditions Investigated for Both Plate Configurations

The results obtained for the first three natural modes of the vibrating plates are shown in Tables 1 and 2. These theoretical results are presented in the form of nondimensional frequencies, and the experimental results are shown for comparison.

Table 1. Natural Frequencies (Ω) of Square Cantilever Plate with Root Completely Clamped

Mode	Reissner Theory				Experiment
	Case 1	Case 2	Case 3	Case 4	
1	3.43	3.15	3.41	3.44	3.33
2	6.13	9.26	6.13	8.21	8.17
3	19.46	17.93	16.86	21.09	19.97

Table 2. Natural Frequencies (Ω) of Square Cantilever Plate with Hub-Pin Support

Mode	Reissner Theory				Experiment
	Case 1	Case 2	Case 3	Case 4	
1	2.51	2.31	1.89	2.67	2.77
2	5.13	5.27	4.90	5.86	5.63
3	13.39	10.46	10.18	15.87	16.0

COMPARISON WITH EXPERIMENT

Experimental Apparatus

The experimental results to be presented in this section were obtained with the apparatus shown in Fig. 5. The device consists of two major components, an electromagnetic exciting system and a capacity transducer system used for measuring deflections of the various plates as they vibrate in their natural modes. The theory and the construction of the device are discussed in ref [9]; only its basic operation will be mentioned here.

$$M_\eta = \begin{bmatrix} (M_\eta)_0 \\ (M_\eta)_1 \\ (M_\eta)_2 \\ (M_\eta)_3 \\ (M_\eta)_4 \\ (M_\eta)_5 \\ (M_\eta)_6 \\ (M_\eta)_7 \\ (M_\eta)_8 \\ (M_\eta)_9 \\ (M_\eta)_{10} \\ (M_\eta)_{11} \\ (M_\eta)_{12} \\ (M_\eta)_{13} \\ (M_\eta)_{14} \\ (M_\eta)_{15} \\ (M_\eta)_{16} \\ (M_\eta)_{17} \\ (M_\eta)_{18} \\ (M_\eta)_{19} \\ (M_\eta)_{20} \\ (M_\eta)_{21} \\ (M_\eta)_{22} \\ (M_\eta)_{23} \\ (M_\eta)_{24} \end{bmatrix} = \begin{bmatrix} 000000000000000 \\ 000000000000000 \\ 000000000000000 \\ 000000000000000 \\ 000000000000000 \\ 100000000000000 \\ 010000000000000 \\ 001000000000000 \\ 000100000000000 \\ 000010000000000 \\ 000001000000000 \\ 000000100000000 \\ 000000010000000 \\ 000000001000000 \\ 000000000100000 \\ 000000000010000 \\ 000000000001000 \\ 000000000000100 \\ 000000000000010 \\ 000000000000001 \\ 000000000000000 \\ 000000000000000 \\ 000000000000000 \\ 000000000000000 \\ 000000000000000 \end{bmatrix} \begin{bmatrix} (M_\eta)_5 \\ (M_\eta)_6 \\ (M_\eta)_7 \\ (M_\eta)_8 \\ (M_\eta)_9 \\ (M_\eta)_{10} \\ (M_\eta)_{11} \\ (M_\eta)_{12} \\ (M_\eta)_{13} \\ (M_\eta)_{14} \\ (M_\eta)_{15} \\ (M_\eta)_{16} \\ (M_\eta)_{17} \\ (M_\eta)_{18} \\ (M_\eta)_{19} \end{bmatrix} = H\bar{M}_\eta$$

$$M_\xi = \begin{bmatrix} (M_\xi)_0 \\ (M_\xi)_1 \\ (M_\xi)_2 \\ (M_\xi)_3 \\ (M_\xi)_4 \\ (M_\xi)_5 \\ (M_\xi)_6 \\ (M_\xi)_7 \\ (M_\xi)_8 \\ (M_\xi)_9 \\ (M_\xi)_{10} \\ (M_\xi)_{11} \\ (M_\xi)_{12} \\ (M_\xi)_{13} \\ (M_\xi)_{14} \\ (M_\xi)_{15} \\ (M_\xi)_{16} \\ (M_\xi)_{17} \\ (M_\xi)_{18} \\ (M_\xi)_{19} \\ (M_\xi)_{20} \\ (M_\xi)_{21} \\ (M_\xi)_{22} \\ (M_\xi)_{23} \\ (M_\xi)_{24} \end{bmatrix} = \begin{bmatrix} 0000000000000000 \\ 1000000000000000 \\ 0100000000000000 \\ 0010000000000000 \\ 0000000000000000 \\ 0000000000000000 \\ 0001000000000000 \\ 0000100000000000 \\ 0000010000000000 \\ 0000000000000000 \\ 0000001000000000 \\ 0000000100000000 \\ 0000000010000000 \\ 0000000001000000 \\ 0000000000000000 \\ 0000000000000000 \\ 0000000000100000 \\ 0000000000010000 \\ 0000000000001000 \\ 0000000000000000 \\ 0000000000000000 \\ 0000000000000100 \\ 0000000000000010 \\ 0000000000000001 \\ 0000000000000000 \end{bmatrix} \begin{bmatrix} (M_\xi)_1 \\ (M_\xi)_2 \\ (M_\xi)_3 \\ (M_\xi)_6 \\ (M_\xi)_7 \\ (M_\xi)_8 \\ (M_\xi)_{10} \\ (M_\xi)_{11} \\ (M_\xi)_{12} \\ (M_\xi)_{13} \\ (M_\xi)_{16} \\ (M_\xi)_{17} \\ (M_\xi)_{18} \\ (M_\xi)_{21} \\ (M_\xi)_{22} \\ (M_\xi)_{23} \end{bmatrix} = G\bar{M}_\xi$$

The U matrix of Eq. 15 is formed from Eq. 38, and is given by

$$U = \begin{bmatrix} 2.00000 & 1.50000 & 1.00000 & 0.50000 & 0.00000 & 2.00000 & 1.50000 & 1.00000 & 0.50000 & 0.00000 & 2.00000 & 1.50000 \\ 2.00000 & 0.18000 & -1.64000 & -3.46000 & -5.28000 & 2.00000 & 0.18000 & -1.64000 & -3.46000 & -5.28000 & 2.00000 & 0.18000 \\ 2.00000 & -0.80000 & -0.80000 & 2.00000 & 7.60000 & 2.00000 & -0.80000 & -0.80000 & 2.00000 & 7.60000 & 2.00000 & -0.80000 \\ 0.00000 & 0.00000 & 0.00000 & 0.00000 & 0.00000 & 0.00000 & 0.00000 & 0.00000 & 0.00000 & 0.00000 & 0.00000 & 0.00000 \\ 0.00000 & 0.00000 & 0.00000 & 0.00000 & 0.00000 & 0.00000 & 0.00000 & 0.00000 & 0.00000 & 0.00000 & 0.00000 & 0.00000 \\ 1.00000 & 1.00000 & 1.00000 & 1.00000 & 1.00000 & 0.50000 & 0.50000 & 0.50000 & 0.50000 & 0.50000 & 0.00000 & 0.00000 \\ 0.00000 & 0.00000 & 0.00000 & 0.00000 & 0.00000 & 0.00000 & 0.00000 & 0.00000 & 0.00000 & 0.00000 & 0.00000 & 0.00000 \\ 0.20833 & 0.20833 & 0.20833 & 0.20833 & 0.20833 & 0.05989 & 0.05989 & 0.05989 & 0.05989 & 0.05989 & 0.00000 & 0.00000 \end{bmatrix}$$

Using the previously defined U, T, and G matrices

$$\iint M_\xi \frac{\partial^2 v}{\partial \xi^2} d\xi d\eta = \frac{\bar{a}^2}{4} \bar{v}\, UTG\bar{M}_\xi$$

$$= \frac{\bar{a}^2}{4} \begin{bmatrix} A_1, & A_2, & A_3, & B_1, & B_2, & B_3, & C_2, & C_3 \end{bmatrix} \begin{bmatrix} 3.00000 & 2.00000 & 1.00000 & 6.00000 & 4.00000 & 2.00000 \\ 0.36000 & -3.28000 & -6.92000 & 0.72000 & -6.56000 & -13.84000 \\ -1.60000 & -1.60000 & 4.00000 & -3.20000 & -3.20000 & 8.00000 \\ 0.00000 & 0.00000 & 0.00000 & 0.00000 & 0.00000 & 0.00000 \\ 0.00000 & 0.00000 & 0.00000 & 0.00000 & 0.00000 & 0.00000 \\ 2.00000 & 2.00000 & 2.00000 & 2.00000 & 2.00000 & 2.00000 \\ 0.00000 & 0.00000 & 0.00000 & 0.00000 & 0.00000 & 0.00000 \\ 0.41667 & 0.41667 & 0.41667 & 0.23958 & 0.23958 & 0.23958 \end{bmatrix}$$

$$
M_{\xi\eta} = \begin{bmatrix} (M_{\xi\eta})_0 \\ (M_{\xi\eta})_1 \\ (M_{\xi\eta})_2 \\ (M_{\xi\eta})_3 \\ (M_{\xi\eta})_4 \\ (M_{\xi\eta})_5 \\ (M_{\xi\eta})_6 \\ (M_{\xi\eta})_7 \\ (M_{\xi\eta})_8 \\ (M_{\xi\eta})_9 \\ (M_{\xi\eta})_{10} \\ (M_{\xi\eta})_{11} \\ (M_{\xi\eta})_{12} \\ (M_{\xi\eta})_{13} \\ (M_{\xi\eta})_{14} \\ (M_{\xi\eta})_{15} \\ (M_{\xi\eta})_{16} \\ (M_{\xi\eta})_{17} \\ (M_{\xi\eta})_{18} \\ (M_{\xi\eta})_{19} \\ (M_{\xi\eta})_{20} \\ (M_{\xi\eta})_{21} \\ (M_{\xi\eta})_{22} \\ (M_{\xi\eta})_{23} \\ (M_{\xi\eta})_{24} \end{bmatrix} = \begin{bmatrix} 00000000000 \\ 00000000000 \\ 00000000000 \\ 00000000000 \\ 00000000000 \\ 10000000000 \\ 01000000000 \\ 00100000000 \\ 00010000000 \\ 00000000000 \\ 00001000000 \\ 00000100000 \\ 00000010000 \\ 00000001000 \\ 00000000000 \\ 00000000000 \\ 00000000100 \\ 00000000010 \\ 00000000001 \\ 00000000000 \\ 00000000000 \\ 00000000000 \\ 00000000000 \\ 00000000000 \\ 00000000000 \end{bmatrix} \begin{bmatrix} (M_{\xi\eta})_5 \\ (M_{\xi\eta})_6 \\ (M_{\xi\eta})_7 \\ (M_{\xi\eta})_8 \\ (M_{\xi\eta})_{10} \\ (M_{\xi\eta})_{11} \\ (M_{\xi\eta})_{12} \\ (M_{\xi\eta})_{13} \\ (M_{\xi\eta})_{16} \\ (M_{\xi\eta})_{17} \\ (M_{\xi\eta})_{18} \end{bmatrix} = J\bar{M}_{\xi\eta}
$$

$$
\begin{bmatrix}
1.00000 & 0.50000 & 0.00000 & 2.00000 & 1.50000 & 1.00000 & 0.50000 & 0.00000 & 2.00000 & 1.50000 & 1.00000 & 0.50000 & 0.00000 \\
-1.64000 & -3.46000 & -5.28000 & 2.00000 & 0.18000 & -1.64000 & -3.46000 & -5.28000 & 2.00000 & 0.18000 & -1.64000 & -3.46000 & -5.28000 \\
-0.80000 & 2.00000 & 7.60000 & 2.00000 & -0.80000 & -0.80000 & 2.00000 & 7.60000 & 2.00000 & -0.80000 & -0.80000 & 2.00000 & 7.60000 \\
0.00000 & 0.00000 & 0.00000 & 0.00000 & 0.00000 & 0.00000 & 0.00000 & 0.00000 & 0.00000 & 0.00000 & 0.00000 & 0.00000 & 0.00000 \\
0.00000 & 0.00000 & 0.00000 & 0.00000 & 0.00000 & 0.00000 & 0.00000 & 0.00000 & 0.00000 & 0.00000 & 0.00000 & 0.00000 & 0.00000 \\
0.00000 & 0.00000 & 0.00000 & -0.50000 & -0.50000 & -0.50000 & -0.50000 & -0.50000 & -1.00000 & -1.00000 & -1.00000 & -1.00000 & -1.00000 \\
0.00000 & 0.00000 & 0.00000 & 0.00000 & 0.00000 & 0.00000 & 0.00000 & 0.00000 & 0.00000 & 0.00000 & 0.00000 & 0.00000 & 0.00000 \\
0.00000 & 0.00000 & 0.00000 & 0.05989 & 0.05989 & 0.05989 & 0.05489 & 0.05989 & 0.20833 & 0.20833 & 0.20833 & 0.20833 & 0.20833
\end{bmatrix}
$$

$$
\begin{bmatrix}
4.00000 & 6.00000 & 4.00000 & 2.00000 & 6.00000 & 4.00000 & 2.00000 & 3.00000 & 2.00000 & 1.00000 \\
4.00000 & 0.72000 & -6.56000 & -13.84000 & 0.72000 & -6.56000 & -13.84000 & 0.36000 & -3.28000 & -6.92000 \\
4.00000 & -3.20000 & -3.20000 & 8.00000 & -3.20000 & -3.20000 & 8.00000 & -1.60000 & -1.60000 & 4.00000 \\
0.00000 & 0.00000 & 0.00000 & 0.00000 & 0.00000 & 0.00000 & 0.00000 & 0.00000 & 0.00000 & 0.00000 \\
0.00000 & 0.00000 & 0.00000 & 0.00000 & 0.00000 & 0.00000 & 0.00000 & 0.00000 & 0.00000 & 0.00000 \\
0.00000 & 0.00000 & 0.00000 & 0.00000 & -2.00000 & -2.00000 & -2.00000 & -2.00000 & -2.00000 & -2.00000 \\
0.00000 & 0.00000 & 0.00000 & 0.00000 & 0.00000 & 0.00000 & 0.00000 & 0.00000 & 0.00000 & 0.00000 \\
0.00000 & 0.00000 & 0.00000 & 0.00000 & 0.23958 & 0.23958 & 0.23958 & 0.41667 & 0.41667 & 0.41667
\end{bmatrix}
\begin{bmatrix} (M_\xi)_1 \\ (M_\xi)_2 \\ (M_\xi)_3 \\ (M_\xi)_6 \\ (M_\xi)_7 \\ (M_\xi)_8 \\ (M_\xi)_{10} \\ (M_\xi)_{11} \\ (M_\xi)_{12} \\ (M_\xi)_{13} \\ (M_\xi)_{16} \\ (M_\xi)_{17} \\ (M_\xi)_{18} \\ (M_\xi)_{21} \\ (M_\xi)_{22} \\ (M_\xi)_{23} \end{bmatrix}
$$

Fig. 5. Experimental Apparatus Showing Hub-Pin-Supported Plate Installed in Mounting Rig.

The exciting system, shown schematically in Fig. *6a*, consists primarily of a regulated power supply and an electronic amplifier which supplies both a constant current for a constant field and a varying current of proper amplitude and phase to produce an alternating field to a pair of electromagnets. The magnetic force produced by these fields causes the plate to vibrate with an amplitude and frequency controlled by an audio oscillator. With this system there is no direct connection between the shaker and the plate, and for nonferrous plates, only a mass of the order of 3 g needs to be attached to the plate.

The capacitance transducer, shown schematically in Fig. *6b*, consists of two probes placed on the opposite sides of the plate at the point where the deflection is desired. The plate and each probe form an air dielectric capacitor, and as the vibrating plate deflects in one direction, the capacitance of one capacitor thus formed increases, while the capacitance of the other capacitor decreases. These two capacitors, whose capacitances change differentially as a function of plate position, are connected in a bridge circuit, and the output of this bridge is connected directly to an oscilloscope to indicate the deflection of the plate.

Results and Conclusion

The main advantage of applying Reissner's principle instead of the Rayleigh-Ritz technique is that the bending and twisting moments are not functions of the deflections. When applying the Rayleigh-Ritz technique, a deflection function

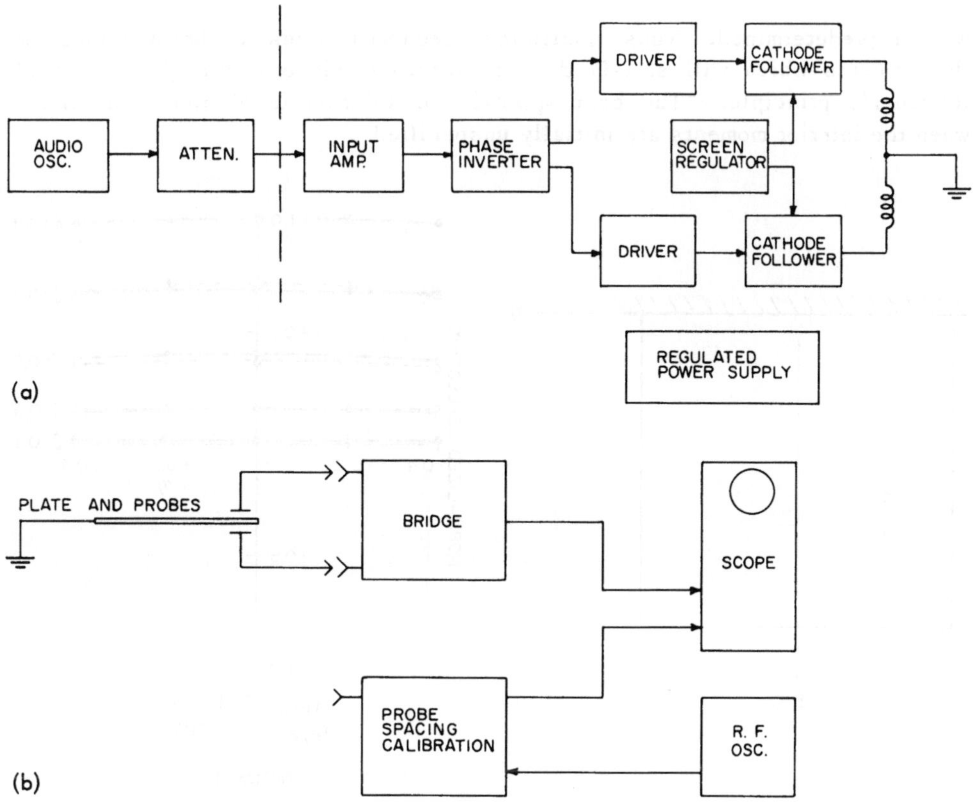

Fig. 6. (*a*) Block Diagram of Exciter; (*b*) Block Diagram of Transducer System.

is chosen, and the moments are linear combinations of the second partial derivatives of the deflection function. Since two differentiations must be performed in order to obtain the moments, it is to be expected that they are much less accurately approximated than the deflections.

The results indicate that approximate solutions to the dynamic problems of square flat plates which are completely or partially supported along the root can be obtained by using the method of finite differences and numerical integration to evaluate the integrals that occur in the modified Reissner principle. Of the four boundary-condition cases investigated, the results obtained for case 4 for both plate configurations are in closest agreement with the frequencies determined experimentally, as can be seen from Tables 1 and 2. In formulating the matrices for case 4, only the boundary moment conditions along the free edges and the twisting moment conditions at the free corners were satisfied; consequently, only the edge moments were initially specified.

In order to satisfy the various shear boundary conditions by finite differences for the other three cases, however, a great many of the interior moments are essentially already specified before Reissner's principle is applied. If the plate under consideration is divided into a small number of parts (i.e., large mesh size), as was the case in this investigation, almost all the interior moments will be predetermined. Thus insufficient freedom remains to obtain values for the moments which will satisfy the equilibrium conditions through the use of Reissner's principle. The best approximate solution is obtained, therefore, when the interior moments are initially unspecified.

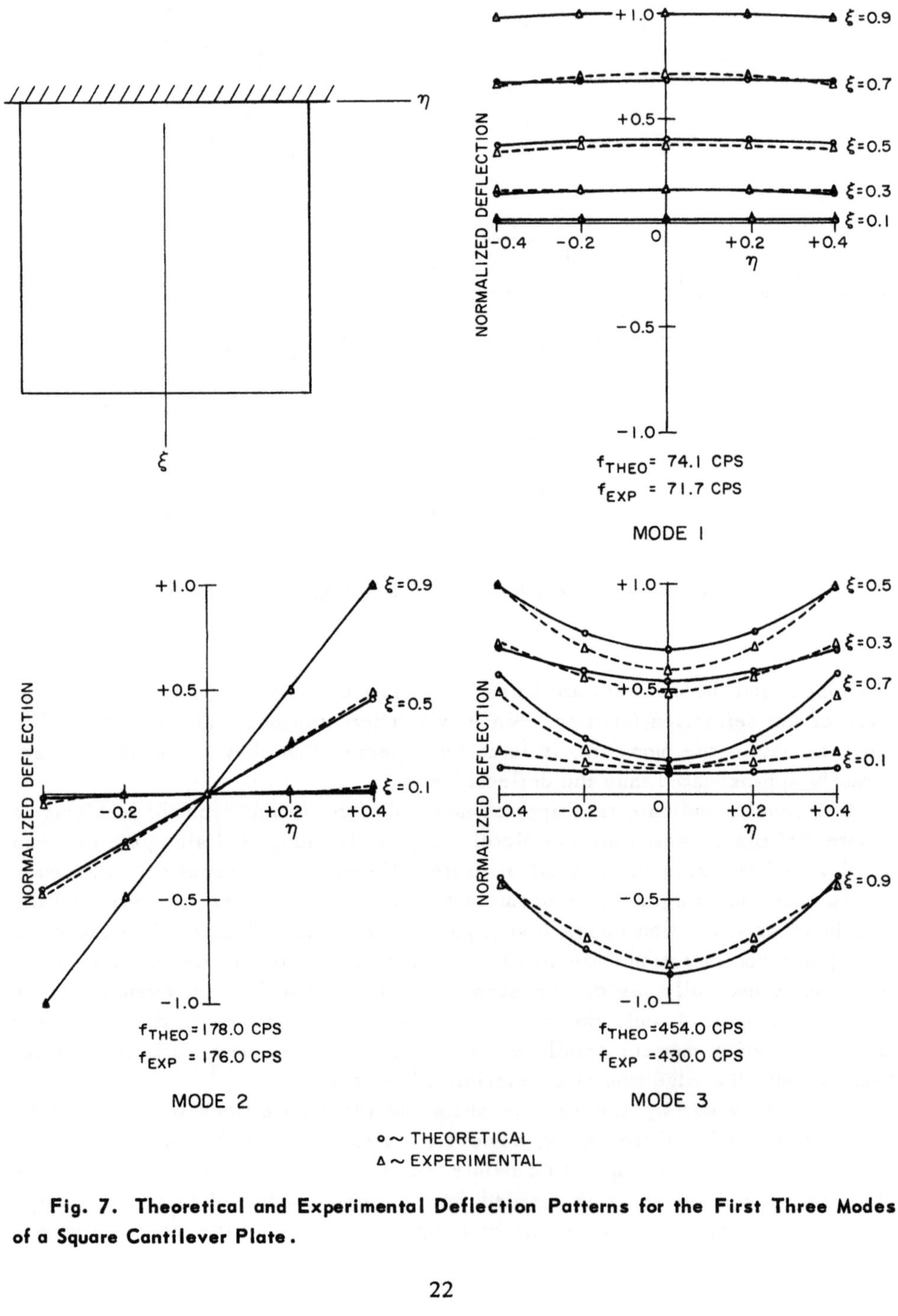

Fig. 7. Theoretical and Experimental Deflection Patterns for the First Three Modes of a Square Cantilever Plate.

Figures 7 and 8 show, respectively, the chordwise deflections for the first three modes of the cantilever plate and the hub-pin plate. Each plate used in the experimental study had an edge dimension of 7.5 in.; the cantilever plate was 0.125 in. thick and the hub-pin plate was 0.250 in. thick. Agreement between these theoretical and experimental results is good when it is considered

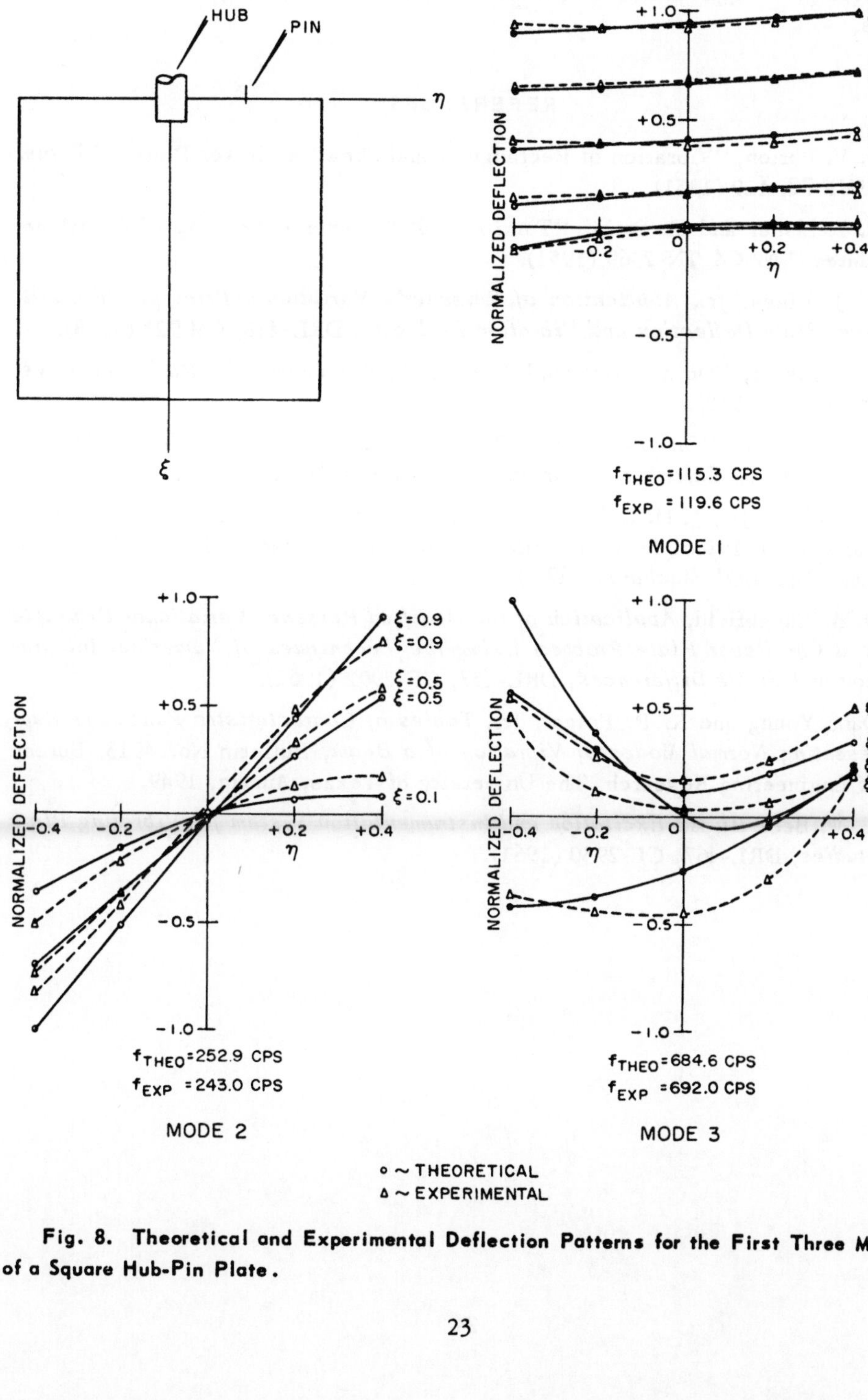

Fig. 8. Theoretical and Experimental Deflection Patterns for the First Three Modes of a Square Hub-Pin Plate.

that the plates were not divided into many parts. Accuracy may also be increased by using deflection expressions containing additional simple functions with unknown coefficients.

One disadvantage in applying Reissner's principle is that the approximate frequencies may be lower or higher than the exact values. The Rayleigh-Ritz technique gives approximate frequencies which are always higher than the exact values.

REFERENCES

[1] M. V. Barton, "Vibration of Rectangular and Skew Cantilever Plates," *Trans. ASME* **73**, 129 (1951).

[2] E. Reissner and M. Stein, "Torsion and Transverse Bending of Cantilever Plates," *NACA* TN 2369 (1951).

[3] H. J. Plass, Jr., *Application of Reissner's Variational Principle to Cantilever Plate Deflection and Vibration Problems*, DRL-418, CM-921 (1958).

[4] E. Reissner, "On a Variational Theorem in Elasticity," *J. Math. and Phys.* **29**, 90–95 (1950).

[5] E. Reissner, "On Variational Principles in Elasticity," *Proceedings of Symposium on Calculus of Variations*, McGraw-Hill, New York, 1958.

[6] H. J. Plass, Jr., J. H. Gaines, and C. D. Newsom, "Application of Reissner's Variational Principle to Cantilever Plate Deflection and Vibration Problems," *J. Appl. Mechanics* **29**, 127 (1962).

[7] D. A. Caughfield, *Application of the Modified Reissner Variational Principle to a Cantilever Plate Problem Using the Techniques of Numerical Integration and Finite Differences*, DRL-457, CF-2902 (1961).

[8] Dana Young and R. P. Felgar, Jr., *Tables of Characteristic Functions Representing Normal Modes of Vibration of a Beam*, Bulletin No. 4913, Bureau of Engineering Research, The University of Texas, Austin, 1949.

[9] C. W. Beck III, *An Excitation and Instrumentation System for Vibrating Plate Studies*, DRL-467, CF-2930 (1961).

STRAIN-WAVE PROPAGATION IN THERMOELASTIC MEDIA

Kuang Liu Cheng
Burroughs Corporation

A complete solution for thermoelastic dilatational waves in homogeneous, isotropic, heat-conducting, thermoelastic media has been obtained for both the steady and transient cases. Discussion of the solution covers a wide range of frequencies, and some of the results may be applied to the explanation of the unusual seismic waves from underground atomic explosions. Some information is also presented on the reflection and refraction of plane waves, attenuative surface waves, spherical waves, and cylindrical waves.

INTRODUCTION

The study of the propagation of elastic waves in solids dates back to the last century. Although the thermal effect on strain, as described by the Duhamel-Neumann law [*1*], has been known for many years, the converse effect – that of strain upon temperature – was actually ignored in the early development of the theory of elasticity. The present (classical) theory assumes that no change of temperature results from a change of strain. The medium undergoing the strain is understood to remain in its prior thermal state. Since, under this assumption, there is no coupling between strain and temperature, the elastic-wave velocities are isothermal. Dilatational elastic-wave velocities in solids have been studied by Jeffreys [*2*], who assumed that elastic waves – like sound velocity in air – are adiabatic.

The attenuation of elastic waves, which cannot be explained by present elasticity theory, can be described by various theories of viscoelasticity [*3*]. These theories, however, do not adequately account for the dissipation of elastic energy and also fail to accommodate the two distinct modes of dilatational waves, which have been observed by Grossling [*4*] and studied in papers by Biot [*5*], Lessen [*6*], Deresiewicz [*7*], and Chadwick and Sneddon [*8*]. The velocities observed by Grossling were those of seismic waves from an underground atomic explosion, and are much higher than the velocities described by the latter authors.

This paper describes a complete solution for the propagation of plane strain waves in homogeneous, isotropic, heat-conducting, thermoelastic media and gives some information on reflection and refraction of plane waves and on the propagation of surface waves, spherical waves, and cylindrical waves.

NOTATION

Two subscripts are used frequently: subscript 1 denotes the elastic mode; subscript 2, the thermal mode.

a_{ij}, a^{ij}	Metric, conjugate tensor
A, A_1, A_2, A_3	Constants
b	Defined by Eq. (16)
B	Base letter of body-force vector
c	Defined by Eq. (18)
C	Specific heat at constant strain
C_S, C_T	Isentropic-phase velocity and isothermal-phase velocity, respectively, of dilatational wave
C_V	Phase velocity of shear wave
C_1^*, C_2^*	Complex-phase velocity of elastic mode and thermal mode, respectively, of *P* wave
e_{ijk}	Permutation symbol without tensor character
f, f_1, f_2, g, g_1, g_2	Real numbers
f_1, f_2	Wave numbers of elastic mode and thermal mode, respectively
g_1, g_2	Coefficients of attenuation of elastic mode and thermal mode respectively
h	Constant
i	$\sqrt{-1}$
$J_{1/2}$, $J_{-1/2}$	Bessel functions
k	Thermal conductivity
m, n	Real numbers
p, q	Constants, complex numbers
P_1, P_2, Q_1, Q_2	Functions defined by Eqs. (35), (37), (36), (38)
r	Radius, a coordinate
S	Entropy of an elastic element of the medium
t	Time
T	Base letter of stress tensor
u	Base letter of displacement vector
U, V	Functions defined by Eqs. (27) and (28)
x, y, z	Rectangular coordinates
α	Coefficient of thermal expansion

$\alpha_a, \alpha_b, \alpha_c, \alpha_d$	Specific angles of incidence, reflection, or refraction
β	Constant defined by Eq. (5)
γ	Base letter of strain tensor
ϵ^{ijk}	Permutation symbol, a relative tensor of weight +1
θ	Temperature above reference temperature Θ
Θ	Reference temperature
λ, μ	Lamé constants of elasticity
ρ	Mass density
Φ, Ψ	Scalar potential and vector potential of displacement respectively

GENERAL EQUATIONS OF THERMOELASTICITY

The general equations of thermoelasticity consist of six equations of strain-displacement relations, six equations of state, three equations of motion, and one equation of heat conduction coupled with change of strain. The sixteen dependent variables are the three components of displacement, the six components of strain, the six components of stress, and temperature. The equations stated below are in tensor form and are based upon the classical theory of elasticity and the assumption of small variation of temperature due to change of strain.

$$\gamma_{ij} = \frac{1}{2}(u_{i|j} + u_{j|i}) . \qquad (1)$$

$$\gamma_{ij} = \alpha\theta a_{ij} + \frac{1}{2\mu} T_{ij} - \frac{\lambda}{2\mu(3\lambda + 2\mu)} T^{n}_{\cdot n} a_{ij} . \qquad (2)$$

$$T^{i}_{j|i} + B_j = \rho \ddot{u}_j . \qquad (3)$$

$$k a^{ij} \theta|_{ij} = C\dot{\theta} + \beta\Theta a^{ij} \dot{u}_{i|j} . \qquad (4)$$

The tensor notation used here and in the following is that of Synge and Schild, as set forth in their book on *Tensor Calculus* [9]. The tensors are noted by base symbols as follows: γ for strain, u for displacement, T for stress, B for body force, and a for the metric or conjugate tensor. The characters of the tensors are indicated by the indices. The invariants are symbolized as follows: α is the coefficient of thermal expansion, θ is temperature above the reference temperature Θ, λ and μ are Lamé constants of elasticity, ρ is mass density, k is thermal conductivity, C is specific heat at constant strain, and β is a constant given by

$$\beta = \alpha(3\lambda + 2\mu)\ . \tag{5}$$

From Eqs. (1), (2), and (3), an equation of motion in terms of displacement may be obtained, of the form

$$\mu a^{ik}(u_{k|ji} + u_{j|ik}) + \lambda a^{ik} u_{k|ji} - \beta\theta|_j + B_j = \rho \ddot{u}_j\ . \tag{6}$$

The covariant tensor of displacement can be resolved into an irrotational part and a solenoidal part, such that

$$u_i = \Phi|_i + \frac{1}{\sqrt{a}}\ a_{ij}\epsilon^{jkm}\Psi_{m|k}\ , \tag{7}$$

where Φ is a scalar potential and Ψ a vector potential of displacement, a is the determinant of the metric tensor a_{ij}, and ϵ^{jkm} is a permutation symbol and is a relative tensor of weight +1. The covariant vector Ψ_i satisfies

$$a^{ij}\Psi_{j|i} = 0\ . \tag{8}$$

With a substitution from Eq. (7), Eq. (4), that for heat conduction, can be written as

$$k a^{ij}\theta|_{ij} = C\dot{\theta} + \beta\Theta a^{ij}\dot{\Phi}|_{ij}\ , \tag{9}$$

and, when body force is negligible, the equation of motion, Eq. (6), can be decomposed into the following two equations:

$$(2\mu + \lambda)a^{ij}\Phi|_{ij} - \rho\ddot{\Phi} = \beta\theta \tag{10}$$

and

$$\mu a^{jk}\Psi_{i|jk} - \rho\ddot{\Psi}_i = 0\ . \tag{11}$$

The phase velocity associated with Ψ_i is of the shear wave

$$C_V = \sqrt{\frac{\mu}{\rho}}\ . \tag{12}$$

The shear wave is independent of temperature, while the dilatational wave associated with Φ is coupled with temperature. The isothermal phase velocity of the dilatational wave is

$$C_T = \sqrt{\frac{2\mu + \lambda}{\rho}}\ . \tag{13}$$

The dependent variable of temperature, θ, may be converted into the entropy of an elastic element of the medium by using the relation [5]

$$S = \frac{C\theta}{\Theta} + \beta u^i|_j \,. \tag{14}$$

Then, if the entropy in Eq. (13) is set equal to a constant, the isentropic-phase velocity of the dilatational wave is seen to be

$$C_S = C_T \sqrt{1+b} \,, \tag{15}$$

where

$$b = \frac{\beta^2\Theta}{\rho C C_T^2} \,. \tag{16}$$

PROPAGATION OF PLANE WAVES IN AN INFINITE MEDIUM

If θ is eliminated from Eqs. (9) and (10), it can be shown that

$$c^2 C_T^2 a^{mn} a^{ij} \Phi|_{ijmn} - a^{ij}(c^2\ddot{\Phi} + C_S^2 \dot{\Phi})|_{ij} + \dddot{\Phi} = 0 \,, \tag{17}$$

where

$$c^2 = \frac{k}{C} \,. \tag{18}$$

Let rectangular Cartesian coordinates be used, and assume a harmonic solution of Φ, of the following complex form:

$$\Phi = A \exp i(pt + qx) \,, \tag{19}$$

where A is a constant, t is time, and x is the coordinate. It is apparent, here, that the constants p and q must satisfy the algebraic equation

$$c^2 C_T^2 q^4 - (c^2 p^2 - i C_S^2 p) q^2 - i p^3 = 0 \,. \tag{20}$$

The constants p and q are complex numbers, such that

$$p = m + in \tag{21}$$

and

$$q = f + ig \,, \tag{22}$$

where m, n, f, and g are real numbers. There are two distinct sets of f and g, corresponding to one set of m and n. One of these sets is

$$f_1^2 = \frac{P_1 + \sqrt{P_1^2 + Q_1^2}}{2} \tag{23}$$

and

$$g_1^2 = \frac{-P_1 + \sqrt{P_1^2 + Q_1^2}}{2} . \tag{24}$$

Here, the functions P_1 and Q_1 are

$$P_1 = \frac{(m^2 - n^2)c^2 + nC_S^2 + \sqrt[4]{U^2 + V^2}\left[m \cos\left(\frac{\tan^{-1} V/U}{2}\right) - n \sin\left(\frac{\tan^{-1} V/U}{2}\right)\right]}{2c^2C_T^2} , \tag{25}$$

$$Q_1 = \frac{2mnc^2 - mC_S^2 + \sqrt[4]{U^2 + V^2}\left[n \cos\left(\frac{\tan^{-1} V/U}{2}\right) + m \sin\left(\frac{\tan^{-1} V/U}{2}\right)\right]}{2c^2C_T^2} , \tag{26}$$

where the functions U and V are

$$U = (m^2 - n^2)c^4 - C_S^4 - 2nc^2(2C_T^2 - C_S^2) \tag{27}$$

and

$$V = 2mc^2[nc^2 + (2C_T^2 - C_S^2)] . \tag{28}$$

The second set of f and g is

$$f_2^2 = \frac{P_2 + \sqrt{P_2^2 + Q_2^2}}{2} \tag{29}$$

and

$$g_2^2 = \frac{-P_2 + \sqrt{P_2^2 + Q_2^2}}{2} , \tag{30}$$

where the functions P_2 and Q_2 are

$$P_2 = \frac{(m^2 - n^2)c^2 + nC_S^2 - \sqrt[4]{U^2 + V^2}\left[m \cos\left(\frac{\tan^{-1} V/U}{2}\right) - n \sin\left(\frac{\tan^{-1} V/U}{2}\right)\right]}{2c^2C_T^2} \tag{31}$$

$$Q_2 = \frac{2mnc^2 - mC_S^2 - \sqrt[4]{U^2 + V^2}\left[n \cos\left(\frac{\tan^{-1} V/U}{2}\right) + m \sin\left(\frac{\tan^{-1} V/U}{2}\right)\right]}{2c^2C_T^2} . \tag{32}$$

There are two distinct phase velocities of the dilatational waves; the squares of these velocities are

$$C_1^2 = \frac{2m^2}{P_1 + \sqrt{P_1^2 + Q_1^2}} \tag{33}$$

and

$$C_2^2 = \frac{2m^2}{P_2 + \sqrt{P_2^2 + Q_2^2}} \, . \tag{34}$$

The velocities C_1 and C_2 are the phase velocities of the elastic and thermal modes of dilatational waves, respectively, since, in the limit as C_S approaches C_T, C_1 approaches the elastic dilatational wave velocity, and C_2 approaches the velocity of the temperature wave. The coefficient of attenuation of the elastic mode is g_1, while that of the thermal mode is g_2.

A brief review of thermoelastic constants (for metallic and nonmetallic materials) listed in a table in Melan and Parkus' book on thermal stresses [*10*] shows that C_T is of the order of 10^5, b is of the order of 10^{-1}, and c is of the order of 1, in cgs units. For frequencies below perhaps 10^6 cps and for the steady state (that is, where $n = 0$), the functions P's and Q's may be expanded into infinite series, such that

$$P_1 = \frac{m^2}{C_T^2(1+b)} + \left(\frac{m^4c^4}{C_T^6}\right)\left[\frac{b(1-b)}{(1+b)^5}\right] + \frac{1}{16}\left(\frac{m^6c^8}{C_T^{10}}\right)\left[\frac{1-b}{(1+b)^9}\right]$$

$$\cdot\,[3(1+b)^4 - 10(1+b)^2(1-b)^2 + 7(1-b)^4] + \ldots, \tag{35}$$

$$Q_1 = \frac{-m^3c^2b}{C_T^4(1+b)^3} - \frac{m^5c^6}{16C_T^8(1+b)^7}[(1+b)^4 - 6(1+b)^2(1-b)^2 + 5(1-b)^4]$$

$$- \frac{m^7c^{10}}{32C_T^{12}(1+b)^{11}}[(1+b)^6 - 15(1+b)^4(1-b)^2 + 35(1+b)^2(1-b)^4$$

$$- 21(1-b)^6] + \ldots, \tag{36}$$

$$P_2 = \frac{m^2b}{C_T^2(1-b)} - \left(\frac{m^4c^4}{C_T^6}\right)\left[\frac{b(1-b)}{(1+b)^5}\right] - \frac{1}{16}\left(\frac{m^6c^8}{C_T^{10}}\right)\left[\frac{1-b}{(1+b)^9}\right]$$

$$\cdot\,[3(1+b)^4 - 10(1+b)^2(1-b)^2 + 7(1-b)^4] + \ldots, \tag{37}$$

$$Q_2 = \frac{m(1+b)}{c^2} + \left[\frac{m^3c^2b}{C_T^4(1+b)^3}\right] + \frac{m^5c^6}{16C_T^8(1+b)^7}[(1+b)^4 - 6(1+b)^2(1-b)^2$$

$$+ 5(1-b)^4] + \frac{m^7c^{10}}{32C_T^{12}(1+b)^{11}}[(1+b)^6 - 15(1+b)^4(1-b)^2$$

$$+ 35(1+b)^2(1-b)^4 - 21(1-b)^6] + \dots. \quad (38)$$

The coefficients of attenuation and the phase velocities of the elastic and thermal modes are

$$g_1^2 = \frac{m^4c^4b^2}{4C_T^6(1+b)^5} - \frac{m^6c^8b^2(2-5b+b^2)}{4C_T^{10}(1+b)^9} + \dots, \quad (39)$$

$$g_2^2 = \frac{m}{2c^2}(1+b) - \frac{m^2b}{2C_T^2(1+b)} + \dots, \quad (40)$$

$$C_1^2 = C_T^2(1+b) - \frac{m^2c^4b(4-3b)}{4C_T^2(1+b)^3} + \dots, \quad (41)$$

$$C_2^2 = \frac{2mc^2}{1+b} - \frac{2m^2c^4b}{C_T^2(1+b)^3} + \dots. \quad (42)$$

At higher frequencies, the series will not converge, and must be expanded by some other means. When the frequency is of the order of 10^{11} to 10^{12} cps and when $m = C_T^2c^{-2}(1+b)$, the phase velocities of the two modes of dilatational waves may be expanded as follows:

$$C_1^2 = C_T^2\left(1 + \frac{b}{2} - \frac{5b^2}{16} - \frac{b^3}{32} - \frac{b^4}{128} + \dots\right) \quad (43)$$

and

$$C_2^2 = 2C_T^2\left(1 - \frac{b^2}{8} + \frac{b^3}{16} - \frac{9b^4}{128} + \dots\right). \quad (44)$$

At extremely high frequencies, the phase velocities of the two modes approach

$$C_1^2 = C_T^2 \quad (45)$$

and

$$C_2^2 = \frac{2mc^2}{3-b}. \quad (46)$$

During the period of rise, particularly when

$$n = -\frac{C_T^2(1-b)}{c^2} \tag{47}$$

and

$$m = \frac{2C_T^2\sqrt{b}}{c^2}, \tag{48}$$

the coefficients of attenuation and the phase velocities of the two modes become the same. Here,

$$g_1^2 = g_2^2 = \frac{[(1-(3b/4)+(3b^2/16)-(b^3/32)+\ldots]}{C_T^2} \tag{49}$$

and

$$C_1^2 = C_2^2 = C_T^2\left(\frac{16}{9} - \frac{4b}{27} + \frac{b^2}{27} + \ldots\right). \tag{50}$$

If the frequency is very low, such that

$$m \ll \frac{2C_T^2\sqrt{b}}{c^2}, \tag{51}$$

then the following approximations may be made:

$$g_1^2 = C_T^2 c^{-4}(1-b-\sqrt{b}), \tag{52}$$

$$g_2^2 = C_T^2 c^{-4}(1-b+\sqrt{b}), \tag{53}$$

$$C_1^2 = \frac{16C_T^2(1-b-\sqrt{b})}{(3-b-2\sqrt{b})^2}, \tag{54}$$

$$C_2^2 = \frac{16C_T^2(1-b+\sqrt{b})}{(3-b+2\sqrt{b})^2}. \tag{55}$$

The high-phase velocities at low frequencies of Eqs. (54) and (55) may be applied to explain the extraordinarily high-phase velocities of the seismic waves recorded from the underground atomic explosion of September 19, 1957, in Nevada [4].

During the period of decay, especially, for example, when

$$n = \frac{C_T^2(1-b)}{c^2} \tag{56}$$

and

$$m \ll \frac{C_T^2(1-b)}{c^2}, \tag{57}$$

the coefficients of attenuation and the phase velocities are such as to be satisfied by the approximations

$$g_1^2 = \frac{C_T^2(1-3b)}{c^4}, \tag{58}$$

$$g_2^2 = \frac{C_T^2(1-b)}{c^4}, \tag{59}$$

$$C_1^2 = \frac{4C_T^2(1-3b)}{(2-3b+2b^2)^2}, \tag{60}$$

$$C_2^2 = \frac{m^2c^4}{C_T^2(1-b)}. \tag{61}$$

REFLECTION AND REFRACTION OF THERMOELASTIC WAVES

In the reflection of thermoelastic waves from a plane boundary, there is a thermal-boundary condition to be satisfied, in addition to the elastic boundary conditions. The reflected waves, therefore, must consist of three distinct kinds, rather than two, as in the reflection of elastic waves. In the reflection and refraction of thermoelastic waves at an interface of two media, there are two thermal continuity conditions, in addition to the elastic continuity conditions. Both the reflected and refracted waves, therefore, must consist of one or more distinct kinds of waves in order to satisfy the two additional continuity conditions. The two distinct modes of thermoelastic dilatational waves provide the additional kinds of waves to satisfy the additional condition or conditions. Thus, when the incident wave is either an SV-wave or one of the two modes of the dilatational waves (P-waves), the reflected waves are an SV-wave, an EP-wave (elastic mode P-wave), and a TP-wave (thermal mode P-wave), and the refracted waves also consist of three kinds of waves. The reflection or refraction of an SH-wave is the same as the reflection or refraction of elastic waves.

Since phase velocities of thermoelastic P-waves depend upon frequency, the reflected or refracted angles also depend upon frequency.

Let rectangular Cartesian coordinates be used and let the medium occupy the half space $x \geqq 0$. If the boundary is thermally insulated and stress-free, the boundary conditions are

$$\theta,_1 = 0, \qquad T_{11} = 0, \qquad T_{12} = 0. \tag{62}$$

Substituting potential functions of the incident and reflected waves into these boundary conditions yields three equations for solving the amplitudes and angles of reflection of the reflected waves. If the incident wave is an EP-wave and the angle of incidence is α_a, then the following equations must be satisfied:

$$\frac{\sin \alpha_a}{C_1^*} = \frac{\sin \alpha_b}{C_1^*} = \frac{\sin \alpha_c}{C_2^*} = \frac{\sin \alpha_d}{C_V}, \tag{63}$$

where α_b, α_c, and α_d are the angles of reflection of the reflected EP-wave, TP-wave, and SV-wave, respectively, and C_1^* and C_2^* are the complex phase velocities of the two modes of P-waves (that is, ratios of the complex frequency p to the complex wave number q for each mode). The angle α_b is equal to α_a, and the angles α_c and α_d are complex numbers, in general. If the incident EP-wave is at a very low frequency and in a steady state, then

$$C_1^* = C_T(1 + b) + \frac{imc^2 b}{2C_T(\sqrt{1 + b})^3} \tag{64}$$

and

$$C_2^* = \frac{\sqrt{2}c}{\sqrt{1 + b}} - \frac{i\sqrt{2}c}{\sqrt{1 + b}}. \tag{65}$$

The angle α_c, then, is approximately zero, and the real part of the angle α_d is much larger than the imaginary part and may be determined from

$$\sin \alpha_d = \frac{C_V \sin \alpha_a}{C_S}. \tag{66}$$

THERMOELASTIC SURFACE WAVES

Where the nature of the disturbing forces is not specified, surface waves can have infinitely many solutions. One type of wave which does not attenuate has been studied by Lockett [11]. An attenuating surface wave is described here. Let the medium occupy the half space $y \geqq 0$, and let there be no displacement in the z direction. Also, let the displacement be represented by two potential functions, as follows:

$$u_i = \Phi_{,i} + e_{ijk} \Psi_{k,j}. \tag{67}$$

Here, the two parts of the displacement will be constrained to have the same phase velocity in the x direction, as well as the same coefficient of attenuation in the x direction. Both potential functions are assumed to be of the form

$$\exp\,(ipt + iqx - hy), \tag{68}$$

where the constant b is allowed to change so that the functions satisfy Eqs. (17) and (11) respectively. The potential functions, then, are

$$\Phi = A_1 \exp(ipt + iqx - b_1 y) + A_2 \exp(ipt + iqx - b_2 y) \tag{69}$$

and

$$\Psi_3 = A_3 \exp(ipt + iqx - b_3 y)\,, \qquad \Psi_1 = \Psi_2 = 0\,, \tag{70}$$

where the b's are constants which must satisfy the conditions

$$q^2 - b_1^2 = P_1 + iQ_1\,, \tag{71}$$

$$q^2 - b_2^2 = P_2 + iQ_2\,, \tag{72}$$

and

$$q^2 - b_3^2 = \frac{p^2}{C_V^2}\,. \tag{73}$$

In Eqs. (71) and (72), P_1, P_2, Q_1, and Q_2 are functions of p and are given by Eqs. (35), (36), (37), and (38).

The potential functions of Eqs. (69) and (70) must also satisfy the boundary conditions. If the boundary is thermally insulated and free from stress, the following three equations may be obtained for the determination of the complex wave number q and any two amplitudes in terms of the third:

$$(2\mu q^2 - \rho p^2)A_1 + (2\mu q^2 - \rho p^2)A_2 + 2i\mu q b_3 A_3 = 0\,, \tag{74}$$

$$2iqb_1 A_1 + 2iqb_2 A_2 - (b_3^2 + q^2)A_3 = 0\,, \tag{75}$$

and

$$[(2\mu + \lambda)(q^2 - b_1^2) - \rho p^2]b_1 A_1 + [(2\mu + \lambda)(q^2 - b_2^2) - \rho p^2]b_2 A_2 = 0\,. \tag{76}$$

Solutions will depend upon frequency and the coefficient of rise or decay. In general, since q is a complex number, a thermoelastic surface wave of this type will attenuate in the x direction. The b's are also complex numbers, in general. Hence, there is a component of the surface wave in the y direction, due to the thermal effect.

PROPAGATION OF SPHERICAL AND CYLINDRICAL THERMOELASTIC WAVES

For spherical symmetrical waves, the solenoidal part of the displacement vanishes. In the special case where $C_S = C_T$,

$$\Phi = \exp(ipt)\left[A_1 r^{-1/2} J_{1/2}\left(\frac{pr}{C_S}\right) + A_2 r^{-1/2} J_{-1/2}\left(\frac{pr}{C_S}\right)\right], \tag{77}$$

where r is radius and $J_{1/2}$ and $J_{-1/2}$ are Bessel functions. When $C_S \neq C_T$, an assumption is made that

$$\Phi = \frac{A}{r} \exp i(pt + qr) . \tag{78}$$

To satisfy the governing equation, Eq. (17), q must, when p is given, satisfy

$$c^2 C_T^2 q^4 - (c^2 p^2 - i C_S^2 p) q^2 - i p^3 = 0 . \tag{79}$$

Equation (79) is identical to Eq. (20), for the plane dilatational wave; hence, the spherical dilatational waves have two distinct modes, with phase velocities precisely those of plane waves, except that amplitude is inversely proportional to radius.

For cylindrical waves, assume that

$$\Phi = \frac{A}{\sqrt{r}} \exp i(pt + qr) . \tag{80}$$

An equation similar to Eq. (79) may then be obtained, of the form

$$c^2 C_T^2 \left(q^2 - \frac{1}{4r^2}\right)^2 - (c^2 p^2 - i C_S^2 p) \left(q^2 - \frac{1}{4r^2}\right) - i p^3 + c^2 C_T^2 \frac{1}{r^4} = 0 . \tag{81}$$

Thus cylindrical waves approach plane dilatational waves as the radius increases.

CONCLUSIONS

The linear theory of thermoelasticity gives a solution of dilatational waves with two distinct modes wherein phase velocities, coefficients of attenuation, and angles of reflection and refraction are all dependent upon frequency and coefficient of rise (or decay). The thermal effect also affects the direction of propagation of an attenuative surface wave.

Spherical dilatational waves propagate in the same manner as plane dilatational waves, except for a diminishing of amplitude with radius. Both cylindrical and spherical dilatational waves approach plane waves asymptotically.

REFERENCES

[1] I. S. Sokolnikoff, *Mathematical Theory of Elasticity*, 2d ed., p 359, McGraw-Hill, New York, 1956.

[2] H. Jeffreys, "The Thermodynamics of an Elastic Solid," *Proc. Cambridge Phil. Soc.* **26**, 101–6 (1930).

[3] H. Kolsky, *Stress Waves in Solids*, Clarendon Press, Oxford, 1953.

[4] B. F. Grossling, "Seismic Waves from the Underground Atomic Explosion in Nevada," *Bull. Seismol. Soc. Amer.* **49**, 1–10 (1959).

[5] M. A. Biot, "Thermoelasticity and Irreversible Thermodynamics," *J. Appl. Phys.* **27**, 240–53 (1956).

[6] M. Lessen, "Thermoelastic Waves and Thermoelastic Shock," *J. Mech. and Phys. Solids* **7**, 77–84 (1959).

[7] H. Deresiewicz, "Plane Waves in a Thermoelastic Solid," *J. Acoust. Soc. Am.* **29**, 204–9 (1957).

[8] P. Chadwick and I. N. Sneddon, "Plane Waves in an Elastic Solid Conducting Heat," *J. Mech. and Phys. Solids* **6**, 223–30 (1958).

[9] J. L. Synge and A. Schild, *Tensor Calculus,* University of Toronto Press, Toronto, 1954.

[10] E. Melan and H. Parkus, *Wärmespannungen,* Springer-Verlag, Vienna, 1953.

[11] F. J. Lockett, "Effect of Thermal Properties of a Solid on the Velocity of Rayleigh Waves," *J. Mech. and Phys. Solids* **7**, 71–75 (1958).

A METHOD FOR COMPUTING BEAM VIBRATION FREQUENCIES BASED ON AN ANALOGY

R. Chicurel
Virginia Polytechnic Institute

The concept of an analogous beam, in which the flexural rigidity and mass per unit length correspond, respectively, to the reciprocal of the mass per unit length and the reciprocal of the flexural rigidity of an original given beam, is introduced. It is shown that both beams possess the same natural frequencies of flexural vibration. For the purpose of an approximate calculation of these frequencies, the continuously distributed mass of the original beam is substituted for a number of concentrated masses. The analogous beam then becomes a chain of rigid links connected by pins and equipped with springs restraining the relative rotation of adjacent links. The equations of motion for the analogous beam can be solved by a procedure which consists in assuming a value for the natural frequency and calculating the deflections successively from one end of the beam to the other. In general, there will be a certain error, and one boundary condition will not be satisfied. The procedure is repeated with different values of the frequency until the error is removed. The method is illustrated by an example of a cantilever beam for which the fundamental frequency is found.

THE ANALOGY

The differential equation describing the free, flexural vibrations of an elastic beam is [*1*]

$$\frac{\partial^2}{\partial x^2}\left(EI\,\frac{\partial^2 y}{\partial x^2}\right) + \rho A\,\frac{\partial^2 y}{\partial t^2} = 0\,. \tag{1}$$

The variable y may be eliminated by introducing the new variable q defined by

$$q = EI\,\frac{\partial^2 y}{\partial x^2}\,. \tag{2}$$

From Eqs. (1) and (2),

$$\frac{\partial^2 q}{\partial x^2} + \rho A\,\frac{\partial^2 y}{\partial t^2} = 0\,. \tag{3}$$

NOTATION

a, b	Coefficients in equations of motion for analogous beam
A	Cross-sectional area of original beam
$\bar{A}$	Cross-sectional area of analogous beam
E	Modulus of elasticity of original beam
$\bar{E}$	Modulus of elasticity of analogous beam
f	Force at joint of analogous beam
F	Amplitude of f
h	Length of each section into which beam is divided
H	Mass moment of inertia of portion of analogous beam between two joints
I	Moment of inertia of cross section of original beam about the neutral axis
$\bar{I}$	Moment of inertia of cross section of analogous beam about the neutral axis
J	Mass moment of inertia of rigid body attached to end of original beam
k	Torque developed at joint of analogous beam per unit rotation of adjacent portions of beam
l	Length of beam
m	Concentrated mass in original beam, or mass of rigid body attached to end of original beam
$\bar{M}$	Bending moment acting on analogous beam
N	Number of links in analogous beam
p	Circular frequency of normal vibration
q	Deflection of analogous beam
Q	Amplitude of q
t	Time
T	Kinetic energy of analogous beam
U	Strain energy of analogous beam
x	Coordinate along beam axis
y	Deflection of original beam
$\bar{\theta}$	Angle of slope of analogous beam
λ	$h^4 \rho A p^2 / EI$
μ	Mass of a link of analogous beam
ρ	Density of original beam
$\bar{\rho}$	Density of analogous beam
i, j	Indices
$(\dot{\ })$	$d/dt\ (\ )$

Dividing both sides of Eq. (3) by ρA, differentiating twice with respect to x, and changing the order of differentiating in the second term [it is assumed that $(\partial^2/\partial x^2)(\partial^2 y/\partial t^2)$ and $(\partial^2/\partial t^2)(\partial^2 y/\partial x^2)$ are continuous],

$$\frac{\partial^2}{\partial x^2}\left(\frac{1}{\rho A}\frac{\partial^2 q}{\partial x^2}\right) + \frac{\partial^2}{\partial t^2}\left(\frac{\partial^2 y}{\partial x^2}\right) = 0 \, ,$$

or

$$\frac{\partial^2}{\partial x^2}\left(\frac{1}{\rho A}\frac{\partial^2 q}{\partial x^2}\right) + \frac{\partial^2}{\partial t^2}\left(\frac{1}{EI} q\right) = 0 \, ,$$

and, since E, I are not functions of t,

$$\frac{\partial^2}{\partial x^2}\left(\frac{1}{\rho A}\frac{\partial^2 q}{\partial x^2}\right) + \frac{1}{EI}\frac{\partial^2 q}{\partial t^2} = 0 \, . \tag{4}$$

A comparison of Eqs. (1) and (4) shows that q may be interpreted as the deflection of a beam, which hereafter will be referred to as the analogous beam, having a rigidity $\overline{EI}$ and a mass per unit length $\overline{\rho A}$ given by

$$\overline{EI} = \frac{1}{\rho A} \, , \qquad \overline{\rho A} = \frac{1}{EI} \, .$$

If y satisfies Eq. (1), then q will satisfy Eq. (4). One implication is that the two beams will have the same natural frequencies of vibration.

BOUNDARY CONDITIONS

The boundary conditions imposed on the original beam can be transformed readily into conditions on the analogous beam. For example, if for all values of t, $y = 0$ at one end, then Eq. (3) indicates that $\partial^2 q/\partial x^2 = 0$ there; also, if for all values of t, $\partial y/\partial x = 0$, then, from Eq. (3),

$$\frac{\partial}{\partial x}\left(\frac{1}{\rho A}\frac{\partial^2 q}{\partial x^2}\right) = \frac{\partial}{\partial x}\left(\overline{EI}\frac{\partial^2 q}{\partial x^2}\right) = 0 \, ;$$

that is, the shear force on the analogous beam is zero; finally, if for all values of t, $EI(\partial^2 y/\partial x^2) = 0$, then $q = 0$ according to Eq. (2); that is, the deflection of the analogous beam is zero. Thus, a simple support at one end of the original beam corresponds to a simple support at the corresponding location in the analogous beam, a built-in support to a free end, and a free end to a built-in support. Also, a simple support in the original beam at a point other than an end results in a hinge at the corresponding location in the analogous beam.

An end condition of frequent occurrence is that of a rigid object attached to the free end of a cantilever. If this object has a mass m and a mass moment of inertia J (about an axis perpendicular to the plane of vibration of the beam and through the mass center of the object), then the shear force and bending moments at the end of the beam are, respectively,

$$m\frac{\partial^2 y}{\partial t^2}=\frac{\partial}{\partial x}\left(EI\frac{\partial^2 y}{\partial x^2}\right), \qquad J\frac{\partial^2\bar{\theta}}{\partial t^2}=J\frac{\partial^3 y}{\partial t^2\partial x}=EI\frac{\partial^2 y}{\partial x^2},$$

$$m\left(\frac{1}{\rho A}\frac{\partial^2 q}{\partial x^2}\right)=\frac{\partial q}{\partial x}, \qquad J\frac{\partial}{\partial x}\left(\frac{1}{\rho A}\frac{\partial^2 q}{\partial x^2}\right)=q,$$

$$\overline{EI}\frac{\partial^2 q}{\partial x^2}=\frac{1}{m}\frac{\partial q}{\partial x}, \qquad \frac{\partial}{\partial x}\left(\overline{EI}\frac{\partial^2 q}{\partial x^2}\right)=\frac{1}{J}q,$$

which indicates that, in the analogous beam, a bending moment proportional to the slope and a shear force proportional to the deflection must be provided by the support; that is, the support consists of a torsional spring of modulus $1/m$ and a linear spring of modulus $1/J$.

COMPUTATION OF NATURAL FREQUENCIES

The natural frequencies of vibration can be found by solving either Eq. (1) or (4). If, as an approximation, the mass of the original beam is assumed to be concentrated at a few points along its length with massless elastic portions connecting these points, then the analogous beam correspondingly will consist of rigid links connected by elastic joints which may be pictured as pinned connections with spiral springs resisting deformation of the chain from the straight configuration. The contrast between the original and the analogous beams in the character of their configurations, as shown in Fig. 1, suggests that the computations necessary to obtain the natural frequencies may naturally follow different courses, depending on whether the original or the analogous beam is considered. The original beam as depicted in Fig. 1 serves as a basis for well-known methods, for example, the method of Myklestad [2] or the method of influence coefficients [3]. The method to be developed here, on the other hand, results from the formulation of equations of motion for the analogous beam as depicted in Fig. 1.

Fig. 1. Original and Analogous Beams.

Figure 2 shows three consecutive rigid portions, or links, of the analogous beam. There are N links and $(N + 1)$ joints; both are numbered successively from left to right starting with number 1. The displacement of the ith joint is q_i. For the sake of simplicity it will be assumed that all links have the same length, h, and that each one is a uniform bar. As the beam vibrates, the links execute plane motions.

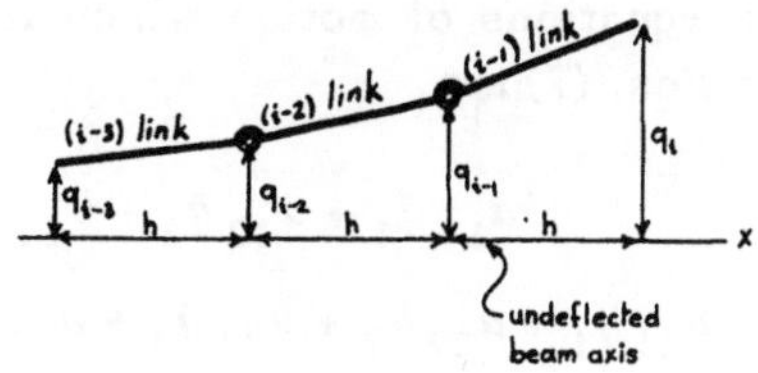

Fig. 2. Three Consecutive Links of the Analogous Beam.

The angle that the ith link makes with the x axis is $(q_{i+1} - q_i)/h$, where the usual small deflection approximation is used. The displacement of the mass center of the ith link is $(q_{i+1} + q_i)/2$. Thus, the kinetic energy in the system is

$$T = \frac{1}{8}\sum_{i=1}^{N} \mu_i(\dot{q}_i + \dot{q}_{i+1})^2 + \frac{1}{2h^2}\sum_{i=1}^{N} H_i(\dot{q}_i - \dot{q}_{i+1})^2 , \tag{5}$$

where μ_i and H_i are, respectively, the mass and mass moment of inertia about the mass center of the ith link. The strain energy is concentrated in the springs acting at the joints. The angular deflection of the spring at the ith joint, that is, the angle of rotation of the ith link relative to the $(i - 1)$st, is, for small deflections, $(q_{i+1} - 2q_i + q_{i-1})/h$, where $i = 2, 3, \ldots, N$. It will be assumed that the angular deflections of the springs at the 1st and $(N + 1)$st joints are, respectively, $(q_2 - q_1)/h$ and $(q_{N+1} - q_N)/h$; these expressions apply at least in the important cases of simple supports, free ends, and built-in ends. If k_i represents the torque developed by the ith spring per radian of deflection at the ith joint, the total strain energy in the system is

$$U = \frac{1}{2h^2}\sum_{i=2}^{N} k_i(q_{i+1} - 2q_i + q_{i-1})^2 + \frac{1}{2h^2} k_1(q_2 - q_1)^2$$

$$+ \frac{1}{2h^2} k_{N+1}(q_{N+1} - q_N)^2 . \tag{6}$$

Interpreting the q's as generalized coordinates in a Lagrangian formulation, the $(N + 1)$ equations of motion are obtained by substituting for T and U according to Eqs. (5) and (6) into the equations

$$\frac{d}{dt}\frac{\partial T}{\partial \dot{q}_j} - \frac{\partial T}{\partial q_j} + \frac{\partial U}{\partial q_j} = f_j , \qquad j = 1, \ldots, N + 1 , \tag{7}$$

where

f_1 = upward force on beam at left end,

f_{N+1} = upward force on beam at right end,

$f_j = 0, j = 2, 3, \ldots, N.$

It has been assumed that external forces act only at the ends of the beam. The formulation can, however, be easily generalized to include other cases.

The equations of motion which result after the indicated substitutions are made in Eqs. (7) are

$$\begin{aligned}
&a_{11}\ddot{q}_1 + a_{12}\ddot{q}_2 + b_{11}q_1 + b_{12}q_2 + b_{13}q_3 = f_1 , \\
&a_{21}\ddot{q}_1 + a_{22}\ddot{q}_2 + a_{23}\ddot{q}_3 + b_{21}q_1 + b_{22}q_2 + b_{23}q_3 + b_{24}q_4 = 0 , \\
&a_{i,i-2}\ddot{q}_{i-2} + a_{i,i-1}\ddot{q}_{i-1} + a_{ii}\ddot{q}_i + a_{i,i+1}\ddot{q}_{i+1} + b_{i,i-2}q_{i-2} + b_{i,i-1}q_{i-1} \\
&\quad + b_{ii}q_i + b_{i,i+1}q_{i+1} + b_{i,i+2}q_{i+2} = 0 , \quad i = 3, 4, \ldots, N-1 , \\
&a_{N,N-2}\ddot{q}_{N-2} + a_{N,N-1}\ddot{q}_{N-1} + a_{NN}\ddot{q}_N + a_{N,N+1}\ddot{q}_{N+1} + b_{N,N-2}q_{N-2} \\
&\quad + b_{N,N-1}q_{N-1} + b_{NN}q_N + b_{N,N+1}q_{N+1} + b_{N,N+2}q_{N+2} = 0 , \\
&a_{N+1,N-1}\ddot{q}_{N-1} + a_{N+1,N}\ddot{q}_N + a_{N+1,N+1}\ddot{q}_{N+1} + b_{N+1,N-1}q_{N-1} \\
&\quad + b_{N+1,N}q_N + b_{N+1,N+1}q_{N+1} = f_{N+1} ,
\end{aligned} \tag{8}$$

where

$$a_{11} = \frac{\mu_1}{4} + \frac{H_1}{h^2} ,$$

$$a_{i,i-2} = 0 , \qquad i = 3, 4, \ldots, N+1 ,$$

$$a_{i,i-1} = \frac{\mu_{i-1}}{4} - \frac{H_{i-1}}{h^2} , \qquad i = 2, 3, \ldots, N+1 ,$$

$$a_{ii} = \frac{\mu_i + \mu_{i-1}}{4} + \frac{H_i + H_{i-1}}{h^2} , \qquad i = 2, 3, \ldots, N ,$$

$$a_{i,i+1} = \frac{\mu_i}{4} - \frac{H_i}{H^2} , \qquad i = 1, 2, \ldots, N ,$$

$$a_{N+1,N+1} = \frac{\mu_N}{4} + \frac{H_N}{H^2} ,$$

$$b_{11} = \frac{k_1 + k_2}{h^2} ,$$

$$b_{12} = -\frac{k_1 + 2k_2}{h^2} ,$$

$$b_{13} = \frac{k_2}{h^2},$$

$$b_{i,i-2} = \frac{k_{i-1}}{h^2}, \qquad i = 3, 4, \ldots, N+1,$$

$$b_{i,i-1} = -\frac{2(k_{i-1} + k_i)}{h^2}, \qquad i = 3, 4, \ldots, N,$$

$$b_{ii} = \frac{k_{i-1} + 4k_i + k_{i+1}}{h^2}, \qquad i = 2, 3, \ldots, N,$$

$$b_{i,i+1} = -\frac{2(k_i + k_{i+1})}{h^2}, \qquad i = 2, 3, \ldots, N-1,$$

$$b_{i,i+2} = \frac{k_{i+1}}{h^2}, \qquad i = 2, 3, \ldots, N-1,$$

$$b_{N,N+1} = -\frac{2k_N + k_{N+1}}{h^2},$$

$$b_{N+1,N} = -\frac{2k_N + k_{N+1}}{h^2},$$

$$b_{N+1,N+1} = \frac{k_N + k_{N+1}}{h^2}. \tag{9}$$

For a normal mode, the solution of Eqs. (8) may be written as

$$q_i = Q_i \cos pt, \qquad i = 1, 2, \ldots, N+1,$$

$$f_{i1} = F_1 \cos pt,$$

$$f_{N+1} = F_{N+1} \cos pt,$$

where Q_i, F_1 are the amplitudes of q_i, f_1, respectively, and p is the circular frequency of the vibration. Inserting this solution in Eqs. (8) and solving for Q_{i+2} in the ith equation ($i = 1, 2, \ldots, N-1$) and for Q_{N+1} in each of the last two,

$$Q_3 = \sum_{j=1}^{2} \frac{p^2 a_{1j} - b_{1j}}{b_{13}} Q_j + F_1, \tag{10}$$

$$Q_4 = \sum_{j=1}^{3} \frac{p^2 a_{2j} - b_{2j}}{b_{24}} Q_j, \tag{11}$$

$$Q_{i+2} = \sum_{j=i-2}^{i+1} \frac{p^2 a_{ij} - b_{ij}}{b_{i,i+2}} Q_j, \qquad i = 3, 4, \ldots, N-1, \tag{12}$$

$$Q_{N+1} = \sum_{j=N-2}^{N} \frac{p^2 a_{Nj} - b_{Nj}}{b_{N,N+1} - p^2 a_{N,N+1}} Q_j , \tag{13}$$

$$Q_{N+1} = \sum_{j=N-1}^{N} \frac{p^2 a_{N+1,j} - b_{N+1,j}}{b_{N+1,N+1} - p^2 a_{N+1,N+1}} Q_j + \frac{F_{N+1}}{b_{N+1,N+1} - p^2 a_{N+1,N+1}} . \tag{14}$$

The simultaneous solution of the homogeneous Eqs. (10)–(14) for the Q's and F's may be carried out in the following manner.

1. A value of p^2 is assumed.
2. If the boundary conditions at the left end of the analogous beam are such that $Q_1 = 0$, then F_1 is set equal to unity. If, on the other hand, $F_1 = 0$, then Q_1 is set equal to unity. In either case, Eq. (10) gives Q_3 in terms of Q_2. Only boundary conditions of these types will be considered, but the method can easily be adapted to others.
3. The expression for Q_3 in terms of Q_2 is substituted in the right side of Eq. (11). Thus, Q_4 is expressed in terms of Q_2 alone. If Q_3 and Q_4 are now eliminated from the first of Eqs. (12) by using the expressions for them in terms of Q_2, an expression for Q_5 in terms of Q_2 will result. Proceeding in this manner successively with each of Eqs. (12), $Q_6, Q_7, \ldots, Q_{N+1}$ are each found in terms of Q_2 alone. In every case, a linear function of Q_2 will result.
4. The expressions for $Q_{N-2}, Q_{N-1}, Q_N, Q_{N+1}$ in terms of Q_2 are now inserted in Eqs. (13) and (14), which can then be regarded as simultaneous linear equations for Q_2 and F_{N+1}. Their solution will, in general, be incompatible with the boundary conditions at the right end of the analogous beam; for instance, the condition $F_{N+1} = 0$.
5. The procedure is repeated starting with step 1 and using a new value of p^2 until the boundary conditions at the right end are satisfied.

CALCULATION OF ANALOGOUS BEAM PARAMETERS

Before the calculations described above can be carried out, it is necessary to know the numerical values of all the μ_i, H_i, and k_i. Consider the original beam with its mass continuously distributed. Since the rigidity of the analogous beam is $\overline{EI} = 1/\rho A$, the curvature $d\bar{\theta}/dx$ that it will acquire due to the bending moment $\bar{M}$ will be

$$\frac{d\bar{\theta}}{dx} = \bar{M}\rho A .$$

Thus,

$$\frac{\bar{M}}{d\bar{\theta}} = \frac{1}{\rho A \, dx} , \tag{15}$$

where $\rho A\, dx$ is the mass of the original beam in the length dx. If the mass is now assumed to be concentrated at a number of discrete points and if the interval between x and dx contains one of these points, then the right side of Eq. (15) becomes $1/m$, m being the concentrated mass. But $\overline{M}/d\overline{\theta}$ is the stiffness of the spring in the analogous beam corresponding to the concentrated mass in the original beam. Thus,

$$k_i = \frac{1}{m_i}. \tag{16}$$

The mass of the ith link of the analogous beam is

$$\mu_i = (\overline{\rho A} h)_i = \left(\frac{h}{EI}\right)_i, \tag{17}$$

and its mass moment of inertia is

$$H_i = \frac{1}{12}\,[(\overline{\rho A} h) h^2]_i = \left(\frac{h^3}{12EI}\right)_i. \tag{18}$$

ILLUSTRATIVE EXAMPLE

The fundamental frequency of a uniform cantilever beam has been found by the method described in this paper. The mass was assumed to be concentrated at five points equally spaced along the beam as depicted in Fig. 3, where the analogous beam is also shown.

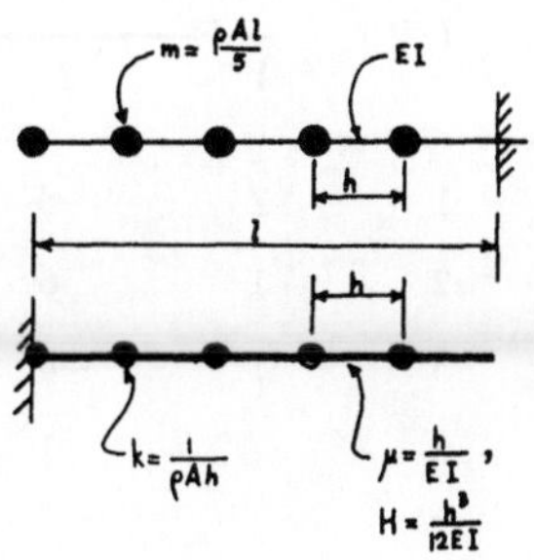

Fig. 3. Original and Analogous Beams of Original Example.

From Eqs. (16)–(18), the following expressions for the analogous beam parameters result:

$$k_1 = k_2 = k_3 = k_4 = k_5 = k_6 = \frac{1}{\rho A h},$$

$$\mu_1 = \mu_2 = \mu_3 = \mu_4 = \mu_5 = \frac{h}{EI}, \tag{19}$$

$$H_1 = H_2 = H_3 = H_4 = H_5 = \frac{h^3}{12EI},$$

from which the constants a_{ij}, b_{ij} are evaluated by using Eqs. (9). These constants, a_{ij} and b_{ij}, can be put in dimensionless forms by multiplying them, respectively, by EI/h and $h^3 \rho A$. These dimensionless constants appear in Tables 1 and 2.

Table 1. Values of the Dimensionless Coefficients $(EI/h)a_{ij}$ for the Beam of the Illustrative Example

i	j = 1	2	3	4	5	6
1	0.3333	0.1667				
2	0.1667	0.6667	0.3333			
3	0	0.3333	0.6667	0.3333		
4		0	0.1667	0.6667	0.1667	
5			0	0.1667	0.6667	0.1667
6				0	0.1667	0.3333

Table 2. Values of the Dimensionless Coefficients $(h^3 \rho A)b_{ij}$ for the Beam of the Illustrative Example

i	j = 1	2	3	4	5	6
1	2	−3	1			
2	1	6	−4	1		
3	1	−4	6	−4	1	
4		1	−4	6	−4	1
5			1	−4	5	−2
6				1	−2	1

If in Eqs. (10)–(14), a_{ij} and b_{ij} are written in terms of their dimensionless counterparts and if the numerator and denominator of each coefficient of Q_j are multiplied by $h^3 \rho A$, one is led to consider the dimensionless quantity $\lambda = h^4 \rho A p^2/EI = l^4 \rho A p^2/625EI$, rather than p^2.

The boundary conditions for the analogous beam can be expressed as $Q_1 = 0$, $F_6 = 0$. By assuming different values of λ and solving Eqs. (10)–(14) successively as described before, corresponding values of the dimensionless force at the right end of the analogous beam, $F_6/(b_{66} - p^2 a_{66})$, were obtained:

λ	$F_6/(b_{66} - p^2 a_{66})$
0.0200	−0.8655
0.0170	−0.1506
0.0160	+0.6095

The interpolated value of λ for which $F_6 = 0$ is 0.0168, which corresponds to a frequency $p = 1.798\sqrt{EI/\rho A l^4}$. The exact value for the numerical coefficient in the expression for p is 1.875 [3].

CONCLUDING REMARKS

There is a parallel in procedure between the method presented in this paper and the method of Myklestad. Essentially, it differs from the latter in that the analogous, rather than the original, beam is used as a basis for formulating equations which apply to successive portions of the beam. In either method, these equations are solved one by one after a value of the frequency has been assumed, the error in this assumption being measured by the departure from a specified boundary condition at the right end of the beam.

In Myklestad's method, 8 quantities, called amplitude coefficients, associated with each station along the beam, are defined. From their values at one station, one computes their values at the next station. This advance from one station to the next involves 16 multiplications and 14 additions. In the method presented here, the corresponding calculations, necessary to advance from one station to the next, consist in solving for a Q_i. This requires 11 multiplications and 9 additions[1] when $i = 5, 6, \ldots, N + 1$, while fewer operations are needed in the case Q_3 and Q_4. Thus, there is a saving in the amount of computational work when this method is used. On the other hand, a certain amount of labor is required in the preliminary step of calculating the constants a_{ij} and b_{ij}; however, this is done only once no matter how many natural frequencies are to be found. The method may, of course, be used to find an approximation to the natural frequency corresponding to any mode. The higher the order of the mode, however, the more stations must be taken, that is, the larger N must be.

---

[1]In verifying this statement it must be assumed that numerical values of $a_{ij}/b_{i,i+2}$ and $b_{ij}/b_{i,i+2}$ have been found initially. Also, it must be remembered that $a_{i,i-2} = 0$, $i = 3, 4, \ldots, N + 1$.

REFERENCES

[1] S. Timoshenko, *Vibration Problems in Engineering*, 2d ed., Van Nostrand, New York, 1937.

[2] N. O. Myklestad, *Fundamentals of Vibration Analysis*, McGraw-Hill, New York, 1956.

[3] L. S. Jacobsen and R. S. Ayre, *Engineering Vibrations*, McGraw-Hill, New York, 1958.

THE PLANE MOTION OF A RIGID BODY CONSTRAINED TO MOVE ON A FLEXIBLE BEAM[1]

C. C. Fretwell and M. Stippes
University of Illinois

This paper analyzes the plane motion of a rigid body constrained to move on a spring-supported flexible beam. Both the inertia of the body and the "translational" inertia of the beam are included in the analysis, but such effects as rotary inertia and shear deformation in the beam are neglected. Furthermore, the motion of the rigid body along the beam is assumed to be known. The method of solution is an extension of a method developed by C. E. Inglis. An expansion of the transverse displacement of the beam in a series of generalized coordinates, whose multipliers are the normalized modes for the spring-supported beam, leads to a system of integral equations. The dependent variables in these equations are the transverse components of acceleration of those parts of the rigid body which are in contact with the beam. A solution of these equations is obtained numerically for a specific set of parameters.

INTRODUCTION

The problem of determining the response of a beam to an excitation caused by the motion of a mass along the beam is extremely complicated by the fact that the forces acting on the beam are not known functions. Considerable attention, however, has been given to this problem, beginning in the work of Willis [*1*] over a century ago. Original studies were directed toward determining the strength of railroad beds and bridges loaded by moving locomotives. Numerous treatments, all approximations, have appeared through the years, culminating in a treatise by Inglis [2] in 1934. Interest in this type of problem has recently been revived because of its applicability to missile launching systems. However, the goals differ. Stress values, needed for proper design of bridges, are of secondary consideration in the launcher problem. The primary aim is determination of the motion of the missile at the time it leaves the launcher, since this affects the range and dispersion of the rocket. By generalization of a method developed by Inglis for treating the motion of a locomotive on a bridge regarded as a simply supported beam, it is possible to obtain a system of integral equations

[1]Sponsored by Rock Island Arsenal, Rock Island, Illinois, under a general contract (DA-11-070-508-ORD-593) with the University of Illinois (this item subcontracted to the University of Iowa).

NOTATION

$R(t)$	Thrust force on mass
W	Weight of mass
m	Mass
θ_E	Angle of elevation of beam
$\xi(t)$	Position coordinate of mass on beam
k_1, k_2	Spring rate constants for beam supports
a, b	Locations of springs with respect to rear of beam
c	Total length of beam
L	Length between supports of the mass
L_1	Length from rear support to center of gravity of the mass
$P(t), Q(t)$	Contact forces between mass and beam
EI	Flexural rigidity of beam
ρA	Mass of beam per unit length
J	Principal transverse moment of inertia of the mass
d_1	Length of travel for two-point contact
d_2	Total length of constrained travel

in a set of kinematical variables which completely defines the plane motion of the rigid body. The method may also be extended to the study of the motion of a rigid body moving on a flat plate.

The model considered here is shown schematically in Fig. 1. In the undeformed state the uniform beam $ABEF$ is straight and makes an angle θ_E with the horizontal. The x axis is taken along the beam. The beam is spring-supported at B and E and suitable constraints act to prevent axial motion of the beam. The behavior of the beam is assumed according to the Bernoulli-Euler theory. The thrust force $R(t)$ moves the rigid mass m along the beam. Devices at C and D on the mass keep it in contact with the beam. At some time T_Q, and at a position coordinate $\xi(T_Q) = d_1$, the constraints at the point D cease to act. At some

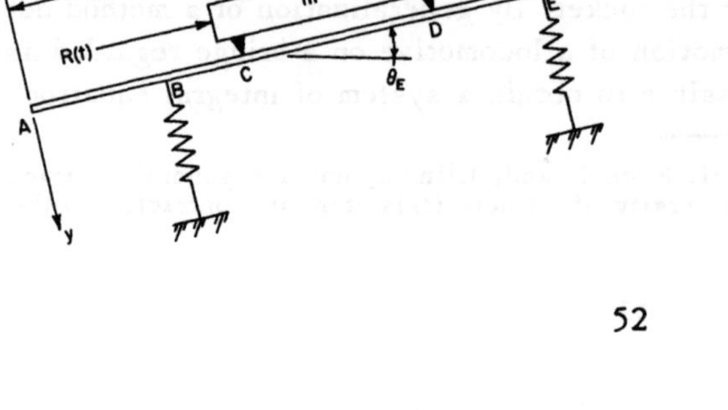

Fig. 1. Schematic Diagram of Model.

tipoff time T_P, and at $\xi(T_P) = d_2$, the constraints at C are removed and the mass is free of the beam. (The time T_P is equal to or greater than T_Q.) Our problem is to determine the rotational velocity of the mass m at tipoff. All motion is assumed to occur in the xy plane.

ANALYSIS

Figure 2 shows the free-body diagrams used in establishing the governing equations. We shall neglect friction and any interaction between the motion of the mass along the beam and changes in slope of the beam.

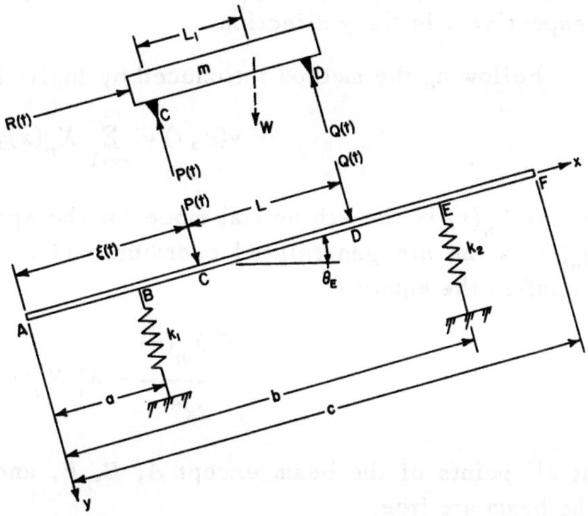

Fig. 2. Free-Body Diagrams.

If P and Q are the time-dependent contact forces between the mass and the beam, then the equation of motion for the beam is

$$EI \frac{\partial^4 y(x\,,t)}{\partial x^4} + \rho A \frac{\partial^2 y(x\,,t)}{\partial t^2} = P\delta(\xi\,, x) + Q\delta(\xi + L\,, x)\,, \qquad (1)$$

where EI is the stiffness of the beam, ρA is its mass per unit length, $y(x\,,t)$ its transverse deflection, and δ is the Dirac delta or unit impulse function. The beam is initially at rest in a displaced position due to the static load of the mass m.

Assuming small rotation and small displacement in the xy plane of the mass,

$$P(t) = \frac{L - L_1}{L} mg \cos\theta_E - \left[\frac{J}{L^2} + m\left(\frac{L - L_1}{L}\right)^2\right] \ddot{y}_P(t)$$

$$+ \left[\frac{J}{L^2} - m\frac{L_1(L - L_1)}{L^2}\right] \ddot{y}_Q(t)\,, \qquad (2)$$

$$Q(t) = \frac{L_1}{L} mg \cos \theta_E + \left[\frac{J}{L^2} - m \frac{L_1(L - L_1)}{L^2} \right] \ddot{y}_P(t) - \left[\frac{J}{L^2} + m \frac{L_1^2}{L^2} \right] \ddot{y}_Q(t) . \quad (3)$$

To obtain $\xi(t)$ we use the approximate equation

$$m\ddot{\xi}(t) = R(t) , \qquad \xi(0) = 0 , \ \dot{\xi}(0) = \xi_0 , \quad (4)$$

which is justifiable if $R(t)$ is large and the Bernoulli-Euler theory is applicable. In Eqs. (2), (3), and (4), L is the distance between C and D, L_1 is the distance from C to the center of gravity of the mass m, J is the transverse moment of inertia of the mass about its center of gravity, and m is the mass of the rigid body. Finally $\ddot{y}_P(t)$ and $\ddot{y}_Q(t)$ are the components of acceleration of C and D respectively in the y direction.

Following the method introduced by Inglis [2], we assume that

$$y(x, t) = \sum_{n=1}^{\infty} X_n(x) q_n(t) , \quad (5)$$

where $X_n(x)$ is the nth normal mode for the spring-supported beam of Fig. 2 and $q_n(t)$ is the nth generalized coordinate [1]. Each function in the set $\{X_n(x)\}$ satisfies the equation

$$\frac{d^4 X_n(x)}{dx^4} - \lambda_n^4 X_n(x) = 0 , \qquad \lambda_n^4 = \frac{\rho A}{EI} \omega_n^2 , \quad (6)$$

at all points of the beam except A, B, E, and F. Since the ends, A and F, of the beam are free,

$$X_n''(0) = X_n'''(0) = X_n''(c) = X_n'''(c) = 0 . \quad (7)$$

At B and E, X_n, X_n', X_n'' must be continuous;

$$EIX_n'''(a^-) - EIX_n'''(a^+) - k_1 X_n(a) = 0 , \quad (8)$$

$$EIX_n'''(b^-) - EIX_n'''(b^+) - k_2 X_n(b) = 0 , \quad (9)$$

where k_1 and k_2 are the spring rate constants for the supports at B and E. The functions in the set $\{X_n(x)\}$ are orthogonal, and we normalize them according to

$$\int_0^c X_n(x) X_n(x) \, dx = 1 . \quad (10)$$

Under these conditions we may write

$$\delta(\xi, x) = \sum_{n=1}^{\infty} X_n(\xi) X_n(x) , \quad (11)$$

$$\delta(\xi + L, x) = \sum_{n=1}^{\infty} X_n(\xi + L) X_n(x) . \quad (12)$$

As a consequence, Eq. (1) becomes

$$\sum_{n=1}^{\infty} [\rho A \ddot{q}_n(t) + EI\lambda_n^4 q_n] X_n(x) = \sum_{n=1}^{\infty} [PX_n(\xi) + QX_n(\xi + L)] X_n(x), \quad (13)$$

and, because $\{X_n(x)\}$ represents a complete set of functions, there results

$$\ddot{q}_n(t) + \omega_n^2 q_n(t) = \frac{P}{\rho A} X_n(\xi) + \frac{Q}{\rho A} X_n(\xi + L), \quad n = 1, 2, 3, \ldots . \quad (14)$$

The general solution of this set of equations is

$$q_n(t) = q_n(0) \cos \omega_n t + \frac{1}{\rho A \omega_n} \int_0^t \{P(\tau) X_n[\xi(\tau)]$$

$$+ Q(\tau) X_n[\xi(\tau) + L]\} \sin \omega_n(t - \tau)\, d\tau, \quad n = 1, 2, 3, \ldots, \quad (15)$$

since the beam is initially at rest. We compute $q_n(0)$ from

$$q_n(0) = \int_0^c X_n(x)\, y(x, 0)\, dx .$$

If we multiply each of Eqs. (15) by $X_n(\xi)$ and add for all n, the left side becomes $y_P(t)$. A similar process yields $y_Q(t)$. We thus obtain two integrodifferential equations involving $y_P(t)$, $y_Q(t)$, $\ddot{y}_P(t)$, and $\ddot{y}_Q(t)$. These equations can be reduced to integral equations in $\ddot{y}_P(t)$ and $\ddot{y}_Q(t)$ by noting that

$$y_P(t) = \int_0^t (t - \tau)\, \ddot{y}_P(\tau)\, d\tau + y_P(0), \quad (16)$$

$$y_Q(t) = \int_0^t (t - \tau)\, \ddot{y}_Q(\tau)\, d\tau + y_Q(0). \quad (17)$$

Performing these operations, we finally obtain

$$y_P(0) = F(t) + \int_0^t \{K_1(t, \tau) + [K_2(t, \tau) - (t - \tau)]\, u_P(\tau) + K_3(t, \tau)\, u_Q(\tau)\}\, d\tau, \quad (18)$$

$$y_Q(0) = G(t) + \int_0^t \{K_4(t, \tau) + K_5(t, \tau)\, u_P(\tau) + [K_6(t, \tau) - (t - \tau)]\, u_Q(\tau)\}\, d\tau, \quad (19)$$

where $u_P(\tau) \equiv \ddot{y}_P(\tau)$, $u_Q(\tau) \equiv \ddot{y}_Q(\tau)$, and $K_i(t, \tau)$ are various nonsymmetric kernels involving $\{X_n(x)\}$ such that $K_i(t, t) = 0$, and $F(t)$ and $G(t)$ are known functions of time. Detailed computations of these kernels are given elsewhere [3].

Equations (18) and (19) are valid for the period $0 \leqq t \leqq T_Q$. To obtain the appropriate equation for the period $T_Q \leqq t \leqq T_P$, we set $Q(t) = 0$ and carry through the same process for $y_P(t)$. Using stars to denote the condition of one-point contact, we obtain

$$(t - T_Q)\,\dot{y}_P(T_Q) + y_P(T_Q) + \frac{1}{2}(t - T_Q)^2\, g \cos\theta_E = F^*(t)$$

$$+ \int_{T_Q}^{t} [K_7^*(t,\tau) + (t-\tau)][g\cos\theta_E - u_P^*(\tau)]\, d\tau\,, \quad (20)$$

where $K_7^*(t,\tau)$ is another nonsymmetric kernel involving $\{X_n(x)\}$ and having the property $K_7^*(t,t) = 0$. It may be worth noting that a discontinuity occurs in $u_P(t)$ at the time of release of the load $Q(t)$.

NUMERICAL ANALYSIS

The integral equations were solved numerically by replacing the integrals with sums according to the trapezoidal rule of integration. As an example consider Eq. (18). If the interval $0 \leqq t \leqq T_Q$ is divided into M equal subintervals of length Δt,

$$y_P(0) = F(N\Delta t) + \frac{1}{2}\Delta t\, K_1(N\Delta t, 0) + (\Delta t) \sum_{j=1}^{N-2} \{K_1(N\Delta t, j\Delta t)$$

$$+ [K_2(N\Delta t, j\Delta t) - (N-j)\Delta t]\, u_P(j\Delta t) + K_3(N\Delta t, j\Delta t)\, u_Q(j\Delta t)\}$$

$$+ (\Delta t)\Big(K_1[N\Delta t, (N-1)\Delta t] + \{K_2[N\Delta t, (N-1)\Delta t] - \Delta t\} u_P[(N-1)\Delta t]$$

$$+ K_3[N\Delta t, (N-1)\Delta t] u_Q[(N-1)\Delta t]\Big)\,, \quad 2 \leqq N \leqq M\,, \quad (21)$$

with similar expressions for Eqs. (19) and (20). These linear equations are recursive in character; therefore there is no need for an inversion of a matrix to obtain their solution. The bulk of computation is involved in evaluating the various kernels at the subdivision points. Since the kernels are represented as infinite series, it was prohibitive to consider more than three terms in each on the digital computer used [3].

A thrust force $R(t)$ whose general character is depicted in Fig. 3 was used to determine $\xi(t)$. Table 1 gives a set of typical numerical data, which was used in the numerical work performed. The first mode frequency $\omega_1/2\pi$ for this system is 4.6 cps; $\omega_2/2\pi$ is 52.6 cps; and $\omega_3/2\pi$ is 156.3 cps. Figures 4 and 5 show graphically the results obtained for θ, the small rotation of the mass, and for $\dot{\theta}$, the corresponding angular velocity. Having determined the values of u_P, u_Q, and $g\cos\theta_E - u_P^*$ from the iteration technique, we use Simpson's integration rule to determine θ and $\dot{\theta}$ from

$$\theta = \theta(0) + \frac{1}{L}\int_0^t (t-\tau)[u_Q(\tau) - u_P(\tau)]\, d\tau\,, \qquad 0 \leqq t \leqq T_Q\,,$$

$$= (t - T_Q)\dot{\theta}(T_Q) + \theta(T_Q) + \frac{d}{J_2 L}\int_{T_Q}^{t} (t-\tau)[g\cos\theta_E - u_P^*(\tau)]\, d\tau\,,$$

$$T_Q \leqq t \leqq T_P\,, \quad (22)$$

$$\dot{\theta} = \frac{1}{L} \int_0^t [u_Q(\tau) - u_P(\tau)]\, d\tau \,, \qquad 0 \leqq t \leqq T_Q \,,$$

$$= \dot{\theta}(T_Q) + \frac{d}{J_2 L} \int_{T_Q}^{t} [g \cos \theta_E - u_P^*(\tau)]\, d\tau \,,$$

$$T_Q \leqq t \leqq T_P \,. \quad (23)$$

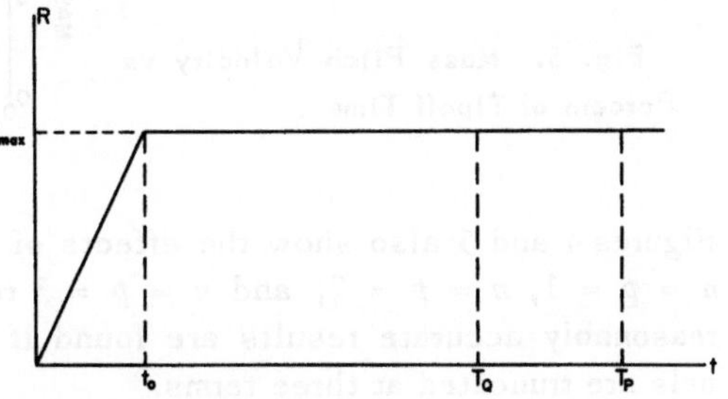

Fig. 3. Thrust Force Diagram.

Table 1. Typical Values of Parameters

Parameter	Value	Parameter	Value
$k_1 c^3/EI$	∞	$W/\rho A c$	690
$k_2 c^3/EI$	6.03	J/mL^2	0.185
a/c	0.140	ξ_0/c	0.0122
b/c	0.488	R_{max}/W	30
$EI/\rho A g c^3$	2163	d_1/c	0.55
θ_E	0.800 rad	d_2/c	0.55
L_1/L	0.694	t_0/T_P	0.0451

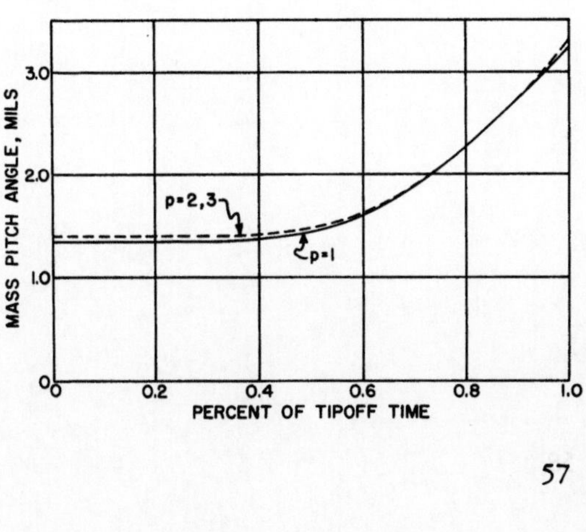

Fig. 4. Mass Pitch Angle vs Percent of Tipoff Time.

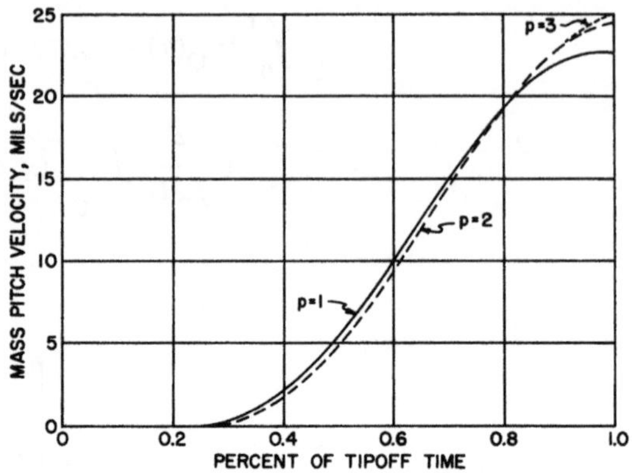

Fig. 5. Mass Pitch Velocity vs Percent of Tipoff Time.

Figures 4 and 5 also show the effects of truncating the series in the kernels at $n = p = 1$, $n = p = 2$, and $n = p = 3$ respectively. For a Δt properly chosen, reasonably accurate results are found if the series representations of the kernels are truncated at three terms.

REFERENCES

[1] S. Timoshenko and D. H. Young, *Vibrations Problems in Engineering*, 3d ed., Van Nostrand, Princeton, N.J., 1955.

[2] C. E. Inglis, *A Numerical Treatise on Vibrations in Railway Bridges*, Cambridge University Press, New York, 1934.

[3] C. C. Fretwell, "A Two Point Mass Moving on a Flexible Beam," Department of Theoretical and Applied Mechanics, University of Illinois Report, 1962.

[4] C. C. Fretwell, " The Lateral Displacements of a Spring Supported Flexible Beam on Which a Mass Moves with Constant Acceleration," Master's thesis, University of Iowa, 1960.

NEW TABLES OF EIGENFUNCTIONS REPRESENTING NORMAL MODES OF VIBRATION OF TIMOSHENKO BEAMS[1]

T. C. Huang and C. S. Kung
University of Florida

The frequency equations of highly transcendental nature for Timoshenko beams have been solved for a wide range of parameters in the bulletin, "Tables of Eigenvalues and Charts of Modifying Quotients for Frequencies and Normal Modes of Beam Vibrations." The present paper introduces new tables of eigenfunctions representing normal modes of vibration of Timoshenko beams to supplement the "Tables of Characteristic Functions Representing Normal Modes of Vibration of a Beam" (University of Texas Publication No. 4913, 1949), since the latter tables are applicable to the vibration of classical beams only. An illustrative example is given in which these new tables are used.

INTRODUCTION

Because of the inadequacy of the classical beam theory, various efforts have been devoted to its improvement, among which the beam equation according to the Timoshenko theory agrees remarkably well with the exact solution. In a recent paper [*1*] the Timoshenko beam equation was treated completely, resulting in two sets of frequency equations and consequently two sets of normal-mode equations. New frequency equations of highly transcendental nature and lengthy normal-mode equations call for tables to facilitate the solutions without the necessity of using high-speed computers in each case.

The Timoshenko beam equation and its solution are briefly outlined in order to provide necessary equations, variables, and parameters to be used later. New frequency and normal-mode equations and their reduction to the classical case are illustrated for a cantilever beam. Tables of eigenvalues and charts of modifying quotients are mentioned, followed by an explanation of the construction of two types of tables for eigenfunctions. Eight representative tables of eigenfunctions of normal modes for a cantilever and an example of application of these tables conclude the paper.

---

[1]Part of a research program sponsored by the National Science Foundation under grant NSF-G20256.

NOTATION

A	Cross-sectional area
C_i, C_i'	Constants (i = 1, 2, 3, 4)
D	Constant
E	Modulus of elasticity
G	Modulus of rigidity
H	Constant
I	Area moment of inertia of cross section
L	Length of beam
Y	Normal function of y
b	Frequency parameter, defined by Eq. (4)
j	$\sqrt{-1}$
k	Numerical shape factor for cross section
p	Angular frequency
q	Modifying quotient, defined by Eq. (18)
r, s	Parameters, defined by Eq. (4)
y	Total deflection of beam
Ψ	Normal function of ψ
α, β	Defined by Eq. (6)
γ	Weight per unit volume
δ, ζ, θ, λ	Defined by Eq. (11)
ξ	Nondimensional length of beam
ψ	Bending slope

TIMOSHENKO BEAM EQUATION AND ITS SOLUTION

The coupled-beam equations for the total deflection y and the bending slope ψ are given by Timoshenko [2] as

$$EI\frac{\partial^2\psi}{\partial t^2} + kAG\left(\frac{\partial y}{\partial x} - \psi\right) - \frac{I\gamma}{g}\frac{\partial^2\psi}{\partial t^2} = 0\,,$$

$$\frac{\gamma A}{g}\frac{\partial^2 y}{\partial t^2} - kAG\left(\frac{\partial^2 y}{\partial x^2} - \frac{\partial\psi}{\partial x}\right) = 0\,, \tag{1}$$

in which E is the modulus of elasticity, G the modulus of rigidity, A the cross-sectional area, I the area moment of inertia of cross section, γ the weight of beam per unit volume, and k the shear coefficient. When the beam vibrates harmonically, we can set

$$y = Ye^{jpt}\,, \qquad \psi = \Psi e^{jpt}\,, \qquad \xi = x/L\,, \tag{2}$$

in which Y is the normal function of y, Ψ the normal function of ψ, p the angular frequency, L the length of the beam, ξ the dimensionless length, and $j = \sqrt{-1}$. Omitting the factor e^{jpt}, Eqs. (1) are reduced to

$$\begin{aligned} &s^2\Psi'' - (1 - b^2r^2s^2)\Psi + Y'/L = 0\,, \\ &Y'' + b^2s^2Y - L\Psi' = 0\,, \end{aligned} \tag{3}$$

in which

$$b^2 = \frac{1}{EI}\frac{\gamma A}{g}L^4p^2\,, \qquad r^2 = \frac{I}{AL^2}\,, \qquad s^2 = \frac{EI}{kAGL^2}\,, \tag{4}$$

and the primes for Y and Ψ represent differentiation with respect to ξ.

The solutions of Eqs. (3), being

$$\begin{aligned} Y &= C_1 \cosh b\alpha\xi + C_2 \sinh b\alpha\xi + C_3 \cos b\beta\xi + C_4 \sin b\beta\xi\,, \\ \Psi &= C_1' \sinh b\alpha\xi + C_2' \cosh b\alpha\xi + C_3' \sin b\beta\xi + C_4' \cos b\beta\xi\,, \end{aligned} \tag{5}$$

where

$$\begin{aligned} \alpha &= \frac{\sqrt{2}}{2}\{-(r^2+s^2)+[(r^2-s^2)^2+4/b^2]^{1/2}\}^{1/2}\,, \\ \beta &= \frac{\sqrt{2}}{2}\{+(r^2+s^2)+[(r^2-s^2)^2+4/b^2]^{1/2}\}^{1/2}\,, \end{aligned} \tag{6}$$

together with appropriate boundary conditions give two sets of natural-frequency equations, according to whether

$$[(r^2 - s^2)^2 + 4/b^2]^{1/2} > (r^2 + s^2) \tag{7}$$

or

$$[(r^2 - s^2)^2 + 4/b^2]^{1/2} < (r^2 + s^2)\,. \tag{8}$$

Only the positive values of the radicals are used. Corresponding to each set of natural frequencies, there is a set of normal-mode equations.

FREQUENCY AND NORMAL-MODE EQUATIONS

Each set of frequency or normal-mode equations deals with six types of uniform beams with simple end conditions. They are (1) supported-supported, (2) free-free, (3) clamped-clamped, (4) clamped-free, (5) clamped-supported, and (6) supported-free. For example, the frequency equation for the cantilever, or clamped-free beam, in the first set is

$$2 - \frac{b(r^2 + s^2)}{\sqrt{1 - b^2 r^2 s^2}} \sinh b\alpha \sin b\beta + [b^2(r^2 - s^2)^2 + 2] \cosh b\alpha \cos b\beta = 0 \,, \quad (9)$$

and the corresponding normal-mode equations are

$$\begin{aligned} Y &= D(\cosh b\alpha\xi - \lambda\zeta\delta \sinh b\alpha\xi - \cos b\beta\xi + \delta \sin b\beta\xi) \,, \\ \Psi &= H\left(\cosh b\alpha\xi + \frac{\theta}{\lambda\zeta} \sinh b\alpha\xi - \cos b\beta\xi + \theta \sin b\beta\xi\right) \,, \end{aligned} \quad (10)$$

where

$$\begin{aligned} \lambda &= \alpha/\beta \,, \\ \zeta &= (\alpha^2 + r^2)/(\alpha^2 + s^2) \,, \\ \delta &= \frac{(1/\lambda) \sinh b\alpha - \sin b\beta}{\zeta \cosh b\alpha + \cos b\beta} \,, \\ \theta &= -\frac{\lambda \sinh b\alpha + \sin b\beta}{(1/\zeta) \cosh b\alpha + \cos b\beta} \,. \end{aligned} \quad (11)$$

The constants D and H are related by the equation of coefficients in ref [1], from which we have

$$D = \frac{L}{b\alpha}[1 - b^2 s^2(\alpha^2 + r^2)]\frac{\theta}{\lambda\zeta} H \,. \quad (12)$$

REDUCTION TO CLASSICAL CASE

The classical frequency and normal-mode equations can be obtained as a special case for which $r = s = 0$. For this case

$$\alpha = \beta = 1/\sqrt{b} \,, \qquad \lambda = \zeta = 1 \,. \quad (13)$$

Again considering the cantilever beam, the frequency equation is simplified to

$$1 + \cosh \sqrt{b} \cos \sqrt{b} = 0 \,, \quad (14)$$

and the normal-mode equations are reduced to

$$\begin{aligned} Y &= \cosh \sqrt{b}\xi - \cos \sqrt{b}\xi - \delta(\sinh \sqrt{b}\xi - \sin \sqrt{b}\xi) \,, \\ \Psi &= \cosh \sqrt{b}\xi - \cos \sqrt{b}\xi + \theta(\sinh \sqrt{b}\xi + \sin \sqrt{b}\xi) \,. \end{aligned} \quad (15)$$

These are familiar classical equations and can be found in most of the vibration texts. Timoshenko [2] uses kl for $\sqrt{b}$, and α_i for δ_i.

TABLES OF EIGENVALUES AND CHARTS OF MODIFYING QUOTIENTS

The eigenvalues b are tabulated in ref [3] for a wide range of combinations of r and s. Part 1 has $r = s = 0$ to 0.1 with an interval of 0.01, and part 2 has $r = 0$ to 0.3 and $s = 0$ to 0.5 with an interval of 0.05. For each set of r and s, the values of b can be obtained directly from these tables or by interpolation. The natural frequencies p can be calculated by using the first equation of (4).

Let b_0 be the classical eigenvalues and p_0, the classical natural frequencies corresponding to b_0. From the first equation of (4) we have

$$b/b_0 = p/p_0 \; . \tag{16}$$

Comparing the mechanism of vibration of classical and Timoshenko beams, we note that the classical beam is equivalent to the Timoshenko beam with rotating and shear constraints. Therefore

$$p_0 \geqq p \tag{17}$$

and

$$b/b_0 = p/p_0 = q \, , \qquad q \leqq 1 \; . \tag{18}$$

The ratio of b/b_0 and p/p_0, denoted by q, will be referred to as the "modifying quotient." The values of b_0 are available in the literature; for example, for the first five modes in Eq. (14),

$$b_0 = 3.5160 \, , \; 22.0344 \, , \; 61.6972, \; 120.9018 \, , \; 199.8595 \; .$$

If the modifying quotients are known, Eq. (18) immediately gives the values of b from which the natural frequencies p can be calculated.

The charts of modifying quotients are also provided in ref [3] for a wide range of r and s. Part 1 has $r = s = 0$ to 0.1, and part 2 has $r = 0$ to 0.3 and $s = 0$ to 0.5.

For example, the values of b for a cantilever beam with four combinations of r and s are:

r	s	b_1	b_2	b_3	b_4	b_5
0.02	0.04	3.49980	21.3496	57.4673	106.579	166.658
0.02	0.05	3.49256	21.0654	55.8354	102.056	156.319
0.03	0.04	3.49579	21.1925	56.5854	104.364	161.386
0.03	0.05	3.48858	20.9167	55.0485	99.9310	152.248

Alternatively, we may obtain b by applying Eq. (18), in which the value of q is obtained from the charts of modifying quotients.

TABLES OF CHARACTERISTIC FUNCTIONS

For each set of parameters r and s, the eigenvalues b can be obtained from tables of eigenvalues or can be calculated from the modifying quotients q and classical values of b. For a given beam the eigenfunctions of normal modes can be calculated from the normal-mode equations and the values of b. These eigenfunctions are tabulated in ref [4] for the corresponding set of r and s as in ref [3]. For example, the tables of eigenfunctions for the normal modes of Y and $L\Psi$ for a cantilever with four combinations of r and s are shown in Tables 1–4 and 5–8 respectively.

Table 1. Eigenfunctions of Deflection; Clamped-Free Timoshenko Beam for $r = 0.02$, $s = 0.04$

ξ	Y_1	Y_2	Y_3	Y_4	Y_5
0.00	0.0000	0.0000	0.0000	0.0000	0.0000
0.05	0.0047	−0.0323	0.0954	−0.1903	0.3123
0.10	0.0174	−0.1045	0.2721	−0.4765	0.6781
0.15	0.0378	−0.2038	0.4742	−0.7179	0.8369
0.20	0.0650	−0.3178	0.6512	−0.8091	0.6648
0.25	0.0986	−0.4344	0.7625	−0.7019	0.2151
0.30	0.1380	−0.5427	0.7817	−0.4136	−0.3165
0.35	0.1825	−0.6329	0.6997	−0.0189	−0.6893
0.40	0.2316	−0.6967	0.5242	0.3736	−0.7331
0.45	0.2847	−0.7279	0.2790	0.6542	−0.4299
0.50	0.3413	−0.7220	−0.0007	0.7445	0.0771
0.55	0.4008	−0.6769	−0.2732	0.6202	0.5500
0.60	0.4628	−0.5926	−0.4937	0.3192	0.7675
0.65	0.5268	−0.4710	−0.6378	−0.0694	0.6289
0.70	0.5924	−0.3158	−0.6709	−0.4301	0.2018
0.75	0.6591	−0.1321	−0.5873	−0.6536	−0.3091
0.80	0.7266	0.0740	−0.3930	−0.6667	−0.6557
0.85	0.7947	0.2961	−0.1075	−0.4519	−0.6600
0.90	0.8631	0.5279	0.2403	−0.0482	−0.2914
0.95	0.9315	0.7638	0.6186	0.4650	0.3278
1.00	1.0000	1.0000	1.0000	1.0000	1.0000

Table 2. Eigenfunctions of Deflection; Clamped-Free Timoshenko Beam for $r = 0.02$, $s = 0.05$

ξ	Y_1	Y_2	Y_3	Y_4	Y_5
0.00	0.0000	0.0000	0.0000	0.0000	0.0000
0.05	0.0049	−0.0361	0.1097	−0.2216	0.3607
0.10	0.0178	−0.1110	0.2942	−0.5142	0.7287
0.15	0.0383	−0.2121	0.4984	−0.7521	0.8582
0.20	0.0657	−0.3269	0.6729	−0.8245	0.6462
0.25	0.0994	−0.4437	0.7784	−0.6940	0.1667
0.30	0.1389	−0.5516	0.7901	−0.3852	−0.3712
0.35	0.1834	−0.6408	0.6998	0.0211	−0.7253
0.40	0.2326	−0.7034	0.5169	0.4136	−0.7354
0.45	0.2857	−0.7329	0.2661	0.6828	−0.3992
0.50	0.3423	−0.7254	−0.0168	0.7540	0.1256
0.55	0.4019	−0.6786	−0.2899	0.6087	0.5943
0.60	0.4638	−0.5927	−0.5120	0.2907	0.7895
0.65	0.5278	−0.4696	−0.6485	−0.1062	0.6215
0.70	0.5932	−0.3132	−0.6763	−0.4646	0.1712
0.75	0.6599	−0.1287	−0.5870	−0.6760	−0.3468
0.80	0.7273	0.0779	−0.3877	−0.6715	−0.6830
0.85	0.7952	0.3000	−0.0988	−0.4392	−0.6657
0.90	0.8634	0.5311	0.2500	−0.0248	−0.2765
0.95	0.9317	0.7658	0.6257	0.4860	0.3491
1.00	1.0000	1.0000	1.0000	1.0000	1.0000

Table 3. Eigenfunctions of Deflection; Clamped-Free Timoshenko Beam for $r = 0.03$, $s = 0.04$

ξ	Y_1	Y_2	Y_3	Y_4	Y_5
0.00	0.0000	0.0000	0.0000	0.0000	0.0000
0.05	0.0047	−0.0322	0.0950	−0.1904	0.3139
0.10	0.0174	−0.1043	0.2720	−0.4798	0.6899
0.15	0.0378	−0.2036	0.4752	−0.7265	0.8623
0.20	0.0650	−0.3176	0.6542	−0.8234	0.6997
0.25	0.0986	−0.4344	0.7681	−0.7199	0.2488
0.30	0.1379	−0.5430	0.7902	−0.4317	−0.2965
0.35	0.1824	−0.6336	0.7106	−0.0332	−0.6913
0.40	0.2315	−0.6979	0.5369	0.3666	−0.7567
0.45	0.2846	−0.7296	0.2922	0.6561	−0.4650
0.50	0.3412	−0.7243	0.0116	0.7547	0.0464
0.55	0.4007	−0.6798	−0.2633	0.6360	0.5387
0.60	0.4627	−0.5959	−0.4910	0.3363	0.7828
0.65	0.5267	−0.4747	−0.6359	−0.0555	0.6664
0.70	0.5923	−0.3196	−0.6735	−0.4230	0.2467
0.75	0.6590	−0.1359	−0.5937	−0.6546	−0.2755
0.80	0.7266	0.0706	−0.4019	−0.6748	−0.6471
0.85	0.7946	0.2932	−0.1172	−0.4637	−0.6783
0.90	0.8630	0.5257	0.2320	−0.0592	−0.3246
0.95	0.9315	0.7627	0.6137	0.4588	0.3011
1.00	1.0000	1.0000	1.0000	1.0000	1.0000

Table 4. Eigenfunctions of Deflection; Clamped-Free Timoshenko Beam for $r = 0.03$, $s = 0.05$

ξ	Y_1	Y_2	Y_3	Y_4	Y_5
0.00	0.0000	0.0000	0.0000	0.0000	0.0000
0.05	0.0049	−0.0360	0.1093	−0.2213	0.3634
0.10	0.0178	−0.1108	0.2939	−0.5205	0.7415
0.15	0.0383	−0.2118	0.4993	−0.7604	0.8837
0.20	0.0657	−0.3267	0.6759	−0.8395	0.6798
0.25	0.0994	−0.4437	0.7840	−0.7142	0.1982
0.30	0.1388	−0.5519	0.7985	−0.4073	−0.3534
0.35	0.1834	−0.6415	0.7107	0.0021	−0.7286
0.40	0.2325	−0.7045	0.5296	0.4023	−0.7587
0.45	0.2856	−0.7347	0.2792	0.6827	−0.4327
0.50	0.3422	−0.7277	−0.0046	0.7656	0.0971
0.55	0.4017	−0.6814	−0.2800	0.6297	0.5850
0.60	0.4637	−0.5960	−0.5057	0.3160	0.8057
0.65	0.5277	−0.4733	−0.6465	−0.0831	0.6587
0.70	0.5931	−0.3170	−0.6787	−0.4496	0.2149
0.75	0.6599	−0.1324	−0.5934	−0.6731	−0.3147
0.80	0.7272	0.0745	−0.3966	−0.6809	−0.6795
0.85	0.7952	0.2971	−0.1085	−0.4572	−0.6844
0.90	0.8634	0.5291	0.2416	0.0446	−0.3095
0.95	0.9317	0.7647	0.6208	0.4728	0.3225
1.00	1.0000	1.0000	1.0000	1.0000	1.0000

Table 5. Eigenfunctions of Bending Slope; Clamped-Free Timoshenko Beam for $r = 0.02$, $s = 0.04$

ξ	$L\Psi_1$	$L\Psi_2$	$L\Psi_3$	$L\Psi_4$	$L\Psi_5$
0.00	0.0000	0.0000	0.0000	0.0000	0.0000
0.05	0.1688	−0.9256	2.2072	−3.5022	4.4648
0.10	0.3255	−1.5975	3.3126	−4.2410	3.7231
0.15	0.4703	−2.0194	3.3882	−2.6809	−0.4876
0.20	0.6031	−2.1997	2.5878	0.3127	−5.4246
0.25	0.7239	−2.1534	1.1464	3.6224	−8.3479
0.30	0.8331	−1.9017	−0.6397	6.1387	−7.6512
0.35	0.9306	−1.4728	−2.4457	7.0414	−3.5297
0.40	1.0168	−0.9003	−3.9601	6.0134	2.1588
0.45	1.0920	−0.2222	−4.9252	3.3179	6.7907
0.50	1.1565	0.5204	−5.1722	−0.2854	8.2215
0.55	1.2109	1.2860	−4.6428	−3.7633	5.8000
0.60	1.2557	2.0343	−3.3952	−6.1084	0.6807
0.65	1.2915	2.7288	−1.5926	−6.6280	−4.7115
0.70	1.3192	3.3390	0.5248	−5.1394	−7.8045
0.75	1.3396	3.8421	2.6800	−2.0148	−7.0747
0.80	1.3536	4.2251	4.6061	1.9387	−2.7672
0.85	1.3624	4.4862	6.0944	5.7380	3.4409
0.90	1.3670	4.6358	7.0368	8.5403	8.9891
0.95	1.3687	4.6972	7.4576	9.9504	12.1007
1.00	1.3690	4.7073	7.5325	10.6369	12.7605

Table 6. Eigenfunctions of Bending Slope; Clamped-Free Timoshenko Beam for $r = 0.02$, $s = 0.05$

ξ	$L\Psi_1$	$L\Psi_2$	$L\Psi_3$	$L\Psi_4$	$L\Psi_5$
0.00	0.0000	0.0000	0.0000	0.0000	0.0000
0.05	0.1683	−0.9026	2.0710	−3.0989	3.6396
0.10	0.3246	−1.5557	3.0836	−3.6275	2.6781
0.15	0.4689	−1.9629	3.1133	−2.0615	−1.1993
0.20	0.6013	−2.1330	2.3144	0.7567	5.4774
0.25	0.7218	−2.0810	0.9175	3.7719	−7.7568
0.30	0.8305	−1.8282	−0.7901	5.9649	−6.7416
0.35	0.9278	−1.4023	−2.4976	6.6106	−2.7687
0.40	1.0137	−0.8366	−3.9095	5.4655	2.3922
0.45	1.0886	−0.1687	−4.7843	2.8243	6.3870
0.50	1.1529	0.5615	−4.9668	−0.5721	7.3873
0.55	1.2071	1.3129	−4.4077	−3.7536	4.9515
0.60	1.2517	2.0465	−3.1687	−5.8003	0.2370
0.65	1.2874	2.7266	−1.4097	−6.1058	−4.5296
0.70	1.3150	3.3245	0.6377	−4.5469	−7.0806
0.75	1.3354	3.8151	2.7089	−1.5111	−6.1437
0.80	1.3493	4.1891	4.5514	2.2280	−2.0088
0.85	1.3580	4.4437	5.9696	5.7545	3.6509
0.90	1.3626	4.5895	6.8643	8.3082	8.6406
0.95	1.3644	4.6493	7.2622	9.5630	11.4055
1.00	1.3646	4.6590	7.3324	9.8032	11.9823

Table 7. Eigenfunctions of Bending Slope; Clamped-Free Timoshenko Beam for $r = 0.03$, $s = 0.04$

ξ	$L\Psi_1$	$L\Psi_2$	$L\Psi_3$	$L\Psi_4$	$L\Psi_5$
0.00	0.0000	0.0000	0.0000	0.0000	0.0000
0.05	0.1687	−0.9248	2.2147	−3.5582	4.6262
0.10	0.3254	−1.5972	3.3349	−4.3436	3.9643
0.15	0.4701	−2.0205	3.4266	−2.7925	−0.3198
0.20	0.6028	−2.2028	2.6371	0.2375	−5.4599
0.25	0.7237	−2.1587	1.1994	3.6190	−8.6237
0.30	0.8328	−1.9092	−0.5921	6.2205	−8.0891
0.35	0.9304	−1.4821	−2.4128	7.1960	−3.9644
0.40	1.0166	−0.9108	−3.9495	6.2055	1.9083
0.45	1.0918	−0.2332	−4.9416	3.4998	6.8388
0.50	1.1564	0.5099	−5.2160	−0.1600	8.5590
0.55	1.2108	1.2768	−4.7101	−3.7255	6.2895
0.60	1.2557	2.0275	−3.4778	−6.1643	1.1115
0.65	1.2916	2.7254	−1.6789	−6.7563	−4.5306
0.70	1.3194	3.3397	0.4484	−5.2959	−7.9589
0.75	1.3399	3.8475	2.6267	−2.1442	−7.4910
0.80	1.3540	4.2354	4.5870	1.8896	−3.2016
0.85	1.3627	4.5013	6.1159	5.8029	3.1619
0.90	1.3674	4.6551	7.0985	8.7266	9.0765
0.95	1.3692	4.7195	7.5509	10.2333	12.5752
1.00	1.3694	4.7308	7.6388	10.5589	13.4122

Table 8. Eigenfunctions of Bending Slope; Clamped-Free Timoshenko Beam for $r = 0.03$, $s = 0.05$

ξ	$L\Psi_1$	$L\Psi_2$	$L\Psi_3$	$L\Psi_4$	$L\Psi_5$
0.00	0.0000	0.0000	0.0000	0.0000	0.0000
0.05	0.1682	−0.9021	2.0805	−3.1534	3.7819
0.10	0.3244	−1.5558	3.1082	−3.7284	2.8670
0.15	0.4687	−1.9645	3.1526	−2.1744	−1.1024
0.20	0.6010	−2.1366	2.3633	0.6757	−5.5739
0.25	0.7215	−2.0867	0.9684	3.7607	−8.0584
0.30	0.8302	−1.8359	−0.7465	6.0434	−7.1592
0.35	0.9275	−1.4116	−2.4699	6.7735	−3.1473
0.40	1.0135	−0.8469	−3.9044	5.6822	2.2069
0.45	1.0884	0.1792	−4.8056	3.0460	6.4781
0.50	1.1528	0.5515	−5.0140	−0.3998	7.7245
0.55	1.2071	1.3045	−4.4764	−3.6753	5.3967
0.60	1.2518	2.0406	−3.2502	−5.8375	0.6004
0.65	1.2876	2.7241	−1.4925	−6.2486	−4.4077
0.70	1.3152	3.3251	0.5665	−4.7542	−7.2531
0.75	1.3356	3.8214	2.6619	−1.7198	−6.5268
0.80	1.3496	4.2002	4.5387	2.0868	−2.4110
0.85	1.3584	4.4596	5.9966	5.7361	3.4510
0.90	1.3630	4.6094	6.9300	8.4380	8.7909
0.95	1.3648	4.6721	7.3578	9.8180	11.8974
1.00	1.3650	4.6830	7.4401	10.1118	12.6299

Tables of eigenfunctions for the vibration of classical beams [5] have been available since 1949. They are a particular case of the new tables when $r = s = 0$.

ILLUSTRATIVE EXAMPLE

The application of new tables of eigenfunctions of normal modes will be illustrated by an example. Consider a steel cantilever beam of 14.4 in. in length and 1 in.^{2} in cross section. From Eq. (4) we have $r = 0.024$ and $s = 0.048$. The application of interpolations from Tables 1–4 and 5–8 gives the normal modes of deflection and bending slope, as shown in Tables 9 and 10 respectively. These results are also shown graphically in Figs. 1 and 2.

Table 9. Eigenfunctions of Deflection; Clamped-Free Timoshenko Beam for $r = 0.024$, $s = 0.048$, $L = 14.4$ in.

ξ	Y_1	Y_2	Y_3	Y_4	Y_5
0.00	0.0000	0.0000	0.0000	0.0000	0.0000
0.05	0.0047	−0.0340	0.1015	−0.2039	0.3346
0.10	0.0177	−0.1096	0.2897	−0.5107	0.7236
0.15	0.0380	−0.2037	0.4852	−0.7363	0.8565
0.20	0.0656	−0.3250	0.6698	−0.8273	0.6635
0.25	0.0990	−0.4385	0.7717	−0.7063	0.2066
0.30	0.1387	−0.5499	0.7918	−0.3994	−0.3530
0.35	0.1829	−0.6367	0.7041	−0.0099	−0.7063
0.40	0.2323	−0.7025	0.5234	0.4014	−0.7443
0.45	0.2851	−0.7307	0.2753	0.6669	−0.4299
0.50	0.3416	−0.7256	−0.0087	0.7566	0.1043
0.55	0.4012	−0.6788	−0.2766	0.6231	0.5656
0.60	0.4636	−0.5939	−0.5061	0.3059	0.7915
0.65	0.5272	−0.4718	−0.6417	−0.0771	0.6405
0.70	0.5810	−0.3152	−0.6762	−0.4523	0.1949
0.75	0.6594	−0.1321	−0.5897	−0.6658	−0.3128
0.80	0.7271	0.0757	−0.3923	−0.6742	−0.6744
0.85	0.7949	0.2966	−0.1075	−0.4530	−0.6700
0.90	0.8633	0.5296	0.2447	0.0367	−0.2927
0.95	0.9316	0.7643	0.6080	0.4695	0.3265
1.00	1.0000	1.0000	1.0000	1.0000	1.0000

Table 10. Eigenfunctions of Bending Slope; Clamped-Free Timoshenko Beam for $r = 0.024$, $s = 0.048$, $L = 14.4$ in.

ξ	Ψ_1	Ψ_2	Ψ_3	Ψ_4	Ψ_5
0.00	0.0000	0.0000	0.0000	0.0000	0.0000
0.05	0.0117	−0.0631	0.1460	−0.2257	1.2683
0.10	0.0226	−0.1086	0.2180	−0.2632	0.2060
0.15	0.0326	−0.1371	0.2211	−0.1549	−0.0703
0.20	0.0418	−0.1492	0.1659	0.0444	−0.3801
0.25	0.0502	−0.1457	0.0683	0.2596	−0.5551
0.30	0.0577	−0.1282	−0.0516	0.4188	−0.4925
0.35	0.0645	−0.0986	−0.1735	0.4695	−0.2137
0.40	0.0704	−0.0593	−0.2720	0.3930	0.1574
0.45	0.0757	−0.0128	−0.3348	0.2089	0.4514
0.50	0.0801	0.0381	−0.3491	−0.0312	0.5363
0.55	0.0822	0.0906	−0.3113	−0.2589	0.3683
0.60	0.0870	0.1418	−0.2255	−0.4082	0.0331
0.65	0.0895	0.1865	−0.1028	−0.4352	−0.3134
0.70	0.0914	0.2311	0.0408	−0.3295	−0.5064
0.75	0.0928	0.2655	0.1864	−0.1173	−0.4504
0.80	0.0938	0.2917	0.3164	0.1473	−0.1611
0.85	0.0944	0.3096	0.4170	0.3994	0.2446
0.90	0.0947	0.3199	0.4809	0.5841	0.6087
0.95	0.0948	0.3242	0.5115	0.6767	0.8153
1.00	0.0948	0.3249	0.5150	0.6955	0.8582

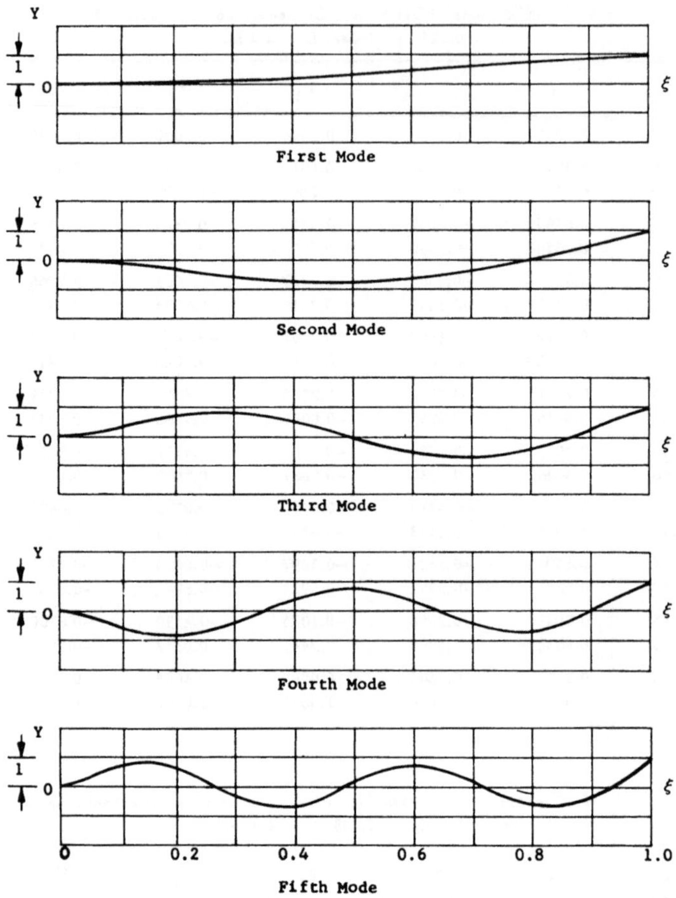

Fig. 1. Eigenfunctions of Deflection; Clamped-Free Timoshenko Beam for $r = 0.024$, $s = 0.048$, $L = 14.4$ in.

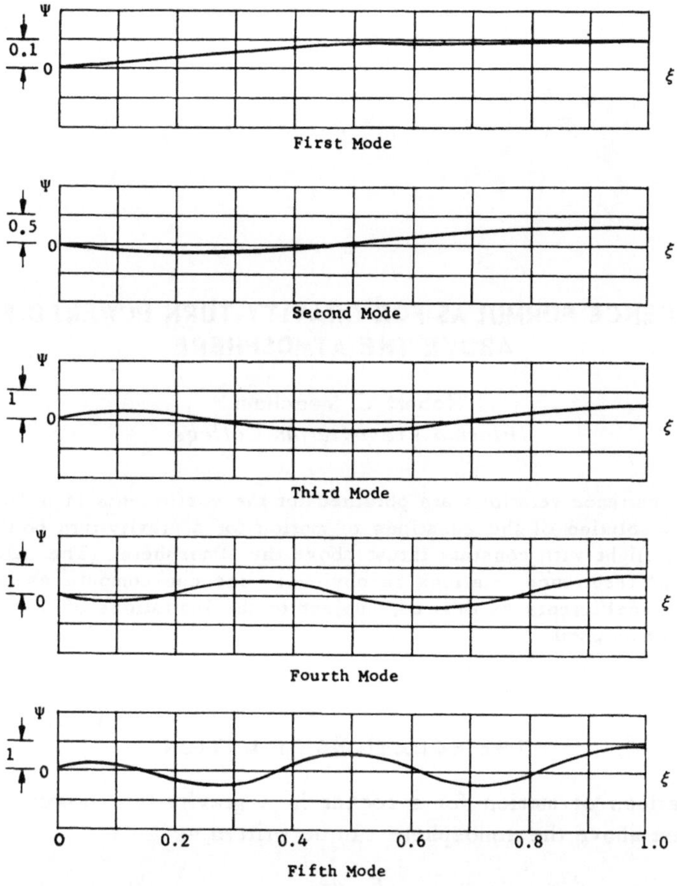

Fig. 2. Eigenfunctions of Bending Slope; Clamped-Free Timoshenko Beam for $r = 0.024$, $s = 0.048$, $L = 14.4$ in.

REFERENCES

[1] T. C. Huang, "The Effect of Rotatory Inertia and of Shear Deformation on the Frequency and Normal Mode Equations of Uniform Beams with Simple End Conditions," *J. Appl. Mechanics*, **28**(4), 579–84 (1961).

[2] S. P. Timoshenko, *Vibration Problems in Engineering*, 3d ed., pp 329–31, Van Nostrand, New York, 1955.

[3] T. C. Huang, "Tables of Eigenvalues and Charts of Modifying Quotients for Frequencies and Normal Modes of Beam Vibrations," to be published as a bulletin.

[4] T. C. Huang, "Tables of Eigenfunctions Representing Normal Modes of Vibration of Timoshenko Beams," to be published as a bulletin.

[5] Dana Young and R. P. Felgar, "Tables of Characteristic Functions Representing Normal Modes of Vibration of a Beam," University of Texas Publication No. 4913, 1949.

RECURRENCE FORMULAS FOR GRAVITY-TURN POWERED FLIGHT ABOVE THE ATMOSPHERE

Robert C. Meacham [1]

Florida Presbyterian College

Recurrence relations are obtained for the coefficients in a Taylor series solution of the equations of motion for a gravity-turn powered rocket flight with constant thrust above the atmosphere. The advantage of recurrence relations is obvious: one can compute as many series coefficients as desired, subject to the limitations of the computer being used.

THE EQUATIONS OF MOTION

The equations of motion for a rocket in a gravity-turn powered flight with constant thrust above the atmosphere can be written as

$$\frac{d^2x}{dt^2} = -\frac{\mu}{r^3}x + \frac{F}{mv}\frac{dx}{dt} + \omega^2 x + 2\omega\frac{dy}{dt},$$

$$\frac{d^2y}{dt^2} = -\frac{\mu}{r^3}y + \frac{F}{mv}\frac{dy}{dt} + \omega^2 y - 2\omega\frac{dx}{dt}, \tag{1}$$

$$\frac{d^2z}{dt^2} = -\frac{\mu}{r^3}z + \frac{F}{mv}\frac{dz}{dt},$$

where an earth-fixed reference system is employed, with the origin at the center of the earth, the z axis coincident with the axis of rotation of the earth, and the x and y axes so chosen that an orthogonal right-handed coordinate system is obtained. As is customary,

$$r = \sqrt{x^2 + y^2 + z^2}. \tag{2}$$

The gravitational field contributes the first term on the right in each of the above equations. This field is taken here to be spherically symmetric. The term μ is the gravitational constant.

---

[1]Consultant, RCA Service Company, Patrick Air Force Base, Florida.

NOTATION

t	Time
ω	Angular velocity of rotation of earth
(x, y, z)	Position coordinates of rocket in geocentric, earth-fixed coordinate system, with z axis coincident with axis of rotation of earth
r	Distance from origin of coordinate system to rocket
v	Speed of rocket
m	Mass of rocket (varies with time)
F	Magnitude of constant-thrust force developed by rocket
μ	Gravitational constant for spherically symmetric gravitational field

The thrust force produced by the rocket engine accounts for the second term on the right side of each equation. This force is taken to be constant in magnitude, F, and is directed tangentially to the rocket trajectory at each instant. The mass of the rocket will be expressed in the form of a Taylor series:

$$m = \sum_{j=0}^{\infty} \frac{1}{j!} m_j t^j = m_0 + m_1 t + \frac{1}{2!} m_2 t^2 + \frac{1}{3!} m_3 t^3 + \dots, \tag{3}$$

where

$$m_j = \left. \frac{d^j m}{dt^j} \right|_{t=0} \qquad \text{for } j = 0, 1, 2, \dots .$$

(For simplicity the Taylor series expansion will be referred to the initial conditions which prevail at time $t = 0$, although by replacing t by the quantity, $t - t_0$, the reference time can be shifted to arbitrary t_0.) The velocity of the rocket has speed v, defined by

$$v = \sqrt{(dx/dt)^2 + (dy/dt)^2 + (dz/dt)^2} . \tag{4}$$

The choice of an earth-fixed – and therefore rotating – coordinate system accounts for the last two terms in each of the first two equations of motion. The term ω is the angular velocity of rotation of the earth about its axis.

THE SOLUTION

The solution of the equations of motion will be sought in the form of a Taylor series:

$$
\begin{aligned}
x &= \sum_{p=0}^{\infty} \frac{1}{p!}\, a_p\, t^p \,, \\
y &= \sum_{p=0}^{\infty} \frac{1}{p!}\, b_p\, t^p \,, \qquad\qquad (5) \\
z &= \sum_{p=0}^{\infty} \frac{1}{p!}\, c_p\, t^p \,.
\end{aligned}
$$

It is the purpose of this paper to demonstrate how the solution coefficients can be computed. Recurrence formulas will be developed which enable one to compute the set (a_n, b_n, c_n) for $n = p$ in terms of the previously computed sets for $n = 0, 1, 2, \ldots, p-1$. Since the computation scheme will be based upon recurrence formulas, computing machines can be programmed to carry out the computation of as many successive solution coefficients as desired (subject to limitations of the particular computer).

The coefficients (a_p, b_p, c_p) are given by the equations

$$
a_p = \left.\frac{d^p x}{dt^p}\right|_{t=0}, \qquad b_p = \left.\frac{d^p y}{dt^p}\right|_{t=0}, \qquad c_p = \left.\frac{d^p z}{dt^p}\right|_{t=0} \qquad (6)
$$

for $p = 0, 1, 2, 3, \ldots$. Hence the initial position and initial velocity account for the first two sets of coefficients:

$$
\begin{aligned}
(a_0, b_0, c_0) &= (x_0, y_0, z_0), \\
(a_1, b_1, c_1) &= \left(\left.\frac{dx}{dt}\right|_{t=0}, \left.\frac{dy}{dt}\right|_{t=0}, \left.\frac{dz}{dt}\right|_{t=0} \right).
\end{aligned} \qquad (7)
$$

The equations of motion, when evaluated at $t = 0$, yield

$$
\begin{aligned}
a_2 &= -\frac{\mu}{r_0^3}\, a_0 + \frac{F}{m_0 v_0}\, a_1 + \omega^2 a_0 + 2\omega b_1 \,, \\
b_2 &= -\frac{\mu}{r_0^3}\, b_0 + \frac{F}{m_0 v_0}\, b_1 + \omega^2 b_0 - 2\omega a_1 \,, \qquad\qquad (8) \\
c_2 &= -\frac{\mu}{r_0^3}\, c_0 + \frac{F}{m_0 v_0}\, c_1 \,.
\end{aligned}
$$

Subsequent coefficients can be obtained by successive differentiation of the equations of motion. In order to carry this out in an orderly fashion and in a manner to reveal the sought-after recurrence relations, it will be convenient to define some auxiliary quantities and to obtain several useful relations among them.

Let quantities u_{pq} be defined as follows:

$$u_{pq} = a_p a_q + b_p b_q + c_p c_q \qquad \text{for } p, q = 0, 1, 2, 3, \ldots . \tag{9}$$

Let

$$r_j = \left.\frac{d^j r}{dt^j}\right|_{t=0} \qquad \text{for } j = 0, 1, 2, 3, \ldots . \tag{10}$$

Then by mathematical induction it can be proved (see Appendix 1) that

$$r_p = \frac{1}{r_0}\left[u_{0p} + \frac{1}{2}\sum_{q=1}^{p-1}\binom{p}{q}(u_{q,p-q} - r_q r_{p-q})\right] \qquad \text{for } p = 0, 1, 2, 3, \ldots , \tag{11}$$

and where $\sum_{q=1}^{P} = 0$ whenever $P < 1$, and where

$$\binom{p}{q} = \frac{p!}{q!(p-q)!} , \tag{12}$$

binomial coefficients.

Let

$$g = \mu/r^3 , \tag{13}$$

and let

$$g_j = \left.\frac{d^j g}{dt^j}\right|_{t=0} \qquad \text{for } j = 0, 1, 2, 3, \ldots . \tag{14}$$

In Appendix 2 it is proved that

$$g_p = \frac{-1}{r_0}\sum_{q=0}^{p-1}\binom{p-1}{q}\left(3 + \frac{q}{p-q}\right) g_q r_{p-q} \qquad \text{for } p = 1, 2, 3, \ldots . \tag{15}$$

Let

$$v_j = \left.\frac{d^j v}{dt^j}\right|_{t=0} \qquad \text{for } j = 0, 1, 2, 3, \ldots . \tag{16}$$

Then by mathematical induction it can be shown (see Appendix 3) that

$$v_p = \frac{1}{v_0}\left[u_{1,p+1} + \frac{1}{2}\sum_{q=1}^{p-1}\binom{p}{q}(u_{q+1,p+1-q} - v_q v_{p-q})\right]$$

for $p = 0, 1, 2, 3, \ldots,$ (17)

and where $\sum_{q=1}^{P} = 0$ whenever $P < 1$.

Let

$$\zeta_j = \left.\frac{d^j}{dt^j}(v^{-1})\right|_{t=0} \qquad \text{for } j = 0, 1, 2, 3, \ldots. \quad (18)$$

In Appendix 4 it is shown that

$$\zeta_p = -\frac{1}{v_0}\sum_{q=0}^{p-1}\binom{p}{q}\zeta_q v_{p-q} \qquad \text{for } p = 1, 2, 3, \ldots. \quad (19)$$

Let

$$\xi_j = \left.\frac{d^j}{dt^j}(m^{-1})\right|_{t=0} \qquad \text{for } j = 0, 1, 2, 3, \ldots. \quad (20)$$

Then it follows, by analogy from Eqs. (18) and (19), that

$$\xi_p = -\frac{1}{m_0}\sum_{q=0}^{p-1}\binom{p}{q}\xi_q m_{p-q} \qquad \text{for } p = 1, 2, 3, \ldots. \quad (21)$$

Recursive formulas have now been given for the Taylor series coefficients for each of the functions g, m^{-1}, and v^{-1} which appear in the equations of motion (1). It remains to give recursive formulas for the solution coefficients (a_n, b_n, c_n).

Before the formulas for (a_n, b_n, c_n) are written down, however, it will be convenient to let

$$f = Fm^{-1}v^{-1}, \quad (22)$$

the coefficient in the thrust-force term on the right side of Eqs. (1), and to define

$$f_j = \left.\frac{d^j}{dt^j}(Fm^{-1}v^{-1})\right|_{t=0} \qquad \text{for } j = 0, 1, 2, 3, \ldots. \quad (23)$$

Since F is constant in the problem under consideration, then Leibniz' rule for the differentiation of the product of two functions can be employed to obtain

$$f_p = F\sum_{q=0}^{p}\binom{p}{q}\xi_q\zeta_{p-q} \qquad \text{for } p = 0, 1, 2, 3, \ldots. \quad (24)$$

In terms of the functions g and f, which were defined in Eqs. (13) and (22), the equations of motion are

$$\frac{d^2x}{dt^2} = -gx + f\frac{dx}{dt} + \omega^2 x + 2\omega\frac{dy}{dt},$$

$$\frac{d^2y}{dt^2} = -gy + f\frac{dy}{dt} + \omega^2 y - 2\omega\frac{dx}{dt}, \tag{25}$$

$$\frac{d^2z}{dt^2} = -gz + f\frac{dz}{dt}.$$

The Taylor series coefficients for g and f were computed above [Eqs. (1) and (24)]. Leibniz' rule for differentiation of the product of two functions yields the following recursion formulas (see Appendix 5):

$$a_{p+2} = -\sum_{q=0}^{p+1} K_{pq} a_q + \omega^2 a_p + 2\omega b_{p+1},$$

$$b_{p+2} = -\sum_{q=0}^{p+1} K_{pq} b_q + \omega^2 b_p - 2\omega a_{p+1}, \tag{26}$$

$$c_{p+2} = -\sum_{q=0}^{p+1} K_{pq} c_q,$$

for $p = 0, 1, 2, 3, \ldots$, and where

$$K_{pq} = \binom{p}{q} g_{p-q} - \binom{p}{q-1} f_{p-q+1}$$

$$\text{for } q = 0, 1, 2, 3, \ldots, p, p+1;\ p = 0, 1, 2, 3, \ldots, \tag{27}$$

and where the additional definition is made:

$$\binom{P}{Q} = 0 \quad \text{when } Q = -1 \text{ or } Q = P + 1. \tag{28}$$

With these formulas one can program a digital computer so that any desired number of coefficients may be computed, subject of course to the limitations of the particular computer. An error analysis is needed next to provide criteria for the rapidity of convergence of the Taylor series so computed.

CONVERGENCE OF THE SOLUTION

To investigate convergence of the Taylor series solution, one can rewrite the equations of motion as a system of six first-order differential equations as follows:

$$
\begin{aligned}
\frac{dx}{dt} &= f_1(x, y, z, \dot{x}, \dot{y}, \dot{z}\,;\, t) = \dot{x}\,, \\
\frac{dy}{dt} &= f_2(x, y, z, \dot{x}, \dot{y}, \dot{z}\,;\, t) = \dot{y}\,, \\
\frac{dz}{dt} &= f_3(x, y, z, \dot{x}, \dot{y}, \dot{z}\,;\, t) = \dot{z}\,, \\
\frac{d\dot{x}}{dt} &= f_4(x, y, z, \dot{x}, \dot{y}, \dot{z}\,;\, t) = -\frac{\mu}{r^2}\cdot\frac{x}{r} + \frac{F}{m}\cdot\frac{\dot{x}}{v} + \omega^2 x + 2\omega\dot{y}\,, \\
\frac{d\dot{y}}{dt} &= f_5(x, y, z, \dot{x}, \dot{y}, \dot{z}\,;\, t) = -\frac{\mu}{r^2}\cdot\frac{y}{r} + \frac{F}{m}\cdot\frac{\dot{y}}{v} + \omega^2 y - 2\omega\dot{x}\,, \\
\frac{d\dot{z}}{dt} &= f_6(x, y, z, \dot{x}, \dot{y}, \dot{z}\,;\, t) = -\frac{\mu}{r^2}\cdot\frac{z}{r} + \frac{F}{m}\cdot\frac{\dot{z}}{v}\,.
\end{aligned} \tag{29}
$$

We have obtained the Taylor series solution of this set of first-order differential equations subject to the initial conditions

$$(x, y, z, \dot{x}, \dot{y}, \dot{z})|_{t=0} = (a_0, b_0, c_0, a_1, b_1, c_1)\,.$$

The solution has been displayed for x, y, and z in Eqs. (5) and (26), and the solution for $\dot{x}$, $\dot{y}$, and $\dot{z}$ is obtained from (5) by differentiation:

$$
\begin{aligned}
\dot{x} &= \sum_{p=0}^{\infty} \frac{1}{p!}\, a_{p+1} t^p\,, \\
\dot{y} &= \sum_{p=0}^{\infty} \frac{1}{p!}\, b_{p+1} t^p\,, \\
\dot{z} &= \sum_{p=0}^{\infty} \frac{1}{p!}\, c_{p+1} t^p\,.
\end{aligned} \tag{30}
$$

The radius of convergence of a differentiated power series is the same as that of the original series, and so the radius of convergence of the six series in (5) and (30) is the same, whatever that may be.

A lower bound for the radius of convergence, ρ, of such a system is given by the following [1]:

$$\rho = a(1 - e^{-b/7Ma}) > b/7M\,, \tag{31}$$

where $|t| < \rho < a$ and where the functions f_i, $i = 1, \ldots, 6$, are analytic in the closed region R defined by the inequalities

$$
\begin{gathered}
|t| \leqq a, \quad |x - a_0| \leqq b, \quad |y - b_0| \leqq b, \quad |z - c_0| \leqq b, \\
|\dot{x} - a_1| \leqq b, \quad |\dot{y} - b_1| \leqq b, \quad |\dot{z} - c_1| \leqq b,
\end{gathered}
\tag{32}
$$

and throughout this closed region the functions f_i are bounded uniformly:

$$
|f_i(x, y, z, \dot{x}, \dot{y}, \dot{z}\,;\, t)| \leqq M \tag{33}
$$

for $(x, y, z, \dot{x}, \dot{y}, \dot{z}\,;\, t)$ in R.

For the system of Eqs. (29), pertaining as they do to powered rocket flight, physical considerations lead readily to estimates for the parameters a, b, and M in Eq. (31):

$$
\begin{aligned}
&a \leqq \text{maximum burning time for the rocket}, \\
&b < r_0/\sqrt{3}\,.
\end{aligned}
\tag{34}
$$

To obtain an estimate for M, notice first that the preceding estimates for a and b guarantee that the functions f_i, $i = 1, \ldots, 6$, are analytic in the region R, and from this it follows that, for $i = 1, \ldots, 6$:

$$
|f_i(x, y, z, \dot{x}, \dot{y}, \dot{z}, \,;\, t)| \leqq \frac{\mu}{r^2} + \frac{F}{m} + \omega^2 r + 2\omega v\,. \tag{35}
$$

The first term on the right is the magnitude of the gravitational acceleration, and the second term is the thrust-to-mass ratio. Practical considerations require that

$$
\frac{\mu}{r^2} \leqq 1g\,, \tag{36}
$$

where g is the gravitational acceleration of the rocket at the surface of the earth. The ratio F/m is, of course, approaching its maximum value just prior to burnout, when the mass of the rocket is least (holding F constant):

$$
\frac{F}{m} \leqq \frac{F}{m_{\text{burnout}}} = ng\,. \tag{37}
$$

The remaining two terms are small in comparison with the first two in the present state of rocket problems, but in any event these two terms can certainly be accounted for, if need be, by taking a larger value for n in Eq. (37):

$$
M = (n + 1)g\,. \tag{38}
$$

Thus it appears that a lower bound for the radius of convergence, ρ, of the Taylor series solution is obtained from (31), (34), and (38):

$$
\rho = \frac{r_0}{7\sqrt{3}(n + 1)g}\,.
$$

Of course this is merely a lower bound for the radius of convergence for the Taylor series solution (5) of the equations of motion (1). The recurrence procedure displayed herein is now being coded for the IBM 7090 by T. R. Lewis. As a consequence of the study of numerical examples, it may be found that much larger radii of convergence are obtained in various problems of interest.

CONCLUSIONS

Herein is displayed a method for generating by recurrence relations the Taylor series solution of the equations of motion for a constant-thrust gravity-turn rocket flight above the atmosphere. The solution coefficients (a_{p+2}, b_{p+2}, c_{p+2}) are computed in terms of the initial conditions and previously computed coefficients – that is, in terms of (a_n, b_n, c_n) for $n = 0, 1, 2, \ldots, p+1$.

The treatment here given is an outgrowth of an earlier report [2] in which some of the needed recurrence relations were only partially developed.

Recursion-formula techniques for the Taylor series solution of the 3-body and n-body problems in which only gravitational forces are present have been developed by Steffensen [3, 4], and Cheney [5].

The advantage of recursion techniques such as these is obvious, assuming that high-speed digital computers are employed to carry out the algebraic manipulations. Numerical studies to investigate the rapidity of convergence of the series in the present paper are now being undertaken. From these studies it is hoped that answers will be found to such questions as "How long a time-span can be taken with truncated Taylor series of various degrees of truncation?" If convergence is rapid enough and if the computation time is also rapid, then it may be that such a technique as this can be employed for the real-time computation of trajectories.

APPENDIX 1

The equation

$$r_p = \frac{1}{r_0}\left[u_{0p} + \frac{1}{2}\sum_{q=1}^{p-1}\binom{p}{q}(u_{q,p-q} - r_q r_{p-q})\right] \tag{11}$$

is easily verified for $p = 0, 1$, and 2. It will now be shown by mathematical induction to be true for all nonnegative integers. Note first that the formula can also be written as

$$r_p = \frac{1}{2r_0}\left[\sum_{q=0}^{p}\binom{p}{q}u_{q,p-q} - \sum_{q=1}^{p-1}\binom{p}{q}r_q r_{p-q}\right]. \tag{A1.1}$$

From this it is evident that one need only prove that

$$\sum_{q=0}^{p} \binom{p}{q} r_q r_{p-q} = \sum_{q=0}^{p} \binom{p}{q} u_{q,p-q} \,. \tag{A1.2}$$

This can be established at once, however, from the equation

$$\sum_{q=0}^{p} \binom{p}{q} \frac{d^q r}{dt^q} \frac{d^{p-q} r}{dt^{p-q}} = \sum_{q=0}^{p} \binom{p}{q} \left(\frac{d^q x}{dt^q} \frac{d^{p-q} x}{dt^{p-q}} + \frac{d^q y}{dt^q} \frac{d^{p-q} y}{dt^{p-q}} + \frac{d^q z}{dt^q} \frac{d^{p-q} z}{dt^{p-q}} \right), \tag{A1.3}$$

provided that one can show this last equation to be true.

It will be *assumed* then that Eq. (A1.3) is true for $p = k$, when k is some natural number greater than or equal to 2:

$$\sum_{q=0}^{k} \binom{k}{q} \frac{d^q r}{dt^q} \frac{d^{k-q} r}{dt^{k-q}} = \sum_{q=0}^{k} \binom{k}{q} \left(\frac{d^q x}{dt^q} \frac{d^{k-q} x}{dt^{k-q}} + \dots \right),$$

where the dots (...) represent terms in y and z which are analogous to those in x. Differentiation of the above equation results in the following on the left side:

$$\sum_{q=0}^{k} \binom{k}{q} \frac{d^q r}{dt^q} \frac{d^{k+1-q} r}{dt^{k+1-q}} + \sum_{q=0}^{k} \binom{k}{q} \frac{d^{q+1} r}{dt^{q+1}} \frac{d^{k-q} r}{dt^{k-q}} \,;$$

analogous expressions occur on the right side after differentiation. By changing the summation index for the second summation, one obtains

$$\sum_{q=0}^{k} \binom{k}{q} \frac{d^q r}{dt^q} \frac{d^{k+1-q} r}{dt^{k+1-q}} + \sum_{q=1}^{k+1} \binom{k}{q-1} \frac{d^q r}{dt^q} \frac{d^{k+1-q} r}{dt^{k+1-q}} \,,$$

or

$$r \frac{d^{k+1} r}{dt^{k+1}} + \sum_{q=1}^{k} \left[\binom{k}{q} + \binom{k}{q-1} \right] \frac{d^q r}{dt^q} \frac{d^{k+1-q} r}{dt^{k+1-q}} + \frac{d^{k+1} r}{dt^{k+1}} r \,.$$

But

$$\binom{k}{q} + \binom{k}{q-1} = \binom{k+1}{q} .$$

Finally then the left side may be written as

$$\sum_{q=0}^{k+1} \binom{k+1}{q} \frac{d^q r}{dt^q} \frac{d^{k+1-q} r}{dt^{k+1-q}} \,,$$

with analogous expressions on the right side. Thus, from the assumption that Eq. (A1.3) is true for $p = k$, it follows that it is also true for $p = k + 1$. Since it is easily verified for $p = 0, 1, 2$, it therefore follows (mathematical induction theorem) that the formula holds for all nonnegative integers, as indicated in Eq. (11).

APPENDIX 2

The equation

$$g_p = -\frac{1}{r_0}\sum_{q=0}^{p-1}\binom{p-1}{q}\left(3+\frac{q}{p-q}\right)g_q r_{p-q} \tag{15}$$

is easily verified for $p = 1$. It will be shown that the formula is true for all natural numbers p. In Eq. (13) we have defined g so that

$$gr^3 = \mu .$$

Differentiation with respect to t yields

$$\frac{dg}{dt}r + 3g\frac{dr}{dt} = 0 .$$

For purposes of exposition this will be written as

$$\frac{dg}{dt}r + ag\frac{dr}{dt} = 0 , \tag{A2.1}$$

where $a = 3$. Each term in this expression is the product of two functions, and so Leibniz' rule for differentiation of the product of two functions can be employed:

$$\sum_{q=0}^{p-1}\binom{p-1}{q}\frac{d^{q+1}g}{dt^{q+1}}\frac{d^{p-1-q}r}{dt^{p-1-q}} + a\sum_{q=0}^{p-1}\binom{p-1}{q}\frac{d^q g}{dt^q}\frac{d^{p-q}r}{dt^{p-q}} = 0 .$$

When this equation is evaluated at $t = 0$, the following equation is obtained:

$$\sum_{q=0}^{p-1}\binom{p-1}{q}g_{q+1}r_{p-1-q} + a\sum_{q=0}^{p-1}\binom{p-1}{q}g_q r_{p-q} = 0 ,$$

or

$$g_p r_0 + \sum_{q=0}^{p-2}\binom{p-1}{q}g_{q+1}r_{p-1-q} + a\sum_{q=0}^{p-1}\binom{p-1}{q}g_q r_{p-q} = 0 . \tag{A2.2}$$

The first summation index can be changed so that this group of terms becomes

$$\sum_{q=1}^{p-1} \binom{p-1}{q-1} g_q r_{p-q} .$$

Hence Eq. (A2.2) may be written as

$$g_p r_0 + \sum_{q=1}^{p-1} \binom{p-1}{q-1} g_q r_{p-q} + a \sum_{q=0}^{p-1} \binom{p-1}{q} g_q r_{p-q} = 0 ,$$

or

$$g_p r_0 + \sum_{q=1}^{p-1} \left[\binom{p-1}{q-1} + a \binom{p-1}{q} \right] g_q r_{p-q} + a \binom{p-1}{0} g_0 r_p = 0 .$$

The coefficient in the summation can be put in the form

$$\binom{p-1}{q} \left(a + \frac{q}{p-q} \right) ,$$

so that the above equation becomes

$$g_p r_0 + \sum_{q=1}^{p-1} \binom{p-1}{q} \left(a + \frac{q}{p-q} \right) g_q r_{p-q} + a \binom{p-1}{0} g_0 r_p = 0 ,$$

or

$$g_p r_0 + \sum_{q=0}^{p-1} \binom{p-1}{q} \left(a + \frac{q}{p-q} \right) g_q r_{p-q} = 0 . \tag{A2.3}$$

When this equation is solved for g_p and a is replaced by 3, Eq. (15) is obtained.

APPENDIX 3

The equation

$$v_p = \frac{1}{v_0} \left[u_{1,p+1} + \frac{1}{2} \sum_{q=1}^{p-1} \binom{p}{q} (u_{q+1,p+1-q} - v_q v_{p-q}) \right] \tag{17}$$

is easily verified for $p = 0$, 1, and 2. It will now be shown to be true for all non-negative integers. Note first that the formula can also be written as

$$v_p = \frac{1}{2v_0} \left[\sum_{q=0}^{p} \binom{p}{q} u_{q+1,p+1-q} - \sum_{q=1}^{p-1} \binom{p}{q} v_q v_{p-q} \right] . \tag{A3.1}$$

From this it is evident that one need only prove that

$$\sum_{q=0}^{p} \binom{p}{q} v_q v_{p-q} = \sum_{q=0}^{p} \binom{p}{q} u_{q+1,p+1-q} \, . \tag{A3.2}$$

When the above phraseology is compared with that in Appendix 1, however, it is seen that Eqs. (11), (A1.1), and (A1.2) are analogous, respectively, to Eqs. (17), (A3.1), and (A3.2). The method of Appendix 1 will therefore carry over in precisely the same manner to the problem of Appendix 3. Instead of repeating the inductive proof of Appendix 1, with the slight but obvious changes to make it apply to Appendix 3, it can at once be concluded that Eq. (17) is valid for all nonnegative integers.

APPENDIX 4

$$\zeta_p = -\frac{1}{v_0} \sum_{q=0}^{p-1} \binom{p}{q} \zeta_q v_{p-q} \tag{19}$$

Let

$$\zeta = v^{-1} \, , \tag{A4.1}$$

so that

$$\zeta_j = \left. \frac{d^j \zeta}{dt^j} \right|_{t=0} \qquad \text{for } j = 0, 1, 2, 3, \ldots . \tag{A4.2}$$

By definition

$$\zeta v = 1 \, .$$

Leibniz' rule for differentiating a product leads to

$$\sum_{q=0}^{p} \binom{p}{q} \zeta_q v_{p-q} = 0 \qquad \text{for } p = 1, 2, \ldots . \tag{A4.3}$$

Hence

$$\zeta_p v_0 + \sum_{q=0}^{p-1} \binom{p}{q} \zeta_q v_{p-q} = 0 \, . \tag{A4.4}$$

If this equation is solved for ζ_p, the desired recurrence relation is obtained.

APPENDIX 5

Recursion formulas (26) for a_{p+2}, b_{p+2}, and c_{p+2} are derived below.

From Eq. (25) one obtains by differentiation, employing Leibniz' rule and evaluating the functions at time $t = 0$,

$$a_{p+2} = -\sum_{q=0}^{p} \binom{p}{q} a_q g_{p-q} + \sum_{q=0}^{p} \binom{p}{q} a_{q+1} f_{p-q} + \omega^2 a_p + 2\omega b_{p+1} . \qquad \text{(A5.1)}$$

By changing the index of summation in the second group of terms, one obtains

$$a_{p+2} = -\sum_{q=0}^{p} \binom{p}{q} a_q g_{p-q} + \sum_{q=1}^{p+1} \binom{p}{q-1} a_q f_{p-q+1} + \omega^2 a_p + 2\omega b_{p+1} . \qquad \text{(A5.2)}$$

With the definition

$$\binom{P}{Q} = 0 \qquad \text{when } Q = -1 \text{ or } Q = P + 1 , \qquad (28)$$

the formulas for a_{p+2}, b_{p+2}, and c_{p+2} can be written as

$$a_{p+2} = -\sum_{q=0}^{p+1} K_{pq} a_q + \omega^2 a_p + 2\omega b_{p+1} ,$$

$$b_{p+2} = -\sum_{q=0}^{p+1} K_{pq} b_q + \omega^2 b_p - 2\omega a_{p+1} , \qquad (26)$$

$$c_{p+2} = -\sum_{q=0}^{p+1} K_{pq} c_q$$

for $p = 0, 1, 2, 3, \ldots$, and where

$$K_{pq} = \binom{p}{q} g_{p-q} - \binom{p}{q-1} f_{p-q+1}$$

for $q = 0, 1, 2, \ldots, p, p + 1; p = 0, 1, 2, \ldots .$ (27)

REFERENCES

[1] E. L. Ince, *Ordinary Differential Equations*, p 284, Longmans, Green and Company, London, 1926.

[2] Robert C. Meacham, *Power Series Solution of Rocket Equations of Motion*, AFMTC-TR 59-18, Sept. 18, 1959.

[3] J. F. Steffensen, "On the Restricted Problem of Three Bodies," *Kgl. Danske Videnskab. Selskab, Mat. fys. Medd.* **30**, No. 18 (1956); *Math. Rev.* **18**, 263 (1957).

[4] J. F. Steffensen, "On the Problem of Three Bodies in the Plane," *Kgl. Danske Videnskab. Selskab, Mat. fys. Medd.* **31**, No. 3 (1957); *Math. Rev.* **19**, 369 (1958).

[5] Ward Cheney, "An Example in Differential Equations: the N-Body Problem," *Amer. Math. Mon.* **67**(5), 456–57 (1960).

ANALOG COMPUTER SOLUTION FOR TRANSVERSE VIBRATIONS OF A UNIFORM BEAM WITH DAMPED, FLEXIBLE, MASSIVE END RESTRAINTS[1]

J. P. Raney
University of Virginia

A method is developed for solving, on an analog computer, for the transverse vibration response of a uniform beam mounted on flexible, damped massive end supports and subjected to concentrated forcing functions and couples.

The method used is that of breaking the beam into a long, uniform center section and end sections. An end section is represented by a system of two linear second-order differential equations, accounting for its flexibility, damping, mass, and geometry. The long, slender center section is represented in terms of its normal modes. Since only a few of the normal modes are required to be generated for any given frequency of a harmonic forcing function and a finite number for any arbitrary forcing function, a computer configuration results which does not require an excessively large number of amplifiers.

The results obtained on the University of Virginia PACE computer by use of this method agree to within 10% of the closed-form solution for the special case of zero end mass and zero spring rate on each end and a harmonic concentrated forcing function at the center of the beam.

INTRODUCTION

Analytical solutions for transverse vibrations of a beam become increasingly difficult to obtain as the complexity of the end restraints is increased [*1*]. Even after solutions are obtained, they must be plotted if an amplitude vs frequency or a displacement vs time display is desired. In addition, the fact that one often wants to vary parameters and obtain new solutions in plotted form indicates the desirability of using a modern analog or digital computer.

It is the purpose of this paper to outline a method for solving, on an analog computer, the general problem of transverse beam vibrations when the beam is mounted with damped, flexible, and massive end restraints. An example for a particular beam will be given.

[1]Supported by the AEC [contract No. AT-(40-1)-1779]; some of the material in this paper was published previously in report No. EP-4422-179-61U, Research Laboratories for the Engineering Sciences, University of Virginia, Charlottesville, in fulfillment of contractual obligations.

NOTATION

J	Moment of inertia of end piece
m	Mass of end piece
a	Distance from center of gravity of end piece to point of measurement of coordinate x_8 or x_{10}
b	Distance from point of attachment of flexible end support to point of measurement of coordinate x_8 or x_{10}
k	End-restraint stiffness constant
c	End-restraint viscous damping constant
x_3, x_5, x_7, x_9	Linear displacements
x_4, x_6, x_8, x_{10}	Angular displacements
F_3, F_5, F_7, F_9	Forces
F_4, F_6, F_8, F_{10}	Torques
M	Symmetric mass matrix
K	Symmetric stiffness matrix
F	Driving-force column matrix
X	Column matrix of linear and angular physical displacements
ϕ, ϕ_k	Normal-mode shape matrix and kth normal mode
$\alpha_{ij} = \alpha_i^{(j)}$	Elements of ϕ
ω_k	Natural frequency of kth normal mode
$Q(t)$	Generalized forcing-function column matrix
$F(t)$	External force applied to beam
$H(t)$	External couple applied to beam
EI	Stiffness parameter for beam
$A\rho$	Mass per unit length of beam
L	Length of beam

DEVELOPMENT OF EQUATIONS OF MOTION

A beam system subjected to a concentrated forcing function, or couple, may be represented as shown in Fig. 1.

If a sign convention is adopted so that forces and displacements are considered positive, as shown in Fig. 1, the following system of equations of motion for the end masses results:

1. for the left end,

$$F_7 = m_1\ddot{x}_7 - m_1 a_1\ddot{x}_8 + c_1\dot{x}_7 - c_1 b_1\dot{x}_8 + k_1 x_7 - k_1 b_1 x_8\,, \tag{1}$$

$$-F_8 = -(J_1 + m_1 a_1^2)\ddot{x}_8 + m_1 a_1\ddot{x}_7 + c_1 b_1\dot{x}_7 - c_1 b_1^2\dot{x}_8 + b_1 k_1 x_7 - k_1 b_1^2 x_8\,; \tag{2}$$

2. for the right end,

$$F_9 = m_2\ddot{x}_9 + m_2 a_2\ddot{x}_{10} + c_2\dot{x}_9 + c_2 b_2\dot{x}_{10} + k_2 x_9 + k_2 b_2 x_{10}\,, \tag{3}$$

$$F_{10} = (J_2 + m_2 a_2^2)\ddot{x}_{10} + m_2 a_2\ddot{x}_9 + c_2 b_2\dot{x}_9 + c_2 b_2^2\dot{x}_{10} + k_2 b_2 x_9 + k_2 b_2^2 x_{10}\,; \tag{4}$$

where

J_1, J_2 = moments of inertia of left- and right-end masses respectively,

m_1, m_2 = masses of left- and right-end masses respectively,

a_1, a_2 = distance from center of gravity of left- and right-end masses to point of measurement of coordinates x_8 and x_{10} respectively,

b_1, b_2 = distance from point of attachment of flexible end support to point of measurement of coordinates x_8 and x_{10} respectively,

k_1, k_2 = left and right linear end restraint flexibility parameters,

c_1, c_2 = left and right linear end restraint viscous damping parameters;

linear displacements and forces are

$$x_3 = x_7\,, \qquad F_3 = -F_7\,,$$
$$x_5 = x_9\,, \qquad F_5 = -F_9\,,$$

angles and torques are

$$x_4 = x_8\,, \qquad F_4 = -F_8\,,$$
$$x_6 = x_{10}\,, \qquad F_6 = -F_{10}$$

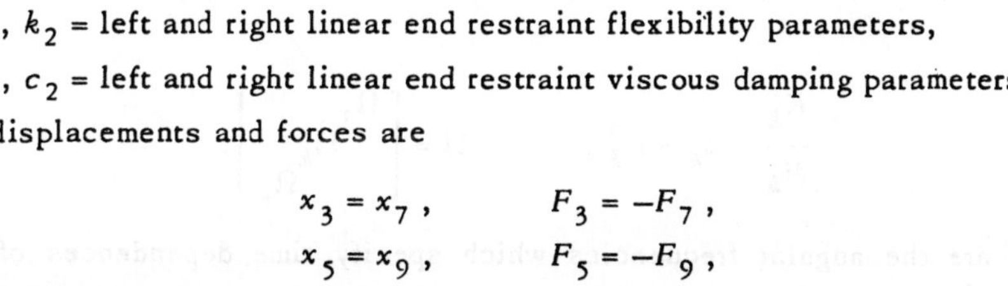

Fig. 1. Beam with Massive, Damped, Flexible End Supports.

It is convenient to express the behavior of the long, slender, center section of the beam in terms of its free-free normal modes; in the presence of suitable geometry, such a description can be transformed into one involving physical coordinates [2, 3]. This general solution for response of the free-free beam to concentrated forces and couples has been justified [4] (see also, the Appendix to this paper).

To see the details, consider a system with column vector, X, of r generalized coordinates. The equation

$$M\ddot{X} + KX = F \tag{5}$$

characterizes the essentials of this system, M and K being symmetric $r \times r$ matrices which are related to mass and stiffness respectively. The column vector F is called the driving force for the system.

Now, there exists a matrix ϕ which characterizes the normal-mode shapes of the system and which diagonalizes M and K; that is,

$$\phi^T M \phi = \begin{bmatrix} M_1 & & \\ & M_k & \\ & & M_r \end{bmatrix}, \qquad \phi^T K \phi = \begin{bmatrix} K_1 & & \\ & K_k & \\ & & K_r \end{bmatrix}.$$

Also,

$$\frac{K_k}{M_k} = \Omega_k = \omega_k^2, \qquad \Omega = \begin{bmatrix} \Omega_1 & & \\ & \Omega_k & \\ & & \Omega_r \end{bmatrix}.$$

The ω_k are the angular frequencies which specify time dependences of these normal modes.

Equation (5) simplifies to

$$\ddot{q} + \Omega q = \begin{bmatrix} \dfrac{1}{M_1} \displaystyle\sum_{k=1}^{r} \alpha_{k1} F_k \\ \vdots \\ \dfrac{1}{M_r} \displaystyle\sum_{k=1}^{r} \alpha_{kr} F_k \end{bmatrix} = \begin{bmatrix} Q_1 \\ \vdots \\ Q_r \end{bmatrix} = Q\ [\phi = (\alpha_{ij})]\ . \tag{6a}$$

The column vector Q is called the generalized forcing function for the system. Clearlv, (6*a*) implies the key relations

$$\begin{Bmatrix} \ddot{q}_1 \\ \ddot{q}_2 \\ \vdots \\ \ddot{q}_r \end{Bmatrix} + \begin{bmatrix} \Omega_1 & & & \\ & \Omega_2 & & \\ & & \ddots & \\ & & & \Omega_r \end{bmatrix} \begin{Bmatrix} q_1 \\ q_2 \\ \vdots \\ q_r \end{Bmatrix} = \begin{Bmatrix} Q_1 \\ Q_2 \\ \vdots \\ Q_r \end{Bmatrix}, \tag{6b}$$

or

$$\ddot{q}_k + \omega_k^2 q_k = Q_k \,, \qquad k = 1, 2, \ldots, r \,, \quad (6c)$$

as derived in the Appendix, where the physical meaning of Eq. (5) should become more clear.

The importance of the system of Eqs. (6*b*) is that they are in a form readily adaptable to analog computation, and it is precisely these equations which will be used to obtain q and, hence, $X(=\phi q)$ for the slender section of the beam.

At this point Plunkett's approach [5] may be utilized, and certain columns of the ϕ matrix discarded. It may be observed that there is no necessity for the dimension r of vectors X and q to be the same, and the discarding of certain columns in Eq. (7) was actually preceded by the tacit discarding of an infinite number of columns. Thus,

$$\begin{Bmatrix} x_1 \\ x_2 \\ x_k \\ x_r \end{Bmatrix} = \begin{bmatrix} 0 & 0 & \alpha_1^{(\beta+1)} & \ldots & \alpha_1^{(r-v)} & 0 & 0 \\ 0 & & & & & & \\ \cdot & & & & & & \\ \cdot & & & & & & \\ 0 & 0 & \alpha_r^{(\beta+1)} & \ldots & \alpha_r^{(r-v)} & 0 & 0 \end{bmatrix} \begin{Bmatrix} q_1 \\ q_2 \\ \cdot \\ \cdot \\ q_r \end{Bmatrix} , \quad (7)$$

where the lowest β and the highest v modes have been neglected, and the second subscript on α is now a superscript. Now,

$$x_k = \sum_{n=\beta+1}^{r-v} \alpha_k^{(n)} q_n \,, \qquad k = 1, 2, \ldots, r \,.$$

In general, the modes discarded will be those making only a negligible contribution to x_k [5]. For response to a harmonic forcing function, all modes except those whose natural frequencies, ω_k, are near the forcing frequency will be neglected, as suggested by part F of the Appendix.

COMPUTER PROGRAM

By the proper combinations of Eqs. (1)–(4), (6), and (7), a computer program can be established. The method is, essentially, as follows:

1. Construct the computer solution of the system of Eqs. (1)–(4), assuming that all displacements and derivatives are available. This results in solutions for F_7, F_8, F_9, and F_{10} (Fig. 2).

2. Form as many mode equations as desired, and assemble the generalized forces Q_k as shown in Eqs. (6) and (7). The Q_k can be formed by use of the F's already available and by use of vibration analysis tables [2] to find appropriate values of α_{ij} (Fig. 3). Notice that, in the figures, a notation more typical to vibration work is used; that is, $\alpha_{ij} = \alpha_i^{(j)}$.

3. Now, solutions for q are available and may be combined by use of $X = \phi q$ to give the physical coordinates and their derivatives (Fig. 4).

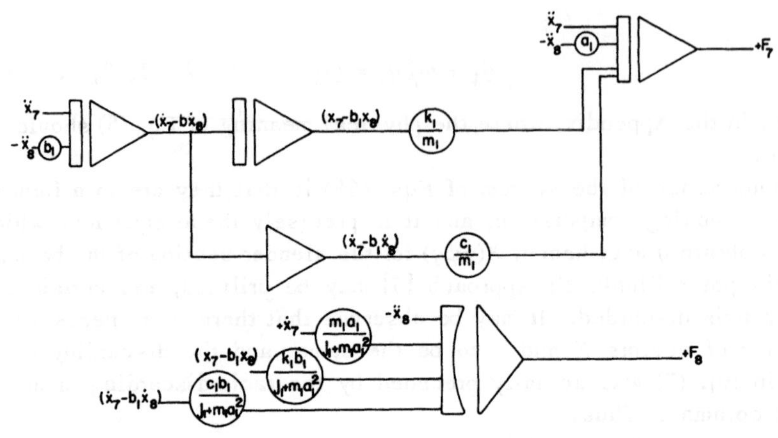

Note: A similar diagram results for the right end mass.

Fig. 2. Computer Diagram for Eqs. (1) and (2) Left-End Mass.

$$\ddot{q}_1 + w_1^2 q_1 = \left[\alpha_1^{(1)}F_1 + \alpha_2^{(1)}F_2 + \alpha_3^{(1)}F_3 + \alpha_4^{(1)}F_4 + \alpha_5^{(1)}F_5 + \alpha_6^{(1)}F_6\right]/M_1$$

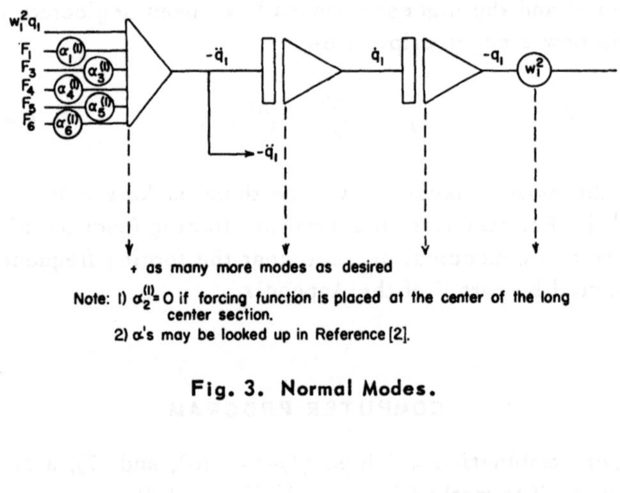

+ as many more modes as desired

Note: 1) $\alpha_2^{(1)} = 0$ if forcing function is placed at the center of the long center section.

2) α's may be looked up in Reference [2].

Fig. 3. Normal Modes.

$$x_i = \sum_{n=1}^{r} \alpha_i^{(n)} q_n$$

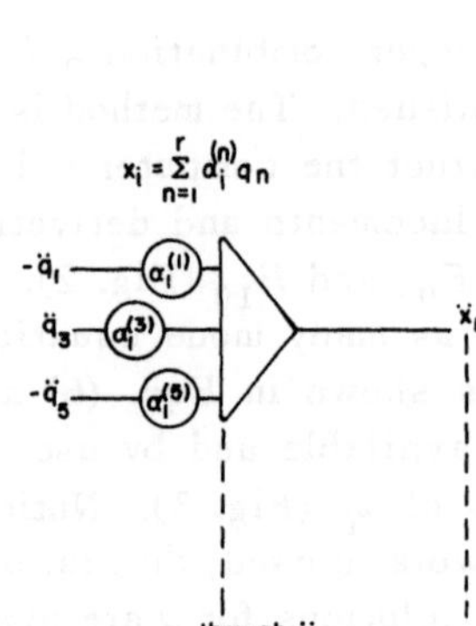

through $\ddot{x}_6$

Note: $\alpha_1^{(2)} = \alpha_1^{(4)}$ for geometry as chosen

Fig. 4. Formation of Physical Coordinates.

4. Finally, the loop is closed by feeding these assembled physical coordinates and their derivatives into the system of Eqs. (1)–(4), which is where the loop was started (Fig. 2).

It is understood that the response of any of the physical coordinates may be observed on a suitable readout device.

Solution of a Typical Problem

Consider a beam with the following specifications (Fig. 1):

Center Section

	Free-Free Natural Frequencies of the First Four Modes
$F(t) = 2.5 \sin \omega t$ at center	$\omega_1 = 2\pi \times 95$ radians/sec
$EI = 7.6 \times 10^5$ lb-in.^{2}	$\omega_2 = 2\pi \times 263$ radians/sec
$A\rho = 0.0293/386$ lb-in.-sec^2	$\omega_3 = 2\pi \times 515$ radians/sec
$L = 65$ in.	$\omega_4 = 2\pi \times 850$ radians/sec

End Sections and Restraints (units are inches, pounds, seconds)

$a_1 = a_2 = 0.75$	$J_1 = 0.96 \times 10^{-4}$ when not 0
$b_1 = b_2 = 1.20$	$J_2 = 0.96 \times 10^{-4}$ when not 0
$m_1 = 0.34 \times 10^{-3}$ when not 0	c_1 = various values
$m_2 = 0.34 \times 10^{-3}$ when not 0	c_2 = various values
	$k_1 = k_2 = 0$

The University of Virginia PACE computer was programmed for the above-given data, and the solutions are shown in Figs. 5 and 6.

Extreme accuracy was not attempted, but, rather, it was desired to demonstrate the usefulness of the method described in this paper. Steady amplitudes (voltages) were read directly from a cathode-ray oscilloscope. Frequency was checked with a Hewlett-Packard 522 electronic counter. The forcing function was generated with an HP 202 low-frequency function generator. A time-scaling factor of 20 was used, thus magnifying any error in frequency, if such were present.

In order to generate the four modes desired, some of the feedback capacitors had to be changed from the usual 1 μf to 0.1 μf, which reduced the accuracy of the solutions.

Frequency response data were taken over a limited range near the third-mode (free-free) frequency in a region near a point at which the response amplitude is independent of damping. This was done because of the availability of plotted closed-form check solutions in these regions.

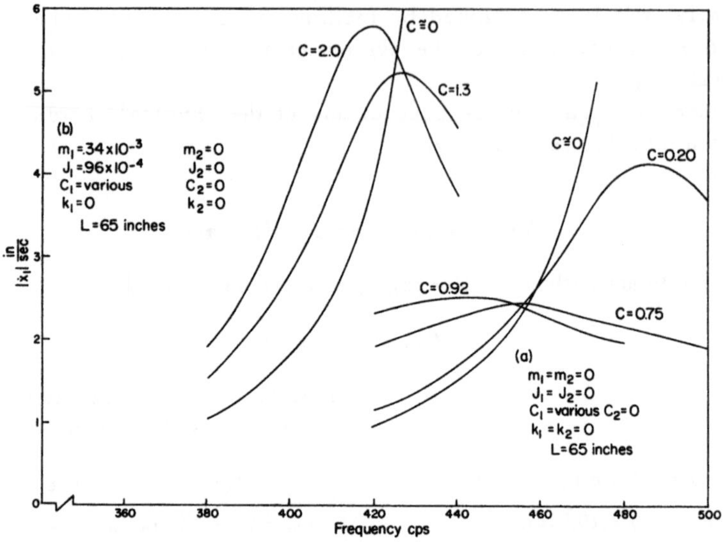

Fig. 5. Analog Computer Solutions.

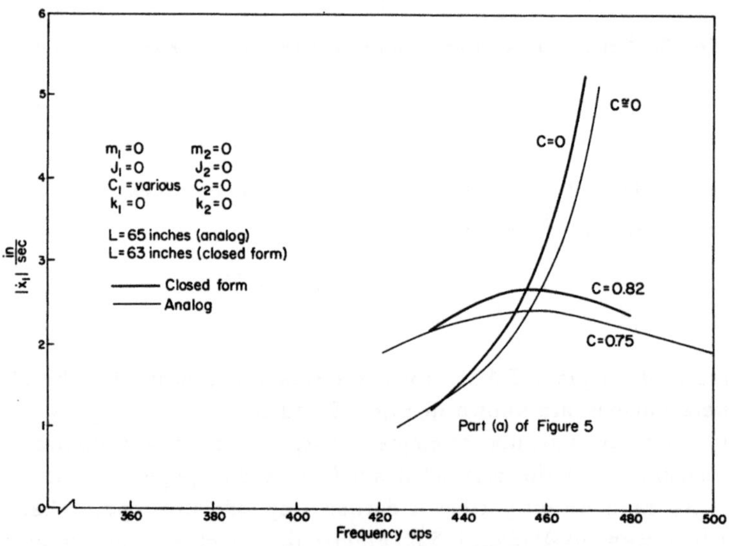

Fig. 6. Analog Computer and Closed-Form Solutions Compared.

DISCUSSION OF RESULTS AND ACCURACY

Analytical closed-form solutions exist for the problem if the end masses and moments of inertias are zero. Both Springfield [1] and Henney [6] plotted certain of these solutions; they also predicted response characteristics in a region near what they referred to as "fixed points" or points through which all response curves pass, irrespective of damping [7].

The analog computer results for zero end mass and moments of inertia (Fig. 6) agree well with the closed-form solutions as computed by the digital computer in the region near the selected fixed point. In order that the frequency error between the two solutions, as shown in Fig. 6, not be unduly magnified, the closed-form solution was adjusted for a beam of length $L = 63$ in. With $L = 65$, the frequency of the invariant point for the closed-form solution was 427 cps, or a deviation of $[(427 - 456)/427]100 = -6.8\%$. Careful tuning of the model loops on the analog computer might help to reduce this discrepancy. Also, with very light damping, the analog solutions produce the antiresonant points near those predicted by the analytical solutions; however, these are not shown.

The error in amplitude of the fixed point was 10%. If one were interested in obtaining the value of damping that gave a zero slope of the response curve at the invariant point [7], the error would be about 9%.

CONCLUSIONS

The normal-mode method of solution of the vibrating-beam problem on the analog computer, even when only a few modes are included, appears to be sufficiently accurate for practical purposes. The advantage of this method lies in the fact that no symmetry is required except that the long, slender center section be uniform in order that the α's may be found in tables; that is, $k_1 \neq k_2$, $c_1 \neq c_2$, and so forth, for the other parameters.

The inclusion of torsional elasticity and torsional damping on the end pieces (Fig. 1) does not alter the basic structure of the system of Eqs. (1)–(4).

Many configurations that would otherwise defy solution can be investigated. In general, any eigenvalue problem, which yields solutions in terms of normal modes, can be solved in this manner.

The arbitrary nature of the forcing function makes the technique of this report valid in the case of shock, as well as in the case of periodic loading; also, the response of any arbitrary point in the system may be determined by choosing to include it in X [8].

ACKNOWLEDGMENT

The author wishes to thank Dr. Robert Plunkett, Professor of Aeronautical Engineering of the University of Minnesota, for first suggesting the technique presented in this report and for offering valuable comments as the work progressed; Mr. R. L. Tomlin, for critically reading and discussing the manuscript; and Mr. Louis Nelson, for assisting with the analog computer work.

APPENDIX

Derivation of Eq. (6c) from the Bernoulli-Euler Beam Equation

Consider a long, uniform, slender, free-free beam in forced transverse vibration, and let

$$EI\frac{\partial^4 x}{\partial y^4} + A\rho\frac{\partial^2 x}{\partial t^2} = F(y\,,t) \tag{8}$$

with

$$\frac{\partial^2 x}{\partial y^2} = \frac{\partial^3 x}{\partial y^3} = 0 \text{ at } y = 0,\ L$$

be the governing equation of motion.

A. From the homogeneous equation, assuming harmonic motion, it is easy to determine functions, $\phi_k(y)$, such that $x(y) = C_k\,\phi_k(y)$, where C_k is a constant dependent on initial conditions, when the beam is vibrating at a characteristic (natural) frequency, ω_k, and such that

$$\int_0^L \phi_k(y)\phi_m(y)\,dy = L, \text{ for } m = k\,,$$
$$= 0, \text{ for } m \neq k\,,$$

an orthogonality property, is also true.[2]

B. Returning now to the nonhomogeneous equation, assume that

$$x(y\,,t) = \sum_{k=1}^{\infty} \phi_k(y)q_k(t) \tag{9}$$

and that

$$F(y\,,t) = \theta(y)\,\psi(t) = \left[\sum_{k=1}^{\infty} A_k\,\phi_k(y)\right]\psi(t)\,. \tag{10}$$

But

$$EI\phi_k^{IV}(y) = A\rho\omega_k^2\,\phi_k(y)$$

if Eq. (9) is substituted in the homogeneous equation (8). Therefore

$$\sum_{k=1}^{\infty} A\rho\phi_k(y)\ddot{q}_k(t) + \sum_{k=1}^{\infty} A\rho\omega_k^2\,\phi_k(y)q_k(t) = \left[\sum_{k=1}^{\infty} A_k\,\phi_k(y)\right]\psi(t)\,. \tag{11}$$

By the orthogonality property,

$$A\rho\ddot{q}_k(t) + A\rho\omega_k^2\,q_k(t) = A_k\,\psi(t)\,,$$

---

[2]Reference [2], pp 318–28.

or

$$\ddot{q}_k(t) + \omega_k^2 q_k(t) = Q_k , \tag{12}$$

where

$$Q_k = \frac{A_k \psi(t)}{A\rho} .$$

Again, by the orthogonality property,

$$\frac{A_k \psi(t)}{A\rho} = \frac{\int_0^L F(y, t)\phi_k(y)\, dy}{A\rho \int_0^L \phi_k^2(y)\, dy} = \frac{1}{A\rho L} \int_0^L F(y, t)\phi_k(y)\, dy . \tag{13}$$

C. To evaluate Eq. (13) for a concentrated forcing function at $y = C$, the following procedure is suitable (see Fig. 7):

$$\frac{A_k \psi(t)}{A\rho} = \lim_{\epsilon \to 0} \frac{1}{A\rho L} \int_{C-\epsilon}^{C+\epsilon} \phi_k(y) \frac{F}{2\epsilon} \psi(t)\, dt$$

$$= \lim_{\epsilon \to 0} \left\{ \frac{[(C+\epsilon) - (C-\epsilon)]}{A\rho L} \frac{F}{2\epsilon} \phi_k(Y)\psi(t) \right\} = \frac{F\phi_k(C)\psi(t)}{A\rho L} .$$

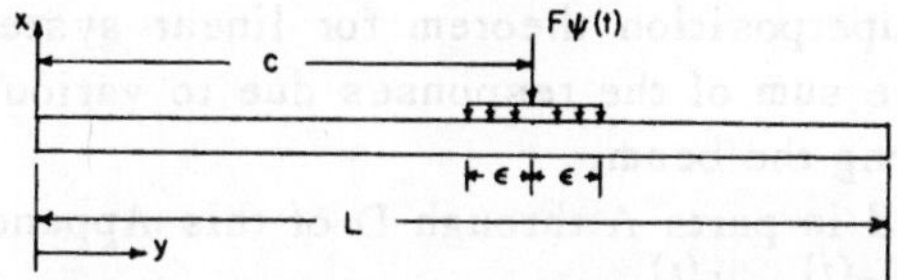

Fig. 7. Beam with Concentrated Forcing Function.

Equation (12) now becomes

$$\ddot{q}_k + \omega_k^2 q_k = \frac{\phi_k(C)F\psi(t)}{M_k} = Q_k , \tag{14}$$

and Eq. (14) is exactly Eq. (6c), where $A\rho L = M_k$ = total mass of the beam.

D. To evaluate Eq. (13) for a couple acting at $y = C$, the procedure of part C above is modified as indicated in Fig. 8. Let

$$F\psi(t) = \frac{1}{\epsilon} H\psi(t) ,$$

where H is a couple. Then

$$\ddot{q} + \omega_k^2 = \lim_{\epsilon \to 0} \left\{ \frac{[(a+\epsilon) - (a-\epsilon)]}{A\rho L} \frac{1}{2\epsilon^2} H\phi_r(Y)\psi(t) \right.$$

$$\left. - \frac{[(b+\epsilon) - (b-\epsilon)]}{A\rho L} \frac{1}{2\epsilon^2} H\phi_r(Y)\psi(t) \right\}$$

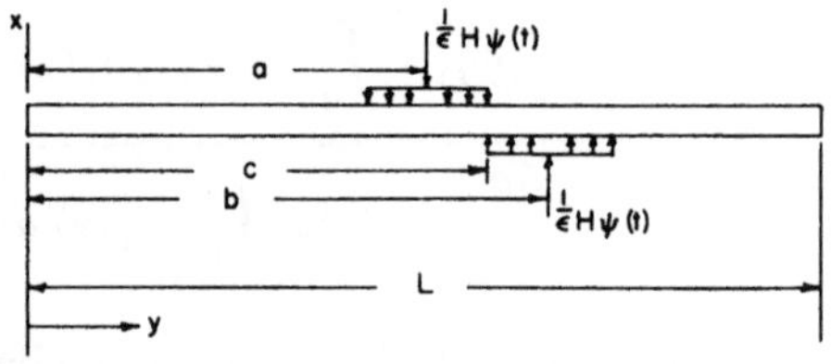

Fig. 8. Beam with Concentrated Couple.

If the limit is taken so that as $\epsilon \to 0$ both forces move toward point $y = C$, then

$$\ddot{q}_k + \omega_k^2 q_k = \frac{\phi_k'(C)H\psi(t)}{M_k} = Q_k \,. \tag{15}$$

Equation (14) is of the same form as Eq. (6c), and Eq. (5) is implicit in Eqs. (9), (14), and (15).

E. The time dependency, $\psi(t)$, has not been specified and can be of an arbitrary nature [8].

By the powerful superposition theorem for linear systems, the response at any point, $x(y, t)$, is the sum of the responses due to various forces and couples applied at any point along the beam.

The $\phi_k(y \text{ or } C)$ used in parts A through D of this Appendix are related to the α_{ij} of the text by $\alpha_{ij} = \alpha_i^{(j)} = \phi_j^{(i)}$.

F. If $\psi(t) = \sin \omega t$, then particular solutions of Eqs. (14) and (15) are

$$q_k = \left[\frac{\phi_k}{A\rho L(\omega_k^2 - \omega^2)} \right] F \sin \omega t \,,$$
$$q_k = \left[\frac{\phi_k'(C)}{A\rho L(\omega_k^2 - \omega^2)} \right] H \sin \omega t \,, \tag{16}$$

$$x(y, t) = \left[\sum_{k=1}^{\infty} \frac{\phi_k(y)\phi_k(C)}{A\rho L(\omega_k^2 - \omega^2)} \right] F \sin \omega t \,,$$
$$x(y, t) = \left[\sum_{k=1}^{\infty} \frac{\phi_k(y)\phi_k'(C)}{A\rho L(\omega_k^2 - \omega^2)} \right] H \sin \omega t \,. \tag{17}$$

Also,

$$
\frac{\partial x}{\partial y} = \left[\sum_{k=1}^{\infty} \frac{\phi_k'(y)\phi_k(C)}{A\rho L(\omega_k^2 - \omega^2)} \right] F \sin \omega t \; ,
$$

$$
\frac{\partial x}{\partial y} = \left[\sum_{k=1}^{\infty} \frac{\phi_k'(y)\phi_k'(C)}{A\rho L(\omega_k^2 - \omega^2)} \right] H \sin \omega t \; , \tag{18}
$$

so that

$$
\begin{pmatrix} x \\ \\ \dfrac{\partial x}{\partial y} \end{pmatrix} = (\text{a receptance}) \begin{pmatrix} \begin{pmatrix} F \sin \omega t \\ H \sin \omega t \end{pmatrix} \\ \begin{pmatrix} F \sin \omega t \\ H \sin \omega t \end{pmatrix} \end{pmatrix}
$$

as given by Bishop [2].

REFERENCES

[1] J. F. Springfield, *Optimization of Parameters of a Vibrating Beam Mounted on Damped, Flexible End Supports,* unpublished M.S. thesis, University of Virginia, Charlottesville, 1961.

[2] R. E. D. Bishop and D. C. Johnson, *The Mechanics of Vibration,* Cambridge University Press, Cambridge, England, 1960.

[3] G. M. L. Gladwell and R. E. D. Bishop, "The Vibration of Rotating Shafts Supported in Flexible Bearings," *J. Mech. Eng. Sci.* **1**(3), 195 (1959).

[4] G. M. L. Gladwell and R. E. D. Bishop, "The Receptances of Uniform and Non-Uniform Rotating Shafts," *J. Mech. Eng. Sci.* **1**(1), 78 (1959).

[5] R. Plunkett, "Semi-Graphical Method for Plotting Vibration Response Curves," p 221 in *Proc. Second Natl. Congr. Appl. Mech., ASME,* New York, 1954.

[6] A. Henney, *The Theoretical Optimization of Damping for Four Configurations of a Vibrating Uniform Beam,* unpublished M.S. thesis, University of Virginia, Charlottesville, 1961.

[7] R. Plunkett, "The Calculation of Optimum Concentrated Damping for Continuous Systems," *J. Appl. Mech.* **80**, 219 (1958).

[8] W. T. Thompson and T. A. Rogers, "Analog Solution for Beams Excited by an Arbitrary Force," *J. Appl. Mech.* **20**, 482–84 (1953).

THE USE OF THE WKB METHOD IN APPLIED MECHANICS[1]

Frank M. White, Jr.
Georgia Institute of Technology

The WKB method, which is used by theoretical physicists in the study of high-frequency oscillations in quantum mechanics, is proposed here as providing satisfactory engineering accuracy for a wide variety of problems in applied mechanics.

The general purposes of this paper are (1) to reacquaint workers in applied mechanics with the WKB method, (2) to show the versatility of the method, and (3) to indicate how the approximating error involved may be estimated. To these ends, a number of examples are provided, which include problems of buckling loads in variable-stiffness columns, free vibrations with a time-variable spring constant, and bending moments in variable-stiffness beam-columns with axial tension loads. The examples chosen are intended to show both the advantages and the disadvantages of the WKB method.

INTRODUCTION

The WKB method is well known to physicists who work in the field of quantum mechanics. The name "WKB" is taken from the initials of the three men – Wentzel [*1*], Kramers [*2*], and Brillouin [*3*] – who independently proposed the method in 1926. The method is used by theoretical physicists to approximate the behavior of the high-frequency modes of oscillation of the Schroedinger wave equation. As frequency increases, the WKB method becomes increasingly accurate.

Possibly because of the emphasis upon very high frequencies, the WKB method is relatively unfamiliar to workers in applied mechanics. However, the method is applicable to a wide variety of problems in mechanics, particularly to buckling and vibration problems. It is the purpose of this paper to once again bring attention to the WKB method and to illustrate its accuracy with a number of different examples.

---

[1]Sponsored by the Academic Research Program of the Georgia Institute of Technology.

NOTATION

a, b, C, C_1, C_2	Constants
δ	Error term in the WKB method
$\bar{\delta}$	Average value of δ in the interval
E, E_0	Moduli of elasticity
f	Special function in beam-column example
G, H	Arbitrarily variable coefficients
I, I_0	Beam section moments of inertia
K	Dimensionless parameter, $K^2 = PL^2/(E_0 I_0)$
K_n	Dimensionless parameter, $K_n^2 = P_n L^2/(E_0 I_0)$
L	Beam length
λ	Dimensionless variable, $\lambda = x/L$
M, M_0	Bending moments
n	Integer number of buckled waves
P	Axial load
P_n	nth order buckling load
π	3.1416
ψ	Phase angle
ξ	Dummy variable
t	Time
x	Independent variable, axial distance
y	Dependent variable, beam deflection
y_a	Approximate WKB solution
z	Dependent variable
z_a	Approximate WKB solution

STATEMENT OF THE METHOD

In its simplest form, the WKB method is concerned with the second-order, linear, homogeneous differential equation in normal form:

$$y'' + G^2(x)y = 0 , \tag{1}$$

where the prime signifies differentiation with respect to x. If the coefficient G^2 is an arbitrarily variable function of x, the general solution to Eq. (1) is not known, and one must resort to numerical or analog techniques. However, the WKB method proposes the following approximate solution to Eq. (1):

$$y_a = \frac{1}{\sqrt{G}} \exp\left[i \int_0^x G \, d\xi \right], \tag{2}$$

where $i = \sqrt{-1}$. Equation (2) is the exact solution to the following approximating differential equation:

$$y_a'' + G^2 y_a (1 - \delta) = 0 \,, \tag{3}$$

where $\delta = \delta(x)$ is the associated error term which indicates the degree of approximation:

$$\delta = \frac{3G'^2}{4G^4} - \frac{G''}{2G^3} \,. \tag{4}$$

It is clear that the function y_a will provide an accurate approximate solution to Eq. (1) whenever $|\delta| << 1$. In quantum mechanics, the requirement that δ be small results in the WKB method being used only for "slowly varying, low wavelength" oscillations, that is, for G very large and G' very small. The WKB solution is actually exact for the two special cases G = a constant and $G = a/(b + x)^2$, where a and b are constants.

In the event that δ is small but not negligible, it may be possible to use an average value $\bar{\delta}$ to improve the WKB solution. The author has made preliminary studies with this end in mind. No general conclusions have been obtained, but specific corrections for particular problems are illustrated in the examples which follow.

The WKB method may be readily extended to partial differential equations. For further details on the general use of the method in theoretical physics, see, for example, refs [4–6].

APPLICATIONS TO MECHANICS PROBLEMS

In applied mechanics, conditions for the use of the WKB method are, in general, less strict than in quantum mechanics. The examples which follow are intended to show that Eq. (2) may provide a reasonable engineering approximation, even when the error term δ is not "small," say, $\delta = 0.3$ or even larger. Further, the examples will illustrate the author's preliminary attempts to use the term $\delta(x)$ to correct Eq. (2) and thereby obtain a better approximation.

Example 1: Buckling of a Variable-Stiffness Column

Consider a pin-ended column of length L and variable stiffness $EI(x)$ subjected to a compression load P. Then the problem of calculating the small-deflection buckling modes of this column is an eigenvalue problem:

$$\frac{d^2y}{d\lambda^2} + \frac{PL^2}{EI} y = 0 , \tag{5}$$

subject to $y(0) = y(1) = 0$, where $\lambda = x/L$. If the stiffness EI is arbitrarily variable, one usually resorts to the Rayleigh-Ritz-Timoshenko method or to numerical matrix methods. However, substitution into Eq. (2) gives the following approximate WKB solution:

$$\begin{aligned} L\sqrt{P_n} \int_0^1 \frac{d\lambda}{\sqrt{EI}} &= n\pi , \\ y_a = C \sin &\left(L\sqrt{P_n} \int_0^\lambda \frac{d\xi}{\sqrt{EI}} \right), \end{aligned} \tag{6}$$

where C is an arbitrary constant, and n is the number of buckled waves. If the error term $\delta(\lambda)$ is positive, Eqs. (6) will overestimate the buckling load P_n. In particular, if δ is small, comparison of Eqs. (6) and (3) indicates that P_n is overestimated by the amount $\bar{\delta}$, where $\bar{\delta}$ is the average value of δ over the length of the beam. Then a better approximation to the buckling load should be P_n, from Eqs. (6), divided by $(1 + \bar{\delta})$. The general accuracy of the method may be shown by the following two specific cases:

Case 1a. – Let $EI = E_0 I_0/(1 + \lambda)$. Substituting into Eq. (6), one obtains

$$\begin{aligned} K_n^2 = P_n L^2/E_0 I_0 &= 0.67302\, n^2\pi^2 \text{ (uncorrected)} , \\ y_a = C \sin &\left\{ \frac{2}{3} K_n [(1 + \lambda)^{3/2} - 1] \right\} . \end{aligned} \tag{7}$$

The average error is given by $\bar{\delta} = 15/128K_n^2$. The approximate values for K_n from Eqs. (7) are compared below with an exact digital computer solution for $n = 1$ and $n = 2$. Also included in the table below are values corrected for the effect of $\bar{\delta}$:

	$n = 1$	$n = 2$
K_n^2 (exact)	6.5484	26.465
K_n^2 (WKB)	6.6424	26.570
% error	1.44	0.40
K_n^2 (WKB)/$(1 + \bar{\delta})$	6.5272	26.453
% error	0.32	0.04

As expected, the WKB method is increasingly accurate at the higher buckling modes, the error decreasing roughly as $1/n^2$. However, the accuracy is satisfactory, even at the lowest mode, $n = 1$. The buckled shape $y_a(\lambda)$ as predicted by Eqs. (7) is also very accurate, as shown in Fig. 1, in which both curves are normalized to have a maximum value of unity. The curve for $y_a(\lambda)$ is so accurate that it may be conveniently substituted into the Rayleigh-Ritz-Timoshenko method:

$$K_n^2 \approx \frac{\int_0^1 y_a'^2 \, d\lambda}{\int_0^1 \frac{y_a^2 E_0 I_0}{EI} \, d\lambda} . \tag{8}$$

The result, by using y_a from Eqs. (7), is $K_1^2 = 6.5484$ and $K_2^2 = 26.465$, correct to five significant figures.

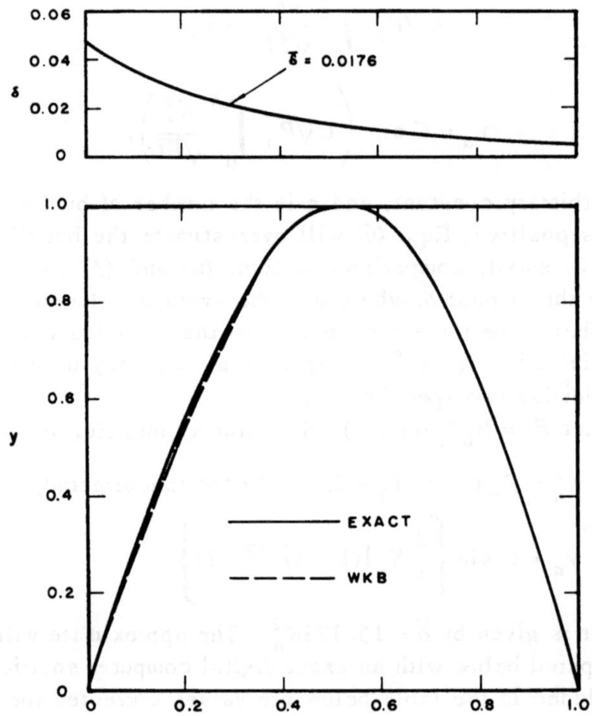

Fig. 1. First-Mode Buckling for $EI = E_0 I_0/(1 + \lambda)$.

Case 1b. – Let $EI = E_0 I_0/\lambda$. Substitution into Eqs. (6) gives the following WKB result:

$$\begin{aligned} K_n^2 &= 2.25 \, n^2 \pi^2 , \\ y_a &= C \sin (\pi \lambda^{3/2}) , \\ \bar{\delta} &= \infty . \end{aligned} \tag{9}$$

For this example, EI is infinite at $\lambda = 0$, and the error term breaks down, giving an apparently infinite $\bar{\delta}$. Thus, one may expect Eqs. (9) to be an inaccurate approximation, although of course the actual error will not be infinite. Comparison of the exact (see ref [7]) and the WKB eigenvalues for this case is given in the table below:

	$n = 1$	$n = 2$	$n = 3$
K_n^2 (exact)	18.956	81.887	189.221
K_n^2 (WKB)	22.207	88.826	199.859
% error	17.15	8.47	5.62

The WKB method in this instance gives unsatisfactory accuracy and should not be used. The approximate shape $y_a(\lambda)$ is badly in error, as shown in Fig. 2. Note that the WKB solution predicts that $y_a'(0) = 0$, which is physically incorrect. Substitution of y_a into the Rayleigh-Ritz-Timoshenko method gives $K_1^2 = 20.554$, which is off by 8.43%. By comparison, the use of the simple guess $y_a = C \sin (\pi\lambda)$ in Eq. (8) gives $K_1^2 = 19.739$, which is only in error by 4.13%.

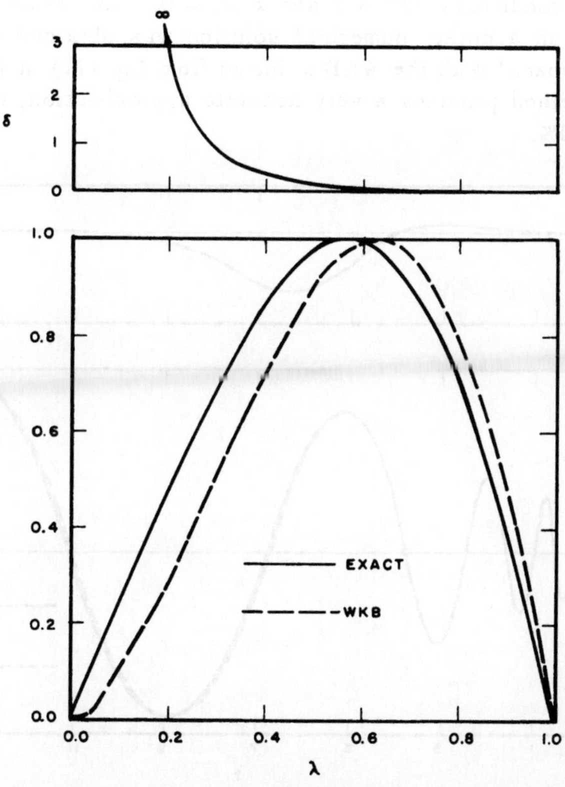

Fig. 2. First-Mode Buckling for $EI = E_0 I_0/\lambda$.

Example 2: Vibrations with Time-Dependent Coefficients

Consider a freely vibrating elastic body with a time-dependent spring constant. The governing differential equation would be

$$\frac{d^2 z}{dt^2} + H^2(t) z = 0 \; . \tag{10}$$

This type of equation governs the oscillatory motion of aircraft and missiles in a decelerating condition (see, e.g., refs [8 and 9]). The WKB method indicates the following approximate solution for the motion:

$$z_a = \frac{C}{\sqrt{H}} \cos \left[\int_0^t H \, d\xi + \psi \right] , \tag{11}$$

where C is a constant and ψ is a phase angle. The error term $\delta(t)$ from Eq. (4) appears to be of only limited usefulness here, since the solution is unbounded in the time domain. To illustrate the use of Eq. (11), let us take the following numerical example:

$$H^2(t) = 50(0.01 + 1.99e^{-t}) \; , \tag{12}$$

with the initial conditions $z(0) = 1$ and $z'(0) = 0$. The exact analytic solution being unknown, an accurate numerical solution was obtained on a digital computer and is compared with the WKB solution from Eq. (11) in Fig. 3. It is seen that the WKB method provides a very accurate approximation, the maximum error being less than 1%.

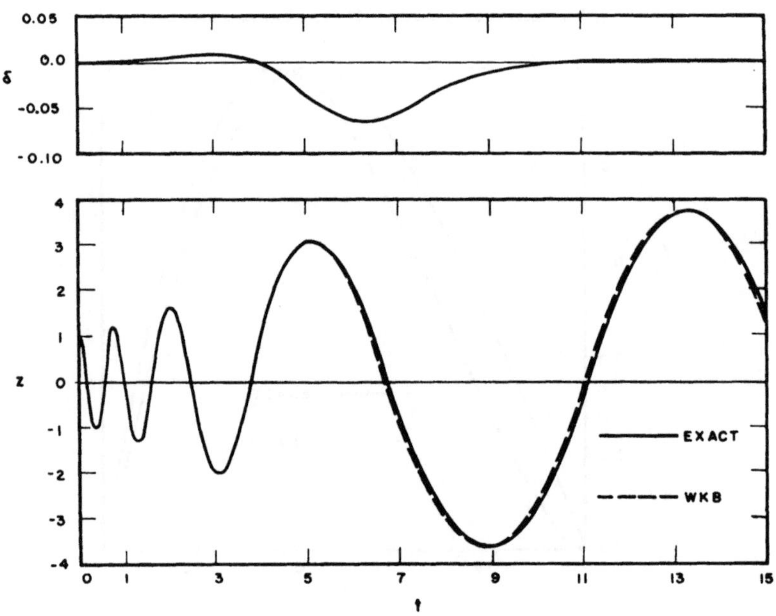

Fig. 3. Oscillations with a Time-Dependent Spring Constant.

Example 3: Beam-Columns with Variable Stiffness

A solution by the WKB method need not be oscillatory in nature. To illustrate this, consider a pin-ended beam-column of length L and variable EI, subjected to an axial tension load P and an applied couple at both ends. The differential equation for small deflections is

$$\frac{d^2M}{d\lambda^2} - \frac{PL^2}{EI} M = 0 \,, \tag{13}$$

where M is the local bending moment. The WKB solution to this problem may be written as

$$M = \left(\frac{EI}{PL^2}\right)^{1/4} \left[C_1 \sinh \int_0^\lambda \sqrt{\frac{PL^2}{EI}}\, d\xi + C_2 \cosh \int_0^\lambda \sqrt{\frac{PL^2}{EI}}\, d\xi \right], \tag{14}$$

where C_1 and C_2 are constants which depend upon the end couples.

To show the accuracy of this approximation, let us take an assumed stiffness $EI = E_0 I_0[1 + \sin(\pi\lambda)]$. Further, let the end conditions be $M(0) = M(1) = M_0$. By using the short notation

$$K^2 = PL^2/E_0 I_0 \,, \qquad f(\lambda) = \int_0^\lambda \frac{d\xi}{\sqrt{1 + \sin(\pi\xi)}} \,,$$

one may solve for the constants C_1 and C_2 and write the solution as follows:

$$M/M_0 = [1 + \sin(\pi\lambda)]^{1/4} \left\{ \left[\frac{1 - \cosh(0.794K)}{\sinh(0.794K)} \right] \sinh(f) + \cosh(f) \right\}, \tag{15}$$

where the numerical value $0.794 = f(1)$. The maximum error in this approximation is very nearly $0.5\overline{\delta}$, if δ is small. A comparison between Eq. (15) and an exact digital computer solution of Eq. (13) is shown in Fig. 4, for two values of the parameter K. The error term $\delta(\lambda)$ is relatively large, since the assumed EI distribution was not "slowly varying." However, it will be noticed from Fig. 4 that good accuracy could have been obtained by correcting the WKB solution for a maximum error of $0.5\overline{\delta}$.

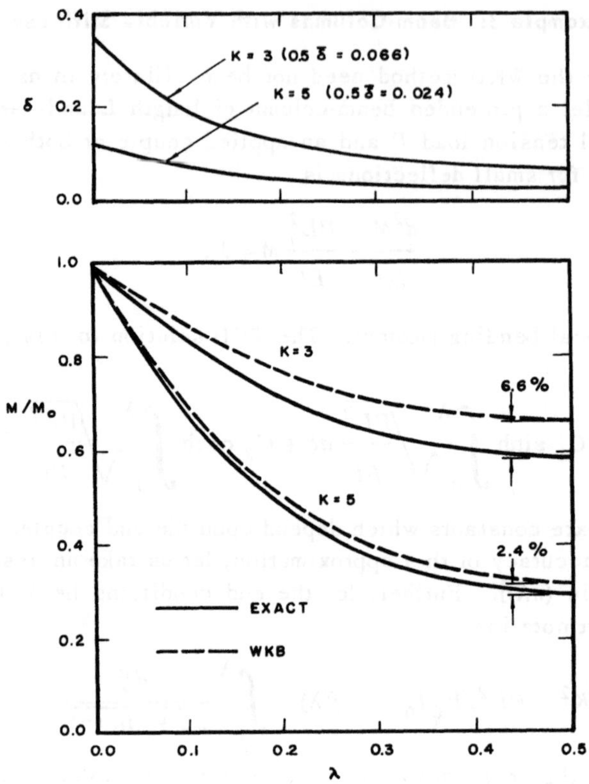

Fig. 4. Bending Moments in an Axial-Tension Beam-Column.

CONCLUSIONS

The examples and discussion given here are intended to show that the WKB method can provide a very useful approximation in applied mechanics problems and that its accuracy is not as limited as is commonly believed. Furthermore, it has been shown that the accuracy can sometimes be considerably improved by careful examination of the error term δ in the method.

The chief advantages of the WKB method are as follows: (1) it provides an analytical tool for study of a class of problems for which there are, in general, no exact solutions; (2) the error involved in the approximation can often be estimated with reasonable precision, particularly if δ is small; and (3) the calculations involved are relatively simple and can easily be performed with pencil and paper.

The chief disadvantages of the WKB method are as follows: (1) it is approximate in nature and does not often give the exact solution to a problem; (2) it is concerned only with the homogeneous second-order equation and does not give much insight into particular solutions to a nonhomogeneous equation; and (3) if the error term δ becomes infinite in the interval studied, the method breaks down and should be discarded in favor of some other technique.

REFERENCES

[1] G. Wentzel, "Eine Verallgemeinerung der Quantumbedingungen fur die Zwecke der Wellenmechanik," *Z. Physik* **38**, 518–29 (1926).

[2] H. A. Kramers, "Wellenmechanik und halbzahlige Quantisierung," *Z. Physik* **39**, 828–40 (1926).

[3] L. Brillouin, "Remarques sur la Mecanique Ondulatoire," *J. Phys. radium.* **7**, 353–68 (1926).

[4] E. Merzbacher, *Quantum Mechanics*, chap. 7, pp 112 ff., Wiley, New York, 1961.

[5] P. Morse and H. Feshbach, *Methods of Theoretical Physics*, vol II, pp 1092 ff., McGraw-Hill, New York, 1951.

[6] J. Slater and N. Frank, *Mechanics*, chap. 10, sec. 2, pp 183 ff., McGraw-Hill, New York, 1947.

[7] Z. Kopal, *Numerical Analysis*, p 248, Chapman & Hall, London, 1955.

[8] H. Allen and A. Eggers, Jr., *A Study of the Motion and Aerodynamic Heating of Missiles Entering the Earth's Atmosphere at High Supersonic Speeds*, National Advisory Committee for Aeronautics, Technical Note 4047 (October 1957).

[9] A. Collar, "On the Stability of Accelerated Motion: Some Thoughts on Linear Differential Equations with Variable Coefficients," *Aeronaut. Quart.* **8**, 309–30 (1957).

Solid Mechanics

PARAMETRIC (DYNAMIC) STABILITY OF ELASTIC SYSTEMS

R. M. Evan-Iwanowski
Syracuse University

The class of problems of the parametric (dynamic) stability is defined. Illustrative examples incorporating the main features of the theory are discussed in detail. A review of pertinent works in the field of parametric stability is given.

INTRODUCTION

The dynamic stability of mechanical systems encompasses a class of problems in mechanics which is on the border line between static stability and vibration problems.

There are two types of static instability, depending upon the shape of the load parameter vs deformation parameter curves, or (p,v) curves. The normal behavior is shown in Fig. 1 by a curve $0N$. For some elastic systems, however, the (p,v) curves may have a branching point, such as shown in Fig. 1 by point B. The load corresponding to this point on the (p,v) curve can therefore support two different equilibrium states – one corresponding to the form of the "normal" branch $0N$ and another corresponding to the $0BB'$ branch. The first form becomes unstable, and the second becomes stable. The loads corresponding to

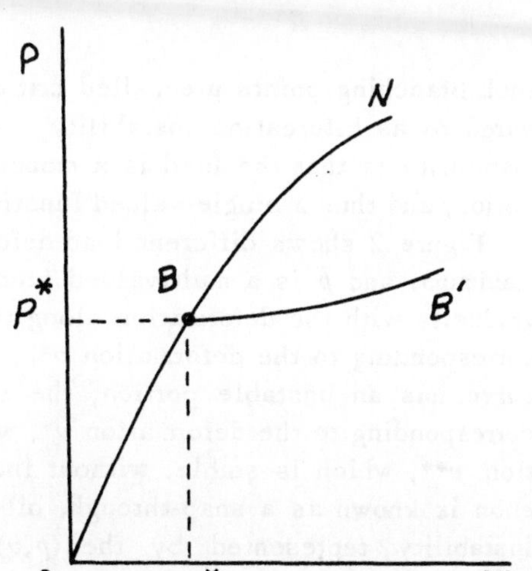

Fig. 1. Bifurcation Type of Instability; *B*, Branch Point; *P, Critical Load.**

NOTATION

x, y, z	Coordinates
t	Time
P	Load, $P = P_0 + P_1 \cos \gamma t$
E, ν	Young's modulus and Poisson's ratio
I	Moment of inertia of a cross section
v	Lateral displacement
w	Transverse displacement
$(')$	Differentiation with respect to x
$(\dot{})$	Differentiation with respect to time t
γ	Frequency of external force
ω	Natural frequency of the system
m	Mass of a column per unit length
l	Length of a column
F	Area of cross section of a column
$A(t)$	$v(x,t) = A(t)X(x)$
P_E	Euler's critical load
$\bar{\omega}$	Defined by Eq. (10)
μ	Defined by Eq. (12)
$f(t)$	Periodic function of time
$p(x,t)$	Time varying load, $p(x,t) = p_0 A(t) \cdot \sin(\pi x/l)$

such branching points are called critical loads. This type of instability is referred to as bifurcation instability. The characteristic feature of this type of instability is that the load is a monotonically increasing function of the deformation, and thus a single-valued function [1].

Figure 2 shows different load-deformation behavior. The (p,v) curve has a maximum, and p is a multivalued function of v. The load increases from zero gradually with the deformation along the solid line $0B$ until point B is reached, corresponding to the deformation v^*. At this point the continuation of the (p,v) curve has an unstable portion; the system thus jumps from the configuration corresponding to the deformation v^*, which is now unstable, to a new configuration v^{**}, which is stable, without increasing the load. This "jump" phenomenon is known as a snap-through, oil-canning, or Durchschlag, and the type of instability represented by the (p,v) curve in Fig. 2 is called transitional instability.

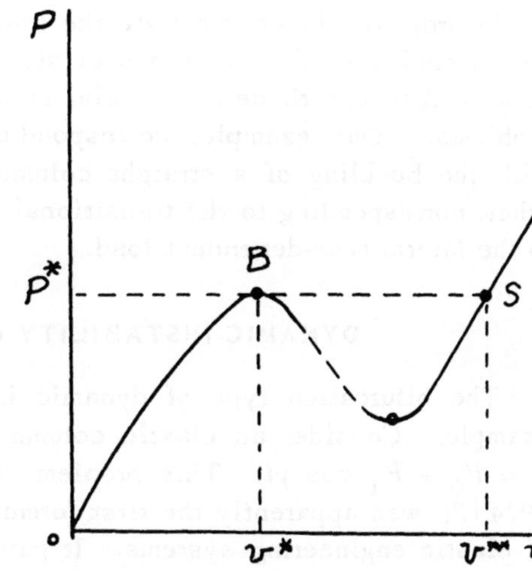

Fig. 2. Transitional Instability. The system jumps from the configuration *B* to the configuration *S*.

Figure 3 actually represents another variant of the transitional instability – namely, when the load jumps from the value p^* to p^{**} for the same deformation. The loads corresponding to the branch points or maxima on the (p,v) curve are called critical loads P_{cr}. The known Euler's critical load corresponds to a linear relationship between p and v, and it is the bifurcation type of instability.

The class of dynamic instability problems finds its counterpart in the static instability in the sense that both the bifurcation and snap-through or transitional instability are possible. It has a common ground with forced vibrations problems in that the instabilities occur at some values of the physical parameters of the systems and frequencies of the exciting force.

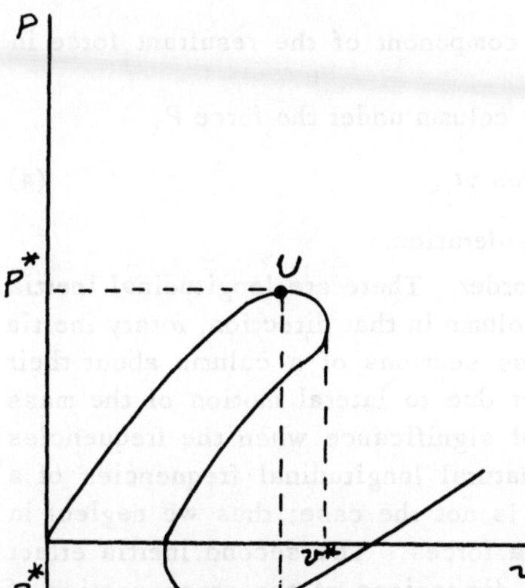

Fig. 3. Transitional Instability. The system jumps from the configuration *U* to the configuration *S*.

In order to closer formulate the problem of dynamic stability, two examples corresponding to the two types of static instability will be discussed in some detail. Although these are special cases, they have all the features of general problems. One example, corresponding to the bifurcation instability, deals with the buckling of a straight column under the time-varying axial load; the other, corresponding to the transitional instability, deals with an arch subjected to the lateral time-dependent load.

DYNAMIC INSTABILITY OF A STRAIGHT COLUMN

The bifurcation type of dynamic instability will now be illustrated in an example. Consider an elastic column subjected to a time-varying axial load, $P = P_0 + P_1 \cos \gamma t$. This problem, which is the one analyzed by Beliaev in 1924 [2], was apparently the first formulation of problems of dynamic instability of elastic engineering systems. It paved the way for impressive developments in this field.

The well-known differential equation governing the behavior of a column under the action of static load P (moment equation) is

$$EIv'' + Pv = 0 , \tag{1}$$

where v is lateral deflection,

$$v = v(x) , \tag{2}$$

x is the length coordinate, the primes denote differentation with respect to x, E is Young's modulus, and I is the moment of inertia of a cross section, assumed to be constant. The equation representing the balance of the transverse shear forces is obtained by differentiating Eq. (1) with respect to x:

$$EIv^{iv} + Pv'' = 0 , \tag{3}$$

which expresses the fact that there is no component of the resultant force in the transverse direction.

To account for the lateral vibrations of a column under the force P,

$$P = P_0 + P_1 \cos \gamma t , \tag{4}$$

the inertia effects have to be taken into consideration.

A short discussion of this matter is in order. There are longitudinal inertia effects, due to the motion of the mass of a column in that direction, rotary inertia effects related to the rotations of the cross sections of a column about their principal axes of inertia, and inertia effect due to lateral motion of the mass of a column. The first inertia effect is of significance when the frequencies of the applied loads γ are close to the natural longitudinal frequencies of a column, which we assume in this example is not the case; thus we neglect in the present analysis the longitudinal inertia forces. The second inertia effect is also of little significance, because the dimensions of the cross section of a column are assumed to be small compared with the length of a column; we thus neglect these inertia forces. Consequently, the only inertia effect is due to the lateral motions and the mass of a column. This inertia force must be added to Eq. (3). Denoting by m the mass of a column per unit length, we get

$$EIv^{iv} + Pv'' + m\ddot{v} = 0 , \tag{5}$$

where the dots indicate differentiation with respect to time t.

It is only natural to seek the solution of Eq. (5) in a form

$$v_\lambda(x, t) = A_\lambda(t) \sin(\lambda\pi x/l) , \tag{6}$$

which obviously satisfies the boundary conditions. The term $A_\lambda(t)$ is an unknown function of time ($\lambda = 1, 2, \ldots, \infty$), and l is the length of a column. Clearly, the function $\sin(\lambda\pi x/l)$ is an eigenfunction of a corresponding static problem.

Substituting Eq. (6) in Eq. (5) results in the following differential equation for the function $A_\lambda(t)$:

$$\ddot{A}_\lambda + \omega_\lambda^2[1 - (P_0 + P_1 \cos\gamma t)/P_{E\lambda}]A_\lambda = 0 , \tag{7}$$

where

$$\omega_\lambda^2 = \lambda^4\pi^4 EI/l^4 m , \tag{8}$$

in which ω_λ is the natural lateral frequency of the column, and $P_{E\lambda}$ is Euler's critical load corresponding to the λ's mode of buckling:

$$P_{E\lambda} = \lambda^2\pi^2 EI/l^2 . \tag{9}$$

The term *parametric instability* is derived from the fact that the time-dependent force P appears as a parameter in Eq. (7). Denoting

$$\bar{\omega}_\lambda = \omega_\lambda(1 - P_0/P_{E\lambda})^{1/2} , \tag{10}$$

which is the natural frequency of a column subject to a constant axial load P_0, we can rewrite Eq. (7) in the following form:

$$\ddot{A}_\lambda + \bar{\omega}_\lambda^2(1 - 2\mu_\lambda \cos\gamma t)A_\lambda = 0 , \tag{11}$$

where μ_λ is called, after Beliaev, the excitation coefficient,

$$\mu_\lambda = \frac{P_1}{2(P_{E\lambda} - P_0)} . \tag{12}$$

We are now in a position to drop the subscript λ, since the same equation, Eq. (11), is valid for all modes of vibration. Equation (7) is a well-known Mathieu differential equation. For a more general case, when

$$P = P_0 + P_1 f(t) ,$$

where $f(t)$ is any periodic function of time, Eq. (7) transforms into

$$\ddot{A}(t) + \bar{\omega}^2[1 - 2\mu f(t)]A(t) = 0 , \tag{13}$$

which is known as Hill's equation.

Thus the problem of the deformation of a straight column under the load P in Eq. (3) reduces to the solution of the Mathieu-Hill type of differential equations.

One of the main features of the class of differential equations represented by Eq. (13) is that for some values of the coefficients the amplitudes of the solution increase indefinitely with increasing t. Such critical values of the coefficients are covering some regions of the $(\bar{\omega},\mu)$ plane, called instability regions.

Investigation of Eq. (11) for very small μ (or P_1) leads to the conclusion that the loss of dynamic stability occurs for the parameter values

$$\frac{\gamma}{2\bar{\omega}} = 1, \frac{1}{2}, \frac{1}{4}, \ldots, \frac{1}{n}. \tag{14}$$

The instability region corresponding to the value 1 in Eq. (14) is called the first or fundamental instability region; that corresponding to $\frac{1}{2}$ is called the second, etc. Beliaev determined one value, namely, $\gamma/2\bar{\omega} = 1$, and later Andronov and Leontovich [3] determined other cases.

A physical argument leads directly to the relationship $\gamma = 2\bar{\omega}$. For each cycle of the transverse vibration of a column, the moving end (top end in Fig. 4) completes two cycles: twice up and down. Thus if the frequency of applied load γ is twice the natural lateral frequency $\bar{\omega}$, then the resonance occurs.

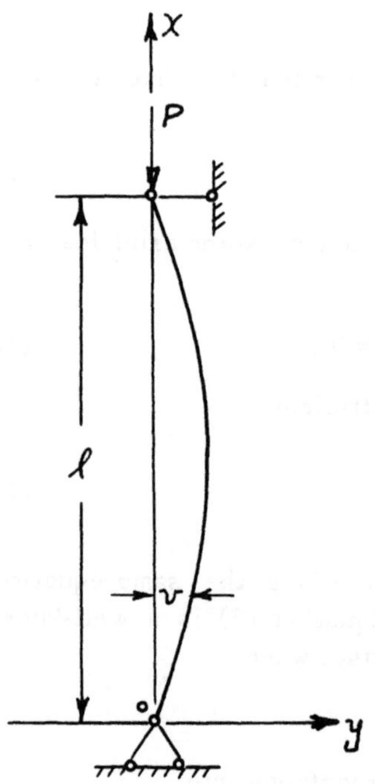

Fig. 4. Schematic Representation of a Column Pinned at Both Ends and Loaded Axially by the Load P.

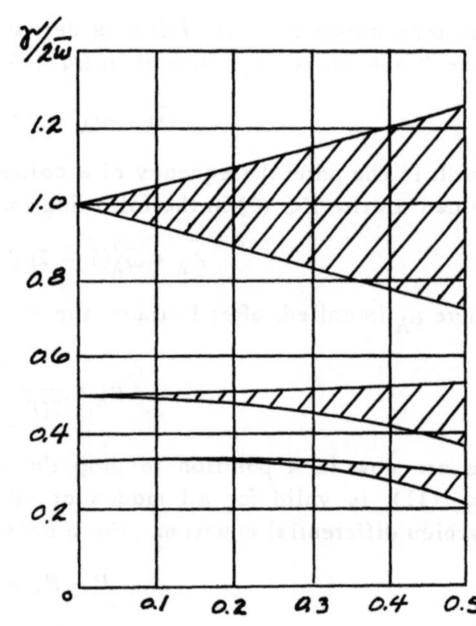

Fig. 5. Regions of Dynamic Instability. Shaded areas represent unstable regions; blank, stable.

Figure 5 represents the regions of dynamic instability for Beliaev's column in the $(\gamma/2\bar{\omega},\mu)$ plane. The shaded areas indicate the regions of dynamic instability, and the blank areas, dynamic stability; the boundaries correspond to periodic solutions. Figure 6 shows the stability diagram in the $(P_0/P_E,\gamma/2\bar{\omega})$ plane for $P_1/P_0 = 0.3$. For large values of P_0/P_E the reference should be made to the expanded Mathieu diagram.

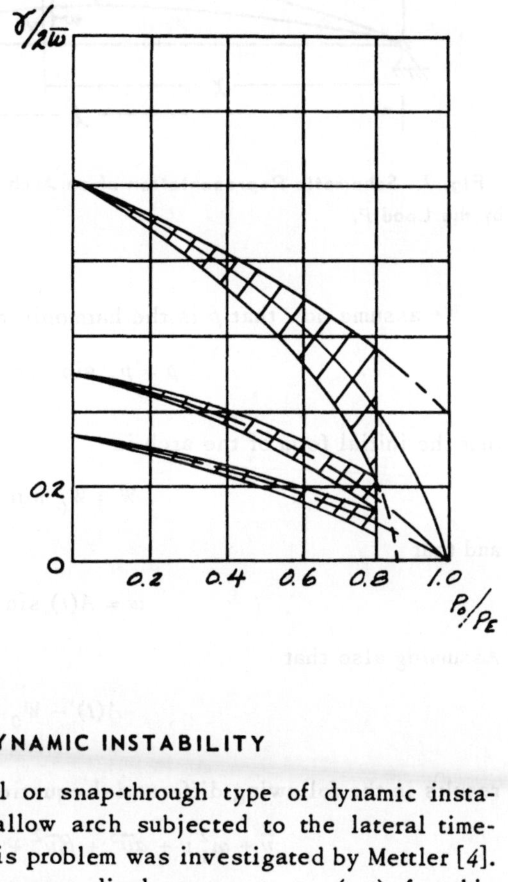

Fig. 6. Regions of Dynamic Instability in $(P_0/P_E,\gamma/2\bar{\omega})$ Plane. Shaded areas, unstable; blank, stable; $P_1/P_0 = 0.3$.

TRANSITIONAL DYNAMIC INSTABILITY

As an example of the transitional or snap-through type of dynamic instability, consider the problem of a shallow arch subjected to the lateral time-dependent load $p(x,t)$ (see Fig. 7). This problem was investigated by Mettler [4]. The differential equation for the transverse displacement $w = w(x,t)$ for this case is

$$EIw^{iv} - EIW^{iv} - Sw'' + p_0 + m\ddot{w} = 0 , \tag{15}$$

where

$$S = \frac{EF}{2l}\int_0^l (w'^2 - W'^2)\, dx , \tag{16}$$

and it represents the horizontal component of the force acting on the cross section of a column; F is the area of the cross section, and W is the initial form of an arch.

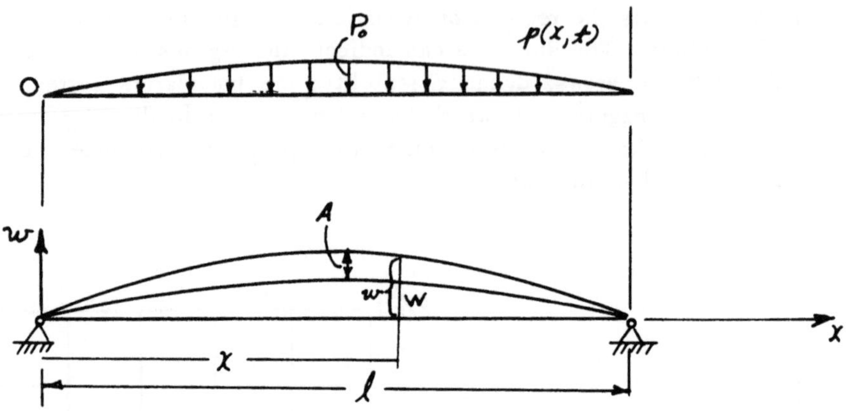

Fig. 7. Schematic Representation of an Arch Pinned at Both Ends and Loaded Laterally by the Load P.

We assume now that p is the harmonic function of the time t,

$$p = p_0 \sin(\pi x/l) \cdot \cos \gamma t , \tag{17}$$

that the initial form of the arch is

$$W = W_0 \sin(\pi x/l) , \tag{18}$$

and that

$$w = A(t) \sin(\pi x/l) . \tag{19}$$

Assuming also that

$$A(t) - W_0 = \bar{v}(t) \tag{20}$$

results in the following differential equation for $\bar{v}(t)$:

$$\ddot{\bar{v}} + \omega_1^2 v + a\bar{v}^3 + \beta\bar{v}^2 + (p_0/m) \cos \gamma t = 0 . \tag{21}$$

Employing Duffing's approximation method for the investigation of this equation, we seek a solution in the form,

$$\bar{v} = a \cos \gamma t + b , \tag{22}$$

where a and b are constants. Substituting Eq. (22) in (21), equating the coefficient of cos γt to zero, and assuming that the vibrations are small, we obtain a relationship between the parameters a, p_0, and γ in the form,

$$a(\gamma^2 - \gamma_1^2) + \alpha a^3 = p_0/m , \tag{23}$$

where

$$\begin{aligned}
\gamma_1^2 &= \pi^4(EI + \tfrac{1}{2}EFW_0^2)/ml^4 , \\
\omega_1 &= \pi^4 E(I - FW_0^2)/l^4 , \\
\alpha &= \beta^2/\omega_1^2 - 1 \text{ (generally} > 1) , \qquad (24) \\
\beta &= 3\bar{\beta}W_0 , \\
\bar{\beta} &= \pi^4 EF/4ml^4 .
\end{aligned}$$

Equation (23) represents a relationship between the three parameters p_0, a, and γ and, thus, a surface in the (p_0, a, γ) space. Graphical representation of such a surface is very complicated, and, instead, the graphs are presented in the planes of pairs of the parameters [contour lines of the surface (24)].

Figure 8 represents diagrammatically the relationship between p_0 for γ = a constant $< \gamma_1$, and Fig. 9 represents the relationship between a and γ/γ_1 for p = a constant. Figure 8 is identical with the frequency diagram (amplitude vs frequency) for a nonlinear oscillator with "soft" spring. This should be expected since Eq. (21) represents a differential equation for such an oscillator.

Characteristic features of the results are represented in Figs. 8 and 9, where stable (solid lines) and unstable (dotted lines) branches, stationary points, and the multivaluedness of the functions are depicted. In the case shown in Fig. 9, when the frequency γ/γ_1 approaches to the maximum point on the lower branch of the curve, the amplitude jumps from the lower branch, which becomes unstable (dotted lines), to the upper branch, which is stable (solid lines). Thus the value of the amplitude increases rapidly. Figure 9 is a dynamic counterpart of a static load-deformation curve as shown in Fig. 2. When p_0, which is increasing from zero, reaches a maximum point U, the equilibrium

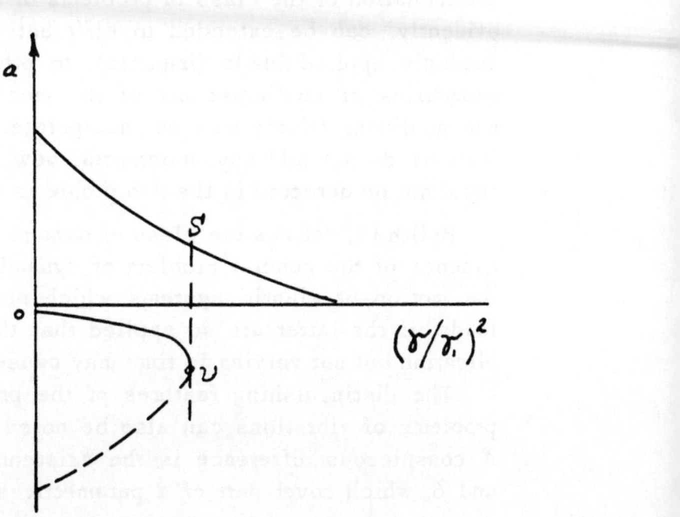

Fig. 8. The $(a, \gamma/\gamma_1)$ Diagram for p_0 = a Constant. Solid line, stable; dashed line, unstable.

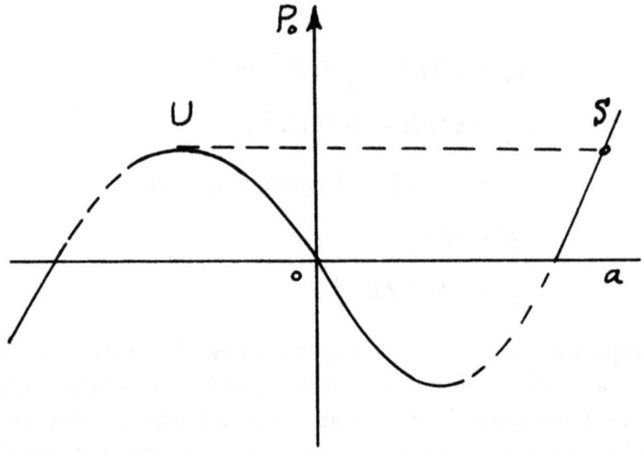

Fig. 9. The (p_0, a) Diagram for γ = a Constant $< \gamma_1$. Solid line, stable; dashed line, unstable.

configuration of a system is unstable, and the (p_0, a) point leaves the unstable branch and jumps to the stable configuration on a stable branch to point S, accompanied by a sudden increase of the amplitude.

The dynamic jumps evidently are manifestation of analogous static phenomenon, namely, snap-through or oil-canning phenomenon. Since this type of instability is called transitional unstability [*1*], a term of "dynamic transitional instability" is appropriate for the dynamic case.

CLASS OF PROBLEMS OF THE DYNAMIC INSTABILITY

The two examples discussed in the preceding sections permit a closer determination of the class of problems of the dynamic instability. The problem, evidently, can be extended to different types of time-varying loads, such as suddenly applied loads (impacts), to other boundary conditions, and to other geometries or configurations of the mechanical systems; the damping effects and nonlinear effects may be incorporated. But these expansions and generalizations do not add any substantial new features of the dynamic stability that could not be detected in the two problems above.

Beilin [*5*] defines the class of dynamic-stability problems as follows: "The essence of the general problem of dynamic instability consists of the study of the motion of elastic systems which are excited by time-dependent external loads, if the latter are so applied that the loads in direction and point of application but not varying in time may cause static loss of stability."

The distinguishing features of the problems of dynamic stability from the problems of vibrations can also be noted from the examples considered above. A conspicuous difference is the existence of the instability regions, Figs. 5 and 6, which cover part of a parametric space. Another distinguishing feature is that the ordinary resonance occurs when the frequency of the exciting force equals the natural frequency of a system; in the parametric resonance the exciting frequencies may be multiples of the natural frequencies. Still another essential difference is a possibility of the occurrence of excited vibrations for the frequencies lower than the principal resonance frequency. For the problems of this type the presence of variable coefficients in the corresponding differential equations is typical. For periodic loading, these coefficients likewise change according to a period law.

The occurrence of the parametric instability has been observed by Melde (1859) in connection with the investigations of the string fixed at one end and attached to the tuning fork at another. Lord Rayleigh proposed a plausible explanation of this phenomenon (1885).[1]

DAMPING EFFECTS

Bolotin [6] set a series of experiments corresponding to Beliaev's theoretical problem (see Fig. 10). The results of these experiments indicate that the boundary points of the fundamental instability region are in good agreement with the theoretical results. It was impossible, however, to obtain experimentally the boundaries of the second or third instability regions. Moreover, even

[1] An electrical analog is generally called "parametric amplification."

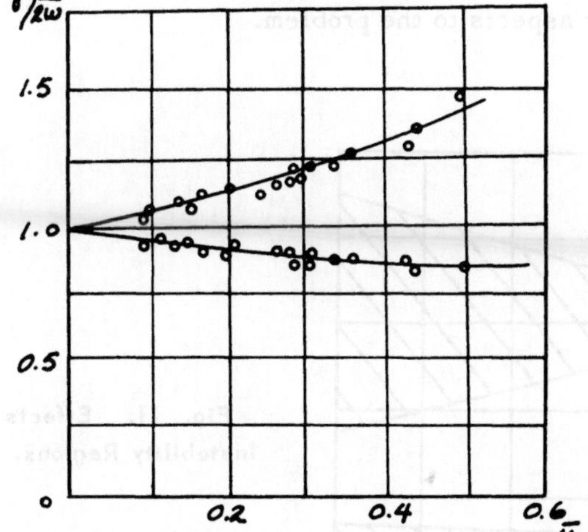

Fig. 10. Experimental Results for the Beliaev Column. Bolotin [6], $\bar{\mu} = \dfrac{\bar{P}_1}{2(P_E - P_0)}$, $\bar{P}_1 = \left(\dfrac{2\bar{\omega}}{\gamma}\right)^2 P_1$.

in these relatively uncomplicated experiments, the amplitudes remained finite, contrary to the results obtained from the small-vibration (linear) analysis.

It is true that initially the measured amplitudes increase exponentially with time, but they do not increase indefinitely. The answer to that question lies naturally in the characteristic of a material column, as opposed to the idealized column. To present a more realistic picture, new factors were introduced in the analysis, such as damping and nonlinear effects.

The damping effects to the first order of approximation could be taken proportional to the velocity of the lateral displacements v. By introduction of such damping effects Eq. (11) is changed to

$$\ddot{A}(t) + 2\epsilon\dot{A}(t) + \bar{\omega}^2(1 - 2\mu \cos \gamma t)A(t) = 0 \,, \tag{25}$$

where ϵ is determined experimentally. The effect of the damping on the instability regions is to move them away from the ordinate (Figs. 10 and 11) and to make them narrower. This means that the dynamic instability is impossible for arbitrary small values of the parameters. It is interesting to note that the fundamental zone is the least removed from the ordinate, and thus it still remains "dangerous," since the instability may occur for relatively small values of the parameters. One may generalize the form of the damping term in Eq. (25), but the essence remains the same as explained above.

We turn now to another factor which alters the dynamic instability regions and adds some other aspects to the problem.

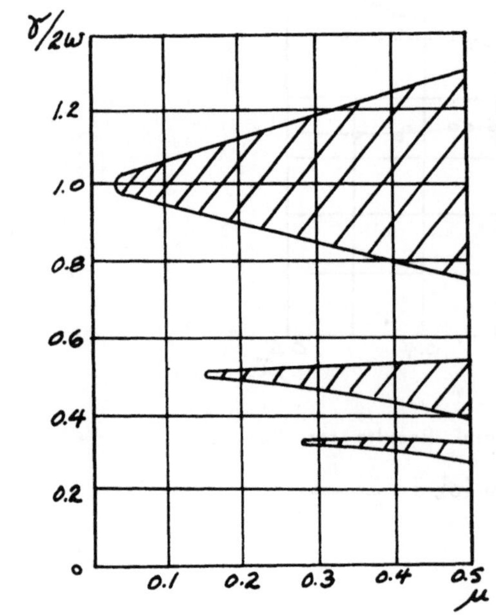

Fig. 11. Effects of Damping on the Instability Regions.

NONLINEAR EFFECTS

It has been mentioned that the experimentally observed amplitude variations do not increase indefinitely with time for the regions of instability, but after following for a short period of time an exponential increase, the character of the oscillatory motion is changed and the rate of the increase of amplitude is decreasing until it reaches a constant (or almost a constant) value.

The linear theories cannot provide an exact description of the oscillatory motion, nor can they determine the values of the amplitudes – they remain undeterminate [7,8]. A correct approach to these problems lies in the application of nonlinear theories. It is interesting to note, however, that a complete picture of the dynamic-instability regions may be obtained from the linear theories [7,8]. In some way it is analogous to the linear results in the static buckling of columns – the critical (Euler's) loads are correct critical values, but the deformations prior to the buckling and in the postbuckling regime remain undeterminate.

The nonlinearity may be introduced in several ways. We may consider "nonlinear elasticity" [6]; that is, we introduce into Eq. (11) those factors which incorporate all static nonlinear factors both of the geometric and physical origin. Equation (11) will have, then, the following form:

$$\ddot{A}(t) + 2\epsilon\dot{A}(t) + \bar{\omega}^2(1 - 2\mu \cos \gamma t)A(t) + g_E(A, \dot{A}, \ddot{A}) = 0\,, \qquad (26)$$

where $g_E(A, \dot{A}, \ddot{A})$ is the nonlinear function corresponding to the nonlinear elasticity.

We may also consider nonlinear inertia effects, both longitudinal as well as lateral. To illustrate this point consider a concentrated mass M on the top of the column in Fig. 4. An additional longitudinal force appears in this case:

$$\Delta P = -M\ddot{\bar{w}}\,, \qquad (27)$$

where $\bar{w}(t)$ is a longitudinal displacement:

$$\bar{w}(t) = \pi^2 A^2/4l + 3\pi^4 A^4/64l^3 + \ldots\,. \qquad (28)$$

Thus, including this new longitudinal force, Eq. (11) becomes

$$\ddot{A}(t) + 2\epsilon\dot{A}(t) + \omega^2\{1 - [P_0 + P_1 \cos \gamma t - M\ddot{\bar{w}}(t)]/P_E\}A(t) = 0\,. \qquad (29)$$

Differentiating Eq. (28) and substituting in Eq. (29) results in the following:

$$\ddot{A} + 2\epsilon\dot{A} + \bar{\omega}^2(1 - 2\mu \cos \gamma t)A + 2\bar{K}_i A[A\ddot{A} + (\dot{A})^2] = 0\,, \qquad (30)$$

where obviously the terms of third and higher orders are dropped. The function

$$g_i(A, \dot{A}, \ddot{A}) = 2\bar{K}A(A\ddot{A} + \dot{A}^2) \qquad (31)$$

represents nonlinear inertia effects, and

$$\bar{K}_i = \pi^4 M/4ml^3\,. \qquad (32)$$

A concentrated mass at the end of a column may not be the only source of inertia forces. Every cross section during the lateral vibrations undergoes small longitudinal displacement, which is of the second order of magnitude compared with the transverse bending,

$$w(x) \doteq \frac{1}{2}\int_0^x (\partial v/\partial x)^2\,dx\ ,$$

where again $v(x,t)$ is the lateral displacement assumed to be $v(x,t) = A(t)\sin(\pi x/l)$.

After some calculations Eq. (11) is transformed into

$$\ddot{A} + \omega^2[1 - (P_0 + P_1\cos\gamma t)/P_E]A + 2\overline{K}_i A(A\ddot{A} + \dot{A}^2) = 0\ , \qquad (33)$$

where

$$\overline{K}_i = \pi^4(\tfrac{1}{3} - \tfrac{5}{8}\pi^2)/4l^2\ .$$

Finally, one may also consider a nonlinear damping; but since the problem of the nonlinear damping did not reach any conclusive stage, we will not discuss it further at this time.

OTHER DEVELOPMENTS IN PARAMETRIC STABILITY

In this section the works pertaining to the dynamic stability will be discussed. In view of the enormous magnitude of the published papers in this area, especially in Russian, only some selected results will be incorporated.

General Considerations

Brachkovskii [9] proved an important theorem which states that the dynamic-stability problems reduce to the Mathieu-type differential equation only in those cases when the modes of free vibrations coincide with the buckling modes. In other cases the separation of variables is impossible, and the determination of equations for $A(t)$, analogous to Eq. (11), requires the usage of approximation methods. Brachkovskii's theorem was further expanded by Dzanelidze [*10*] and Bolotin [6].

Gol'denblat [7, *8*] proved that linear analysis is sufficient to determine the instability regions for the rods, but it is insufficient to determine the amplitude in the resonance regions; he supported this by some experimental work.

Kovalenko indicated an apparent paradox in the dynamic stability: from the linear analysis it follows that the instability for Beliaev's column (instability regions of Mathieu equations) occurs for arbitrary, small harmonic loads. He indicated that this paradox can be removed if a proper damping is introduced [*11*].

Columns, Rods, Beams

In 1933, Köenig and Taub [*12, 13*] conducted an extensive study of the stability of columns under the impact load.

Krylov and Bugolubov [*14*] in 1935 considered boundary conditions different from those of Beliaev, and the resulting equations do not admit solutions by means of separation of variables in the form $y = T(t)X(x)$. They also extended the solutions to the case of multiharmonic axial forces.

Bondarenko [*15*] considered some simple problems when in addition to the axial force a small transverse force was present.

The effect of the rate of the shortening of the end points of initially curved columns on the critical loads was discussed by Hoff [*16, 17*] in a number of publications and verified experimentally by Hoff and his associates [*18*]. It was found, among other things, that the short columns snap through under the rapidly applied axial loads, and the long columns deflect gradually.

The analysis of buckling under loads increasing linearly with time was also treated by Katsuta [*19*, 20] and Suharo [*21*].

Makushin [*22*] considered stability of columns under the action of axial, piecewise constant, periodic loadings. Smirnov [*23*] treated a problem of the axially distributed time-varying load. He showed that for $\mu > 0.8$ the linear theory must be replaced by the nonlinear theory.

Gastev [*24*] considered the time-varying force of the form

$$P(t) = P_1[\sigma_1(t) + \sigma_1(t - \tau) + \sigma_1(t - 2\tau) + \ldots] ,$$

where $\sigma(t)$ is the unit step function.

Extensive and detailed work on stability of columns under the time-varying loads was carried on by Mettler [*4, 25*] and his associate Weidenhammer [*16, 26*].

Gol'denblat's [*7, 8*] contribution is significant. He analyzed thin-walled beams, rods, effects of elastic media, and resonance of suspension bridges. He also conducted experiments on the Beliaev column.

An interesting result was obtained for the authors in ref [*27*], where it was demonstrated that with an impact load which exceeds the *n*th critical static load, there is a stable equilibrium mode having *n* half-waves. The same effect was verified in ref [*27*] experimentally.

Chelomei [*28*] considered periodic loads distributed along the length of a rod.

Snitko [*29*] reported the loss of stability under the impact loads, whose magnitude was lower than the critical static load – the faster the applied load, the lower its critical value.

The effects of damping were considered by Andronov and Leontovich [*3*]. Bolotin [*6*] also investigated in detail the influence of the damping effects, especially for the nonlinear damping.

Bolotin [*6*] studied in detail the nonlinear effects. He considered three types of nonlinearity: elastic, inertia, and damping (see "Nonlinear Effects").

Smirnov [*23*] studied matrix methods of solutions. He also demonstrated the necessity of incorporating the nonlinear effects when P_0 is near P_E.

Arches and Rings

Bolotin [6] analyzed extensively a group of problems of arches loaded by the time-dependent loads at the apex.

Mettler extended his work [25] in cooperation with Weidenhammer [30] and analyzed the case of an arbitrary time-dependent force acting on an arch.

Beilin [5], Leiderman [31], and Larkin [32] treated problems of arches subjected to a variety of transverse and longitudinal forces, while Salion [33] considered stability of arches under the action of periodic moments. Bondar' [34] analyzed stability of parabolic arches. Makushin [35] in 1947 considered stability of rings.

An extensive analysis of rings was carried out by Dzanelidze and Radtsig [36].

Plates and Shells

A thorough investigation of stability of plates under dynamic loads was done by Khalilov [37], Chelomei [38], Bodner [39], and Leiderman [40].

The loss of stability of shells under dynamic loads was studied experimentally and analytically by a few workers in the field. Schmitt [41] performed a series of tests on aluminum shells. The first analytical investigation of dynamic stability of both cylindrical and spherical shells was performed by Bolotin in 1956 [6]. Using linear theories, Suhara [42] studied the snapping of shallow spherical shells under impact loads. Some other contributions are: Agamirov and Vol'mir [43], cylindrical shells, using finite deflection theories; Vol'mir [44], cylindrical panels. Experimental work is almost lacking on this subject; thus it is difficult to assess the validity of the theoretical works.

Varia

Leonov [45] initiated the studies of the stability of elastic systems under the quasi-harmonic excitations. He was followed by Kovalenko [46].

It was already mentioned that Gol'denblat and Bolotin experimented on the Beliaev column. A weak point of these experimental works is that the dynamic component of the external force, $P_1 \cos \gamma t$, could not be varied independently of γ. This drawback does not permit, for instance, the investigation of the influence of the amplitude P_1 on the stability at constant frequencies and, conversely, the influence of γ at a constant amplitude.

REFERENCES

[1] C. B. Biezeno and R. Grammel, *Engineering Dynamics,* vol 2, Van Nostrand, Princeton, New Jersey, 1957.

[2] N. M. Beliaev, "Stability of Prismatic Rods Subject to Variable Longitudinal Forces," in *Collection Engineering Construction and Structural Mechanics,* Leningrad, Put', 1924.

[3] A. A. Andronov and M. A. Leontovich, "On the Vibration of Systems with Periodically Varying Parameters," *Zhur. Russ. Fiz.-Khim. Obschestva.* **59** (1927).

[4] E. Mettler, "Allgemeine Theorie der Stabilität erzwungener Schwingungen elasticher Körper," *Ingr.-Arch.* **17**, 418 (1949).

[5] E. A. Beilin, "On the Dynamic Stability of Circular Arches with Internal Damping," *Trudy Leningrad. Inzhener.-Strotilel. Inst.* **17** (1954).

[6] V. V. Bolotin, "On the Parametric Excitation of Transverse Vibrations," *Poperechnye Kolebania i Kriticheskie skorosti,* Moscow.

[7] I. I. Gol'denblat, "On Some Problems of Structural Dynamics," *Stroitel. Prom.* **10**, 11 (1944).

[8] I. I. Gol'denblat, *Dynamic Stability of Structures,* Stroiizdat, Moscow, 1948.

[9] B. Z. Brachkovskii, "On the Dynamic Stability of Elastic Systems," *Prikladnaia Matematika i Mekhaniks* **6**(1), 87–88 (1942).

[10] G. Y. Dzanelidze, "The General Problem of Dynamic Stability of Elastic Systems," *Usp. math. nauk* **10**(4), 66 (1955).

[11] K. R. Kovalenko, *Summary of the Doctoral Dissertation,* Odeskii Institut morskogo flota, 1950.

[12] C. Köenig and J. Taub, "Impact Buckling of Thin Bars in the Elastic Range Hinged at Both Ends," NACA-TM 748 (1934); *Trans. Luftfahrtforschung* **10**, 2 (1933).

[13] J. Taub, "Impact Buckling of Thin Bars in the Elastic Range for Any End Condition," NACA-TM 749 (1934); *Trans. Luftfahrtforschung* **10**, 2 (1933).

[14] N. M. Krylov and N. N. Bogolubov, "Investigation of Phenomena of Resonance in Transverse Vibrations of Rods, Found When Periodic Normal Forces Are Applied to One of the Ends of the Rod," in *Collection Researches in Vibration of Structures,* Kiev-Khar'kov, Gos. Nauchno-tekh. izd. Ukrainy, 1935.

[15] G. V. Bondarenko, *Hill's Equation and Its Application in the Field of Engineering Dynamics,* Moscow-Leningrad, izd. akad. Nauk SSR, 1936.

[16] F. Weidenhammer, "Der eingespannte, axial pulsierend belastete Stab als Stabilitä tsproblem," *Ingr.-Arch.* **19**, 162–91 (1951).

[17] N. J. Hoff, "The Dynamics of the Buckling of Elastic Column," *J. Appl. Mechanics* **18**, 68–74 (1951).

[*18*] N. J. Hoff and V. G. Bruce, "Dynamic Analysis of the Buckling of Laterally Loaded Flat Arches," *J. Math. Phys.* **32**(4), 276–88 (1954).

[*19*] Chitoshi Katsuta, "On Buckling of an Elastic Long Column Under High Speed Loading," paper read before a meeting of the Architectural Institute of Japan, Nov. 23, 1947, Library of Congress Reproduction, PB94937.

[*20*] Chitoshi Katsuta, "High Speed Buckling of Mild Steel Columns," *Proc. First Japan National Congress for Applied Mechanics,* p 65, Japan National Committee for Theoretical and Applied Mechanics, Science Council of Japan, May 1952.

[*21*] Jiro Suhara, "Dynamic Stability of Column Compressed by Axial Load Increasing with Time," paper read before the Research Committee on Applied Mechanics, Faculty of Engineering, University of Kyusha, May 1948, Library of Congress Reproduction, PB94945.

[*22*] V. M. Makushin, "Dynamic Stability of Elastic Rectilinear Rods," *Trudy 2-i nauchno tekhn. Konferentsii,* Mosk. vysh. tekh. uchil., 1946.

[*23*] A. F. Smirnov, *The Static and Dynamic Stability of Structures,* Transzeldozizdat, Moscow, 1947.

[*24*] V. A. Gastev, "Transverse Vibrations and Stability of Rods Subjected to Periodic Repeated Impulses," *Trudy Leningrad. inst. avia. priborostroeniia* **1** (1949).

[*25*] E. Mettler, "Erzwungene nichtlineare Schwingungen elastischer Körper," *IXe Congr. Internat. Mec. App., Actes, t.* **V**, Bruxelles, 5–26, 1957.

[*26*] F. Weidenhammer, "Nichtlineare Biegeschwingungen des axial- pulsierend belasteten Stabe," *Ingr.-Arch.* **20**, 315 (1952).

[*27*] M. A. Lavrent'ev and A. I. Ishlinskii, "Dynamic Forms of Loss of Stability of Elastic Systems," *Doklady Akad. Nauk SSSR* **14**, 6 (1949).

[*28*] V. N. Chelomei, "On the Stability of Rods Subject to the Action of Longitudinal Periodically Varying Forces Distributed Along the Length," *Trudy Kieosk. aviats. inst.* **8** (1938).

[*29*] N. K. Snitko, "The Stability of Slightly Bent Compressed Steel Rods Under Vibrating Loads," *Trudy voenno-transp. akad.* **2** (1949).

[*30*] E. Mettler and F. Weidenhammer, "Kinetisches Durchschlagen des schwach gekrümmnten Stabes," *Ingr.-Arch.* **29**(5), 301–15 (1960).

[*31*] Y. R. Leiderman, "On the Parametric Resonance of an Elastic Arch Subjected to Longitudinal Transverse Bending," *akad. nauk Uzbekoskoi SSR, Doklady* **9** (1954).

[*32*] R. P. Larkin, *Dynamic Stability of Curved Rod,* Dissertation, Leningrad Polyt. Inst., 1955.

[*33*] V. E. Salion, "The Dynamic Stability of a Curved Arch Under the Action of Periodic Moments," *Akad. Nauk Dinnik Anniv Volume,* Akad. Nauk ukr. SSR, 1956.

[34] N. G. Bondar', "Dynamic Stability and Vibrations of Hingeless Parabolic Arches," *Akad. Nauk SSSR, Inzhenernyi Sbornik* **13** (1953).

[35] V. M. Makushin, "Dynamic Stability of the Deformed Rods," Notes, Mosk. vysh. tekh. uchi'l, 1947.

[36] G. I. Dzanelidze and M. A. Radtsig, "Dynamic Stability of Rings Subject to Normal Periodic Forces," *Prikladnaia Matematicks i Mekhaniks* **5** (5–6), 55–60 (1940).

[37] Z. I. Khalilov, "Dynamic Stability of Plates," *Baku. Azerbaidzhanskii gosudarstvennyi universitet* **1** (1949).

[38] V. K. Chelomei, "The Dynamic Stability of Plates," *Trudy Kiev. aviats.* **8** (1938).

[39] V. A. Bodner, "Stability of Plates Subjected to Longitudinal Perpendicular Forces," *Prikladnaia Matematicks i Mekhaniks* **2**(1), 87–104 (1938).

[40] Y. R. Leiderman, "On the Parametric Resonance of Plastic Plates," *Doklady, Akad. nauk Uzbekoskoi SSR* **3** (1955).

[41] A. F. Schmitt, "Dynamic Buckling Tests of Aluminum Shells," *Aeronaut. Eng. Rev.* **15**(9), 54–58 (1956).

[42] J. Suhara, *Snapping of Shallow Spherical Shells Under Static and Dynamic Loading,* AFOSR TN-60-831 (1950).

[43] V. L. Agamirov and A. S. Vol'mir, "Behavior of a Cylindrical Shell Subject to the Dynamic Loadings of Pressure Around the Circumference," *Akad. Nauk, SSSR Investia* **1**(3), 78–83 (1959).

[44] A. S. Vol'mir, "Stability of Cylindrical Shells Subject to Dynamic Loading," *Doklady Akad. Nauk SSSR* **123**, 806–8 (1958).

[45] M. Y. Leonov, "On Quasi Harmonic Stability," *Prikladnaia Matematicks i Mekhaniks* **10**, 575–80 (1946).

[46] K. R. Kovalenko, *The Basic Assumption of A. M. Liapunov's Theory of Instability Regions and Their Applications to the Dynamic Stability of Rods,* Dissertation, Odessk. inst. inzh. Marsk. Flota., 1950.

BIBLIOGRAPHY

Beliaev, N. M., "On Dynamic Coefficients of Struts in Tension and Compression," Otdel. inzh. issled., Sb., 11, Moscow, Transpechat', 1927.

Benz, G., "Die mechanische Bedeutung des instabilen Zweiges der Frequenz – Amplituden Kurve bei parametererregten Schwingungen," *Zeitschrift fur angewandte Mathematik und Mechanik* **36**, 273–4 (1956).

Bolotin, V. V., *Dynamic Stability of Elastic Systems,* Gos. Izdat. Tekh.-Teor. Lit, Moscow, 1956.

Bolotin, V. V., "Dynamic Stability of Symmetric Modes of Oscillation of Compressed and Bent Arches," *Doklady Akad. Nauk SSSR,* **83**(4) (1952).

Bolotin, V. V., "Transverse Vibrations and Critical Velocities," *Akad. Nauk SSSR, Knigi Izdatel' Stva.* **1** (1951).

Duffing, G., *Erzwungene Schwingungen bei veränderlicher Eigenfrequenz und ihre technische Bedeutung*, Braunschweig, 1918.

Erickson, B., S. V. Nardo, S. A. Patel, and N. J. Hoff, "An Experimental Investigation of the Maximum Loads Supported by Elastic Columns in Rapid Compression Tests," *Proc. Society Experimental Stress Analysis*, vol XIV, No. 1, pp 13–20, 1956.

Hoff, N. J., "The Dynamics of the Buckling Process," La Ricerga Scientifica, Revista del Consiglio Nazionale delle Ricerche, vol 20, No. 12, p 1919, Rome, Italy, 1950.

Malkina, R. L., "Stability of Curved Arches Subject to Longitudinal Periodic Forces," *Akad. Nauk SSSR, Inzhenernyi Sbornik* **14** (1953).

Oniashvili, O. D., "Analysis of Earthquake-Proof Shallow Shells," *Akad. Nauk Gruz. SSR, Inst. Stroitel-nogo dela* **3** (1950).

Oniashvili, O. D., "Dynamic Stability of Shallow Shells," *Akad. Nauk Gruz. SSR, Soobshcheniia* **11**(3) (1950).

Panovko, Y. G., "Dynamic Analysis of Structures," *Structural Mechanics in the USSR, 1917–57*, Academy of Sciences USSR, 1957.

Salerno, V. L., F. Bauer, and James Sheng, "The Behavior of Simply Supported Column Under Constant or Varying End Load with Transverse Displacement of One Point of Support," *Proc. First U.S. Nat'l Congress of Applied Mech.*, p 425, Am. Soc. Mech. Eng., New York, 1952.

Utida, J., and K. Terawa, "Dynamical Stability of a Column Under Periodic Longitudinal Forces," *Tokyo Imperial University Aeronaut. Research. Inst. Report* **15**, 1939 (1940).

LARGE DEFORMATIONS OF THIN SHELLS OF REVOLUTION BY INTERNAL-PRESSURE LOADING[1]

E. R. Chubbuck
Louisiana State University

A method is developed for the calculation of large displacements of thin shells of revolution under internal-pressure loading. The original meridian curve of the shell may be given either as a function of polar angle, or as the *x-y* coordinates of 100 points. Membrane theory is used. The method may be used for the plastic as well as for the elastic range of the biaxial stress-strain relations of the shell material, although the presently available procedure of finding biaxial stress-strain relations from uniaxial tests should be verified by experiment.

INTRODUCTION

Missile tanks are often made of thin (0.015 to 0.030 in.) sheet metals formed into a circular cylinder onto which a cap is attached at each end. These tanks are pressurized for various reasons, and there is a consequent deformation of tank and tank closure (cap). As far as material utilization is concerned, the hemispherical closure is ideal, but savings can be made in weight and total tank length by making "flatter" closures (such as a surface of revolution formed by an ellipse revolved about its minor axis). The purpose of this investigation was to develop a method whereby the deformations of such pressurized thin shells of revolution could be determined, even for the large deformations for which the linear methods are inaccurate. Also desired was a comparison between the true deformations when large and those calculated by the linear methods.

The problem, as formulated herein, allows the shell of revolution to expand freely at the circle where it is joined to the tank, and thus vertical meridian tangents remain vertical along that circle. Also, since the shell thickness is quite small compared with the principal radii of curvature, zero moments were assumed at the circle of juncture to the tank.

There are two sources of nonlinearity in the present problem: geometry changes and nonlinear stress-strain relations. The investigation as originally

[1]The problem presented in this paper originated as a topic investigated by the author during the summer of 1961 while employed by the Stress Group of General Dynamics/Astronautics in San Diego, Calif.

NOTATION

ϕ	Polar angle (Figs. 1 and 2)
θ	Longitude (Fig. 1)
R_1	Principal radius of curvature in plane perpendicular to meridian (Figs. 1 and 2)
R_2	Principal radius of curvature in plane of meridian (Fig. 2)
R_0	Perpendicular distance from axis of revolution to point – same as x (Fig. 2)
a, b	Characteristic lengths for particular shell of revolution (Fig. 2)
R_1'; R_2'	R_1/a, nondimensional R_1; R_2/a, nondimensional R_2
x_n, y_n	Rectangular coordinates of nth point (Fig. 2)
x_n', y_n'	x/a, y/a, the nondimensional rectangular coordinates of the nth point
u, v, w	Displacements in θ, ϕ, and z directions respectively
v', w'	v/a, w/a, nondimensional tangential and normal displacements
ϵ_ϕ^t	$\epsilon_\phi^e + \epsilon_\phi^p$, total nominal strain in the ϕ direction
ϵ_θ^t	$\epsilon_\theta^e + \epsilon_\theta^p$, total nominal strain in the θ direction
ϵ_i^e, ϵ_i^p	Elastic and plastic nominal strain
N_ϕ, N_θ	Membrane forces in ϕ and θ directions
δN_ϕ, δN_θ	Increments in membrane forces from a pressure increment δp
p_z	Inward component of distributed unit loading
p	Internal pressure on shell
p'	p/E, nondimensional pressure
δp	Increment of p
$\delta p'$	$\delta(p/E)$, nondimensional pressure increment
δ	Increment of quantity following
F_c	Total vertical force on shell above latitude line
E	Young's modulus
ν	Poisson's ratio
δB	Increment of plastic flow
σ_ϕ, σ_θ	Stress in the ϕ and θ directions
σ_y	True yield stress in uniaxial tension
$\bar{\sigma}$	$\sigma_\phi^2 + \sigma_\theta^2 - \sigma_\phi \sigma_\theta$, equivalent octahedral shear stress

$\bar{\sigma}'$	$\dfrac{\bar{\sigma}}{E} = \dfrac{p'R'}{2h'R_2'}\sqrt{3R_2'^2 - 2R_1'R_2' + (R_1')^2}$, nondimensional equivalent octahedral shear stress
$F(\bar{\sigma})$	Function of $\bar{\sigma}$ used in determining δB
σ_t	True stress for axial tension
ϵ_t^p	True plastic strain for axial tension, $\ln(1 + \epsilon_0^p)$
ϵ_0^p	Nominal plastic strain for axial tension
$f(\phi)$	$\dfrac{R_2'\delta\epsilon_\phi^t - R_1'\delta\epsilon_\theta^t}{\sin\phi}$
G_n	$S_n\phi^2 + T_n\phi + U_n$, second-degree curve used in modified Simpson's rule for numerical integration
S_n, T_n, U_n	Coefficients for G_n, given by Eq. (41)
ΔA_n	$\dfrac{S_n}{3}(\phi_{n+1}^3 - \phi_n^3)$ and $\dfrac{T_n}{2}(\phi_{n+1} - \phi_n) + U_n(\phi_{n+1} - \phi_n)$
ΔA_{n-1}	$\dfrac{S_n}{3}(\phi_n^3 - \phi_{n-1}^3) + \dfrac{T_n}{2}(\phi_n^2 - \phi_{n-1}^2) + U_n(\phi_n - \phi_{n-1})$
y_n'	$M_n x'^2 + N_n x' + P_n$ for $0 \leqq n \leqq 50$
x_n'	$Q_n y'^2 + W_n y' + Z_n$ for $50 < n \leqq 100$
h_0'	h_0/a, original nondimensional shell thickness
h_0, h	Original and current shell thickness
h'	h/a, current nondimensional shell thickness
$\delta v, \delta w$	Tangential and normal displacements due to δp
$\delta v', \delta w'$	Nondimensional δv and δw
$\delta x, \delta y$	Rectangular displacements due to δp
$\delta x', \delta y'$	Nondimensional δx and δy
I_n	$\displaystyle\int_{\phi_n}^{\pi/2}\left(\frac{R_2'\delta\epsilon_\phi^t - R_1'\delta\epsilon_\theta^t}{\sin\phi}\right)d\phi$

conducted was divided into two parts: (1) a so-called "elastic" part, wherein only those nonlinearities due to geometry change were considered, and (2) a so-called "plastic" part, in which both types of nonlinearities were involved. There were so many duplications of procedure, however, that the two parts will be handled simultaneously in this paper, with proper explanation of the differences in procedure wherever necessary. Thus, if no division into elastic and plastic categories is indicated, the procedure is identical for that step regardless of the type of stress-strain relation.

Figure 1 shows a three-dimensional view of a part of a surface of revolution and indicates ϕ and θ directions for stresses and strains. The element shown is positioned between two lines of latitude and two meridians.

Figure 2 is an intersection of the shell of revolution with any plane containing the y axis; thus, the curve is a meridian.

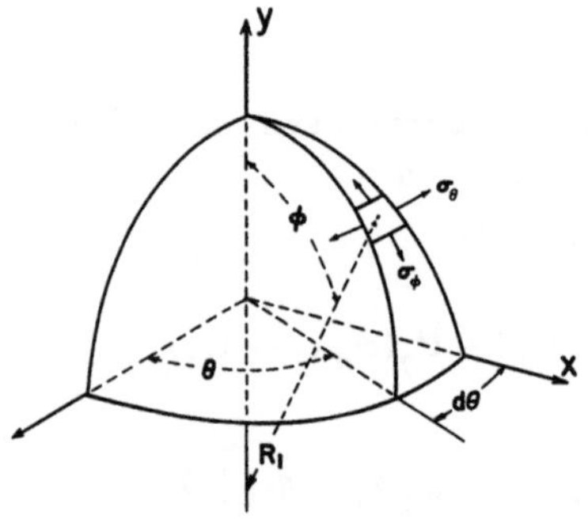

Fig. 1. Sketch of Part of Shell of Revolution Showing Some of the Nomenclature.

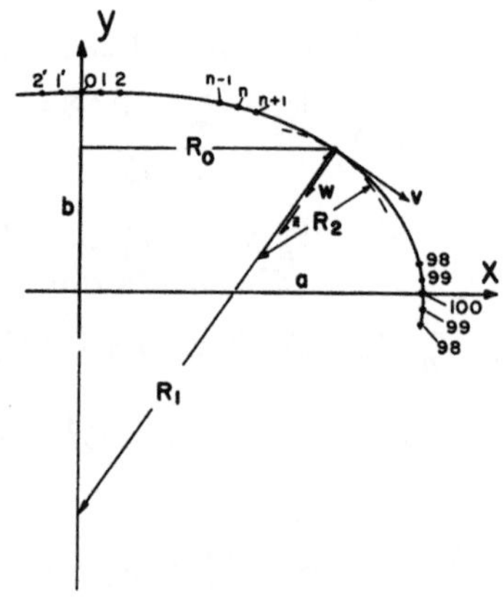

Fig. 2. Meridian Curve of a Surface of Revolution Showing Point Numbering System.

BASIC METHOD OF ANALYSIS

Membrane Theory

Since the wall thickness of the sheet materials considered here is small (0.015 to 0.035 in.) compared with the principal radii of curvature (4 to 10 ft or more), the use of the membrane theory, instead of the much more complex moment theory of shells, appears entirely justified. The membrane theory was used exclusively in this work, an important advantage of which is that the equilibrium equations are statically determinate.

Stepwise Computation with Geometry Correction

The availability of modern, high-speed computers makes feasible the computation of difficult problems by using elementary methods a great number of times. For this problem, the basic scheme was to use linear analysis to determine the deformations of the shell after each increment of pressure is applied and then correct the geometry at each point of the shell before another pressure increment is added. Thus, the final configuration of a given pressure step is the initial configuration for the succeeding step. This process continues until the desired total pressure is reached. The number of pressure steps to use is not predetermined; the number of steps is increased by doubling (starting with the entire pressure applied in one step) until the deflection of a particular point of the shell changes less than a certain percentage upon doubling the number of pressure steps.

Such a procedure would obviously be out of the question were it not for the modern computers, since we must perform corrections to several entities at each of, say, 100 points after each of several pressure steps.

DETERMINATION OF DISPLACEMENTS

It can be shown that the strains in a shell of revolution in the ϕ and θ directions, ϵ^t_ϕ and ϵ^t_θ, can be expressed for small geometry changes in terms of v and w as follows:

$$\epsilon^t_\phi = \frac{1}{R_2}\left(\frac{\partial v}{\partial \phi} - w\right) - \frac{z}{R_2}\frac{\partial}{\partial \phi}\left(\frac{v}{R_2} + \frac{1}{R_2}\frac{\partial w}{\partial \phi}\right), \tag{1}$$

$$\epsilon^t_\theta = \frac{v \cot \phi}{R_1} - \frac{w}{R_1} - \frac{z}{R_0 R_2}\left(v + \frac{\partial w}{\partial \phi}\right)\cos \phi\,, \tag{2}$$

where v and w are the displacements along and perpendicular to the meridian, and z is the distance from the middle surface at which the strains are measured.

In the thin shells whose middle surfaces are surfaces of revolution, there is no dependence upon θ, and because the shells are thin, we shall ignore the second term on the right of each of Eqs. (1) and (2). Thus, the remainder of the analysis will treat the shell as if it were entirely located at the middle surface; all expressions for strain, displacement, etc., will refer to the middle surface. Equations (1) and (2) become

$$\epsilon^t_\phi = \frac{1}{R_2}\left(\frac{dv}{d\phi} - w\right), \tag{3}$$

$$\epsilon^t_\theta = \frac{1}{R_1}(v \cot \phi - w)\,. \tag{4}$$

If Eq. (3) is multiplied by R_2, Eq. (4) multiplied by R_1, and Eq. (4) subtracted from Eq. (3), the following equation results:

$$\left(\frac{dv}{d\phi} - w\right) - (v \cot \phi - w) = R_2\,\epsilon^t_\phi - R_1\,\epsilon^t_\theta\,,$$

or

$$\frac{dv}{d\phi} - v \cot \phi = R_2\,\epsilon^t_\phi - R_1\,\epsilon^t_\theta\,. \tag{5}$$

The expression on the left of Eq. (5) is an exact derivative; thus,

$$\sin\phi\frac{d(v/\sin\phi)}{d\phi}=\sin\phi\left(\frac{\sin\phi\,dv/d\phi-v\cos\phi}{\sin^2\phi}\right)=\frac{dv}{d\phi}-v\cot\phi\,.$$

Therefore, Eq. (5) becomes

$$\sin\phi\frac{d(v/\sin\phi)}{d\phi}=R_2\,\epsilon_\phi^t-R_1\,\epsilon_\theta^t\,. \tag{6}$$

Integrating,

$$\left.\frac{v}{\sin\phi}\right]_{\phi=\pi/2}^{\phi}=\int_{\phi=\pi/2}^{\phi}\left(\frac{R_2\,\epsilon_\phi^t-R_1\,\epsilon_\theta^t}{\sin\phi}\right)d\phi\,, \tag{7}$$

$$w=v\cot\phi-R_1\,\epsilon_\theta^t\,. \tag{8}$$

Equations (7) and (8) are the desired relations for the displacements v and w (the displacement u is zero at each point in a shell of revolution loaded axisymmetrically). These equations are valid only for small geometry changes, but the entire basis of approach is to apply the total pressure in small steps and thus produce a small geometry change at each step.

For the linearly elastic problem, we could now substitute the values of ϵ_ϕ^t and ϵ_θ^t in terms of membrane forces, but this cannot be done for the plastic problem; therefore, we shall leave the expressions for v and w in terms of strains and later express these strains as necessary for the two cases.

EQUATIONS OF EQUILIBRIUM

If an element of the shell of revolution in Fig. 1 is placed in equilibrium, there are three equilibrium equations resulting from force summations in the ϕ, θ, and z directions. It can be shown (and it is obvious physically) that equilibrium in the θ direction is automatically obtained for a shell of revolution and axisymmetric loading. The other two force summations result in equilibrium equations as follows:

$\Sigma F_\phi=0$:

$$R_0\frac{dN_\phi}{d\phi}+N_\phi\frac{dR_0}{d\phi}-N_\theta\frac{dR_0}{d\phi}+R_0R_2\,p_\phi=0\,, \tag{9}$$

where N_ϕ and N_θ are the membrane forces, and p_ϕ is the ϕ component of the axisymmetric loading; in this problem, $P_\phi=0$.

$\Sigma F_z=0$:

$$\frac{N_\theta}{R_1}+\frac{N_\phi}{R_2}+p_z=0\,. \tag{10}$$

If Eq. (9) is manipulated and integrated, we obtain the familiar form which can also be obtained by an equilibration of the portion of the shell above the parallel for angle ϕ; thus,

$$2\pi R_0 N_\phi \sin\phi + F_c = 0 , \tag{11}$$

where F_c = total vertical force on the cap above the ϕ position. In our problem with internal-pressure loading, we have $p_\phi = p_\theta = 0$, and $p_z = -p$, since we measure positive p_z in the same direction as positive w, or "inward." Thus, for the case of internal pressure, we may calculate F_c as

$$F_c = -\pi R_0^2 p = -\pi (R_1 \sin\phi)^2 p , \tag{11a}$$

where p = internal pressure. Equation (11) thus becomes, after using Eq. (11a),

$$N_\phi = \frac{-(-\pi R_1^2 \sin\phi) p}{2\pi (R_1 \sin\phi)\sin\phi} = \frac{pR_1}{2} . \tag{12}$$

From Eq. (10) we obtain, after using Eq. (12),

$$N_\theta = -R_1 \left(\frac{N_\phi}{2} + p_z \right) = pR_1 \left(1 - \frac{R_1}{2R_2} \right) . \tag{13}$$

Equations (12) and (13) show that the membrane forces are statically determinate at each point of the shell of revolution, if the pressure and principal radii of curvature are known. These radii are to be corrected after each pressure increment is applied. Since our total pressure is to be applied in increments, we will obtain increments of membrane forces, δN_ϕ and δN_θ, which, in turn, give increments of strain; thus, we will need expressions for δN_ϕ and δN_θ. If we express each of N_ϕ and N_θ before and after the application of a pressure increment δp and subtract, we will obtain values for δN_ϕ and δN_θ as follows:

$$N_\phi + \delta N_\phi = \frac{(p+\delta p)(R_1 + \delta R_1)}{2} = \frac{pR_1 + \delta p R_1 + p\delta R_1 + \delta p \delta R_1}{2} ,$$

$$\delta N_\phi = \frac{\delta p R_1}{2} + \frac{p \delta R_1}{2} , \tag{14}$$

after subtracting Eq. (12) and neglecting $\delta p \delta R_1$ as a higher-order differential for any given step of pressure. Similarly for δN_θ, we obtain, after more algebra,

$$\delta N_\theta = \delta p R_1 \left[1 - \frac{R_1}{2(R_2 + \delta R_2)} \right] + p\delta R_1 \left(1 - \frac{R_1}{R_2 + \delta R_2} \right) . \tag{15}$$

Of course, δR_1 and δR_2 are not known yet, because the δN_ϕ and δN_θ currently being calculated would be necessary to compute the δR's. However, a reasonable estimate of the values of the δR's would be better than to ignore them; therefore, we shall use as an estimate their values for the previous pressure increment, except for the first step for which the values of the δR's will be taken as zero.

STRESS-STRAIN RELATIONS

Up to now, we have been able to develop our relations without regard to the type of stress-strain relations that may exist, because Eqs. (7) and (8) arise strictly from geometry, and Eqs. (14) and (15) are statically determinate. In order to obtain values for the strains, ϵ_ϕ^t and ϵ_θ^t, in Eqs. (7) and (8), we must convert increments of membrane forces into increments of strain. To do this, we must divert our discussion of the stress-strain relations into two parts labeled "elastic" and "plastic."

Elastic Case

Making use of Hooke's law for biaxial stress, we have

$$\epsilon_\phi^t = \frac{1}{Eh}(N_\phi - \nu N_\theta) , \tag{16}$$

$$\epsilon_\theta^t = \frac{1}{Eh}(N_\theta - \nu N_\phi) . \tag{17}$$

Since we will apply the pressure loading in increments, we will want expressions for increments of strain, $\delta\epsilon_\phi^t$ and $\delta\epsilon_\theta^t$, caused by the pressure increment δp. Thus, from Eqs. (16) and (17), we obtain

$$\delta\epsilon_\phi^t = \frac{1}{Eh}(\delta N_\phi - \nu\delta N_\theta) , \tag{18}$$

$$\delta\epsilon_\theta^t = \frac{1}{Eh}(\delta N_\theta - \nu\delta N_\phi) , \tag{19}$$

where we consider the shell thickness h as constant. After substituting expressions for δN_ϕ and δN_θ from Eqs. (14) and (15) into Eqs. (18) and (19), we obtain

$$\delta\epsilon_\phi^t = \frac{1}{Eh}\left[\left(\frac{1}{2} - \nu\right)(R_1\,\delta p + p\delta R_1) + \nu\left(\frac{R_1^2\,\delta p}{2R_2} + \frac{pR_1\,\delta R_1}{R_2}\right)\right] , \tag{20}$$

$$\delta\epsilon_\theta^t = \frac{1}{Eh}\left[\left(1 - \frac{\nu}{2}\right)(R_1\,\delta p + p\delta R_1) - \frac{R_1\,\delta p}{2R_2} - \frac{pR_1\,\delta R_1}{R_2}\right] , \tag{21}$$

after dropping δR_2, which is small compared with R_2. Equations (20) and (21) will actually be used in a dimensionless form, as will all relations in this paper, so as to reduce significantly the number of problems that must be solved with computers. We place Eqs. (20) and (21) in dimensionless form by multiplying and dividing by the characteristic length a to obtain

$$\delta\epsilon_\phi^t = \frac{1}{h'}\left[\left(\frac{1}{2} - \nu\right)(R_1'\delta p' + p'\delta R_1') + \nu\left(\frac{R_1'^2\delta p'}{2R_2'} + \frac{p'R_1'\delta R_1'}{R_2'}\right)\right] , \tag{22}$$

$$\delta\epsilon_\theta^t = \frac{1}{h'}\left[\left(1 - \frac{\nu}{2}\right)(R_1'\delta p' + \delta R_1') - \frac{R_1'^2\delta p'}{2R_2'} - \frac{p'R_1'\delta R_1'}{R_2'}\right], \tag{23}$$

where $h' = h/a$, $R_1' = R_1/a$, $R_2' = R_2/a$, $p' = p/E$, and $\delta p' = \delta p/E$.

Plastic Case

By plastic behavior, we mean in this paper the behavior of a material in the region beyond the proportional limit for strain-hardening materials such as are commonly used for missile tanks. Though the uniaxial stress-strain relations for the plastic range are available for the interesting materials, there is no general method at present for using uniaxial test results for predicting biaxial stress-strain relations with accuracy. Some biaxial tests [*1–3*] have been made, mostly on materials other than the ones of present interest, but predicted results usually are not in good agreement with test results. Plastic strains appear to be so dependent upon stress history, as well as the current stress state, that theoretical predictions are quite difficult.

Two types of plastic stress-strain theories have been used: the deformation theory and the flow or incremental theory. Deformation theory promulgates that strains can be found from current stress state alone, with no consideration of stress history. In the incremental theory, both current stress state and stress history are used to arrive at the strains. The disadvantage of the incremental theories is that they are considerably more difficult to formulate mathematically.

Though the principal stress ratio varies with ϕ for the shells of revolution, the stress ratio of a given element would not change radically during loading; therefore, biaxial test data for constant stress ratios could be applied directly if such data were available for sufficiently closely spaced ratios. However, this data is not presently available, and we are forced to rely on uniaxial data as a basis for finding biaxial strains by some incremental theory, even though we recognize the shortcomings of the procedure.

Two incremental theories which appear tractable were found in the literature. One method was developed by Marin, Ulrich, and Hughes [*4*] and is based on a flow theory in which the incremental changes in the principal shear stresses are proportional to the incremental changes in the principal shear strains. The other method was developed by Hwang [*5*] and is also a flow theory, but the basis is the assumption that equivalent stress depends only on the plastic work done. It was found by the author that the two above methods were equivalent for biaxial conditions; therefore, only the method of Marin, Ulrich, and Hughes will be used here. In order to illustrate the use of the method, it will be applied for a specific material (CA 0-71004 at −320°F) of the type used for missile tanks.

The results of the method of Marin *et al.* are shown below as modified for the case of biaxial stress and with the notation for our shell problem. It should be noted that true strains are used in the uniaxial relations, even though nominal strains are dealt with in the biaxial relations. However, for the small strains for which this method is valid, there is very little difference in nominal and true strain.

Total nominal strain increments are given by

$$\delta\epsilon^t_\phi = \frac{1}{E}(\delta\sigma_\phi - \nu\delta\sigma_\theta) + \delta B\left(\sigma_\phi - \frac{1}{2}\sigma_\theta\right)$$

$$= \delta\epsilon^e_\phi + \delta\epsilon^p_\phi\,, \tag{24}$$

$$\delta\epsilon^t_\theta = \frac{1}{E}(\delta\sigma_\theta - \nu\delta\sigma_\phi) + \delta B\left(\sigma_\theta - \frac{1}{2}\sigma_\phi\right)$$

$$= \delta\epsilon^e_\theta + \delta\epsilon^p_\theta\,, \tag{25}$$

where

δB = increment of plastic flow,

$\delta\epsilon^e_i$ = increment of nominal elastic strain,

$\delta\epsilon^p_i$ = increment of nominal plastic strain,

σ_i = true stress.

The onset of plastic flow is determined by the Von Mises criterion, which is based on the distortion energy theory (or the equivalent octahedral shear stress) and is given for our case by

$$\sigma_y^2 = \sigma_\phi^2 + \sigma_\theta^2 - \sigma_\phi\sigma_\theta\,, \tag{26}$$

where σ_y = true yield stress in axial tension.

We now assume isotropic expansion of the yield surface as plastic deformation proceeds and that the function of Eq. (26), which defines the onset of plastic flow, also defines subsequent plastic flow. In other words, the term $\bar\sigma$, where

$$\bar\sigma^2 = \sigma_\phi^2 + \sigma_\theta^2 - \sigma_\phi\sigma_\theta$$

$$= \text{equivalent octahedral shear stress}\,, \tag{27}$$

is assumed to be the variable upon which δB in Eqs. (24) and (25) depends. Thus,

$$\delta B = F(\bar\sigma)\delta\bar\sigma\,,\ \text{for } \delta\bar\sigma > 0 \text{ (loading)}\,,$$

$$\delta B = 0\,,\ \text{for } \delta\bar\sigma \leqq 0 \text{ (unloading or nonloading)}\,. \tag{28}$$

We now further assume that the function $F(\bar\sigma)$ must be the same for plastic flow for any stress condition, including axial loading; therefore, we shall determine $F(\bar\sigma)$ for uniaxial tension and assume that the functional form is valid for multiaxial plastic stress conditions. From Eq. (24) or (25) for the uniaxial case,

$$\delta\epsilon^p_t = \delta B\sigma_t\,, \tag{29}$$

where $\delta\epsilon_t^p$ = increment of true plastic strain for axial test, and σ_t = true stress for axial test.

From the first part of Eq. (28), $\delta B = F(\overline{\sigma})\delta\overline{\sigma}$, and for uniaxial conditions, $\overline{\sigma} = \sigma_t$; therefore,

$$\delta\epsilon_t^p = \overline{\sigma}F(\overline{\sigma})\delta\overline{\sigma} ,$$

or

$$F(\overline{\sigma}) = F(\sigma_t) = \frac{1}{\overline{\sigma}} \cdot \frac{\delta\epsilon_t^p}{\delta\overline{\sigma}} = \frac{1}{\sigma_t} \cdot \frac{\delta\epsilon_t^p}{\delta\sigma_t} = \frac{1}{\sigma_t} \cdot \frac{d\epsilon_t^p}{d\sigma_t} . \tag{30}$$

We see from Eq. (30) that we determine the functional form of $F(\overline{\sigma})$ by the true stress–true strain curve for our material in axial tension. Figure 3 is a mean curve of nominal stress vs nominal strain compiled from a fairly large number of axial test results for the example material, stainless steel CA 0-71004 at −320°F. The term "mean" here refers to the fact that the material does not behave exactly the same in the rolling direction of the sheet as it does in a direction transverse to rolling. No attempt was made to include this difference in the present analysis. Figure 4 is a plot of true stress vs true plastic strain as calculated from Fig. 3, in which true plastic strain is given by

$$\epsilon_t^p = \ln (1 + \epsilon_0^p) , \tag{31}$$

where

ϵ_t^p = true plastic strain for uniaxial case,

ϵ_0^p = nominal plastic strain for uniaxial case.

In obtaining the curve of Fig. 4 from Fig. 3, we could have used nominal stresses and strains with very little error because of the rather small plastic strains involved. A functional form for the curve of Fig. 4 is expressed as indicated in two parts rather than one more-involved equation for the entire curve.

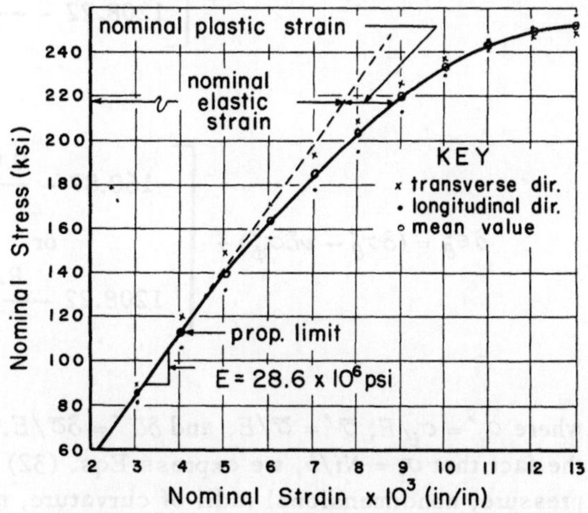

Fig. 3. Nominal Stress-Strain Curve for Stainless Steel CA 0-71004 at −320°F.

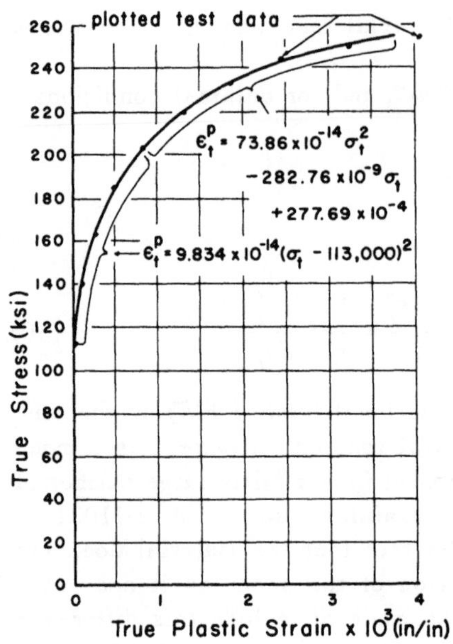

Fig. 4. Axial Tensile True Stress vs True Plastic Strain for Stainless Steel CA 0-71004 at –320°F.

Applying the above method to the specific example material, we obtain in dimensionless form the following equations for $E = 28.6 \times 10^6$ psi:

$$\delta\epsilon_\phi^t = (\delta\sigma_\phi' - \nu\delta\sigma_\theta') + F(\bar{\sigma}')\left(\sigma_\phi' - \frac{1}{2}\sigma_\theta'\right)\delta\bar{\sigma} ,$$

$$= (\delta\sigma_\phi' - \nu\delta\sigma_\theta') + \begin{bmatrix} 160.87 - \dfrac{0.63561}{\bar{\sigma}'} \\ \text{or} \\ 1208.22 - \dfrac{8.08794}{\bar{\sigma}'} \end{bmatrix} \left(\sigma_\phi' - \frac{1}{2}\sigma_\theta'\right)\delta\bar{\sigma}', \quad (32)$$

$$\delta\epsilon_\theta^t = (\delta\sigma_\theta' - \nu\delta\sigma_\phi') + \begin{bmatrix} 160.87 - \dfrac{0.63561}{\bar{\sigma}'} \\ \text{or} \\ 1208.22 - \dfrac{8.08794}{\bar{\sigma}'} \end{bmatrix} \left(\sigma_\theta' - \frac{1}{2}\sigma_\phi'\right)\delta\bar{\sigma}', \quad (33)$$

where $\sigma_i' = \sigma_i/E$, $\bar{\sigma}' = \bar{\sigma}/E$, and $\delta\bar{\sigma}' = \delta\bar{\sigma}/E$. Using Eqs. (14) and (15) along with the fact that $\sigma_i = Ni/h$, we express Eqs. (32) and (33) in terms of nondimensional pressure, nondimensional radii of curvature, nondimensional shell thickness, and appropriate increments of them as follows:

$$\delta\epsilon^t_\phi = \frac{1}{h'}\left[\left(\frac{1}{2}-\nu\right)(R_1'\,\delta p' + p'\delta R_1') + \nu\left(\frac{R_1'^2\delta p'}{2R_2'} + \frac{p'R_1'\delta R_1'}{R_2'}\right)\right]$$

$$+\frac{p'R_1'}{4h'}\begin{bmatrix} 160.87 - \dfrac{0.63561}{\bar{\sigma}'} \\ \text{or} \\ 1208.22 - \dfrac{8.08794}{\bar{\sigma}'} \end{bmatrix}\frac{R_1'}{R_2'}\delta\bar{\sigma}', \qquad (34)$$

$$\delta\epsilon^t_\phi = \delta\epsilon^e_\phi + \delta\epsilon^p_\phi\,,$$

$$\delta\epsilon^t_\theta = \frac{1}{h'}\left[\left(1-\frac{\nu}{2}\right)(R_1'\delta p' + p'\delta R_1') - \frac{R_1'^2\delta p'}{2R_2'} - \frac{p'R_1'\delta R_1'}{R_2'}\right]$$

$$+\frac{p'R_1'}{4h'}\begin{bmatrix} 160.87 - \dfrac{0.63651}{\bar{\sigma}'} \\ \text{or} \\ 1208.22 - \dfrac{8.08794}{\bar{\sigma}'} \end{bmatrix}\left(3-\frac{2R_1'}{R_2'}\right)\delta\bar{\sigma}'$$

$$= \delta\epsilon^e_\theta + \delta\epsilon^p_\theta\,, \qquad (35)$$

where

$$\bar{\sigma}' = \frac{P'R_1'}{2h'R_2'}\sqrt{(3R_2')^2 - 2R_1'R_2' + R_1'^2}\,,$$

and we use the upper option of Eqs. (34) and (35) if $3.95 \times 10^{-3} < \bar{\sigma}' \leqq 7.12 \times 10^{-3}$, or the lower option if $7.12 \times 10^{-3} < \bar{\sigma}' \leqq 8.88 \times 10^{-3}$. The reason for not expressing $\bar{\sigma}'$ and $\delta\bar{\sigma}'$ in terms of p', $\delta p'$, R_1', R_2', $\delta R_1'$, and $\delta R_2'$ is that we need the value of $\bar{\sigma}'$ and the sign of $\delta\bar{\sigma}'$ to determine the appropriate option and to establish loading or unloading for the current increment of pressure.

NUMERICAL INTEGRATION OF DISPLACEMENT EQUATION

In order to nondimensionalize our problem, we divide both sides of Eq. (7) by a, some characteristic length, to obtain

$$\frac{v'}{\sin\phi} = \int_{\phi=\pi/2}^{\phi}\left(\frac{R_2'\epsilon^t_\phi - R_1'\epsilon^t_\theta}{\sin\phi}\right)d\phi\,, \qquad (36)$$

where $v' = v/a$, $R' = R_2/a$, and $R_1' = R_1/a$. In Eq. (36), v' is the nondimensional displacement along the meridian of a point having current polar angle ϕ; ϵ_ϕ^t and ϵ_θ^t are the strains in the ϕ and θ directions respectively. All these quantities are caused by the increment of nondimensional pressure $\delta p'$. In order to distinguish these quantities from accumulated strains, and displacements from several increments of pressure, let us, hereafter in this paper, denote the displacements and strains from one pressure increment $\delta p'$ as $\delta v'$, $\delta w'$, $\delta\epsilon_\phi^t$, and $\delta\epsilon_\theta^t$. Equation (36) thus becomes

$$\frac{\delta v'}{\sin\phi} = \int_{\phi=\pi/2}^{\phi} \left(\frac{R_2'\delta\epsilon_\phi^t - R_1'\delta\epsilon_\theta^t}{\sin\phi} \right) d\phi \,, \tag{37}$$

and Eq. (8) becomes, after nondimensionalizing,

$$\delta w' = \delta v' \cot\phi - R_1'\delta\epsilon_\theta^t \,. \tag{38}$$

In determining displacements from Eqs. (37) and (38), we use expressions for $\delta\epsilon_\phi^t$ and $\delta\epsilon_\theta^t$ from Eqs. (22) and (23) for the elastic case, or from Eqs. (34) and (35) for the plastic case.

Point Numbering System

Figure 2 indicates the method of numbering points, starting with 0 on the y axis and proceeding on to 100 on the x axis. Thus, our points increase in number as their x coordinates increase, even though our integration of Eq. (37) proceeds from right to left. This direction of integration was chosen because there is no indeterminacy of the term $v'/\sin\phi$ for $\phi = \pi/2$; whereas there is trouble at $\phi = 0$.

Integration of Equation (37) by Modified Simpson's Rule

In order to avoid the possibility of overlooking the fact that $d\phi$ in Eq. (37) is negative, let us write this equation as follows:

$$\frac{\delta v'}{\sin\phi} = -\int_{\phi}^{\pi/2} \left(\frac{R_2'\delta\epsilon_\phi^t - R_1'\delta\epsilon_\theta^t}{\sin\phi} \right) d\phi \,. \tag{39}$$

Assuming that we can calculate the various terms of the integrand, which are functions of ϕ, we can make a qualitative plot of $f(\phi)$ vs ϕ, where $f(\phi)$ is the integrand of Eq. (39); this is indicated in Fig. 5. We do not need to repeat the entire summation for each point but can merely add on the increments of area. Therefore,

$$\int_{\phi_n}^{\phi=\pi/2} \left(\frac{R_2'\delta\epsilon_\phi^t - R_1'\delta\epsilon_\theta^t}{\sin\phi} \right) d\phi = \int_{\phi_{n+1}}^{\phi=\pi/2} \left(\frac{R_2'\delta\epsilon_\phi^t - R_1'\delta\epsilon_\theta^t}{\sin\phi} \right) d\phi + \Delta A_n \,, \tag{40}$$

where ΔA_n is the area under the $f(\phi)$ curve between the verticals at ϕ_n and ϕ_{n+1}

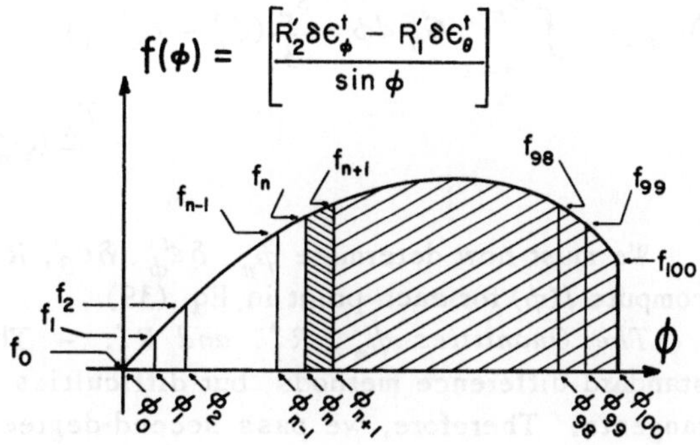

Fig. 5. Qualitative Plot of $f(\phi)$ vs ϕ Showing Method of Indexing.

Since Simpson's rule for numerical integration requires equal intervals of the independent variable, we are unable to use the method in the usual form since our intervals of ϕ will not be equal after the first increment of pressure. Therefore, let us pass a second-degree curve of the form

$$G_n = S_n \phi^2 + T_n \phi + U_n \tag{41}$$

through the three points (ϕ_{n-1}, f_{n-1}), (ϕ_n, f_n), (ϕ_{n+1}, f_{n+1}), after which we integrate from ϕ_n to ϕ_{n+1} to obtain ΔA_n and from ϕ_{n-1} to ϕ_n to obtain ΔA_{n-1}. We thus obtain three equations in the three unknowns S_n, T_n, and U_n, the solutions of which are

$$S_n = \frac{f_{n-1}(\phi_{n+1} - \phi_n) - f_n(\phi_{n+1} - \phi_{n-1}) + f_{n+1}(\phi_n - \phi_{n-1})}{\phi_{n-1}^2(\phi_{n+1} - \phi_n) - \phi_n^2(\phi_{n+1} - \phi_{n-1}) + \phi_{n+1}^2(\phi_n - \phi_{n-1})},$$

$$T_n = \frac{f_{n-1}(\phi_{n+1}^2 - \phi_n^2) - f_n(\phi_{n+1}^2 - \phi_{n-1}^2) + f_{n+1}(\phi_n^2 - \phi_{n-1}^2)}{\phi_{n-1}^2(\phi_{n+1} - \phi_n) - \phi_n^2(\phi_{n+1} - \phi_{n-1}) + \phi_{n+1}^2(\phi_n - \phi_{n-1})},$$

(42)

$$U_n = \frac{f_{n-1}(\phi_{n+1}^2 \phi_n - \phi_n^2 \phi_{n+1}) - f_n(\phi_{n+1}^2 \phi_{n-1} - \phi_{n-1}^2 \phi_{n+1})}{\phi_{n-1}^2(\phi_{n+1} - \phi_n) - \phi_n^2(\phi_{n+1} - \phi_{n-1}) + \phi_{n+1}^2(\phi_n - \phi_{n-1})}$$

$$+ \frac{f_{n+1}(\phi_n^2 \phi_{n-1} - \phi_{n-1}^2 \phi_n)}{\phi_{n-1}^2(\phi_{n+1} - \phi_n) - \phi_n^2(\phi_{n+1} - \phi_{n-1}) + \phi_{n+1}^2(\phi_n - \phi_{n-1})},$$

where we use only odd n (n = 99, 97, ..., 1), since we need S_n, T_n, and U_n only for pairs of increments. Therefore,

$$\Delta A_n = \int_{\phi_n}^{\phi_{n+1}} G_n \, d\phi = \frac{S_n}{3}(\phi_{n+1}^3 - \phi_n^3) + \frac{T_n}{2}(\phi_{n+1}^2 - \phi_n^2) + U_n(\phi_{n+1} - \phi_n), \tag{43}$$

$$\Delta A_{n-1} = \int_{\phi_{n-1}}^{\phi_n} G_n \, d\phi = \frac{S_n}{3}(\phi_n^3 - \phi_{n-1}^3) + \frac{T_n}{2}(\phi_n^2 - \phi_{n-1}^2) + U_n(\phi_n - \phi_{n-1}) \,. \quad (44)$$

We must now determine ϕ_n, $\delta\epsilon_\phi^t$, $\delta\epsilon_\theta^t$, R_1', and R_2' for each point in order to compute $f(\phi)$ for each point in Eq. (39).

The Quantities ϕ_n, R_1', *and* R_2'. – These quantities could be found by standard difference methods, but difficulties would be experienced near vertical tangents. Therefore, we pass second-degree best-fit curves through each point and its two neighbors on each side from which ϕ_n, R_1', and R_2' are obtained functionally. In addition to providing a local functional curve for each point, this procedure smoothes the deformed points, thus reducing round-off errors. These curves are of the form

$$\bar{y}_n' = M_n x'^2 + N_n x' + P_n \,, \quad \text{for } 0 \leqq n \leqq 50 \,, \quad (45)$$

$$\bar{x}_n' = Q_n y'^2 + W_n y' + Z_n \,, \quad \text{for } 50 < n \leqq 100 \,. \quad (46)$$

It should be noted that there is a curve of the form of Eq. (45) or Eq. (46) for each point of the shell; also, symmetrically located points 2′, 1′, 99′, and 98′ are used for determining the curves for 0, 1, 99, and 100. Once the local curves for each point are computed (local curves are computed with coordinates unsmoothed during the calculations for the current-pressure increment), the deformed coordinates will be corrected prior to the next pressure increment according to the following scheme:

For $0 \leqq n \leqq 50$, replace (x_n', y_n') with $(x_n', \bar{y}_n')$.

For $50 < n \leqq 100$, replace (x_n', y_n') with $(\bar{x}_n', y_n')$.

Now from Eq. (45) or Eq. (46) we find for each point:

For $0 \leqq n \leqq 50$,

$$\phi_n = \tan^{-1}\left(-\frac{d\bar{y}_n'}{dx'}\right)_n = \tan^{-1}(-2M_n x_n' - N_n) \,. \quad (47)$$

For $50 < n \leqq 100$,

$$\phi_n = \frac{\pi}{2} - \tan^{-1}\left(-\frac{d\bar{x}_n'}{dy'}\right)_n = \frac{\pi}{2} - \tan^{-1}(-2Q_n y_n' - W_n) \,. \quad (48)$$

Since $x' = R_1' \sin\phi$, we may write

$$\begin{aligned} R_{1_0}' &= R_{2_0}' \,, \quad \text{for } n = 0 \,, \\ R_{1_n}' &= x_n'/\sin\phi_n \,, \quad \text{for } 0 < n \leqq 50 \,, \\ R_{1_n}' &= \bar{x}_n'/\sin\phi_n \,, \quad \text{for } 50 < n \leqq 100 \,. \end{aligned} \quad (49)$$

We can determine R'_{2_n} by the ordinary method of calculating the radius of curvature, since R_2 is merely the radius of curvature of the meridian.

Strain Increments $\delta\epsilon^t_\phi$ *and* $\delta\epsilon^t_\theta$. – At this point, we must again use different equations for the elastic case than we do for the plastic case.

For the elastic case, we use Eqs. (22) and (23) to compute the strain increments $\delta\epsilon^t_\phi$ and $\delta\epsilon^t_\theta$; here h' is the current nondimensional shell thickness. With small error, this thickness could be used as the constant original thickness; however, a bit more accuracy can be obtained easily by finding h' by the formula

$$h' = h'_0[1 - \nu(\epsilon^e_\phi + \epsilon^e_\theta)] , \tag{50}$$

where $h'_0 = h_0/a$, and ϵ^e_ϕ, ϵ^e_θ = the elastic strains. In the elastic case, of course, the elastic strains ϵ^e_ϕ and ϵ^e_θ are merely the summation of the strain increments $\delta\epsilon^t_\phi$ and $\delta\epsilon^t_\theta$ over the total number of pressure increments.

For the plastic case, we use Eqs. (34) and (35) to calculate the strain increments; here there is an elastic part, ϵ^e_ϕ or ϵ^e_θ, and a plastic part, ϵ^p_ϕ or ϵ^p_θ, once the criterion for plastic behavior has been reached by Eq. (26). Again, the current shell thickness h' can be computed a bit more accurately by the relation

$$h' = h'_0[1 - \nu(\epsilon^e_\phi + \epsilon^e_\theta) - (\epsilon^p_\phi + \epsilon^p_\theta)] , \tag{51}$$

where ϵ^p_ϕ, ϵ^p_θ = the plastic strains.

RECTANGULAR DISPLACEMENTS FROM NORMAL AND TANGENTIAL DISPLACEMENTS

Since the slope of the middle surface will change as the several pressure increments are applied, we cannot merely add up all the increments $\delta v'$ and $\delta w'$ for each pressure increment $\delta p'$; instead, we convert each $\delta v'$ and $\delta w'$ to rectangular displacements $\delta x'$ and $\delta y'$, which can then be added algebraically after all pressure increments are applied for the final cycle. Figure 6 is a sketch indicating the geometry involved. In the nondimensional form we have for each pressure increment $\delta p'$ the incremental rectangular displacements as follows:

$$\delta x' = \delta v' \cos\phi - \delta w' \sin\phi , \tag{52}$$

$$\delta y' = -(\delta v' \sin\phi + \delta w' \cos\phi) . \tag{53}$$

After using Eq. (38) in Eqs. (52) and (53), we obtain for the nth point,

$$\delta x'_n = R'_{1_n} \sin\phi_n \, \delta\epsilon^t_{\theta_n} , \tag{54}$$

$$\delta y'_n = R'_{1_n} \cos\phi_n \, \delta\epsilon^t_{\theta_n} - \frac{\delta v'_n}{\sin\phi_n} , \tag{55}$$

where we obtain $\epsilon^t_{\theta_n}$ from Eq. (23) or (35), $\delta v'_n/\sin\phi_n$ from Eq. (39) [or from Eq. (57)], and R'_{1_n} from Eq. (49).

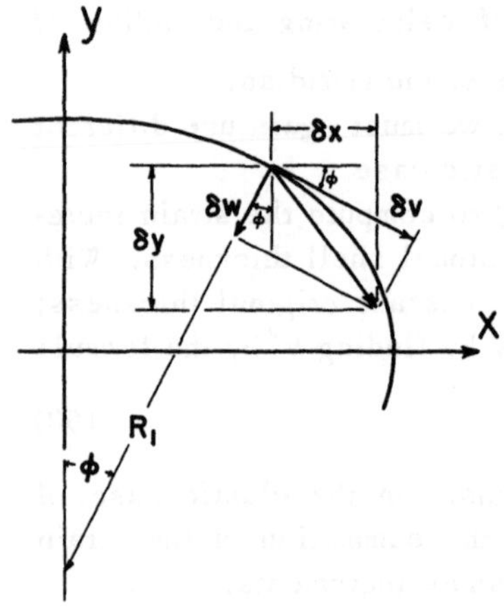

Fig. 6. Geometry for Converting Normal and Tangential Displacements to Rectangular Displacements.

SEQUENCE OF STEPS FOR COMPUTER SOLUTION

The basic order of procedure is identical for both the elastic case and the plastic case and is outlined below.

A. Locate Starting Points

In order to reduce the debugging time, enter the problem with x', y', R_1', and R_2' for each of the 105 points; even if the undeformed meridian curve is of functional form, a separate small program to calculate the starting values is recommended. Also needed to proceed with the problem are values of ν, h_0/a, and p'_{max}, the maximum nondimensional pressure to which the shell will be subjected.

B. Determine Best-Fit Curve for Each Point

After each pressure increment, we calculate the best-fit curve of the form of Eq. (45) or Eq. (46) for each point by a least-squares method using the x'-y' coordinates found by adding $\delta x'$ and $\delta y'$ to the values of x' and y' before $\delta p'$ was applied. We thus obtain values for M_n, N_n, P_n, if $0 \leqq n \leqq 50$, or for Q_n, W_n, Z_n, if $50 < n \leqq 100$, to be used in Eqs. (47) and (48).

C. Determine Polar Angle (ϕ) and Radii of Curvature (R_1' and R_2') for Each Point

Here we use Eq. (47) or (48) to compute ϕ_n for each point after each application of $\delta p'$. Radii of curvature R_1' and R_2' are computed at each point by use of Eq. (49) for R_1' and by standard procedure from Eq. (45) or (46) for R_2'.

D. Compute Strain Increments $\delta\epsilon_\phi^t$ and $\delta\epsilon_\theta^t$ at Each Point

After each pressure increment $\delta p'$, we compute the total strain increments $\delta\epsilon_\phi^t$ and $\delta\epsilon_\theta^t$ by use of Eqs. (22) and (23) for the elastic case or by use of Eqs. (34) and (35) for the plastic case.

E. Compute $f(\phi)$ for Each Point

The integrand of Eq. (39) has been referred to in this paper as $f(\phi)$; thus, at each point, we shall compute $f_n = f(\phi_n)$ from

$$f_n = f(\phi_n) = \frac{R'_{2n}\,\delta\epsilon_\phi^t - R'_{1n}\,\delta\epsilon_{\theta_n}^t}{\sin\phi_n}\,, \text{ for } n > 0\,, \tag{56}$$

$$f_0 = f(\phi_0) = f(0) = 0\,.$$

F. Compute Increments ΔA_n

The increments ΔA_n and ΔA_{n-1} under the $f(\phi)$ curve are calculated in pairs by use of Eqs. (43) and (44), with the values of S_n, T_n, and U_n from Eq. (42).

G. Compute $\delta v'/\sin\phi$ for Each Point

From Eqs. (39) and (40), we have

$$\frac{\delta v'_n}{\sin\phi_n} = -I_n = -(I_{n+1} + \Delta A_n)\,, \tag{57}$$

where

$$I_n = \int_{\phi_n}^{\pi/2}\left(\frac{R'_2\,\delta\epsilon_\phi^t - R'_1\,\delta\epsilon_\theta^t}{\sin\phi}\right)d\phi\,,$$

$$I_{n+1} = \sum_{i=99}^{n+1} \Delta A_n\,,$$

$$I_{100} = 0\,.$$

H. Locate Points of Deformed Shell in x'-y' Coordinates

Once $\delta v'_n/\sin\phi_n$ is available from Eq. (57), R'_{1_n} from Eq. (49), and $\delta\epsilon_{\theta_n}^t$ from Eq. (23) or (35), we may determine the rectangular incremental displacements $\delta x'$ and $\delta y'$ due to $\delta p'$ from Eqs. (54) and (55). Then the new x'-y' coordinates are found by adding $\delta x'$ and $\delta y'$ to the values of x' and y' before $\delta p'$.

I. Loop Back to Step B, Change Cycle, or Proceed to Step K

We do not predetermine the final number of pressure increments but increase this number by doubling until the magnitude of the displacements of some specific point changes less than, say, 1% by such doubling when we move to the next cycle. By "cycle" we mean the set of calculations involved for a given number of increments of the same total pressure.

J. Enter Next Cycle

Here we double the number of increments in the total pressure, p'_{max}, that was used in the previous cycle. We then loop back to step B and begin a computing sequence again.

K. Outputs

We will want to print out the original inputs as well as the various final calculated values of ϕ, x', y', R_1', R_2', $\bar{\sigma}'$, ϵ_ϕ^e, ϵ_θ^e, ϵ_ϕ^p, ϵ_θ^p, and h' for each point.

CONCLUSIONS

By application of simple procedures a large number of times, large deformations of thin shells of revolution under internal pressure loading can be computed with the aid of modern digital computers for the plastic range as well as the elastic range of the stress-strain relations. If comprehensive biaxial stress-strain data were available for the plastic range, a more accurate computation of deformations could be accomplished by the direct use of this information during the step-by-step calculations than is possible with the presently available method based on uniaxial test results.

REFERENCES

[1] H. E. Davis and E. R. Parker, "Behavior of Steel Under Biaxial Stress as Determined by Tests in Tubes," *J. Appl. Mechanics* **15**, 201 (1948).

[2] W. R. Osgood, "Combined-Stress Tests on 24S-T Aluminum-Alloy Tubes," *J. Appl. Mechanics* **14**, A-147 (1947).

[3] Aris Phillips, "Combined Tension-Torsion Tests for Aluminum Alloy 2S-0," *J. Appl. Mechanics* **19**, (1952).

[4] Joseph Marin, B. H. Ulrich, and W. P. Hughes, "Plastic Stress-Strain Relations for 75S-T6 Aluminum Alloy," *NACA* TN 2425, Washington, August, 1951.

[5] Chintsun Hwang, "Incremental Stress-Strain Law Applied to Work-Hardening Plastic Materials," *Trans. ASME* **81**, 594 (1959).

BIBLIOGRAPHY

Batdorf, S. B., and Bernard Budianski, "Polyaxial Stress-Strain Relations of a Strain-Hardening Metal," *Trans. ASME* **76**, 323 (1954).

Budianski, Bernard, "A Reassessment of Deformation Theories of Plasticity," *Trans. ASME* **81**, 259 (1959).

Cunningham, D. M., E. G. Thomsen, and J. E. Dorn, "Plastic Flow of a Magnesium Alloy Under Biaxial Stresses," *Am. Soc. Testing Materials, Proc.* **47**, 546 (1947).

Dorn, J. E., and A. J. Latter, "Stress-Strain Relations for Finite Elastoplastic Deformations," *J. Appl. Mechanics* **15**, 234 (1948).

Flügge, W., *Stresses in Shells*, Springer-Verlag, Berlin, Germany, 1960.

Fraenkel, S. J., "Experimental Studies of Biaxially Stressed Mild Steel in the Plastic Range," *J. Appl. Mechanics* **15**, 193 (1948).

Gleyzal, A., "Plastic Deformation of a Circular Diaphragm Under Pressure," *J. Appl. Mechanics* **15**, 288 (1948).

Hu, L. W., and Joseph Marin, "Anisotropic Loading Functions for Combined Stresses in the Plastic Range," *J. Appl. Mechanics* **22**, 77 (1955).

Hu, L. W., and Joseph Marin, "Determination of Theoretical Plastic Stress-Strain Relations for Variable Combined Stress Ratios," *J. Appl. Mechanics* **19**, 485 (1952).

Marin, Joseph, and L. W. Hu, "Plastic Stress-Strain Relations for Biaxial Tension and Variable Stress Ratios," *Am. Soc. Testing Materials, Proc.* **52**, 1098 (1952).

Nadai, Arpad, *Theory of Flow and Fracture of Solids*, 2d ed., McGraw-Hill, New York, 1950.

Naghdi, P. M., F. Essenburg, and W. Koff, "An Experimental Study of Initial and Subsequent Yield Surfaces in Plasticity," *J. Appl. Mechanics* **25**, 201 (1958).

Naghdi, P. M., J. C. Rowley, and C. W. Beadle, "Experiments Concerning the Yield Surface and the Assumption of Linearity in the Plastic Stress-Strain Relations," *J. Appl. Mechanics* **22**, 416 (1955).

Prager, William, "Theory of Plastic Flow *versus* Theory of Plastic Deformation," *J. Appl. Phys.* **19**, 540 (1948).

Prager, William, "The Stress-Strain Laws of the Mathematical Theory of Plasticity – a Survey of Recent Progress," *J. Appl. Mechanics* **15**, 226 (1948).

Timoshenko, S., and S. Woinowsky-Krieger, *Theory of Plates and Shells*, 2d ed., McGraw-Hill, New York, 1959.

SOME PROBLEMS IN THE BUCKLING OF THICK PLATES[1]

Daniel Frederick
Virginia Polytechnic Institute

The governing equations for the buckling of thick plates including the effects of shear deformation, normal pressure, and middle-surface force resultants are presented in rectangular and polar coordinates, following the theory developed by Reissner. Solutions of these equations are presented which permit the satisfaction of three boundary conditions on each edge instead of two as in the classical theory.

Solutions and numerical results are presented for (*a*) the buckling of a uniformly compressed rectangular plate simply supported along two opposite sides perpendicular to the direction of compression and having various edge conditions on the other two sides, and (*b*) the buckling of a uniformly compressed circular plate with various edge conditions. The solutions are compared with classical theory. Effects of shear deformation and normal stress become important at a certain value of a thickness-length parameter which can be ascertained from the numerical examples. In addition, the freeing of one edge of a simply supported rectangular plate of twisting moment is significant.

INTRODUCTION

Historically, problems in the bending of plates were of interest before the elements of the linear theory of elasticity were discovered by Navier and Cauchy in 1822. In 1828, Poisson developed a theory for the bending of plates which stipulated three boundary conditions on an edge. This led to some controversy, which was resolved by Kirchhoff (1850) and Thompson and Tait (1867), who showed that only two boundary conditions were needed in the classical theory for thin plates.

Although the classical theory of plates (which is not exact from the standpoint of elasticity theory) is adequate for many engineering problems, there are cases where a more exact theory is needed. This led to the development of a number of thick-plate theories, which included correction terms in the thickness direction of the plate. Most of these are too complicated to yield simple or workable mathematical solutions. In 1945, Reissner [*1–3*] developed a simpler

[1]Sponsored by the Virginia Engineering Experiment Station and based partly on a Ph.D. thesis submitted to the University of Michigan.

NOTATION

$\sigma_x, \sigma_y, \sigma_z$	Normal stresses
$\tau_{xy}, \tau_{xz}, \tau_{yz}$	Shearing stresses
$\epsilon_x, \epsilon_y, \epsilon_z$	Normal strains
$\epsilon_{xy}, \epsilon_{xz}, \epsilon_{yz}$	Shearing strains
u, v, w	Displacements
P_x, P_y, P_{xy}	In-plane stress resultants, rectangular plate
M_x, M_y, M_{xy}	Moment stress resultants, rectangular plate
V_x, V_y	Shear stress resultants, rectangular plate
$\bar{u}, \bar{v}, \bar{w}$	Average displacement resultants, rectangular plate
α, β	Average rotations, rectangular plate
h	Plate thickness
q_0	Pressure on upper surface
E	Young's modulus
G	Shearing modulus
ν	Poisson's ratio
D	Flexural rigidity
ϕ	Stress function
P, P_c	Critical loads
a, b	Edge dimensions, rectangular plate
$M_r, M_\theta, M_{r\theta}$	Moment stress resultants, circular plate
V_r, V_θ	Shear stress resultants, circular plate
α_r, α_θ	Average rotations, circular plate
J_n, Y_n	Bessel functions of order n
I_n, K_n	Modified Bessel functions of order n
a, b	Radii, circular plate
$A_{1n}, A_{2n}, C_1 \ldots C_4, C_{1n} \ldots C_{12n}$	Constants of integration

thick-plate theory including the effects of shear deformation and stress normal to the middle surface. Since the publication of Reissner's theory, a large number of solutions have been presented using it.

The purpose of this paper is to present the governing equations for the buckling of rectangular and circular plates based upon Reissner's theory for the case of small deflections and to present some solutions and examples based on it. In the case of buckling, a new mode of deformation called a shear mode arises. The results are compared numerically with the classical theory, showing where the corrections for shear and normal pressure are important. Examples presented are (a) the buckling of a uniformly compressed rectangular plate simply supported[2] along two opposite sides perpendicular to the direction of compression and having various edge conditions on the other two sides, and (b) the buckling of a uniformly compressed circular plate with various edge conditions. All results are for a plate of constant thickness, h.

NOTATION AND DEFINITIONS

An element of the plate with top and bottom faces parallel to each other is shown in Fig. 1 in rectangular Cartesian coordinates. The origin is in the middle surface of the plate, and the positive directions of the axes are shown in the figure.

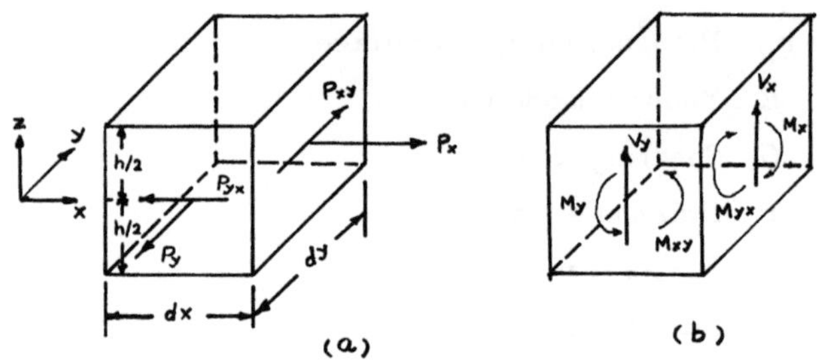

Fig. 1. Element of Rectangular Plate.

Since the derivation in this paper will include both bending and extension of plates, force resultants as well as moment and shear resultants will be introduced. The two normal-stress resultants P_x and P_y and the shear-stress resultant P_{xy}, all lying in the plane of the plate, are defined per unit length as:

$$P_x = \int_{-h/2}^{h/2} \sigma_x \, dz \,, \qquad P_y = \int_{-h/2}^{h/2} \sigma_y \, dz \,, \qquad P_{xy} = \int_{-h/2}^{h/2} \tau_{xy} \, dz \,, \tag{1}$$

where σ_x, σ_y, and τ_{xy} are the corresponding stresses. Furthermore, from the classical theory, the moment resultants M_x and M_y, the twisting-moment resultant M_{xy}, and the shear-stress resultants V_x and V_y are defined as:

$$M_x = \int_{-h/2}^{h/2} \sigma_x z \, dz \,, \qquad M_y = \int_{-h/2}^{h/2} \sigma_y z \, dz \,, \qquad M_{xy} = \int_{-h/2}^{h/2} \tau_{xy} z \, dz \,,$$
$$V_x = \int_{-h/2}^{h/2} \tau_{xz} \, dz \,, \qquad V_y = \int_{-h/2}^{h/2} \tau_{yz} \, dz \,, \tag{2}$$

[2]Simply supported and clamped edges may be defined in two ways, since there are three boundary conditions to be specified in the Reissner theory.

where τ_{xz} and τ_{yz} are shear stresses in the vertical direction. In applying the variational principle developed by Reissner [3] to derive the governing equations, it will be necessary to consider average-displacement resultants and average-rotation resultants. The average-displacement resultants are defined as:

$$\bar{u}(x,y) = \frac{1}{h}\int_{-h/2}^{h/2} u\,dz\,, \qquad \bar{v}(x,y) = \frac{1}{h}\int_{-h/2}^{h/2} v\,dz\,, \qquad \bar{w} = \int_{-h/2}^{h/2} w\,dz\,, \tag{3}$$

where u, v, and w are the usual displacements considered as functions of x, y, and z. Average rotations of line elements normal to the middle surface about the y and x axes, respectively, are

$$\alpha = \frac{12}{h^3}\int_{-h/2}^{h/2} uz\,dz\,, \qquad \beta = \frac{12}{h^3}\int_{-h/2}^{h/2} vz\,dz\,. \tag{4}$$

Reissner developed a theory based upon linear relations between the strains (normal strains are ϵ_x, ϵ_y, ϵ_z, and shearing strains are ϵ_{xy}, ϵ_{xz}, ϵ_{yz}) and the displacements. For the buckling problem the additional strain in the middle surface due to the deflection $\bar{w}$ must be considered, since this part of the strain in combination with the finite forces P_x, P_y, and P_{xy} may add terms to the total energy of the same magnitude as those considered by Reissner. The six components of finite strain to be used in the variational principle instead of the infinitesimal ones are given below. In writing these it is assumed that all derivatives of u and v and the derivative of w in the z direction are all small. Hence, the strain components become:

$$\begin{aligned}
\epsilon_x &= \frac{\partial u}{\partial x} + \frac{1}{2}\left(\frac{\partial w}{\partial x}\right)^2, & \epsilon_{xy} &= \frac{\partial u}{\partial y} + \frac{\partial v}{\partial x} + \frac{\partial w}{\partial x}\frac{\partial w}{\partial y}\,, \\
\epsilon_y &= \frac{\partial v}{\partial y} + \frac{1}{2}\left(\frac{\partial w}{\partial y}\right)^2, & \epsilon_{xz} &= \frac{\partial u}{\partial z} + \frac{\partial w}{\partial x}\,, \\
\epsilon_z &= \frac{\partial w}{\partial z}\,, & \epsilon_{yz} &= \frac{\partial v}{\partial z} + \frac{\partial w}{\partial y}\,.
\end{aligned} \tag{5}$$

THE GOVERNING EQUATIONS IN RECTANGULAR COORDINATES

By using the above strains and stress resultants in the variational principle[3] of Reissner, the governing equations for the buckling of a rectangular plate will arise. Since this paper is concerned mainly with the applications, the lengthy manipulations required to arrive at the governing equations are not presented.

[3] This principle states that the difference of the strain energy expressed in a mixed form and the work done by surface forces must be stationary.

These are given in detail in ref [5]. The governing equations in the standard form are:

$$D\nabla^4\overline{w} = \underline{\frac{h^2}{10}\frac{2-\nu}{1-\nu}\nabla^2 q_0} - q_0 \,,$$

$$\underline{\frac{h^2}{10}\nabla^2 V_x} - V_x = D\frac{\partial}{\partial x}\nabla^2\overline{w} - \underline{\frac{h^2}{10(1-\nu)}\frac{\partial q_0}{\partial x}} \,,$$

$$\underline{\frac{h^2}{10}\nabla^2 V_y} - V_y = D\frac{\partial}{\partial y}\nabla^2\overline{w} - \underline{\frac{h^2}{10(1-\nu)}\frac{\partial q_0}{\partial y}} \,, \tag{6}$$

$$M_x = D\left[\underline{\frac{6(1-\nu)}{5Gh}\frac{\partial V_x}{\partial x}} - \frac{\partial^2\overline{w}}{\partial x^2} + \nu\frac{\partial^2\overline{w}}{\partial y^2} + \underline{\frac{3\nu}{5Gh}q_0}\right] \,,$$

$$M_y = D\left[\underline{\frac{6(1-\nu)}{5Gh}\frac{\partial V_y}{\partial y}} - \frac{\partial^2\overline{w}}{\partial y^2} + \nu\frac{\partial^2\overline{w}}{\partial x^2} + \underline{\frac{3\nu}{5Gh}q_0}\right] \,,$$

$$M_{xy} = \frac{Gh^3}{12}\left[\underline{\frac{6}{5Gh}\left(\frac{\partial V_x}{\partial y} + \frac{\partial V_y}{\partial x}\right)} - 2\frac{\partial^2\overline{w}}{\partial x\,\partial y}\right] \,,$$

where

$$q_0 = -P_x\frac{\partial^2\overline{w}}{\partial x^2} - 2P_{xy}\frac{\partial^2\overline{w}}{\partial x\,\partial y} - P_y\frac{\partial^2\overline{w}}{\partial y^2} \,,$$

$D = Eh^3/12(1-\nu^2)$ is the flexural rigidity of the plate,

E = Young's modulus,

G = shearing modulus,

ν = Poisson's ratio.

The boundary conditions for this problem also emanate from the variational equation. Using as an example an edge of the plate perpendicular to the x axis, these conditions are, with primed quantities as prescribed functions:

$$\begin{aligned} M_x &= M_x' \text{ or } \alpha \text{ prescribed} \,, \\ M_{xy} &= M_{xy}' \text{ or } \beta \text{ prescribed} \,, \\ V_x &= V_x' \text{ or } \overline{w} \text{ prescribed} \,. \end{aligned} \tag{7}$$

It should be noted that three boundary conditions must be satisfied instead of two as in the classical theory. Consistent with this, the governing equations are a sixth-order system that yields three constants of integration for each edge. The classical governing equations [which may be obtained from Eqs. (6) by omitting the underlined, or h^2, terms] are a fourth-order system with two constants for each edge.

SOLUTIONS OF THE GOVERNING EQUATIONS IN RECTANGULAR COORDINATES

The solutions for the system of Eqs. (6) can be found from the first three equations on $\overline{w}$, V_x, and V_y, assuming that P_x, P_y, and P_{xy} have been solved from the extensional problem or else are prescribed. The first equation is then a homogeneous one for $\overline{w}$. Assuming $\overline{w}$ has been found, then the second and third are the governing equations for V_x and V_y. Their solutions can be written as the sum of particular and homogeneous parts:

$$V_x = V_{xH} + V_{xP} ,$$
$$V_y = V_{yH} + V_{yP} . \tag{8}$$

If the function $\phi(x, y)$ is introduced, such that

$$V_{xH} = \frac{\partial \phi}{\partial y} , \qquad V_{xH} = -\frac{\partial \phi}{\partial x} , \tag{9}$$

then the governing equation for ϕ is

$$\frac{h^2}{10} \nabla^2 \phi - \phi = 0 . \tag{10}$$

The particular solutions are then given by

$$V_{xP} = V_x^* + \frac{\partial}{\partial x}(-D\nabla^2 \overline{w}) ,$$
$$V_{yP} = V_y^* + \frac{\partial}{\partial y}(-D\nabla^2 \overline{w}) , \tag{11}$$

where V_x^* and V_y^* must satisfy the following equations by virtue of the second and third equations of Eqs. (6):

$$\frac{h^2}{10}\nabla^2 V_x^* - V_x^* = \frac{h^2}{10}\frac{2-\nu}{1-\nu}\left[\frac{h^2}{10}\nabla^2\frac{\partial q_0}{\partial x} - \frac{\partial q_0}{\partial y}\right] ,$$
$$\frac{h^2}{10}\nabla^2 V_y^* - V_y^* = \frac{h^2}{10}\frac{2-\nu}{1-\nu}\left[\frac{h^2}{10}\nabla^2\frac{\partial q_0}{\partial y} - \frac{\partial q_0}{\partial y}\right] . \tag{12}$$

Since the constants of integration arise only for the homogeneous solutions, they are derived from the first equation of Eqs. (6) and Eq. (10).

BUCKLING OF A RECTANGULAR PLATE SIMPLY SUPPORTED ON FOUR EDGES BY A UNIFORM AXIAL LOAD IN THE x DIRECTION

Solution with $M_{xy} \neq 0$ on the Edges

The axes are chosen as shown in Fig. 2. Terms including P_y and P_{xy} in the equations of the paragraph above have been dropped, and for convenience P_x is

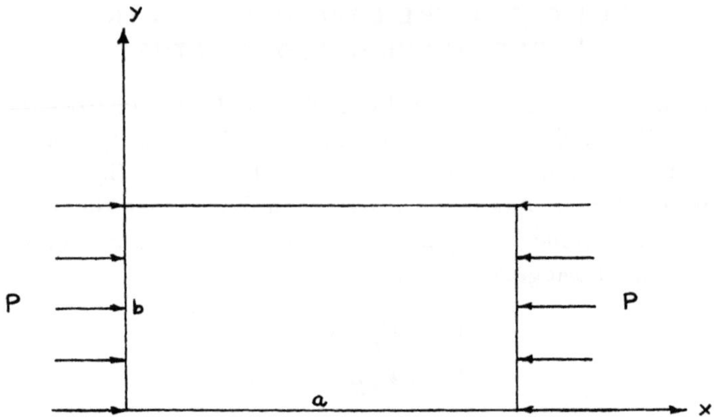

Fig. 2. Rectangular Plate Under Compressive Load.

taken as $-P$. Specifically the governing equation on $\bar{w}$ becomes

$$(1 + C)\frac{\partial^4 \bar{w}}{\partial x^2} + (2 + C)\frac{\partial^4 \bar{w}}{\partial x^2\, \partial y^2} + \frac{\partial^4 \bar{w}}{\partial y^4} + \frac{P}{D}\frac{\partial^2 \bar{w}}{\partial x^2} = 0 , \tag{13}$$

where

$$C = -\frac{P}{D}\frac{b^2}{10}\frac{2 - \nu}{1 - \nu} .$$

By using the separation of variables technique,[4] the solutions for $\bar{w}$, V_x, and V_y can be written as

$$\bar{w} = \sum_{n=1}^{\infty} (C_{1n} \sinh \alpha_n y + C_{2n} \cosh \alpha_n y + C_{3n} \sin \beta_n y + C_{4n} \cos \beta_n y) \sin \lambda_n x ,$$

$$V_x = \sum_{n=1}^{\infty} (A_{1n}\gamma_n \cosh \gamma_n y + A_{2n}\gamma_n \sinh \gamma_n y + B_{1n} \sinh \alpha_n y$$

$$+ B_{2n} \cosh \alpha_n y + B_{3n} \sin \beta_n y + B_{4n} \cos \beta_n y) \cos \lambda_n x ,$$

$$Vy = \sum_{n=1}^{\infty} (\lambda_n A_{1n} \sinh \gamma_n y + \lambda_n A_{2n} \cosh \gamma_n y + D_{1n} \sinh \alpha_n y$$

$$+ D_{2n} \cosh \alpha_n y + D_{3n} \sin \beta_n y + D_{4n} \cos \beta_n y) \sin \lambda_n x ,$$

[4]This is the same as starting with a Levy-type solution. In ref [9], it is shown that both Levy- and Navier-type solutions are also valid in the Reissner theory.

where

$$\alpha_n = \left[\left(1+\frac{C}{2}\right)\lambda_n^2 + \sqrt{\frac{C^2}{4}\lambda_n^4 + \frac{P}{D}\lambda_n^2}\right]^{1/2},$$

$$\beta_n = \left[-\left(1+\frac{C}{2}\right)\lambda_n^2 + \sqrt{\frac{C^2}{4}\lambda_n^4 + \frac{P}{D}\lambda_n^2}\right]^{1/2},$$

$$(B_{1n}, B_{2n}) = \frac{-\lambda_n^3\left[D - \dfrac{h^2}{10}\dfrac{P}{1-\nu}\right] + D\lambda_n\alpha_n^2}{\dfrac{h^2}{10}(\alpha_n^2 - \lambda_n^2) - 1}(C_{1n}, C_{2n}),$$

$$(B_{3n}, B_{4n}) = \frac{-\lambda_n^3\left[D - \dfrac{h^2}{10}\dfrac{P}{1-\nu}\right] + D\lambda_n\beta_n^2}{-\dfrac{h^2}{10}(\beta_n^2 + \lambda_n^2) - 1}(C_{3n}, C_{4n}), \qquad (14)$$

$$(D_{1n}, D_{2n}) = \frac{-\lambda_n^2\alpha_n\left[D - \dfrac{h^2}{10}\dfrac{P}{1-\nu}\right] + D\alpha_n^3}{\dfrac{h^2}{10}(\alpha_n^2 - \lambda_n^2) - 1}(C_{2n}, C_{1n}),$$

$$(D_{3n}, D_{4n}) = \frac{\lambda_n^2\beta_n\left[D - \dfrac{h^2}{10}\dfrac{P}{1-\nu}\right] + D\beta_n^3}{-\dfrac{h^2}{10}(\beta_n^2 + \lambda_n^2) - 1}(C_{4n}, -C_{3n}),$$

$$\gamma_n = \left[\frac{10}{h^2} + \lambda_n^2\right]^{1/2},$$

and where $A_{1n}, A_{2n}, C_{1n} \ldots C_{4n}$ are independent constants of integration.

A simply supported edge in the classical theory is defined by zero deflection and zero moment. In the present theory, a third condition must be specified. It may be $M_{xy} = 0$, or the rotation of line elements in the face of the edge may be taken to be zero. Choosing the latter, then $M_{xy} \neq 0$ on the simply supported edge just as in the case in the classical theory.

Therefore the boundary conditions of the four edges are

$$\begin{aligned} \bar{w}(0, y) &= \bar{w}(a, y) = 0 , \\ M_x(0, y) &= M_x(a, y) = 0 , \\ \beta(0, y) &= \beta(a, y) = 0 , \\ \bar{w}(x, 0) &= \bar{w}(x, b) = 0 , \\ M_y(x, 0) &= M_y(x, b) = 0 , \\ \alpha(x, 0) &= \alpha(x, b) = 0 . \end{aligned} \tag{15}$$

The first six boundary conditions will be satisfied by taking $\lambda_n = n\pi/a$ ($n = 1, 2, \ldots$), and the last six will yield six homogeneous algebraic equations for the six unknown constants. Upon setting the determinant of the coefficients of these six equations equal to zero, the following characteristic equation arises:

$$\sinh \alpha_n b \sinh \gamma_n b \sin \beta_n b \left\{ \frac{6(1-\nu)}{5Gh} \left[\lambda_n (b_{3n} - b_{1n}) + (\beta_n d_{4n} - \alpha_n d_{2n}) \right] - (\beta_n^2 - \alpha_n^2) \right\}^2 = 0 , \tag{16}$$

where

$$b_{1n} = \frac{B_{1n}}{C_{1n}} , \qquad b_{3n} = \frac{B_{3n}}{C_{3n}} , \qquad d_{2n} = \frac{D_{2n}}{C_{1n}} , \qquad d_{4n} = \frac{D_{4n}}{C_{3n}} .$$

After considerable manipulation, the term in the braces reduces to

$$\left\{ P + \frac{100D(1-\nu)}{h^4 \lambda_n^2} \right\} ,$$

and since this cannot be zero unless P is tensile, then

$$\sin \beta_n b = 0 ,$$

and the critical loads are therefore

$$P = \frac{\dfrac{Da^2\pi^2}{n^2} \left[\left(\dfrac{m}{b}\right)^2 + \left(\dfrac{n}{a}\right)^2 \right]^2}{1 + \dfrac{h^2}{10} \dfrac{2-\nu}{1-\nu} \left[\left(\dfrac{m\pi}{b}\right)^2 + \left(\dfrac{n\pi}{a}\right)^2 \right]} . \tag{17}$$

The corresponding equation from the classical theory is

$$P_c = \frac{\pi^2 a^2 D}{n^2} \left(\frac{n^2}{a^2} + \frac{m^2}{b^2} \right)^2 . \tag{18}$$

For the first mode, $m = n = 1$, and the ratio of the classical load to the new buckling load for a square plate of side b is

$$\frac{P_c}{P} = 1 + \frac{\pi^2}{5}\,\frac{2-\nu}{1-\nu}\,\frac{h^2}{b^2}\,. \tag{19}$$

With $\nu = 1/3$, the curve of P_c/P vs h/b is plotted in Fig. 3. It is to be noted that the correction term becomes important for $h/b > \sim 10$. For higher modes of buckling the difference between P_c and P is even greater.

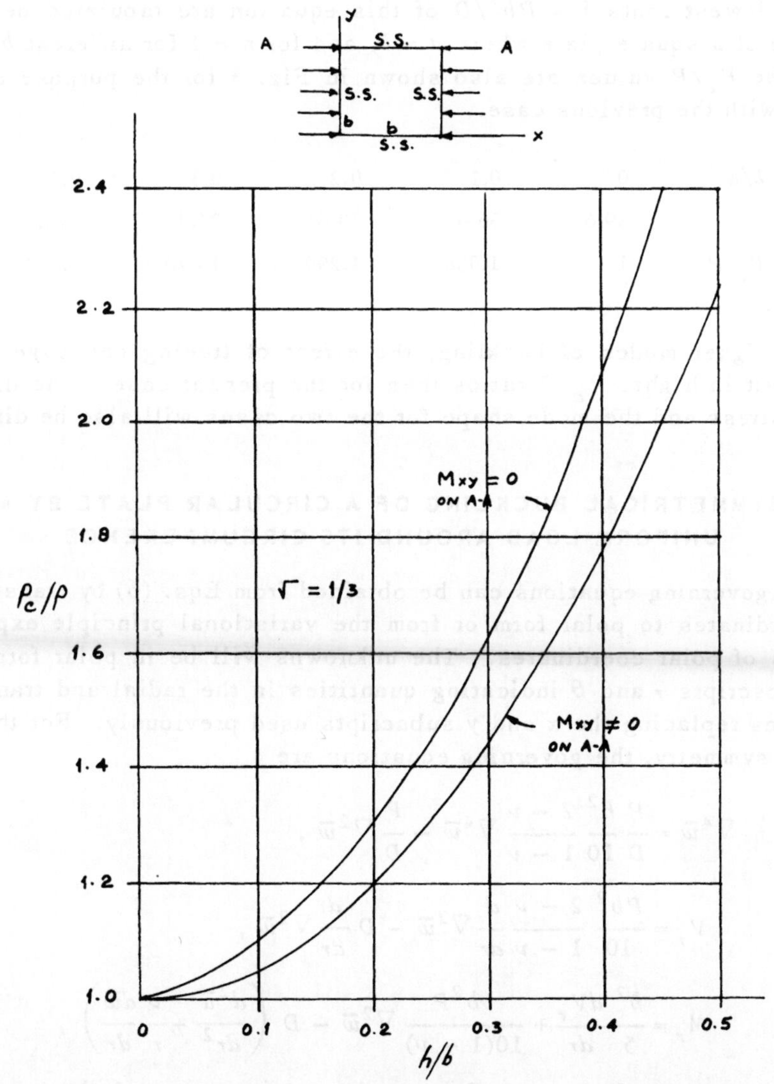

Fig. 3. Critical Loads for Simply Supported Edges.

Solution of the Same Problem with One Edge Free of M_{xy}

The boundary conditions given above in Eqs. (15) will be the same except that $\alpha(x, b) = 0$ will be replaced by $M_{xy}(x, b) = 0$. By using the solutions above with the new boundary conditions, the determinant for determining the critical load is

$$\sin \beta_n b \cosh \gamma_n b \sinh \alpha_n b \, [(\gamma_n^2 + \lambda_n^2)(\beta_n^2 + \alpha_n^2)]$$
$$+ \sinh \alpha_n b \cos \beta_n b \sinh \alpha_n b(2\lambda_n^2 \beta_n \gamma_n)$$
$$- \sin \beta_n b \cosh \alpha_n b \sinh \gamma_n b(2\gamma_n^2 \alpha_n \gamma_n) = 0 \, . \qquad (20)$$

The lowest roots $k = Pb^2/D$ of this equation are tabulated below for the case of a square plate where $a = b$, and for $n = 1$ for different h/b values. The P_c/P values are also shown in Fig. 3 for the purpose of comparison with the previous case.

h/a	0	0.1	0.2	0.3	0.5
k	39.6	36.4	30.65	24.4	14.8
P_c/P	1	1.088	1.290	1.620	2.680

For higher modes of buckling, the effect of freeing one edge of M_{xy} will result in higher P_c/P ratios than for the present case. The distribution of stress and the mode shape for the two cases will also be different.

SYMMETRICAL BUCKLING OF A CIRCULAR PLATE BY A UNIFORM LOAD AROUND ITS CIRCUMFERENCE

The governing equations can be obtained from Eqs. (6) by transforming the coordinates to polar form or from the variational principle expressed in terms of polar coordinates. The unknowns will be in polar form also, with subscripts r and θ indicating quantities in the radial and transverse directions replacing the x and y subscripts used previously. For the case of axial symmetry, the governing equations are

$$\nabla^4 \overline{w} = \frac{P}{D}\frac{h^2}{10}\frac{2-\nu}{1-\nu}\nabla^4 \overline{w} - \frac{P}{D}\nabla^2 \overline{w} \, ,$$

$$V_r = \frac{Ph^2}{10}\frac{2-\nu}{1-\nu}\frac{d}{dr}\nabla^2 \overline{w} - D\frac{d}{dr}\nabla^2 \overline{w} \, ,$$

$$M_r = \frac{h^2}{5}\frac{dV_r}{dr} + \frac{\nu h^2 P}{10(1-\nu)}\nabla^2 \overline{w} - D\left(\frac{d^2 \overline{w}}{dr^2} + \frac{\nu}{r}\frac{d\overline{w}}{dr}\right) ,$$

$$M_\theta = \frac{h^2}{5}\frac{V_r}{r} + \frac{\nu h^2 P}{10(1-\nu)}\nabla^2 \overline{w} - D\left(\frac{1}{r}\frac{d\overline{w}}{dr} + \nu\frac{d^2 \overline{w}}{dr^2}\right) ,$$

$$\alpha_r + \frac{d\bar{w}}{dr} = \frac{6}{5\mathrm{Gh}} V_r , \tag{21}$$

where

$$\nabla^2 = \frac{1}{r}\frac{d}{dr} + \frac{d^2}{dr^2} .$$

The equation on $\bar{w}$ can be written as

$$\nabla^4 \bar{w} + k^2 \nabla^2 \bar{w} = 0 , \qquad \text{or} \qquad \nabla^2(\nabla^2 + k^2)\bar{w} = 0 , \tag{22}$$

where

$$b^2 = \frac{h^2}{10}\frac{2-\nu}{1-\nu} , \qquad \text{and} \qquad k^2 = \frac{P}{D - b^2 P} .$$

Its general solution is found by setting each of the factors operating on $\bar{w}$ equal to zero independently and integrating each. Adding these two solutions, the general solution on $\bar{w}$ is

$$\bar{w} = C_1 + C_2 \ln r + C_3 J_0(kr) + C_4 Y_0(kr) , \tag{23}$$

where C_1, C_2, C_3, and C_4 are constants of integration, and J_0 and Y_0 are Bessel functions of zero order of the first and second kind respectively. For a solid circular plate with no hole, which is being considered here, the solutions $\ln r$ and $Y_0(kr)$ will be discarded since they become infinite when r is equal to zero, and the solution sought should be finite and continuous everywhere.

The Clamped Plate. – With the clamped edge ($r = a$) defined by the boundary conditions

$$\bar{w}(a) = 0 , \qquad \alpha_r(a) = 0 ,$$

and the solution for $\bar{w}$ as

$$\bar{w} = C_1 + C_3 J_0(kr) ,$$

the characteristic equation for determining the critical loads becomes

$$J_1(ka)\left[\frac{6}{5Gh}(b^2 P - D)k^2 + 1\right] = 0 . \tag{24}$$

The procedure for determining the above equation is the same as for the case of the rectangular buckling loads.

For the vanishing of Eq. (24), there are two possibilities, either

$$J_1(ka) = 0 ,$$

or

$$-\frac{6P}{5Gh} + 1 = 0 .$$

By using the first possibility, the roots of $J_1(ka) = 0$ are

$$3.83\,, \qquad 7.02\,, \qquad 10.2\,, \qquad 13.3\,,$$

and the critical loads and mode shapes can be found from these. Use of lowest root gives the first mode and the first critical load P. For $\nu = 1/3$, this can be written as

$$\frac{P_c}{P} = 1 + 3.67\frac{h^2}{a^2}\,,$$

where $P_c = (3.83)^2 D/a^2$ is the corresponding critical load as calculated from the classical theory. It can be seen that the critical load P_c is always greater than that calculated by including the effect of shear deformation. The ratio P_c/P for the first and second modes is shown in Fig. 4.

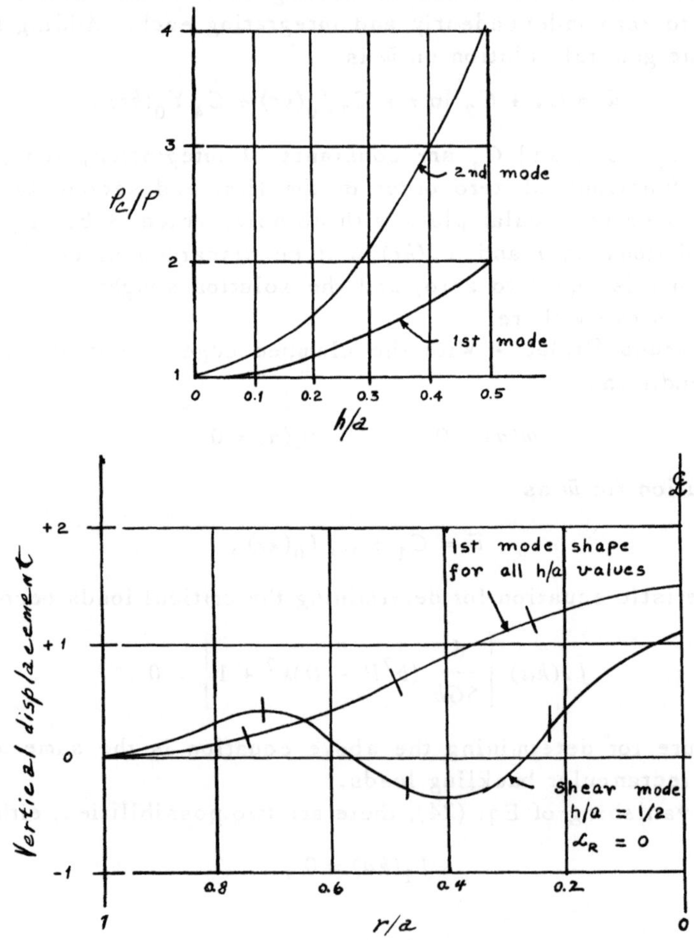

Fig. 4. Mode Shapes and Displacement, Clamped Circular Plate.

The second possibility $P = 5Gh/6$ gives rise to a mode of buckling which is not evident in the classical theory. This will be called a *shear mode of buckling.* For the sake of comparison, the shear-mode shape and the ordinary first-mode shape are shown in Fig. 4, where the vertical displacement is plotted against r/a. A consideration of α_r gives a clearer picture of the mode shapes, since α_r for the shear mode is zero everywhere, while for the normal mode it is zero only at the edge and center of the plate. The value of V_r for these two modes is also different. In fact, the ordinary modes take place with no shear on the boundary, while the shear mode has a nonzero value there. The distributions of V_r are also shown in Fig. 5. From a practical point of view, the shear mode is unlikely to occur, because its critical load is much higher than the buckling loads for the lower normal modes.

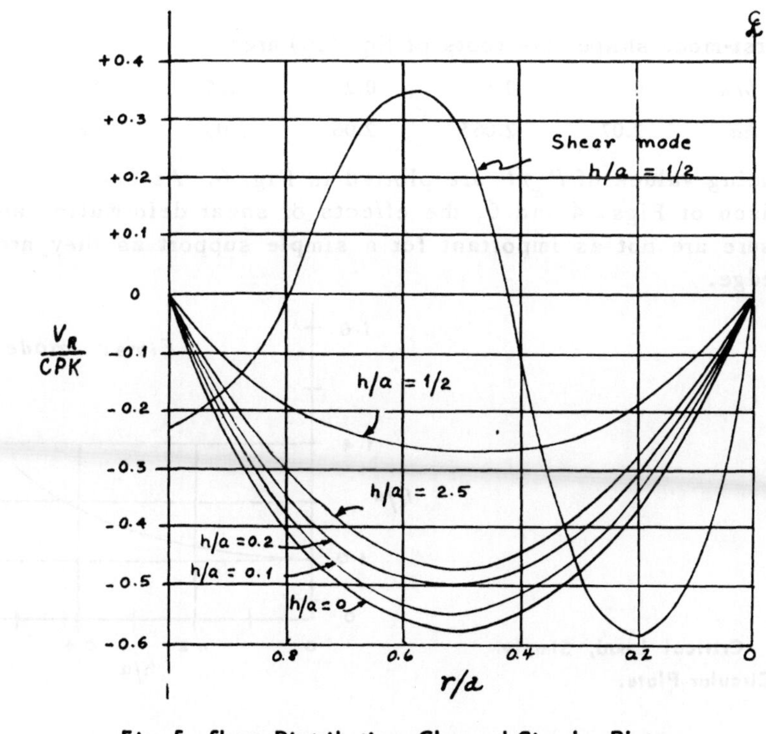

Fig. 5. Shear Distribution, Clamped Circular Plate.

The Simply Supported Plate. – By using the same procedure again, with boundary conditions

$$w(a) = 0 \quad \text{and} \quad M_r(a) = 0 ,$$

the characteristic equation for determining the buckling loads becomes

$$\left[\frac{h^2}{5}\frac{P}{D}ka - \frac{\nu^2 h^2}{10(1-\nu)}\frac{P}{D}ka + ka\right] J_0(ka)$$
$$+ \left(-\frac{h^2}{5}\frac{P}{D} - 1 + \nu\right) J_1(ka) = 0 . \qquad (25)$$

Or, with $\nu = 1/3$, this can be written as

$$\left(\frac{3}{20}\frac{Ph^2}{D} + 1\right) kaJ_0(ka) - \left(\frac{Ph^2}{5D} + \frac{2}{3}\right) J_1(ka) = 0 , \qquad (26)$$

where

$$\frac{Ph^2}{D} = \frac{1}{\frac{1}{4} + \frac{1}{(ka)^2}\left(\frac{a}{h}\right)^2} .$$

For the first-mode shape, the roots of Eq. (26) are:

h/a	0	0.1	0.2	0.25	0.5
ka	2.07	2.065	2.06	2.05	2.025

Corresponding values of P_c/P are plotted in Fig. 6. As can be seen from a comparison of Figs. 4 and 6, the effects of shear deformation and normal pressure are not as important for a simple support as they are for a clamped edge.

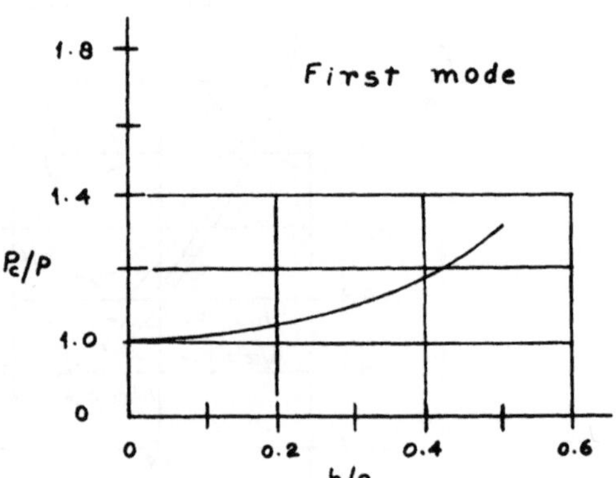

Fig. 6. Critical Load, Simply Supported Circular Plate.

The Circular Plate Clamped on the Outer Edge and Fixed Against Rotation on the Inner Edge. – The inner core is free to move in a vertical direction, but the inner edge is always fixed against rotation as shown in Fig. 7. The boundary conditions are

$$\bar{w}(a) = 0 , \qquad \alpha_r(a) = 0 ,$$
$$v_r(b) = 0 , \qquad \alpha_r(b) = 0 .$$

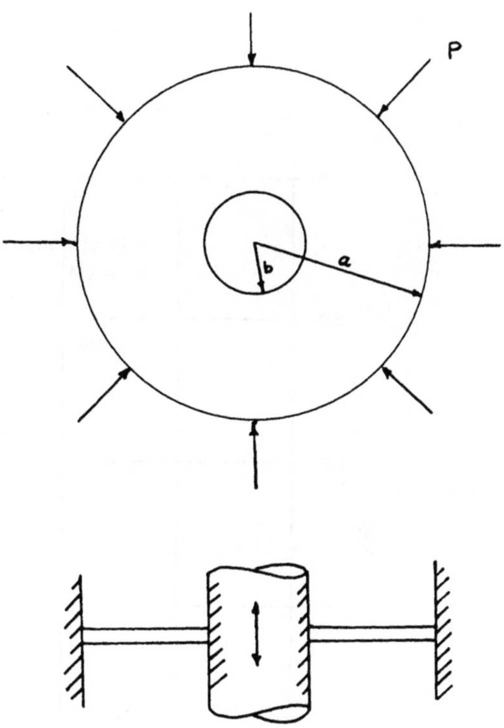

Fig. 7. Circular Plate, Movable Rigid Insert.

For this problem, the complete solution for $\overline{w}$ in Eq. (23) will be used since the radius does not approach the origin. Upon setting the fourth-order determinant which arises from the boundary conditions equal to zero, the following characteristic equation arises:

$$\left[\frac{6}{5Gh}(b^2P - D)k^3 - k\right]\left[\frac{6}{5Gh}(b^2P - D)k^3\right][J_1(ka)Y_1(ka) - J_1(kb)Y_1(ka)] = 0 \,. \qquad (27)$$

If the first bracketed term vanishes, the critical load for the shear mode arises, or $P = 5Gh/6$. Vanishing of the second bracketed term gives the trivial case, $P = 0$. Finally, the roots of the third term, or

$$J_1(ka)Y_1(kb) - J_1(kb)Y_1(ka) = 0 \,, \qquad (28)$$

give the critical loads. The roots of (28) are given by Jahnke and Emde [*10*] for the six modes. The values of P_c/P are plotted in Fig. 8 for the first mode only. As the ratio a/b approaches 1, the shear effects are very important and the critical values of P as calculated from the classical theory are greatly in error. As $a/b \to \infty$, the curve approaches the one for the clamped plate without a hole.

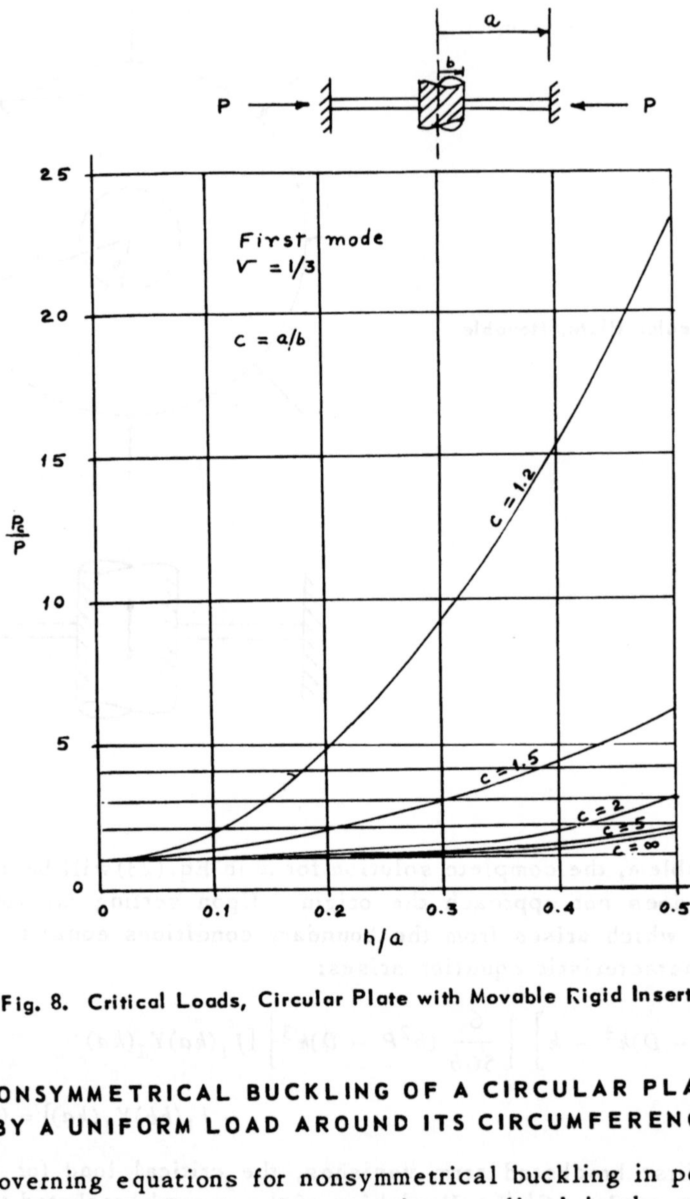

Fig. 8. Critical Loads, Circular Plate with Movable Rigid Insert.

NONSYMMETRICAL BUCKLING OF A CIRCULAR PLATE BY A UNIFORM LOAD AROUND ITS CIRCUMFERENCE

The governing equations for nonsymmetrical buckling in polar coordinates were derived following the procedure outlined in the previous article. They are

$$\nabla^4 \overline{w} = \frac{b^2 P}{D} \nabla^4 \overline{w} - \frac{P}{D} \nabla^2 \overline{w} \, ,$$

$$\frac{b^2}{10} \nabla^2 V_r - V_r + \frac{b^2}{10} \left(\frac{2}{r} \frac{\partial V_r}{\partial r} + \frac{V_r}{r^2} \right) = D \frac{\partial}{\partial r} \nabla^2 \overline{w}$$

$$+ \frac{b^2 P}{10(1-\nu)} \frac{\partial}{\partial r} \nabla^2 \overline{w} + \frac{b^2 D}{5r} \nabla^2 \overline{w} \, ,$$

$$\frac{h^2}{10}\nabla^2 V_\theta - V_\theta + \frac{h^2}{10}\left(\frac{2}{r^2}\frac{\partial V_r}{\partial \theta} - \frac{V_\theta}{r^2}\right) = D\frac{1}{r}\frac{\partial}{\partial \theta}\nabla^2 \overline{w} + \frac{h^2 P}{10(1-\nu)}\frac{1}{r}\frac{\partial}{\partial \theta}\nabla^2 \overline{w} ,$$

$$M_r = \frac{h^2}{5}\frac{\partial V_r}{\partial r} + \frac{\nu h^2 P}{10(1-\nu)}\nabla^2 \overline{w} - D\left(\frac{\partial^2 \overline{w}}{\partial r^2} + \frac{\nu}{r}\frac{\partial \overline{w}}{\partial r} + \frac{\nu}{r^2}\frac{\partial^2 \overline{w}}{\partial \theta^2}\right),$$

$$M_\theta = \frac{h^2}{5}\left(\frac{1}{r}\frac{\partial V_\theta}{\partial \theta} + \frac{V_r}{r}\right) + \frac{\nu h^2 P}{10(1-\nu)}\nabla^2 \overline{w} - D\left(\frac{1}{r}\frac{\partial \overline{w}}{\partial r} + \frac{1}{r^2}\frac{\partial^2 \overline{w}}{\partial \theta^2} + \nu\frac{\partial^2 \overline{w}}{\partial r^2}\right),$$

$$M_{r\theta} = \frac{h^2}{10}\left[\frac{1}{r}\frac{\partial V_r}{\partial \theta} + \frac{\partial}{\partial r}\left(\frac{V_\theta}{r}\right)\right] - D(1-\nu)\frac{\partial}{\partial r}\left(\frac{1}{r}\frac{\partial \overline{w}}{\partial \theta}\right),$$

$$\alpha_r + \frac{\partial \overline{w}}{\partial r} = \frac{6}{5Gh}V_r ,$$

$$\alpha_\theta + \frac{1}{r}\frac{\partial \overline{w}}{\partial \theta} = \frac{6}{5Gh}V_\theta , \tag{29}$$

where

$$\nabla^2 = \frac{1}{r}\frac{\partial}{\partial r}\frac{\partial^2}{\partial r^2} + \frac{1}{r^2}\frac{\partial^2}{\partial \theta^2} .$$

The solutions to be sought here are of the Levy form, which will permit the expansion of boundary conditions in a Fourier series in the variable θ. Following the techniques[5] employed in previous sections (see "Solutions of the Governing Equations in Rectangular Coordinates" and "Buckling of a Rectangular Plate Simply Supported on Four Edges by a Uniform Axial Load in the x Direction"), the most general periodic solutions for the functions $\overline{w}$, V_r, and V_θ are

$$\overline{w} = \sum_{n=1}^{\infty}[C_{1n}r^n + C_{2n}r^{-n} + C_{3n}J_n(kr) + C_{4n}Y_n(kr)]\sin n\theta$$

$$+ \sum_{n=0}^{\infty}[C_{5n}r^n + C_{6n}r^{-n} + C_{7n}J_n(kr) + C_{8n}Y_n(kr)]\cos n\theta ,$$

$$V_r = \frac{1}{r}\frac{\partial \phi}{\partial r} + \sum_{n=1}^{\infty} B\left[\left(-\frac{nJ_n}{kr} + J_{n+1}\right)C_{3n} + \left(-\frac{nY_n}{kr} + Y_{n+1}\right)C_{4n}\right]\sin n\theta$$

$$+ \sum_{n=0}^{\infty} B\left[\left(-\frac{nJ_n}{kr} + J_{n+1}\right)C_{7n} + \left(\frac{nY_n}{kr} + Y_{n+1}\right)C_{8n}\right]\cos n\theta ,$$

---

[5]These are explained in detail in ref [8].

$$V_\theta = -\frac{\partial\phi}{\partial r} + \sum_{n=1}^{\infty} B\left(\frac{nJ_n}{kr} C_{7n} + \frac{nY_n}{kr} C_{8n}\right) \sin n\theta$$

$$+ \sum_{n=1}^{\infty} B\left(-\frac{nJ_n}{kr} C_{3n} - \frac{nY_n}{kr} C_{4n}\right) \cos n\theta \;, \quad (30)$$

where

$$k^2 = \frac{P}{D - b^2 P} \;, \qquad B = \frac{k^3\left[D - \dfrac{Pb^2}{10(1-\nu)}\right]}{\left(\dfrac{b^2}{10} k^2 + 1\right)} \;,$$

$$\phi = \sum_{n=0}^{\infty} \left[C_{9n} I_r\left(\frac{\sqrt{10}}{b} r\right) + C_{10n} K_n\left(\frac{\sqrt{10}}{b} r\right)\right] \cos n\theta$$

$$+ \sum_{n=1}^{\infty} \left[C_{11n} I_n\left(\frac{\sqrt{10}}{b} r\right) + C_{12n} K_n\left(\frac{\sqrt{10}}{b} r\right)\right] \sin n\theta \;,$$

and where $C_{1n} \ldots C_{12n}$ are constants of integration. In the expression for ϕ, I_n and K_n are modified Bessel functions of the first and second kind respectively. Note that there are now six constants of integration for each edge, three to be used with the sine terms and three with the cosine terms.

First Antisymmetric Mode of a Simply Supported Plate with $\alpha_\theta = 0$

Consider a circular plate with no hole of radius a, where the simple support will be defined by the boundary conditions

$$\overline{w}(a) = 0 \;, \qquad M_r(a) = 0 \;, \qquad \text{and} \qquad \alpha_\theta(a) = 0 \;.$$

Using only the part of the solution corresponding to $\sin\theta$ and following the usual procedure, the characteristic equation is

$$J_1(ka)\left\{\frac{Bb^2}{5a^2k^3} - \left[\frac{P\nu b^2}{10(1-\nu)} - D\nu\right]\right\} - J_1''(ka)\left[\frac{b^2 B}{5k} + D(1-\nu)\right] = 0 \;, \quad (31)$$

where the primes indicate differentiations with respect to the argument of the function. Or, with $\nu = 1/3$ and the usual identities for expressing the second derivative of the Bessel function in terms of the zero and first-order functions, Eq. (31) can be written as

$$\frac{J_0(ka)}{J_1(ka)} = \frac{2}{ka} - ka + (ka)\,\frac{\dfrac{h^2}{5a^2} + \dfrac{1}{3}\left[\dfrac{h^2}{10a^2}(ka)^2 + 1\right]}{\dfrac{h^2}{30a^2}(ka)^2 - \dfrac{2}{3}}\,. \qquad (32)$$

The lowest roots of this equation for different h/a ratios are given in the table below along with the corresponding P_c/P values.

h/a	0	0.1	0.2	0.3	0.5
ka	3.635	3.635	3.640	3.645	3.660
P_c/P	1	1.0329	1.128	1.290	1.810

First Antisymmetric Mode of a Simply Supported Edge Defined by $M_{r\theta} = 0$

In order to determine the effect of $M_{r\theta}$ on a simple support, the example above is considered again with the boundary condition $\alpha_\theta = 0$ replaced by $M_{r\theta} = 0$. The characteristic equation becomes

$$J_0(ka)(MG + HN) = J_1(ka)\left[\frac{2}{ka}\,GM + HN\left(\frac{2}{ka} - ka\right) - \frac{H}{3}\left(\frac{h^2k^2}{10} + 1\right)(ka)\right], \qquad (33)$$

where

$$M = -\frac{2}{3} + \frac{7}{30}\,h^2k^2\,,$$

$$N = \frac{2}{3} - \frac{1}{30}\,h^2k^2\,,$$

$$G = \frac{\sqrt{10}h}{a}\,I_1'\left(\frac{\sqrt{10}}{h}\,a\right),$$

$$H = -\left(10 + \frac{4h^2}{a^2}\right) I_1\left(\frac{\sqrt{10}}{h}\,a\right) + \frac{2\sqrt{10}\,h}{a}\,I_0\left(\frac{\sqrt{10}}{h}\,a\right).$$

The lowest roots and corresponding P_c/P values are given below.

h/a	0	0.1	0.2	0.3	0.5
ka	3.635	3.635	3.630	3.630	3.680
P_c/P	1	1.0329	1.135	1.300	1.807

These results differ from those of the example above by less than 1% for all b/a values. Although the buckling loads are nearly the same, the distribution of stress and the buckling modes will be different.

Problems in the buckling of ring-shaped plates can be treated by the equations of this section without any additional difficulty.

CONCLUSION

The thick-plate theory developed by Reissner, which includes the effects of shear deformation and normal pressure, has been used to develop the governing equations for the buckling of plates in rectangular and polar coordinates. Several solutions of these are presented, and numerical results for specific examples are compared with the classical theory.

The effects of shear deformation and normal stress become important at a certain value of a thickness-length parameter which can be obtained from the numerical examples. In addition, the freeing of one edge of a simply supported rectangular plate of twisting moment is significant.

REFERENCES

[1] Eric Reissner, "On the Theory of Bending of Elastic Plates," *J. Math. and Phys.* **23**, 184–91 (1944).

[2] Eric Reissner, "The Effect of Transverse Shear Deformation on the Bending of Elastic Plates," *J. Appl. Mechanics* **12**(2), 69–77 (1945).

[3] Eric Reissner, "On a Variational Theorem in Elasticity," *J. Math. and Phys.* **29**, 90–95 (1950).

[4] S. P. Timoshenko and K. S. Woinowsky, *Theory of Plates and Shells*, McGraw-Hill, New York, 1959.

[5] Daniel Frederick, "On Some Problems in Bending of Thick Plates," Ph.D. Dissertation, University of Michigan, 1955.

[6] *The Table of Bessel Functions* $J_0(z)$ *and* $J_1(z)$ *for Complex Arguments*, prepared by the Mathematical Tables Project of the Federal Works Agency, Columbia University Press, 1943.

[7] *The Table of Bessel Functions* $Y_0(z)$ *and* $Y_1(z)$ *for Complex Arguments*, prepared by the Computation Laboratory, National Bureau of Standards, Columbia University Press, 1950.

[8] Daniel Frederick, "On Some Problems in the Bending of Thick Circular Plates on an Elastic Foundation," *J. Appl. Mechanics* **23**(2), 195 (1956).

[9] Daniel Frederick, "Thick Rectangular Plates on an Elastic Foundation," *Trans. Am. Soc. Civil Engrs.* **122**, 1069 (1957).

[10] Jahnke and Emde, *Tables of Functions*, Dover Publications, 1945.

STRESSES AND DEFORMATIONS INDUCED IN COMPOSITE CYLINDERS BY TRANSVERSE BODY FORCES[1]

Gerald F. Gillis
Rohm & Haas Company
Redstone Arsenal Research Division

An analysis was made of the stresses and displacements in two concentric, hollow, circular cylinders which were intimately bonded at their interface, loaded by transverse body forces, and supported by a normal load uniformly distributed over an angle of 2α on the exterior surface of the outer cylinder. The analysis was made by using the complex variable method of N. I. Muskhelishvili and assuming that the materials of both cylinders were elastic, homogeneous, isotropic, and in a state of plane strain.

As a practical application of the solution, the stresses are determined for a solid-propellant rocket motor lying on its side and supported by a load distributed over a very small angle.

INTRODUCTION

A theoretical analysis is presented of the stresses induced in two composite cylinders by transverse body forces. It was assumed that the body consisted of two infinitely long, concentric, hollow, circular cylinders perfectly bonded together at the interface and supported by a normal load uniformly distributed over an angle of 2α along the entire length of the unit (Fig. 1). In the limit as α approaches zero, this support corresponds to a line load parallel to the longitudinal axes of the cylinders. It was further assumed that the theory of infinitesimal deformations was valid and that both cylinders could be characterized as being composed of elastic, homogeneous, and isotropic materials.

Each cylinder was analyzed separately by representing the unknown loading along the interface with an infinite complex Fourier series with unknown coefficients and employing the circular ring solution of Muskhelishvili [*1*]. The series coefficients were then evaluated by satisfying the boundary conditions for the composite unit and by equating displacements and normal and shearing stresses along the common boundary.

[1]Sponsored by the Army Ordnance Corps under contract No. DA-01-021-ORD-12024.

NOTATION

r, θ	Polar coordinates
a_k, b_k, A_k, B_k	Undetermined complex constants
E_k	Undetermined real constant
G	Modulus of elasticity in shear
g_k	$\dfrac{a_k}{\delta} R_2^{k-1}$
i	As a subscript, denotes values pertaining to the inner cylinder
j_{-k}	$\dfrac{a_{-k}}{\delta} R_2^{-k-1} \zeta^{-2k+2}$
L_1, L_2	Inner and outer boundaries respectively
o	As a subscript, denotes values pertaining to the outer cylinder. This symbol should not be confused with the number 0 used to indicate values of series coefficients when the summation index k is equal to zero
q	Intensity of uniformly distributed load
R_1	Inner radius of inner cylinder
R_2	Outer radius of inner cylinder and inner radius of outer cylinder
R_3	Outer radius of outer cylinder
V_r, V_θ	Radial and circumferential body forces respectively
u, v	Radial and circumferential components of displacements respectively
*	As a superscript, denotes stresses and displacements in problems where the body forces are zero
λ	Lame's constant
ν	Poisson's ratio
κ	$3 - 4\nu$
σ_r, σ_θ	Normal stress components
$\tau_{r\theta}$	Shearing stress component
δ, ρ	Density (lbf/in.^{3}) of the inner and outer cylinder respectively
β	G_i/G_o
ζ	R_1/R_2
η	R_3/R_2
ξ	R_1/r
2α	Angle over which the supporting load is distributed
$\gamma_1, \gamma_2, \gamma_3, \gamma_4$	Complex constants of integration

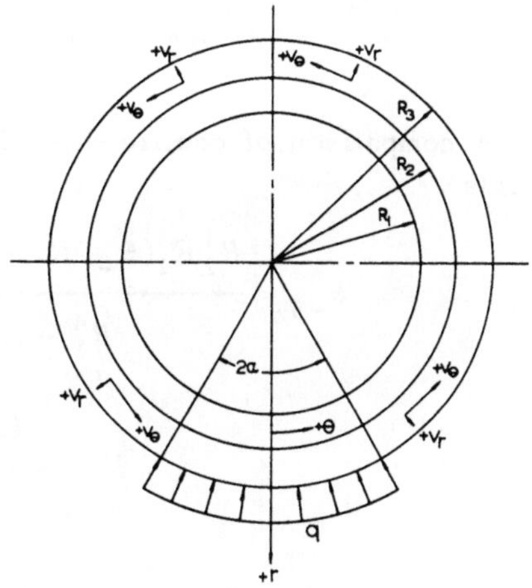

Fig. 1. Origin of Coordinates and Sign Conventions.

MUSKHELISHVILI'S SOLUTION FOR THE CIRCULAR RING

Muskhelishvili proves that, in the absence of body forces, the stresses and displacements in the plane theory of elasticity may be expressed as

$$2G(u^* + iv^*) = a_0 r(\eta - 1) + \bar{b}_{-2} r^{-1} + e^{i\theta}\left(\frac{\eta a_1 r^2}{2} - \bar{a}_{-1} + \frac{\bar{b}_{-3} r^{-2}}{2}\right)$$

$$+ e^{-i\theta}\left[(\eta a_{-1} - \bar{b}_{-1}) \ln \frac{r}{R_2} + i\theta(\eta a_{-1} + \bar{b}_{-1}) - \bar{a}_1 r^2 + \gamma_1 \eta - \bar{\gamma}_2\right]$$

$$+ \sum_{2}^{\infty} e^{ik\theta}\left(\frac{\eta a_k r^{k+1}}{1+k} - \bar{a}_{-k} r^{1-k} + \frac{\bar{b}_{-k-2} r^{-k-1}}{1+k}\right)$$

$$+ \sum_{2}^{\infty} e^{-ik\theta}\left(\frac{\eta a_{-k} r^{1-k}}{1-k} - a_k r^{1+k} + \frac{\bar{b}_{k-2} r^{k-1}}{1-k}\right), \quad (1)$$

$$\sigma_r^* - i\tau_{r\theta}^* = 2a_0 - b_{-2} r^{-2} + e^{i\theta}(\bar{a}_{-1} r^{-1} - b_{-1} r^{-1})$$

$$+ e^{-i\theta}(2a_{-1} r^{-1} + \bar{a}_1 r - b_{-3} r^{-3}) + \sum_{2}^{\infty} e^{ik\theta}[(1-k)a_k r^k - b_{k-2} r^{k-2} + \bar{a}_{-k} r^{-k}]$$

$$+ \sum_{2}^{\infty} e^{-ik\theta}[(1+k)a_{-k} r^{-k} - b_{-k-2} r^{-k-2} + \bar{a}_k r^k], \quad (2)$$

where the bar over a symbol denotes the complex conjugate of that quantity. The asterisk was used in the above equations as a reminder that they apply to problems in which the body forces are negligible.

On L_1 and L_2, the stress boundary conditions may be expressed as infinite complex Fourier series; that is

$$\sigma_r^* - i\tau_{r\theta}^* = \sum_{-\infty}^{\infty} A_k e^{ik\theta}, \quad \text{on } L_1, \quad (3)$$

$$\sigma_r^* - i\tau_{r\theta}^* = \sum_{-\infty}^{\infty} B_k e^{ik\theta}, \qquad \text{on } L_2 . \tag{4}$$

A comparison of coefficients of $e^{ik\theta}$ between Eq. (2) and Eqs. (3) and (4) yields

$$b_{-3} = \frac{R_2^3 R_1^3 (A_{-1} R_2 - B_{-1} R_1)}{R_1^4 - R_2^4} + \frac{2\bar{A}_1 R_1^3 R_2^2}{(1+\eta)(R_1^2 + R_2^2)}, \tag{5}$$

$$b_{-2} = \frac{(B_0 - A_0) R_2^2 R_1^2}{R_2^2 - R_1^2}, \tag{6}$$

$$b_{-1} = \frac{-\eta A_1 R_1}{1+\eta}, \tag{7}$$

$$a_{-1} = \frac{\bar{A}_1 R_1}{1+\eta}, \tag{8}$$

$$a_0 = \frac{R_2^2 B_0 - R_1^2 A_0}{2(R_2^2 - R_1^2)}, \tag{9}$$

$$a_1 = \frac{\bar{A}_{-1} R_1^3 - \bar{B}_{-1} R_2^3}{R_1^4 - R_2^4} - \frac{2A_1 R_1}{(R_1^2 + R_2^2)(1+\eta)}, \tag{10}$$

$$a_k = \frac{(1+k)(R_2^2 - R_1^2) C_k - (R_2^{-2k+2} - R_1^{-2k+2}) \bar{C}_{-k}}{(1-k^2)(R_2^2 - R_1^2)^2 - (R_2^{-2k+2} - R_1^{-2k+2})(R_2^{2k+2} - R_1^{2k+2})},$$

$$k = \pm 2, \pm 3, \pm \ldots, \tag{11}$$

$$b_{k-2} = (1-k) R_1^2 a_k + \bar{a}_{-k} R_1^{-2k+2} - A_k R_1^{-k+2}, \qquad k = \pm 2, \pm 3, \pm \ldots, \tag{12}$$

$$A_1 R_1 = B_1 R_2, \tag{13}$$

where

$$C_k = B_k R_2^{-k+2} - A_k R_1^{-k+2}, \qquad k = \pm 2, \pm 3, \pm \ldots . \tag{14}$$

The solution for a particular problem involving a circular ring geometry may be obtained by comparing the actual boundary conditions to the general ones expressed by Eqs. (3) and (4). Substitution of the A_k's and B_k's so determined into Eqs. (5)–(14) and Eqs. (1) and (2) and separation of the real and imaginary parts will yield the stresses σ_r^* and $\tau_{r\theta}^*$ and the displacements u^* and v^*. The σ_θ^* stress may be obtained from the equilibrium equations.

SOLUTION FOR THE INNER CYLINDER

The complex-variable method of Muskhelishvili is particularly applicable to two-dimensional problems without body forces; however, the method may be extended to include problems involving body forces by making a change of variable which will eliminate them from the equilibrium equations and thus define a modified problem. After solution of the modified problem, one may easily revert to the original problem.

The stresses for the problem being considered in this analysis are caused entirely by body forces. For the inner cylinder, these forces are

$$V_r = \delta \cos \theta , \tag{15}$$

$$V_\theta = -\delta \sin \theta , \tag{16}$$

respectively, where the sign convention illustrated by Fig. 1 was employed. These body forces may be eliminated from the equilibrium equations when written either in terms of stresses or displacements by defining a modified problem such that

$$\sigma_r^* = \sigma_r + \delta r \cos \theta , \tag{17}$$

$$\sigma_\theta^* = \sigma_\theta + \delta r \cos \theta , \tag{18}$$

$$\tau_{r\theta}^* = \tau_{r\theta} , \tag{19}$$

$$u^* = u + \frac{\delta r^2 \cos \theta}{4(\lambda + G)} , \tag{20}$$

$$v^* = v + \frac{\delta r^2 \sin \theta}{4(\lambda + G)} , \tag{21}$$

where the asterisk denotes the stresses and displacements for the modified problem.

A boundary-loading equivalent to the actual stresses along the interface must be applied to the inner cylinder if it is to be analyzed as a separate body. The exact stresses along this boundary are unknown; however, symmetry of the body about the chosen origin (Fig. 1) requires

$$\sigma_r(r, \theta) = \sigma_r(r, -\theta) , \tag{22}$$

$$\tau_{r\theta}(r, \theta) = -\tau_{r\theta}(r, -\theta) , \tag{23}$$

where θ is any angle measured from the origin. Consequently, σ_r, for a given r, must be a function of only the cosine of θ plus a possible constant term, and $\tau_{r\theta}$ must be a function of only the sine of θ. With these facts in mind, the boundary condition along the outer surface of the inner cylinder can be represented by an infinite complex Fourier series with unknown coefficients; that is,

$$\sigma_r - i\tau_{r\theta} = \sum_{-\infty}^{\infty} E_k e^{ik\theta} , \qquad r = R_2 , \tag{24}$$

where E_k is real. The inner surface of the cylinder is stress free; hence

$$\sigma_r - i\tau_{r\theta} = 0\ , \qquad r = R_1\ . \tag{25}$$

The boundary conditions for the modified problem may be obtained by substituting Eqs. (17)–(19) into Eqs. (24) and (25); hence

$$\sigma_r^* - i\tau_{r\theta}^* = \frac{\delta R_1}{2}(e^{i\theta} + e^{-i\theta})\ , \qquad r = R_1\ , \tag{26}$$

$$\sigma_r^* - i\tau_{r\theta}^* = \frac{\delta R_2}{2}(e^{i\theta} + e^{-i\theta}) + \sum_{-\infty}^{\infty} E_k e^{ik\theta}\ , \qquad r = R_2\ . \tag{27}$$

The A_k's and B_k's may now be evaluated by a comparison of coefficients of $e^{ik\theta}$ between Eqs. (3), (4), (26), and (27). Substitution of these values into Eqs. (5)–(14) and the a_k's and b_k's so determined into Eqs. (1) and (2) will yield the stresses and displacements for the modified problem. To revert to the original problem, these stresses and displacements must then be substituted into Eqs. (17)–(21). Such a procedure yields

$$\begin{aligned}
2G_i(u + iv) = {} & \frac{E_0 R_2^2[2R_1^2 + r^2(\eta_i - 1)]}{2r(R_2^2 - R_1^2)} + e^{i\theta}\left[\frac{E_{-1} R_2^3(R_1^4 + r^4\eta_i)}{2R^2(R_2^4 - R_1^4)}\right. \\
& \left. + \frac{\delta R_1^2(R_1^2 R_2^2 - r^4\eta_i - r^2R_1^2 - r^2R_2^2)}{2r^2(1 + \eta_i)(R_1^2 + R_2^2)} + \frac{\delta r^2\eta_i}{4} - \frac{\delta r^2 G_i}{2(\lambda_i + G_i)}\right] \\
& + e^{-i\theta}\left[\frac{\delta R_1^2\eta_i \ln(r/R_2)}{1 + \eta_i} - \frac{\delta r^2}{2} - \frac{E_{-1}R_2^3 r^2}{R_2^4 - R_1^4} + \frac{\delta R_1^2 r^2}{(1 + \eta_i)(R_1^2 + R_2^2)} + \gamma_1\eta_i - \bar{\gamma}_2\right] \\
& + \sum_2^{\infty} e^{ik\theta}\left[\frac{a_k r^{-k-1}}{1 + k}(R_1^{2k+2} + \eta_i r^{2k+2}) + a_{-k}r^{-k-1}(R_1^2 - r^2)\right] \\
& + \sum_2^{\infty} e^{-ik\theta}\left[\frac{a_{-k} r^{k-1}}{1 - k}(R_1^{-2k+2} + \eta_i r^{-2k+2}) + a_k r^{k-1}(R_1^2 - r^2)\right]\ ,
\end{aligned} \tag{28}$$

$$\begin{aligned}
\sigma_r - i\tau_{r\theta} = {} & \frac{E_0 R_2^2(r^2 - R_1^2)}{r^2(R_2^2 - R_1^2)} + e^{i\theta}\frac{\delta(R_1^2 - r^2)}{2r} \\
& + e^{-i\theta}\left[\frac{E_{-1}R_2^3(r^4 - R_1^4)}{r^3(R_2^4 - R_1^4)} + \frac{\delta R_1^2(r^2 - R_1^2)(R_2^2 - r^2)}{r^3(1 + \eta_i)(R_1^2 + R_2^2)}\right] \\
& + \sum_2^{\infty} e^{ik\theta}\,[(1 - k)a_k r^{k-2}(r^2 - R_1^2) + a_{-k}r^{k-2}(r^{-2k+2} - R_1^{-2k+2})] \\
& + \sum_2^{\infty} e^{-ik\theta}[(1 + k)a_{-k}r^{-k-2}(r^2 - R_1^2) + a_k r^{-k-2}(r^{2k+2} - R_1^{2k+2})]\ ,
\end{aligned} \tag{29}$$

where a_k is defined by Eq. (11) and the subscript i was used to denote values pertaining to the inner cylinder.

SOLUTION FOR THE OUTER CYLINDER

The outer cylinder is supported externally by a pressure q uniformly distributed over an angle of 2α and extending the length of the cylinder. This load may be represented by a complex Fourier series as

$$\sigma_r = \sum_{-\infty}^{\infty} D_k e^{ik\theta}, \qquad r = R_3, \tag{30}$$

where

$$D_k = \frac{1}{2\pi}\int_0^{2\pi} f(\theta) e^{ik\theta}\, d\theta, \tag{31}$$

$$f(\theta) = -q, \; -\alpha \leqq \theta \leqq \alpha, \tag{32}$$

$$f(\theta) = 0 \text{ for all other } \theta. \tag{33}$$

Thus

$$\sigma_r = \frac{-q\alpha}{\pi} - \frac{2q}{\pi}\sum_{1}^{\infty} \frac{\cos k\theta \sin k\alpha}{k}. \tag{34}$$

Equation (34), coupled with the requirement that the shearing stress be zero on the outer surface, yields the boundary condition that

$$\sigma_r - i\tau_{r\theta} = \frac{-q\alpha}{\pi} - \frac{2q}{\pi}\sum_{1}^{\infty} \frac{\cos k\theta \sin k\alpha}{k}, \qquad r = R_3. \tag{35}$$

The boundary conditions along the surface $r = R_2$ will be the same as for the inner cylinder; that is,

$$\sigma_r - i\tau_{r\theta} = \sum_{-\infty}^{\infty} E_k e^{ik\theta}, \qquad r = R_2. \tag{36}$$

The body forces for the outer cylinder are identical in form to those for the inner cylinder; hence they may be eliminated from the equilibrium equations by redefining the stresses and displacements as

$$\sigma_r^* = \sigma_r + \rho r \cos\theta, \tag{37}$$

$$\sigma_\theta^* = \sigma_\theta + \rho r \cos\theta, \tag{38}$$

$$\tau_{r\theta}^* = \tau_{r\theta}, \tag{39}$$

$$u^* = u + \frac{\rho r^2 \cos\theta}{4(\lambda_o + G_o)}, \tag{40}$$

$$v^* = v + \frac{\rho r^2 \sin\theta}{4(\lambda_o + G_o)}, \tag{41}$$

where the asterisk denotes stresses and displacements for the modified problem, and the subscript o denotes elastic properties of the outer cylinder.

The boundary conditions for the modified problem, obtained by substituting Eqs. (37)–(39) into (35) and (36), are

$$\sigma_r^* - i\tau_{r\theta}^* = \frac{\rho R_2(e^{i\theta} + e^{-i\theta})}{2} + \sum_{-\infty}^{\infty} E_k e^{ik\theta}, \qquad r = R_2, \tag{42}$$

$$\sigma_r^* - i\tau_{r\theta}^* = \frac{-q\alpha}{\pi} + \frac{\rho R_3(e^{i\theta} + e^{-i\theta})}{2} - \frac{q}{\pi}\sum_{1}^{\infty}\frac{(e^{ik\theta} + e^{-ik\theta})\sin k\alpha}{k}, \qquad r = R_3. \tag{43}$$

A procedure similar to the one outlined in the previous section yields

$$qR_3 \sin\alpha = -E_1 R_2 \pi + \frac{\rho\pi}{2}(R_3^2 - R_2^2), \tag{44}$$

$$\sigma_r - i\tau_{r\theta} = \frac{q\alpha R_3^2(R_2^2 - r^2)}{\pi r^2(R_3^2 - R_2^2)} + \frac{E_0 R_2^2(R_3^2 - r^2)}{r^2(R_3^2 - R_2^2)} + e^{i\theta}\left[\frac{E_1 R_2}{r} + \frac{\rho(R_2^2 - r^2)}{2r}\right]$$

$$+ e^{-i\theta}\left[\frac{2E_1 R_2(r^2 - R_2^2)(R_3^2 - r^2)}{r^3(1 + \eta_o)(R_3^2 + R_2^2)} + \frac{E_{-1} R_2^3(r^4 - R_3^4)}{r^3(R_2^4 - R_3^4)}\right.$$

$$\left. + \frac{\rho R_2^2(r^2 - R_2^2)(R_3^2 - r^2)}{r^3(1 + \eta_o)(R_3^2 + R_2^2)} + \frac{qR_3^3(r^4 - R_2^4)\sin\alpha}{\pi r^3(R_2^4 - R_3^4)}\right]$$

$$+ \sum_{2}^{\infty} e^{ik\theta}\left[(1 - k)d_k r^{k-2}(r^2 - R_3^2)\right.$$

$$\left. + d_{-k} r^{k-2}(r^{-2k+2} - R_3^{-2k+2}) - \frac{qr^{k-2} R_3^{-k+2}\sin k\alpha}{\pi k}\right]$$

$$+ \sum_{2}^{\infty} e^{-ik\theta}\left[(1 + k)r^{-k-2}(r^2 - R_3^2)d_{-k} + r^{-k-2}(r^{2k+2} - R_3^{2k+2})d_k\right.$$

$$\left. - \frac{qR_3^{k+2} r^{-k-2}\sin k\alpha}{\pi k}\right], \tag{45}$$

$$2G_o(u + iv) = \frac{-E_0 R_2^2[2R_3^2 + r^2(\eta_o - 1)]}{2r(R_3^2 - R_2^2)} - \frac{q\alpha R_3^2[2R_2^2 + r^2(\eta_o - 1)]}{2\pi r(R_3^2 - R_2^2)}$$

$$+ e^{+i\theta}\left[\frac{E_{-1} R_2^3(R_3^4 + \eta_o r^4)}{2r^2(R_2^4 - R_3^4)} - \frac{E_1 R_2(r^2 R_2^2 + r^2 R_3^2 + \eta_o r^4 - R_2^2 R_3^2)}{(1 + \eta_o)(R_2^2 + R_3^2)r^2}\right.$$

$$
+\frac{qR_3^3(R_2^4+\kappa_o r^4)\sin\alpha}{2\pi r^2(R_2^4-R_3^4)}-\frac{\rho R_2^2(\kappa_o r^4+r^2R_3^2+r^2R_2^2-R_2^2R_3^2)}{2r^2(1+\kappa_o)(R_3^2+R_2^2)}
$$

$$
+\frac{\rho\kappa_o r^2}{4}-\frac{G_o\rho r^2}{2(\lambda_o+G_o)}\Bigg]+e^{-i\theta}\Bigg[\gamma_3\kappa_o-\bar{\gamma}_4+\left(\ln\frac{r}{R_2}\right)\Bigg(\frac{2E_1R_2\kappa_o}{1+\kappa_o}
$$

$$
+\frac{\rho R_2^2\kappa_o}{1+\kappa_o}\Bigg)-\frac{E_{-1}R_2^3r^2}{R_2^4-R_3^4}+\frac{2E_1R_2r^2}{(R_2^2+R_3^2)(1+\kappa_o)}-\frac{qR_3^3r^2\sin\alpha}{\pi(R_2^4-R_3^4)}-\frac{\rho r^2}{2}
$$

$$
+\frac{\rho r^2R_2^2}{(R_2^2+R_3^2)(1+\kappa_o)}\Bigg]+\sum_2^{\infty}e^{ik\theta}\Bigg[\frac{d_kr^{-k-1}}{1+k}(R_3^{2k+2}+\kappa_o r^{2k+2})
$$

$$
+d_{-k}r^{-k-1}(R_3^2-r^2)+\frac{qR_3^{k+2}r^{-k-1}\sin k\alpha}{\pi k(1+k)}\Bigg]
$$

$$
+\sum_2^{\infty}e^{-ik\theta}\Bigg[d_kr^{k-1}(R_3^2-r^2)+\frac{d_{-k}}{1-k}r^{k-1}(R_3^{-2k+2}+\kappa_o r^{-2k+2})
$$

$$
+\frac{qR_3^{-k+2}r^{k-1}\sin k\alpha}{\pi k(1-k)}\Bigg]. \tag{46}
$$

SOLUTION FOR THE TWO-BODY PROBLEM

The uniqueness of the solution theorem guarantees that there is only one solution corresponding to given surface and body forces. Thus the stress and displacement solutions given by Eqs. (28), (29), (45), and (46) are unique, even though they still contain undetermined constants due to the arbitrariness of the boundary condition used for the surface $r = R_2$. If the assumption is made that the two cylinders are perfectly bonded together along the interface, corresponding stresses and displacements may be equated at this boundary and the undetermined constants evaluated. Substitution of these constants into Eqs. (28), (29), (45), and (46) will yield the stresses and displacements for a composite cylindrical body subjected to transverse body forces with a stress-free inner boundary and a uniform load symmetrically distributed about $\theta = 0$ on the outer boundary. This, of course, is the solution sought in the present analysis. By performing the operations indicated above, one may show that

$$
\frac{E_1}{\delta R_2}=\frac{1}{2}(\zeta^2-1), \tag{47}
$$

$$
E_0=\frac{-q\alpha\eta^2(1-\zeta^2)(1+\kappa_o)\beta}{\pi[(\eta^2-1)(2\zeta^2+\kappa_i-1)+\beta(1-\zeta^2)(2\eta^2+\kappa_o-1)]}, \tag{48}
$$

$$\frac{E_{-1}}{\delta R_2} = \left[\frac{(1-\eta^4)(1-\zeta^4)}{\beta(1-\zeta^4)(\eta^4+\varkappa_o)+(\eta^4-1)(\zeta^4+\varkappa_i)}\right]$$

$$\times\left[\frac{\beta(\zeta^2-1)}{1+\eta^2}-\frac{\zeta^2}{1+\zeta^2}+\frac{1}{2}+\frac{\rho\beta(1-\eta^2)}{2\delta(\eta^2+1)}-\frac{4q\eta^3\beta(1-\nu_o)\sin\alpha}{\pi(1-\eta^4)\delta R_2}\right] \qquad (49)$$

$$\frac{\bar{\gamma}_2}{\delta R_2^2}+\frac{\gamma_3\varkappa_o\beta}{\delta R_2^2}-\frac{\gamma_1\varkappa_i}{\delta R_2^2}-\frac{\bar{\gamma}_4\beta}{\delta R_2^2}=-\frac{1}{2}+\frac{\zeta^2}{(1+\varkappa_i)(\zeta^2+1)}-\frac{\beta(\zeta^2-1)}{(1+\varkappa_o)(1+\eta^2)}$$

$$+\frac{\rho\beta}{\delta}\left[\frac{1}{2}-\frac{1}{(1+\eta^2)(1+\varkappa_o)}\right]-\frac{E_{-1}}{\delta R_2}\left[\frac{(1-\eta^4)-\beta(1-\zeta^4)}{(1-\zeta^4)(1-\eta^4)}\right]$$

$$+\frac{q\eta^3\beta\sin\alpha}{\delta R_2\pi(1-\eta^4)}, \qquad (50)$$

$$(1-k)(1-\zeta^2)a_k+R_2^{-2k}(1-\zeta^{-2k+2})a_{-k}-(1-k)(1-\eta^2)d_k$$

$$-R_2^{-2k}(1-\eta^{-2k+2})d_{-k}=\frac{-qR_3^{-k}\eta^2\sin k\alpha}{\pi k}, \qquad k\geqq 2, \qquad (51)$$

$$(1-\zeta^{2k+2})R_2^{2k}a_k+(1+k)(1-\zeta^2)a_{-k}-R_2^{2k}(1-\eta^{2k+2})d_k$$

$$-(1+k)(1-\eta^2)d_{-k}=\frac{-qR_3^k\eta^2\sin k\alpha}{\pi k}, \qquad k\geqq 2, \qquad (52)$$

$$(\zeta^{2k+2}+\varkappa_i)R_2^{2k}a_k+(1+k)(\zeta^2-1)a_{-k}-R_2^{2k}(\eta^{2k+2}+\varkappa_o)\beta d_k$$

$$-(1+k)(\eta^2-1)\beta d_{-k}=\frac{qR_3^k\eta^2\beta\sin k\alpha}{\pi k}, \qquad k\geqq 2, \qquad (53)$$

$$(\zeta^2-1)(1-k)a_k+(\zeta^{-2k+2}+\varkappa_i)R_2^{-2k}a_{-k}-\beta(\eta^2-1)(1-k)d_k$$

$$-\beta R_2^{-2k}(\eta^{-2k+2}+\varkappa_o)d_{-k}=\frac{qR_3^{-k}\eta^2\beta\sin k\alpha}{\pi k}, \qquad k\geqq 2. \qquad (54)$$

The constants for a line load support may be obtained by substituting Eq. (44) into Eqs. (48)–(54) and taking the limit as α approaches zero.

Simultaneous solution of Eqs. (51)–(54), substitution of the results into Eqs. (29) and (45), and separation of real and imaginary parts will yield equations for the stresses at any point in the body. The displacements may be determined, apart from the rigid-body components contained in γ_1, γ_2, γ_3, and γ_4, in a like manner. These rigid-body components may be evaluated by fixing the displacements at appropriate points in the body and using the relation obtained together with Eq. (50) to determine them.

It should be noted that the equations for the stresses and displacements obtained from the above procedure will contain a semi-infinite series. Muskhelishvili proves that the series involved in his general solution for the circular ring are absolutely and uniformly convergent if the expressions for the boundary tractions on L_1 and L_2 have second-order derivatives with respect to θ which satisfy the Dirichlet conditions, or the somewhat lesser restriction that they be of bounded variation. These conditions are satisfied by the solution for a uniformly distributed load support but not by a line load support solution. In other words, as α approaches zero, q must approach infinity at the same rate if the uniformly distributed load is to reduce to a finite point load; hence Eq. (4) will not be of bounded variation on L_2. The series will still converge for the concentrated-load solution, with the possible exception of points along the boundary on which the load is applied.

Perhaps the easiest method of proving the latter would be to consider a circular ring with a stress-free inner boundary and a loading similar to that of Eq. (34) on the outer boundary. After solving the problem, one may show that the ratio of the $n + 1$ term to the n term of the stress solution is less than unity for all points where $r < R_3$.

NUMERICAL CALCULATIONS FOR A SOLID-PROPELLANT ROCKET MOTOR

The solution previously presented is directly applicable to the slump problem for a solid-propellant rocket motor lying on its side. In particular, the solution may be used to determine if the stresses and strains created in solid-propellant rocket motors during storage are great enough to cause a failure in the bond between the case and the propellant and/or cracking of the propellant. This, of course, is an area of concern that will become even more important with the trend toward thin, lightweight rocket motor cases.

The radial and shearing stresses in the propellant (inner cylinder) may be obtained by separating real and imaginary parts of Eq. (29). These two components and the equations of equilibrium determine the circumferential stress; hence

$$\frac{\sigma_r}{\delta R_2} = \left(\frac{E_0}{\delta R_2}\right)\left(\frac{1-\xi^2}{1-\zeta^2}\right) + \cos\theta \left\{\frac{E_{-1}}{\delta R_2}\left[\frac{\zeta(1-\xi^4)}{\xi(1-\zeta^4)}\right] - \frac{\zeta(1-\xi^2)}{2\xi} + \frac{\zeta(\xi^2-\zeta^2)(1-\xi^2)}{\xi(1+\eta_i)(1+\zeta^2)}\right\} + \sum_2^\infty \cos k\theta \left\{g_k\left(\frac{\zeta}{\xi}\right)^k [2 - k - \xi^2(1-k) - \xi^{2k+2}] + j_{-k}\left(\frac{\zeta}{\xi}\right)^{k-2}[(2+k)\xi^{2k-2} - \xi^{2k}(1+k) - 1]\right\}, \quad (55)$$

$$\frac{\tau_{r\theta}}{\delta R_2} = \sin\theta \left\{\frac{(1-\xi^2)\zeta}{2\xi} + \frac{E_{-1}}{\delta R_2}\left[\frac{\zeta(1-\xi^4)}{\xi(1-\zeta^4)}\right] + \frac{\zeta(\xi^2-\zeta^2)(1-\xi^2)}{\xi(1+\zeta^2)(1+\eta_i)}\right\}$$

$$+\sum_{2}^{\infty} \sin k\theta \left\{ g_k \left(\frac{\zeta}{\xi}\right)^{k} [\xi^2(1-k)+k-\xi^{2k+2}] \right.$$

$$\left. + j_{-k}\left(\frac{\zeta}{\xi}\right)^{k-2} [k\xi^{2k-2}-\xi^{2k}(1+k)+1] \right\}, \quad (56)$$

$$\frac{\sigma_\theta}{\delta R_2} = \left(\frac{E_0}{\delta R_2}\right)\left(\frac{1+\xi^2}{1-\zeta^2}\right) + \cos\theta \left\{ \frac{(1-\xi^2)\zeta}{2\xi} + \frac{E_{-1}}{\delta R_2}\left[\frac{\zeta(3+\xi^4)}{\xi(1-\zeta^4)}\right] \right.$$

$$\left. + \frac{\zeta(\xi^4-3\zeta^2+\xi^2+\zeta^2\xi^2)}{\xi(1+\nu_i)(1+\zeta^2)} \right\} + \sum_{2}^{\infty} \cos k\theta \left\{ g_k \left(\frac{\zeta}{\xi}\right)^{k} [(1-k)\xi^2+2+k+\xi^{2k+2}] \right.$$

$$\left. + j_{-k}\left(\frac{\zeta}{\xi}\right)^{k-2} [1+(2-k)\xi^{2k-2}+\xi^{2k}(1+k)] \right\}, \quad (57)$$

where

$$g_k(1-k)\{(1-\zeta^2)[\beta(1+\nu_o)-(\beta-1)(1-\eta^{-2k+2})]$$
$$+(\eta^2-1)[(\beta+\nu_i)-(\beta-1)\zeta^{2k+2}]\}+j_{-k}\{\beta(1+\nu_o(\zeta^{2k-2}-1)$$
$$+(\beta-1)(1-k^2)(\eta^2-1)(1-\zeta^2)\zeta^{2k-2}$$
$$-(1-\eta^{-2k+2})[\zeta^{2k-2}(\beta+\nu_i)-(\beta-1)]\}$$
$$= -\frac{\beta\Delta(1+\nu_o)\eta^{-k+1}\sin k\alpha}{k\sin\alpha}, \quad k \geqq 2, \quad (58)$$

$$g_k\{\beta(1+\nu_o)(1-\zeta^{2k+2})-(1-\eta^{2k+2})[\beta+\nu_i-(\beta-1)\zeta^{2k+2}]$$
$$+(1-k^2)(\eta^2-1)(1-\zeta^2)(\beta-1)\}+j_{-k}(1+k)\{\beta(1+\nu_o)(1-\zeta^2)\zeta^{2k-2}$$
$$-(1-\zeta^2)(\beta-1)(1-\eta^{2k+2})\zeta^{2k-2}+(\eta^2-1)[\zeta^{2k-2}(\nu_i+\beta)-\beta+1]\}$$
$$= -\frac{\Delta\beta(1+\nu_o)\eta^{k+1}\sin k\alpha}{k\sin\alpha}, \quad k \geqq 2, \quad (59)$$

$$\Delta = \frac{1}{2}\left[(1-\zeta^2)+\frac{\rho}{\delta}(\eta^2-1)\right]. \quad (60)$$

Equations (58) and (59) were obtained by eliminating two of the unknowns from Eqs. (51)–(54).

The rigid-body components of the displacements were evaluated by requiring that

$$v_i = v_o = 0\ , \qquad \theta = 0\ , \qquad \pi\ , \tag{61}$$

$$u_i = u_o = 0\ , \qquad \theta = 0\ , \qquad r = R_2\ , \tag{62}$$

$$(u + iv)_o = (u + iv)_i\ . \tag{63}$$

Substitution of the values of the rigid-body components determined from the above relations into Eq. (28) and separation of real and imaginary parts yielded

$$\frac{2G_i u}{\delta R_2^2} = \frac{E_0}{\delta R_2}\left[\frac{\zeta(2\xi^2 + \eta_i - 1)}{2\xi(1-\zeta^2)}\right] + \cos\theta\left\{-\frac{E_0}{\delta R_2}\left[\frac{2\zeta^2 + \eta_i - 1}{2(1-\zeta^2)}\right]\right.$$

$$+\frac{E_{-1}}{\delta R_2}\left[\frac{(\xi^2-\zeta^2)(2+\xi^2\zeta^2-\eta_i)}{2\xi^2(1-\zeta^4)}\right] + \frac{\zeta^2\eta_i}{1+\eta_i}\ln\frac{\zeta}{\xi} + \frac{\zeta^2(\xi^2-\zeta^2)(\xi^2+\eta_i-2)}{2\xi^2(1+\zeta^2)(1+\eta_i)}$$

$$\left.+\frac{1}{4\xi^2}(\xi^2-\zeta^2)\right\} - \cos\theta\left\{\sum_2^\infty \frac{g_k}{1+k}[\zeta^{2k+2} + \eta_i - (1+k)(1-\zeta^2)]\right.$$

$$\left.+\frac{j_{-k}}{1-k}[1 + \eta_i\zeta^{2k-2} - (1-k)(1-\zeta^2)\zeta^{2k-2}]\right\}$$

$$+\sum_2^\infty \cos k\theta\left\{\frac{g_k}{1+k}\left(\frac{\zeta}{\xi}\right)^{k+1}[\xi^{2k+2} + \eta_i - (1+k)(1-\xi^2)]\right.$$

$$\left.+\frac{j_{-k}}{1-k}\left(\frac{\zeta}{\xi}\right)^{k-1}[1 + \xi^{2k-2}\eta_i - \xi^{2k-2}(1-k)(1-\xi^2)]\right\}\ , \tag{64}$$

$$\frac{2G_i v}{\delta R_2^2} = \sin\theta\left\{\frac{E_{-1}}{\delta R_2}\left[\frac{(\xi^2+\zeta^2)(\xi^2\zeta^2+\eta_i) - 2(\xi^2-\zeta^2)}{2\xi^2(1-\zeta^4)}\right] + \frac{1}{4\xi^2}(3\zeta^2-\xi^2)\right.$$

$$\left.-\frac{\zeta^2\eta_i}{1+\eta_i}\ln\frac{\zeta}{\xi} + \frac{\zeta^2[\xi^2(\xi^2-\zeta^2) - \eta_i(\xi^2+\zeta^2) - 2\zeta^2]}{2\xi^2(1+\eta_i)(1+\zeta^2)} + \frac{E_0}{\delta R_2}\left[\frac{2\zeta^2+\eta_i-1}{2(1-\zeta^2)}\right]\right\}$$

$$+\sin\theta\left\{\sum_2^\infty \frac{g_k}{1+k}[\zeta^{2k+2} + \eta_i - (1+k)(1-\zeta^2)] + \frac{j_{-k}}{1-k}[1 + \eta_i\zeta^{2k-2}\right.$$

$$\left.-(1-k)(1-\zeta^2)\zeta^{2k-2}]\right\} + \sum_2^\infty \sin k\theta\left\{\frac{g_k}{1+k}\left(\frac{\zeta}{\xi}\right)^{k+1}[\xi^{2k+2} + \eta_i\right.$$

$$+ (1 + k)(1 - \xi^2)] - \frac{j_{-k}}{1 - k}\left(\frac{\zeta}{\xi}\right)^{k-1} [1 + n_i \xi^{2k-2}$$

$$+ (1 - k)(1 - \xi^2)\xi^{2k-2}] \Bigg\} . \qquad (65)$$

Equations (55)–(60), (64), and (65) were programmed for an electronic digital computer, so that the effect of the geometric parameters and material properties on the stresses and displacements could be determined. In performing the calculations, a "standard" problem was defined and the effect of departures from these values investigated. The numerical values used in the standard problem were

$$\zeta = 0.3 , \qquad n_i = 1.02 ,$$

$$\eta = 1.05 , \qquad n_o = 1.68 ,$$

$$\rho/\delta = 4.73333 , \qquad \alpha = 0.1^\circ .$$

$$\beta = 1 \times 10^{-5} ,$$

The variation in the stresses and the displacement at $r = R_1$ and $r = R_2$ with the relative flexibility of the case β, all other data remaining the same, is shown in Figs. 2–9. The effect of varying the case thickness parameter η, all other data remaining the same, is shown in Figs. 10–13.

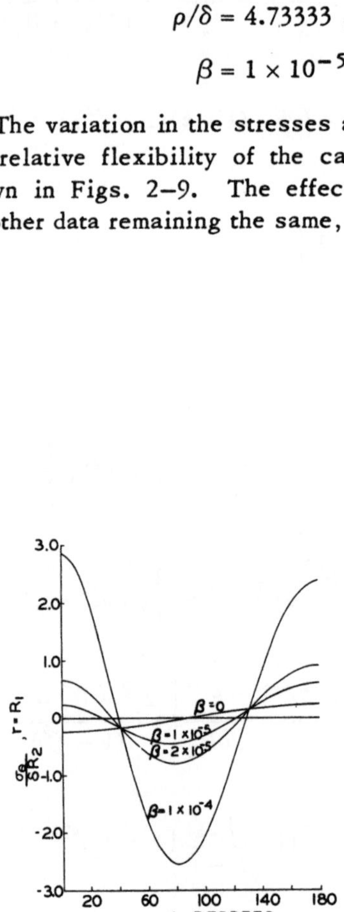

Fig. 2. Variation of σ_θ Along $r = R_1$ with β.

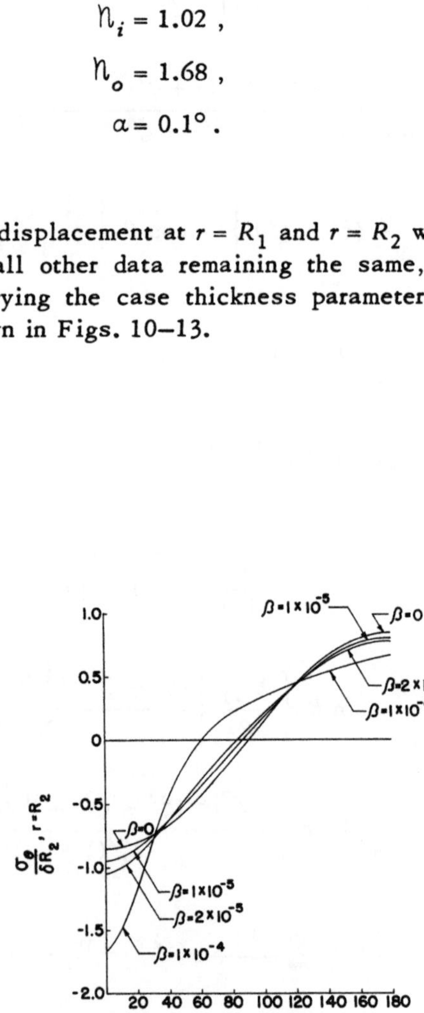

Fig. 3. Variation of σ_θ Along $r = R_2$ with β.

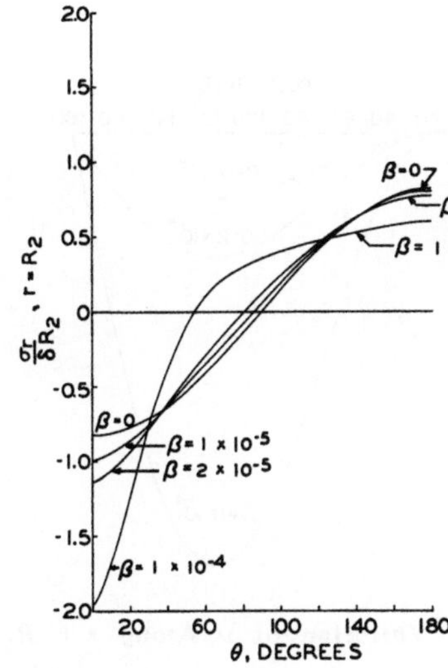

Fig. 4. Variation of σ_r Along $r = R_2$ with β.

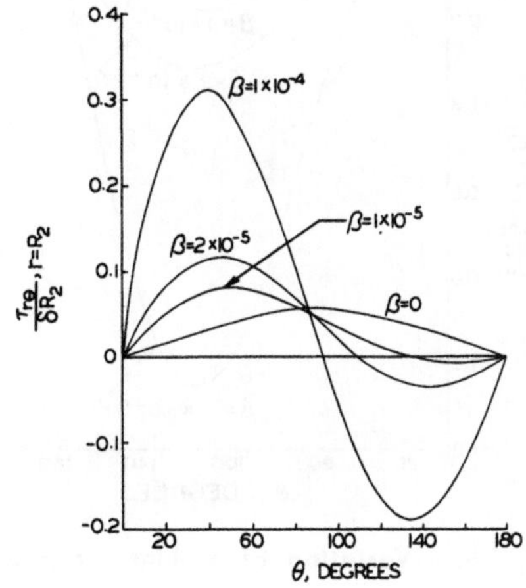

Fig. 5. Variation of $\tau_{r\theta}$ Along $r = R_2$ with β.

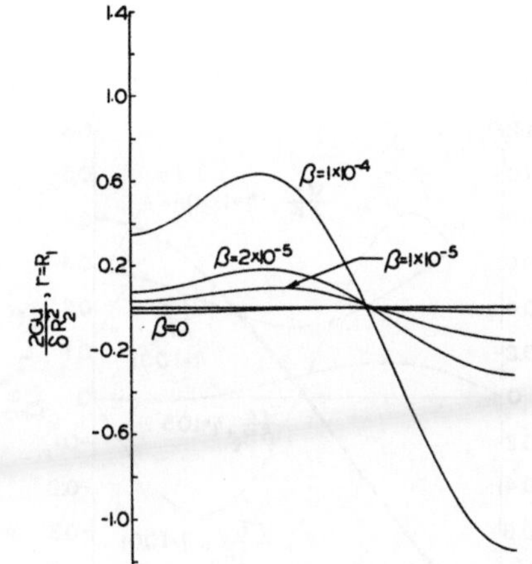

Fig. 6. Variation of u Along $r = R_1$ with β.

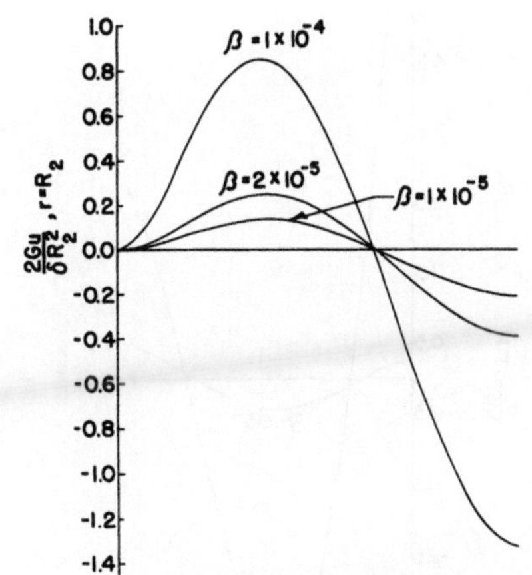

Fig. 7. Variation of u Along $r = R_2$ with β.

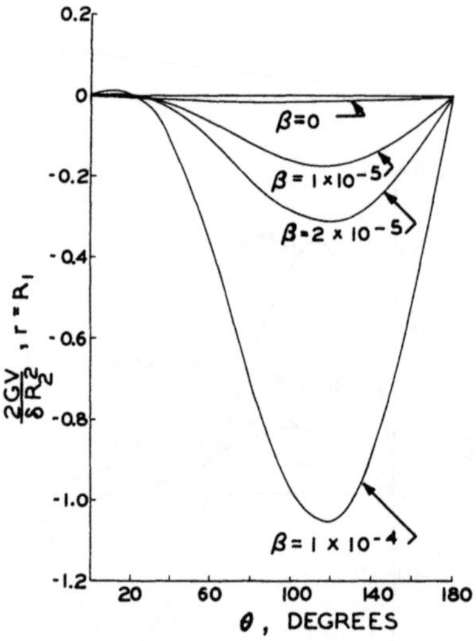

Fig. 8. Variation of v Along $r = R_1$ with β.

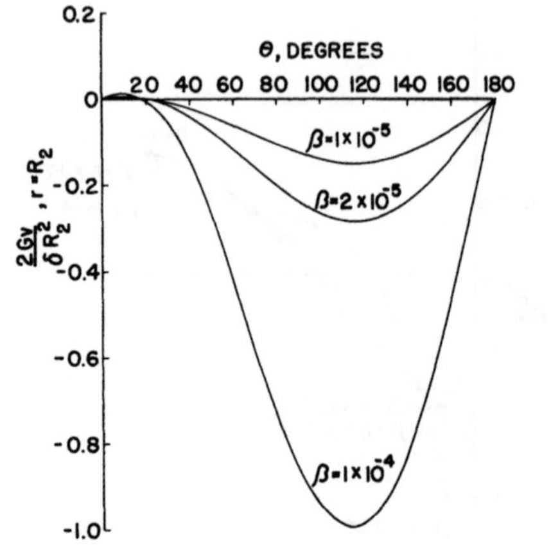

Fig. 9. Variation of v Along $r = R_2$ with β.

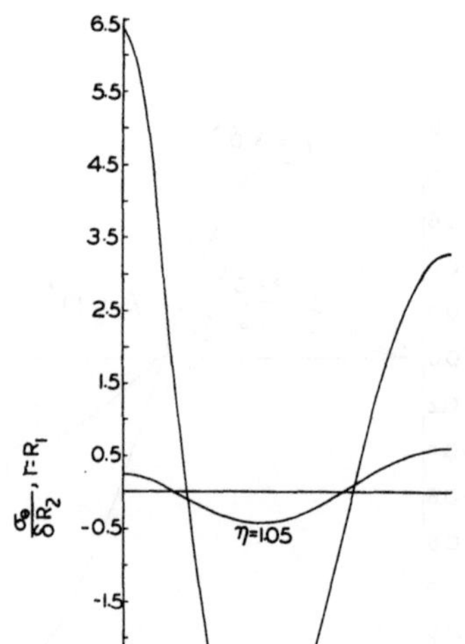

Fig. 10. Variation of σ_θ Along $r = R_1$ with η.

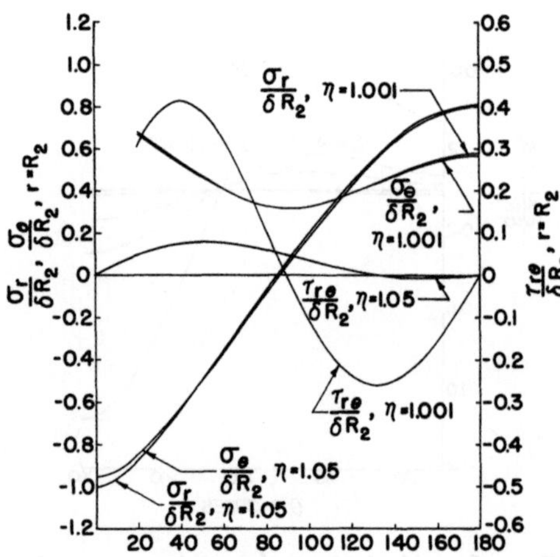

Fig. 11. Variation of σ_r, σ_θ, and $\tau_{r\theta}$ Along $r = R_2$ with η.

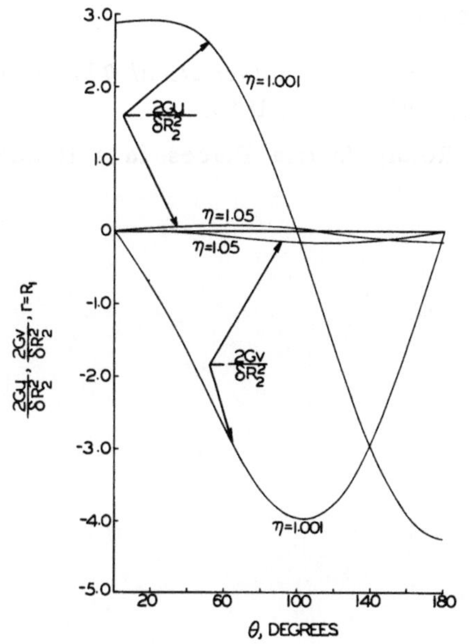

Fig. 12. Variation of u and v Along $r = R_1$ with η.

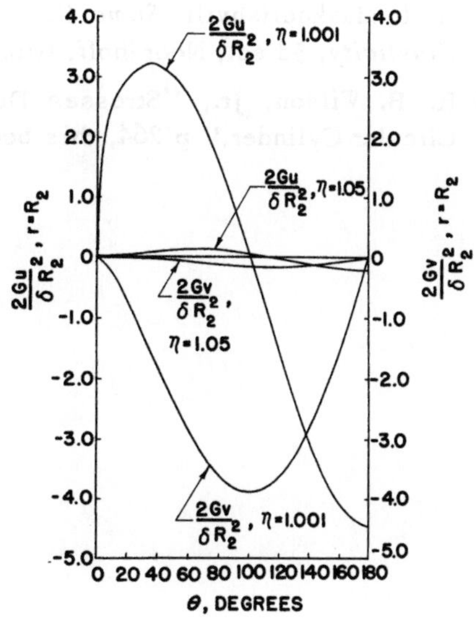

Fig. 13. Variation of u and v Along $r = R_2$ with η.

CONCLUSIONS

It is apparent that changes in both of these parameters result in significant changes in the stresses and displacements. Perhaps the most important point brought out by the figures is that the common practice of assuming the motor case to be rigid ($\beta = 0$) in stress-strain analyses of solid-propellant rocket motors may lead to erroneous results for nonsymmetrically loaded motors. Not only does this assumption rule out any variation in stresses and displacements with case flexibility, but the solutions will also be independent of case thickness.

An investigation was also made of the effect of variations in the geometric parameters and material properties on the rate of convergence of the series. It was found that the thickness of the inner cylinder, within the range $0.2 \leqq \zeta \leqq 0.5$, had little effect. Likewise, values of β within the range investigated, $1 \times 10^{-7} \leqq \beta \leqq 1 \times 10^{-4}$, produced little variation. For $\beta = 0$, g_k and j_{-k} are zero and the stress and displacement solutions contain only a finite number of terms. It was also noted that as long as the case was fairly thick ($\eta = 1.05$), convergence at $r = R_2$ was about as rapid as at $r = R_1$; however, for thin cases ($\eta = 1.001$) there was a definite dependence on r. This, of course, results from the fact that the effect of the support, which is essentially a line load, extends into the region occupied by the inner cylinder. Variations in the thickness of the case had the greatest effect on series convergence. For example, approximately six times as many terms were required to achieve the same degree of accuracy for $\eta = 1.001$ as were required for $\eta = 1.05$. These observations were found to be true in general; however, particular combinations of the various parameters could result in departures from these findings.

REFERENCES

[1] N. I. Muskhelishvili, *Some Basic Problems of the Mathematical Theory of Elasticity*, 3d ed., Noordhoff, Groningen, Netherlands, 1953.

[2] H. B. Wilson, Jr., "Stresses Due to Rotary Inertia Forces in a Hollow Circular Cylinder," p 264, this book.

COMBINED EFFECT OF ROTATION AND EDGE FORCE ON LARGE DEFLECTIONS OF THIN ANNULAR PLATES

A. Koehli
International Business Machines Corporation
and
R. M. Evan-Iwanowski
Syracuse University

Large deflections of thin, elastic, annular plates due to the combined loading of the rotary inertia and uniformly distributed force around and normal to the edge of the plate have been analyzed. The resulting nonlinear equations were solved by means of the power series.

An application of thin annular plates to the clutching devices led to the consideration of this problem. Numerical examples are presented for the values of physical parameters in the range of applicability to clutching devices.

INTRODUCTION

In the linear theories the superposition method provides a direct means to account for combined effects resulting from multiple causes. For thin plates, however, when the deflections exceed half the thickness of a plate, the linear theories cease to yield satisfactory results and nonlinear theories must be employed [*1*]. In the nonlinear theories the superposition method is obviously not applicable, and the effects due to the combined causes have to be assessed by accounting for the coupling of the causes.

In this paper, the combined effects of rotary inertia and the edge force on the large deflections of thin annular plates have been investigated.

Basic differential equations for large deflections of thin plates were developed by von Kármán [*2*] and later by Novozhilov [*3*]. The von Kármán equations are particularly suitable for moderately large deflections, when only the additional effects of the vertical displacements on the radial strain are taken into account. In the present analysis the von Kármán equations are used, which are sets of two nonlinear equations. These resulting equations are solved by use of power-series expansion of the unknown functions. This method was first applied successfully to the plates problem by Way [*4*], to the shallow spherical shells by Simons [*5*], and to the annular plates by Wempner and Schmidt [*6*]. In

NOTATION

r, θ, z	Cylindrical coordinates
a, b	Inner and outer radii
h	Thickness of the plate
E, ν	Elastic constants
ρ	Density
ω	Angular velocity
$\bar{p}$	Uniform load at the edge of the plate per unit length
$\bar{u}, \bar{w}$	Radial and axial displacements
β	$d\bar{w}/dr$
$\bar{N}_r, \bar{N}_\theta$	Radial and circumferential membrane forces per unit length
ψ	$r\bar{N}_r$
$\bar{Q}_r$	Shear per unit length
$\bar{M}_r, \bar{M}_\theta$	Radial and circumferential moments per unit length
D	$Eh/12(1-\nu^2)$
α, α_0	$r/b, a/b$
γ	h/b
u, w	$\bar{u}/b, \bar{w}/b$
N_r	$(1-\nu^2)\bar{N}_r/Eh$
N_θ	$(1-\nu^2)\bar{N}_\theta/Eh$
Q_r	$(1-\nu^2)\bar{Q}_r/Eh$
M_r	$12(1-\nu^2)\bar{M}_r/Eh^2\gamma$
M_θ	$12(1-\nu^2)\bar{M}_\theta/Eh^2\gamma$
p	$(1-\nu^2)\bar{p}/Eh$

ref [4] a method was proposed for the determination of the leading coefficients by interpolating between the approximate values. Simons [5] develops a systematic way to estimate the leading coefficients. In ref [6] the authors discuss also the problem of convergence and evaluation of the error in the resulting series.

GOVERNING EQUATIONS

Consider an annular thin plate, Fig. 1, rotating with constant angular velocity ω about the axis through the center of the plate, perpendicular to its plane, and loaded along its inner edge by a uniformly distributed load $\bar{p}$.

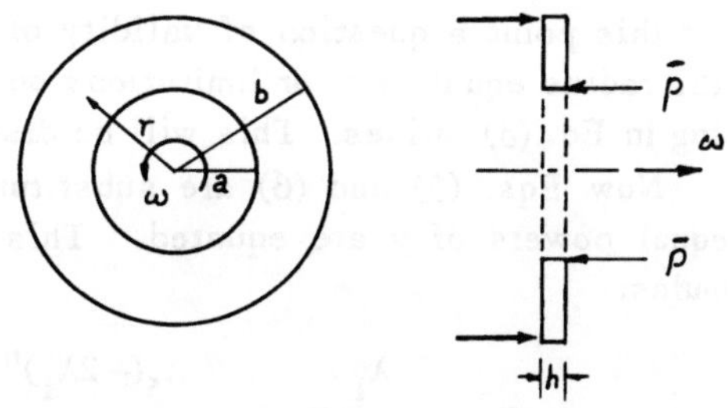

Figure 1.

The von Kármán equations, when applied to this type of loading, are found to have the following form:

$$L(r\psi) = -\frac{1-\nu^2}{2}\left[\frac{\beta^2}{r} + \frac{2(3+\nu)}{E}\rho\omega^2 r\right],$$

$$L(r\beta) = \frac{12}{\gamma^2}\left(\frac{\psi\beta}{r} - \frac{p}{r}\right), \tag{1}$$

where the differential operator (linear) L is

$$L(rf) = \frac{d}{dr}\left[\frac{1}{r}\frac{d}{dr}(rf)\right], \tag{2}$$

β is the variation of the meridianal slope, referred to as "rotation," and ψ is the meridianal resulting force.

It is convenient, however, for the calculations to represent Eq. (1) in a nondimensional form as follows (see ref [6]):

$$L(\alpha\psi) = -\frac{1-\nu^2}{2}\left[\frac{\beta^2}{\alpha} + \frac{2(3+\nu)}{E}\rho\omega^2 b^2 \alpha\right],$$

$$L(\alpha\beta) = \frac{12}{\gamma^2}\left(\frac{\beta\psi}{\alpha} - \frac{p}{\alpha}\right). \tag{3}$$

In view of the application of the power-series solution, the independent variable α is changed to y by the transformation

$$\alpha = e^{Ly}, \quad L = \log \alpha_0 . \tag{4}$$

Now the variable y varies from 0 to 1.

Next, the unknown functions β and ψ in Eq. (3) are expanded into the power series of y:

$$\beta = e^{-Ly} \sum_{n=0}^{\infty} B_n y^n , \tag{5}$$

$$\psi = e^{-Ly} \sum_{n=0}^{\infty} F_n y^n \tag{6}$$

At this point a question of validity of such representation in the y region with the radius equal to 1, or limitations on the values of various parameters appearing in Eq. (3), arises. This will be discussed later.

Now Eqs. (5) and (6) are substituted into Eq. (3), and the coefficients of equal powers of y are equated. This results in the following recurrence formulas:

$$F_n = -\frac{\lambda_1}{n} F_{n-1} - \frac{\lambda_5(-2\lambda_1)^{n-2}}{n!} - \frac{\lambda_2}{n(n-1)} \sum_{i=0}^{n-2} B_i B_{n-2-i} , \tag{7}$$

$$B_n = -\frac{\lambda_1}{n} B_{n-1} - \frac{\lambda_3(-\lambda_1)^{n-2}}{n!} + \frac{\lambda_4}{n(n-1)} \sum_{i=0}^{n-2} F_i B_{n-2-i} , \tag{8}$$

where

$$\lambda_1 = -2L , \quad \lambda_2 = \frac{1-\nu^2}{2} L^2 ,$$

$$\lambda_3 = \frac{12p}{\gamma^2} L^2 , \quad \lambda_4 = \frac{12}{\gamma^2} L^2 , \tag{9}$$

$$\lambda_5 = \frac{(1-\nu^2)(3+\nu)}{E} L^2 \rho\omega^2 b^2 .$$

The relationships (7) and (9) determine F_n and B_n for $n \geqq 2$ in terms of F_0, F_1, B_0, and B_1. The last four constants cannot be determined from the recurrence formulas. They must be selected in such a way that the prescribed boundary conditions are satisfied.

BOUNDARY CONDITIONS

For a plate with clamped inner and free outer edges the boundary conditions require that

$$N_r = M_r = 0 \qquad \text{for } y = 0 \text{ (outer edge)} ,$$

$$\beta = u = 0 \qquad \text{for } y = 1 \text{ (inner edge)} .$$

The first condition results in

$$F_0 = 0 . \tag{10}$$

The radial moment M_r under the transformation (4) becomes

$$M_r = \frac{1}{L} e^{-2Ly} \sum_{n=0}^{\infty} nB_n y^{n-1} - e^{-2Ly} \sum_{n=0}^{\infty} B_n y^n + \nu e^{-2Ly} \sum_{n=0}^{\infty} B_n y^n ,$$

which for $y = 0$ results in

$$B_1 = L(1 - \nu)B_0 \,. \tag{11}$$

The condition $\beta = 0$ at $y = 1$ yields

$$\sum_{n=0}^{\infty} B_n = 0 \,. \tag{12}$$

Applying Eq. (4), u becomes

$$u = \frac{e^{Ly}}{1 - \nu^2}\left[(1 - \nu)e^{-2Ly}\sum_{n=0}^{\infty} F_n y^n + \frac{1}{L} e^{-2Ly}\sum_{n=0}^{\infty} nF_n y^{n-1}\right.$$

$$\left. - 2e^{-2Ly}\sum_{n=0}^{\infty} F_n y^n + \lambda_6 e^{2Ly}\right] ,$$

where

$$\lambda_6 = \frac{1 - \nu^2}{E}\rho\omega^2 b^2 \,. \tag{13}$$

For $y = 1$, the above expression for u yields

$$L\alpha_0^4\lambda_6 + \sum_{n=1}^{\infty} [n - L(1 + \nu)]F_n = 0 \,. \tag{14}$$

The other boundary conditions can be treated similarly. Since, however, the numerical examples discussed here concern the free-clamped boundaries, the attention is limited to these conditions. The derivation of the relationships similar to (12) and (14) is straightforward.

NUMERICAL PROCEDURES

Boundary conditions (10), (11), (12), and (14) together with the recurrence formulas (7) and (8) result in a set of two equations for F_1 and B_0. The exact determination of F_1 and B_0 from these equations is extremely difficult. Thus, one must resort to some approximating values of these constants, and hence to approximate solutions of Eqs. (5) and (6).

Any approximation of F_1 and B_0 physically means that the boundary conditions are only approximately satisfied; that is, if (12) and (14) are different from zero some residual radial moment and slope remain around the boundary. It is argued further that if these residual values are small they affect only slightly the distribution of stress, strain, moments, and other physical quantities inside the plate. Thus, if approximate values of F_1 and B_0 are found such that the boundary conditions are satisfied to a desired degree of accuracy, then it can be concluded that a satisfactory solution to the problem is found (see refs [5] and [6]).

Let $F_n^{(k)}$ and $B_n^{(k)}$ denote the kth approximation of the coefficients F_n and B_n. The $(k+1)$st approximation can be written as

$$F_n^{(k+1)} = F_n^{(k)} + \Delta F_n^{(k)} ,$$
$$B_n^{(k+1)} = B_n^{(k)} + \Delta B_n^{(k)} . \tag{15}$$

Substituting Eqs. (15) into the recurrence formulas (7) and (8), and considering $\Delta F_n^{(k)}$ and $\Delta B_n^{(k)}$ to be small quantities so that the nonlinear terms can be omitted, we obtain

$$F_n^{(k+1)} = F_n^{(k)} + \Delta F_n^{(k)} ,$$

$$\Delta F_n^{(k)} = -\frac{\lambda_1}{n} \Delta F_{n-1}^{(k)} - \frac{\lambda_2}{n(n-1)} \sum_{i=0}^{n-2} [\Delta B_i^{(k)} B_{n-2-i}^{(k)} + \Delta B_{n-2-i}^{(k)} B_i^{(k)}] . \tag{16}$$

A similar expression can be formed for $\Delta B_n^{(k)}$. Repeating Eq. (16) $(n-1)/2$ times results in the (linear) expressions for $\Delta F_n^{(k)}$ and $\Delta B_n^{(k)}$ of the form:

$$\Delta F_n^{(k)} = a_n^{(k)} \Delta F_1^{(k)} + b_n^{(k)} \Delta B_0^{(k)} ,$$
$$\Delta B_n^{(k)} = c_n \Delta F_1^{(k)} + d_n \Delta B_0^{(k)} , \tag{17}$$

where the coefficients $a_n^{(k)}$, $b_n^{(k)}$, $c_n^{(k)}$, and $d_n^{(k)}$ are functions of $F_j^{(k)}$ and $B_j^{(k)}$ $(j = 0 \ldots \infty)$.

Moreover, the boundary conditions result in the following:

$$\sum_{n=0}^{\infty} B_n^{(k+1)} = \sum_{n=0}^{\infty} [B_n^{(k)} + \Delta B_n^{(k)}] = 0 , \quad \sum_{n=0}^{\infty} B_n^{(k)} = -\sum_{n=0}^{\infty} \Delta B_n^{(k)} . \tag{18}$$

Similarly,

$$\sum_{n=0}^{\infty} [n - L(1+\nu)]F_n^{(k)} + L\alpha_0^4 \lambda_6 = -\sum_{n=0}^{\infty} [n - L(1+\nu)]\Delta F_n^{(k)} . \tag{19}$$

By using Eqs. (17), Eqs. (18) and (19) can be written as

$$\sum_{n=0}^{\infty} [n - L(1+\nu)]F_n^{(k)} = -\Delta F_1^{(k)} \sum_{n=0}^{\infty} [n - L(1+\nu)]a_n^{(k)}$$
$$- \Delta B_0^{(k)} \sum_{n=0}^{\infty} [n - L(1+\nu)]b_n^{(k)} = R_2[\Delta F_1^{(k)}, \Delta B_0^{(k)}] \ldots , \tag{20}$$

$$\sum_{n=0}^{\infty} B_n^{(k)} = -\Delta F_1^{(k)} \sum_{n=0}^{\infty} c_n^{(k)} - \Delta B_0^{(k)} \sum_{n=0}^{\infty} d_n^{(k)} = R_2[\Delta F_1^{(k)}, \Delta B_0^{(k)}] \ldots . \tag{21}$$

The left side of these relationships represents the residual radial moment and slope around the boundary resulting from the use of the kth approximation. On the other hand, these equations represent the procedure for obtaining successive approximations. If, for instance, the kth approximation is found by some means, then Eqs. (20) and (21) provide the means for finding $\Delta F_1^{(k)}$ and $\Delta B_0^{(k)}$, which, in turn, determine $F_n^{(k+1)}$ and $B_n^{(k+1)}$. Repeating the same steps yields new approximations, and so on.

It is clear that the usage of Eqs. (20) and (21) for the calculational purposes is impractical, since the index n varies from 0 to ∞. Thus the series are truncated, and the following form is used:

$$\sum_{n=0}^{q} [n - L(1+\nu)]F_{nq}^{(k)} = -\Delta F_{1q}^{(k)} \sum_{n=0}^{q} [n - L(1+\nu)]a_{nq}^{(k)}$$
$$- \Delta B_{0q}^{(k)} \sum_{n=0}^{q} [n - L(1+\nu)]b_{nq}^{(k)} = R_1[\Delta F_{1q}^{(k)}, \Delta B_{0q}^{(k)}] , \quad (22)$$

$$\sum_{n=0}^{q} B_{nq}^{(k)} = -\Delta F_{1q}^{(k)} \sum_{n=0}^{q} C_{nq} - \Delta B_{0q}^{(k)} \sum_{n=0}^{q} d_n^{(k)} = R_2[\Delta F_{1q}^{(k)}, \Delta B_{0q}^{(k)}] , \quad (23)$$

where the index q has to be appropriately chosen. Because of the truncation, all terms in Eqs. (22) and (23) are given the additional index q.

The corrections, $\Delta F_{1q}^{(k)}$ and $\Delta B_{0q}^{(k)}$, are assumed to be very small; consequently, for sufficiently large q the expressions (22) and (23) can be represented as follows:

$$R_{1q}[\Delta F_{1q}^{(k)}, \Delta B_{0q}^{(k)}] = -\frac{\partial R_{1q}}{\partial F_{1q}^{(k)}} \Delta F_{1q}^{(k)} - \frac{\partial R_{1q}}{\partial B_{0q}^{(k)}} \Delta B_{0q}^{(k)} ,$$
$$R_{2q}[\Delta F_{1q}^{(k)}, \Delta B_{0q}^{(k)}] = -\frac{\partial R_{2q}}{\partial F_{1q}^{(k)}} \Delta F_{1q}^{(k)} - \frac{\partial R_{2q}}{\partial B_{0q}^{(k)}} \Delta B_{0q}^{(k)} , \quad (24)$$

where

$$\frac{\partial R_{1q}}{\partial F_{1q}^{(k)}} = \sum_{n=0}^{q} \frac{\partial F_{nq}^{(k)}}{\partial F_{1q}^{(k)}} , \text{ etc.}$$

For $n = 0, 1$,

$$\frac{\partial F_{0q}^{(k)}}{\partial F_{1q}^{(k)}} = \frac{\partial F_0^{(k)}}{\partial B_{0q}^{(k)}} = \frac{\partial B_{1q}^{(k)}}{\partial F_{1q}^{(k)}} = 0 , \quad \frac{\partial B_{1q}^{(k)}}{\partial B_{0q}^{(k)}} = L(1-\nu) ; \quad (25)$$

and for $n \geqq 2$,

$$\frac{\partial F_{nq}^{(k)}}{\partial F_{1q}^{(k)}} = -\frac{\lambda_1}{n}\frac{\partial F_{(n-1)q}^{(k)}}{\partial F_{1q}^{(k)}} - \frac{\lambda_2}{n(n-1)}\sum_{i=0}^{n-2}\left[\frac{\partial F_{iq}^{(k)}}{\partial F_{1q}^{(k)}} B_{(n-2-i)q}^{(k)} + F_{1q}^{(k)}\frac{\partial B_{(n-2-i)q}^{(k)}}{\partial F_{1q}^{(k)}}\right], \text{ etc.} \tag{26}$$

see Eqs. (7), (8), (10), and (11)]. Thus Eqs. (24) constitute sufficient information for the determination of new corrections $\Delta F_{1q}^{(k)}$ and $\Delta B_{0q}^{(k)}$.

Present calculations were made possible by the use of the high-speed computing machine IBM 7090.

NUMERICAL EXAMPLE

The procedures described in the preceding sections are now applied to a numerical example.

The numerical calculations were carried to the point such that the radial bending moment M_r, slope β, radial force N_r, and radial displacement u at the inner edge did not exceed 0.01% of their respective maximum values inside the plate.

First Estimate of the Leading Coefficients

The success of the above method may hinge upon a first approximation of the leading coefficients F_1 and B_0, that is, upon the values $F_1^{(0)}$ and $B_0^{(0)}$. It is proposed to estimate these coefficients based upon previous reasonable knowledge of the approximate behavior of the plate under the specified loadings. We start with a nonrotating plate. Maximum deflection w_0 is computed first by linear theory [7]. Curves given by Way [4] for moderately large deflections of circular plates give an approximation of the amount the linear value is in error. It is then corrected accordingly. Assuming that the plate deforms into a cone, which implies that $\beta = \beta_0$ = constant, where

$$\beta_0 = \tan^{-1}\frac{w_0}{b-a},$$

we thus set $B = \beta_0$ as a first approximation.

From the expression for radial displacement u, the radial displacement u_0 at the outer edge $y = 0$ is found to be for the nonrotating plate:

$$u_0 = \frac{F_1}{(1-\nu^2)L},$$

which provides a first approximation for F_1 if an appropriate value of u_0 is assumed. Once the actual constants F_1 and B_0 for the nonrotating plate have been computed, they, in turn, are taken as first approximations for calculating values for the rotating plate.

Numerical Results

Numerical results were obtained for Mylar plates with the following dimensions and engineering constants:

$$b = 1\,, \quad \gamma = 0.003\,, \quad \alpha_0 = 0.5\,,$$

$$E = 5 \times 10^5 \text{ psi}\,, \quad \nu = 0.3\,,$$

$$\rho = 1.3 \times 10^{-4} \text{ lb in.}^{-4} \text{ sec}^{-2}\,,$$

with free outer edge and clamped inner edge.

Calculations were made for three axial loads:

$$P = 7 \text{ lb}\,, \qquad p = 0.4057 \times 10^{-4}\,;$$

$$P = 9 \text{ lb}\,, \qquad p = 0.5217 \times 10^{-4}\,;$$

$$P = 11 \text{ lb}\,, \qquad p = 0.6377 \times 10^{-4}\,;$$

where P is the total force applied at the inner edge; and for ω = 0, 500, 1000, 2000, 3000, and 4000 radians/sec.

The coefficients F_n and B_n were computed for q = 60, for the cases with angular velocity ω = 0, 500, 1000, and 2000 radians/sec, and q = 80 for the cases with ω = 3000 and 4000 radians/sec.

To illustrate the degree of accuracy attained, a particular case is quoted. For $p = 0.4057 \times 10^{-4}$ and $\omega = 0$, the calculated values at the outer edge were:

$$M_r = 0.1478 \times 10^{-7}\,, \qquad \beta = -0.3725 \times 10^{-8}\,,$$
$$N_r = 0.1767 \times 10^{-10}\,, \qquad u = -0.1809 \times 10^{-8}\,.$$

These values are less than 0.01% of their respective maximal values inside the plate.

The calculation results are represented in Figs. 2–7. By using the graphs in Figs. 2–7, the stresses and the displacements at any point of the plate may now be found. To indicate the order of magnitude of induced stresses and displacements, the following is given:

$$r = 0.5\,, \quad z = -h/2\,, \quad p = 0.4057 \times 10^{-4}\,;$$

for $\omega = 0$ the total stresses are:

$$\sigma_r = 1714.0 \text{ psi}\,, \qquad \sigma_\theta = 514.2 \text{ psi}\,,$$

$$\bar{w} = w/b = -0.0249\,, \qquad \bar{u} = u/b \simeq 0\,;$$

for $\omega = 4000$ the total stresses are:

$$\sigma_r = 2036.0 \text{ psi}\,, \quad \sigma_\theta = 611.64 \text{ psi}\,,$$

$$w = -0.0163\,, \qquad u \simeq 0\,.$$

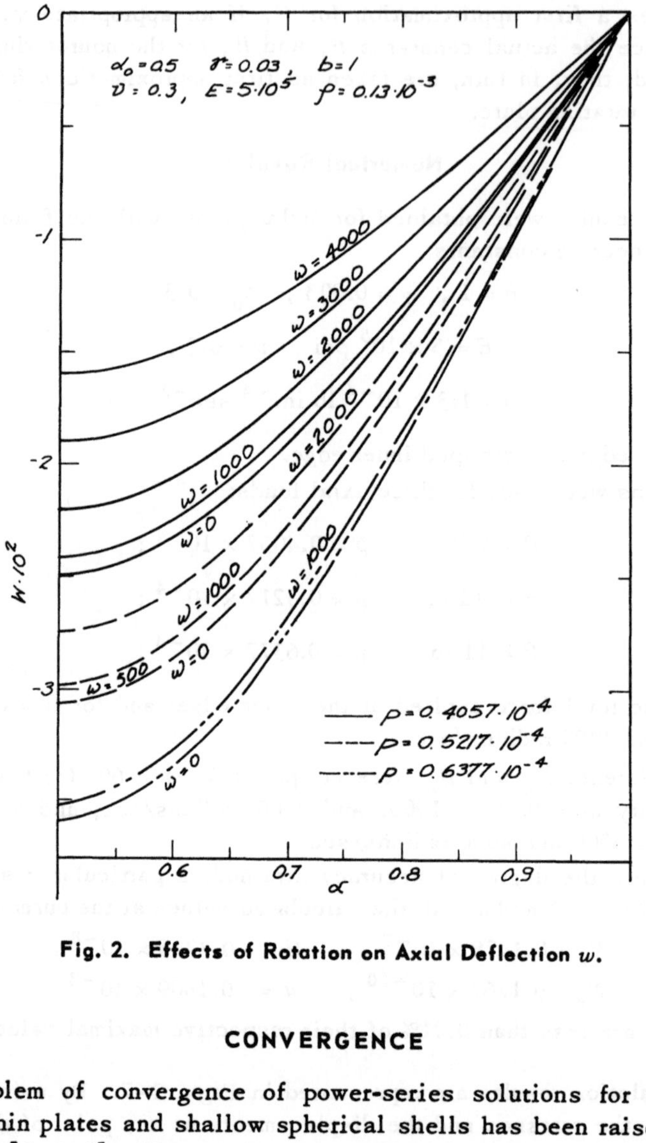

Fig. 2. Effects of Rotation on Axial Deflection w.

CONVERGENCE

The problem of convergence of power-series solutions for the nonlinear theories of thin plates and shallow spherical shells has been raised by various investigators [6, 8, 9]. The convergence obviously depends to a great measure on the values of the set of parameters (α_0, γ, p, and ω) entering in the basic relationships. The situation encountered in this investigation is represented in Table 1, which summarizes the investigated cases. It indicates a rather dramatic and delicate problem of convergence. Quite similar situations were reported by Budiansky [9].

Weinitschke [8] put forth a plausible proposition that for certain parameter values the complex singularities exist within the assumed region of convergence of the power series. To avoid such a possibility he develops the series about a few points within the radius of the shell with smaller radii of convergence. (It has also a beneficial calculation effect, since the series are more rapidly convergent.)

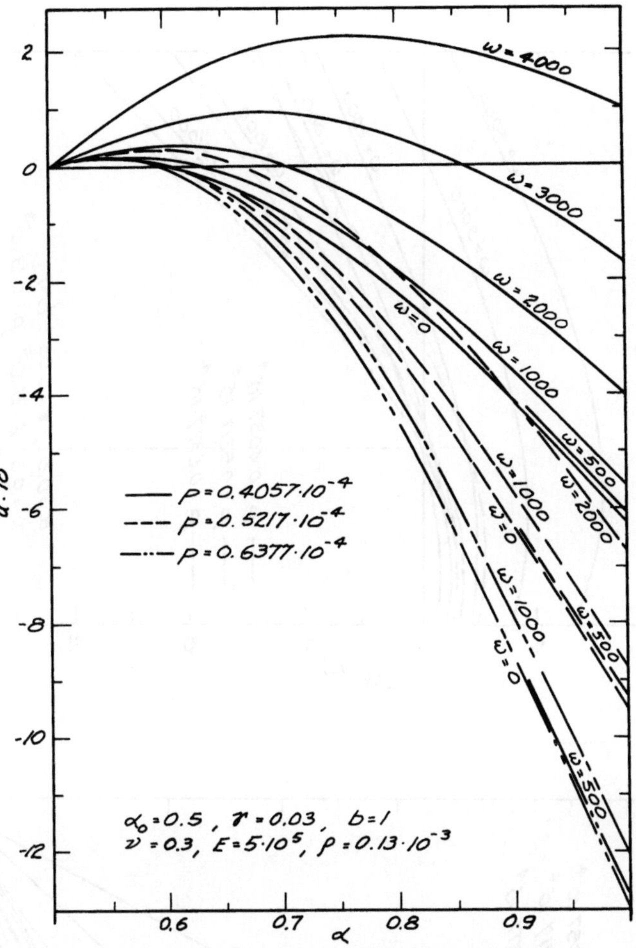

Fig. 3. Effects of Rotation on Radial Displacement u.

In ref [6] a sufficient condition for convergence was discussed, and based upon this condition an estimation of the errors was given.

CONCLUSION

The results given in Figs. 2–8 illustrate the effects of rotation on the stress and the displacement components. In the particular case calculated, these effects are negligible for angular velocities $\omega < 500$ radians/sec.

Axial displacements and bending stresses decrease with increasing angular velocity, while the membrane stresses increase and the radial displacement curves shift in the positive direction.

The example given may be considered a limit of the cases to which the described method may be safely applied. The method will fail when α_0 and γ are taken smaller and ω and p are taken larger than their respective values of the given example.

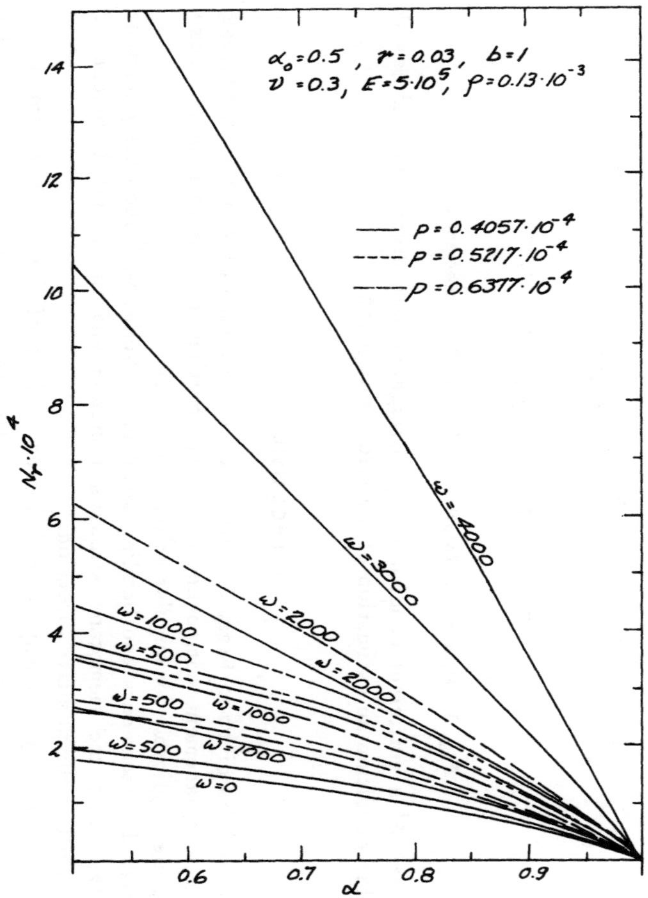

Fig. 4. Effects of Rotation on Radial Membrane Tension N_r.

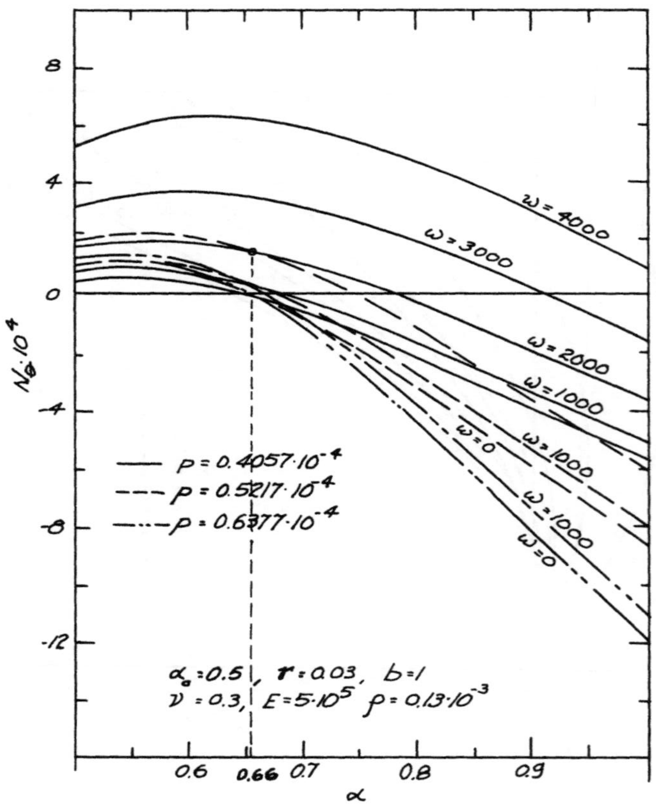

Fig. 5. Effects of Rotation on Circumferential Membrane Tension N_θ.

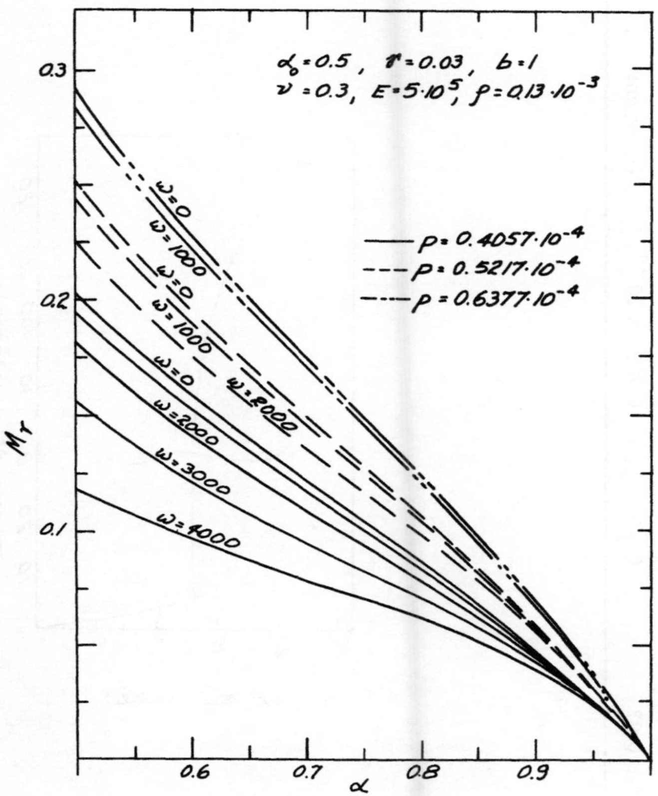

Fig. 6. Effects of Rotation on Radial Bending Moment M_r.

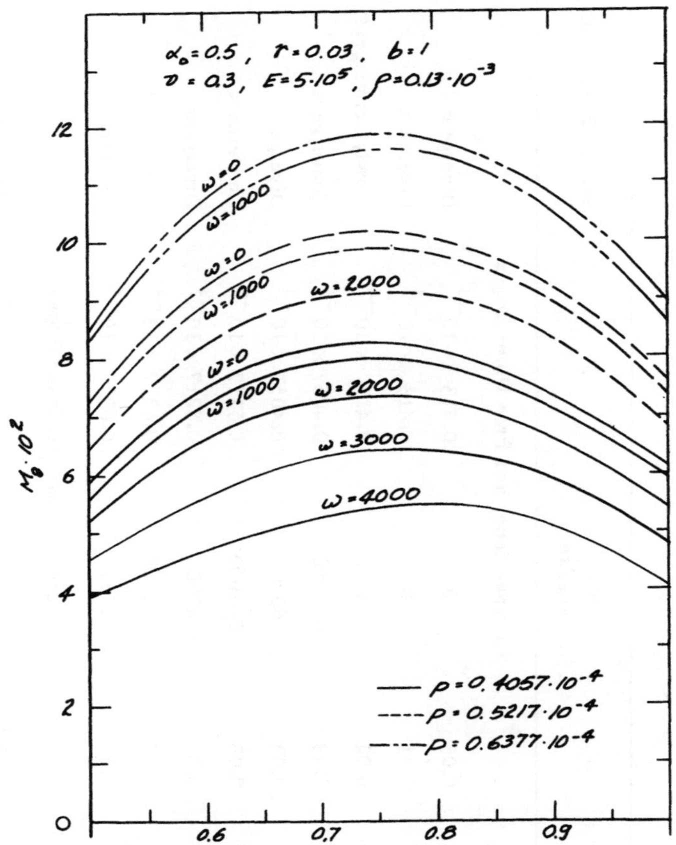

Fig. 7. Effects of Rotation on Circumferential Moment M_θ.

Table 1. Résumé of Computed Cases

($E = 5 \times 10^5$, $\nu = 0.3$, $\psi = 0.13 \times 10^{-4}$ lb in.$^{-4}$ sec^{-2})

a_0	γ	ω (radians/sec)	P	Series, Eqs. (5, 6)
		Clamped Inner and Free Outer Edges		
0.3	0.05	0	0.413×10^{-4}	Divergent
0.5	0.01	0	0.413×10^{-4}	Divergent
0.5	0.02	0	0.413×10^{-4}	Slowly convergent
0.5	0.03	0–4000	0.4057×10^{-4}	Convergent
0.5	0.03	6000	0.4057×10^{-4}	Divergent
0.5	0.03	0–2000	0.5217×10^{-4}	Convergent
0.5	0.03	4000	0.5217×10^{-4}	Divergent
0.5	0.03	0–1000	0.6377×10^{-4}	Convergent
0.5	0.03	2000	0.6377×10^{-4}	Divergent
		Clamped Inner and Outer Edges		
0.5	0.05	0	0.413×10^{-4}	Convergent
0.5	0.04	0	0.2117×10^{-3}	Divergent
0.2	0.02	0	0.289×10^{-4}	Divergent

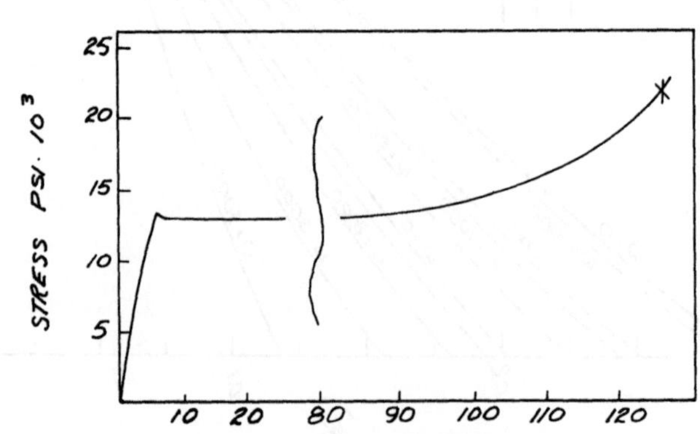

Fig. 8. Tensile Stress-Strain Curve for Mylar Polyester Film.

REFERENCES

[1] E. Reissner, "On Axisymmetrical Deformations of Thin Shells of Revolution," pp 27–51 in *Proceedings of Symposia in Applied Mathematics*, vol 3, American Mathematical Society, 1950.

[2] Theodor von Kármán, "Festigkeitsprobleme im Maschinenbau," *Enzyklopedie der matematischen Wissenschaften*, pp 348–52, vol 4 (4), Teubner, Leibzig, 1910.

[3] V. V. Novozhilov, *Foundations of the Nonlinear Theory of Elasticity*," Graylock, Rochester, N.Y., 1953.

[4] S. Way, "Bending of Circular Plates with Large Deflection," *Trans. Am. Soc. Mech. Engrs.* **56**, 627–36 (1934).

[5] R. M. Simons, "A Power Series Solution of the Nonlinear Equation for Axisymmetrical Bending of Shallow Spherical Shells," *J. Math. and Phys.* **15**, 164–76 (1956).

[6] G. A. Wempner and R. Schmidt, "Large Symmetric Deflections of Annular Plates," *J. Appl. Mechanics* **25**, 449–52 (1958).

[7] R. J. Roark, *Formulas for Stress and Strain*, p 201, McGraw-Hill, New York, 1954.

[8] H. Weinitschke, "On the Stability Problem for Shallow Spherical Shells," *J. Math. and Phys.* **38**(4), 209–31 (1960).

[9] B. Budiansky, "Buckling of Clamped Shallow Spherical Shells," pp 64–94 in *Proceedings of Symposia on the Theory of Thin Elastic Shells*, Delft, 1959, 1960.

ELASTIC ANALYSIS OF A COMPOSITE CIRCULAR PLATE ROTATING ABOUT AN ARBITRARY POINT

Thomas P. Mitchell and William E. Warren
Cornell University

The theory of plane stress is used to determine series expressions for the stress and displacement fields in an elastic ring and insert assembly which rotates with constant angular velocity about an arbitrary point. The generality of the analysis is such that it includes many limiting configurations of interest. Some of these cases are examined, and closed-form expressions for the stresses are presented.

INTRODUCTION

A method of Muskhelishvili in conjunction with the theory of plane stress is used here to determine analytic expressions for the stress and displacement fields in a rotating, thin homogeneous coplanar elastic ring and insert assembly. The angular velocity of the assembly is constant, and the axis of rotation, which is perpendicular to the common plane, passes through an arbitrary point of this plane. The assembly is rigidly connected to the axis of rotation from a point in either the insert or the ring. Thus, in general, a concentrated reaction force and moment are required at this mounting point to maintain equilibrium. The stresses and displacements of the ring and insert are assumed to be identical at the ring-insert interface, and the outside boundary of the ring is taken to be stress free. The case in which the assembly is mounted in the insert is treated first and in detail. The results for that in which the assembly is mounted in the ring are found by similar methods and accordingly are presented with a minimum of explanatory comment.

Basically, the solution is effected by assuming series expressions for the necessary complex potential functions in the ring and insert, applying the boundary conditions, and evaluating the coefficients of the assumed series. The notation of Muskhelishvili [*1*] is used throughout.

The generality of this analysis is such that many special configurations of interest are obtained as simple limiting cases of it. Some of these configurations are: (1) the rotation about an arbitrary point of an elastic ring on a rigid insert; a particular case of which is treated in ref [*2*]; (2) the rotation of a disk about an arbitrary axis; a particular case of this is treated in ref [*3*]; (3) the rotation about an arbitrary axis of an elastic disk mounted in a constraining rigid ring; and (4) an elastic ring rotating about an arbitrary point in its plane.

NOTATION

a_m, b_m, c_m, d_m	mth coefficient in a Laurent series
k	Ratio given by R_2/R_1
m	Variable of series summation
n	Constant equal to $(3 - \nu)/(1 + \nu)$
p	Abscissa of axis of rotation
q	Areal density
u, v	Cartesian components of displacement
x, y	Cartesian coordinates
z	Complex variable $x + iy$
z_0	Point of mounting
A_m, B_m, C_m	mth coefficient of Fourier series expression of a boundary condition
F, F'	Resultant force on a boundary as defined
G	Modulus of rigidity
H	Parameter as defined
M	Concentrated reaction moment at mounting
N	Ratio given by G_1/G_2
R_1, R_2	Radii as defined
X	Concentrated reaction force at mounting
X_x, Y_y, X_y	Cartesian stress components
Y_m, Y'_m, Z_m	Series constants as defined
ρ, θ	Polar coordinates in z plane
$\widehat{\rho\rho}, \widehat{\theta\theta}, \widehat{\rho\theta}$	Curvilinear stress components
ν	Poisson's ratio
ω	Angular velocity
$\Phi(z), \Psi(z), \Omega(z), \chi(z)$	Complex potential functions

CASE 1: ASSEMBLY FASTENED IN INSERT

Consider a thin homogeneous circular plane elastic ring of outer radius R_1 and inner radius R_2 containing a plane homogeneous circular elastic insert of radius R_2. The ring and insert are not assumed to be made of the same material.

The ring and insert assembly is fastened at the point $z_0 = x_0 + iy_0$ of the insert and rotates with a constant angular velocity ω about an axis perpendicular to the plane of the ring and insert and passing through the point p as shown in Fig. 1. The ring and insert are assumed to be bonded together at the interface radius R_2 in such a way that no residual or fitting stresses are present. This yields the conditions that the displacements of the insert and ring at the interface are identical, and that the normal and shear components of stress in the ring and insert at the interface are identical, the result corresponding physically to an insert welded into a ring. It is required to determine the stress and displacement fields in both materials.

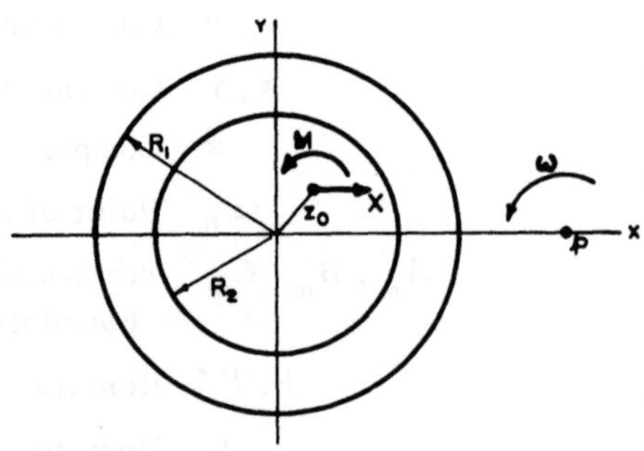

Fig. 1. Rotating Elastic Ring with Circular Elastic Insert. Assembly mounted in insert.

Clearly, one loses nothing in generality by assuming p to lie on the real axis. At the point of attachment z_0 there will be, in general, a concentrated reaction moment M and a concentrated reaction force X, which is required to keep the system in equilibrium. In all that follows, a subscript 1 refers to the material of the ring and a subscript 2 refers to the material of the insert. The reaction force X and the reaction moment M acting at the point z_0 are given by

$$X = \pi q_1 \omega^2 p(R_1^2 - R_2^2) + \pi q_2 \omega^2 p R_2^2 \, , \tag{1}$$

$$M = X\left(\frac{z_0 - \bar{z}_0}{2i}\right) . \tag{2}$$

The force transmitted by the insert to the ring at the interface radius R_2 will be designated by F and is

$$F = \pi q_1 \omega^2 p(R_1^2 - R_2^2) \, , \tag{3}$$

where q is the density of the material.

The Insert

The equations of equilibrium and compatibility, which must be satisfied in the presence of the inertia forces, are:

$$\frac{\partial X_x}{\partial x} + \frac{\partial X_y}{\partial y} = -q_2 \omega^2 (x - p) \, ,$$

$$\frac{\partial X_y}{\partial x} + \frac{\partial Y_y}{\partial y} = -q_2\omega^2 y ,$$

$$\nabla^2(X_x + Y_y) = -2(1 + \nu_2)q_2\omega^2 . \tag{4}$$

A particular set of complex-stress combinations in terms of the complex variables z and $\bar{z}$ which satisfy Eq. (4) are:

$$X_x + Y_y = -\frac{(1 + \nu_2)q_2\omega^2}{2}(z - p)(\bar{z} - p) ,$$

$$Y_y - X_x + 2iX_y = \frac{(1 - \nu_2)q_2\omega^2}{4}(\bar{z} - p)^2 , \tag{5}$$

and the associated complex displacement is given by

$$u + iv = -\frac{(1 - \nu_2)q_2\omega^2}{16G_2}(\bar{z} - p)(z - p)^2 , \tag{6}$$

where ν is Poisson's ratio and G is the modulus of rigidity.

If polar coordinates ρ and θ are now introduced into the z plane, the stress components $\widehat{\rho\rho}$, $\widehat{\theta\theta}$, and $\widehat{\rho\theta}$ corresponding to these coordinates are:

$$(\widehat{\rho\rho} + \widehat{\theta\theta})^{(1)} = -\frac{(1 + \nu_2)q_2\omega^2}{2}(z - p)(\bar{z} - p) , \tag{7}$$

$$(\widehat{\rho\rho} - i\widehat{\rho\theta})^{(1)} = -\frac{(1 + \nu_2)q_2\omega^2}{4}(z - p)(\bar{z} - p) - \frac{(1 - \nu_2)q_2\omega^2 e^{i2\theta}}{8}(\bar{z} - p)^2 . \tag{8}$$

In order to facilitate the establishment of the boundary conditions, the complex displacement (6) is differentiated with respect to θ to give

$$\frac{\partial(u + iv)^{(1)}}{\partial\theta} = -\frac{iz(1 - \nu_2)q_2\omega^2}{16G_2}\left[2(z - p)(\bar{z} - p) - \frac{\bar{z}}{z}(z - p)^2\right] . \tag{9}$$

The complex stress combinations and differentiated complex displacement field due to the concentrated reaction force X are

$$(\widehat{\rho\rho} + \widehat{\theta\theta})^{(2)} = \frac{-X}{\pi(1 + n_2)}\left(\frac{1}{z - z_0} + \frac{1}{\bar{z} - \bar{z}_0}\right), \tag{10}$$

$$(\widehat{\rho\rho} - i\widehat{\rho\theta})^{(2)} = \frac{-X}{2\pi(1 + n_2)}\left[\frac{1 + n_2 e^{i2\theta}}{(z - z_0)} + \frac{1}{(\bar{z} - \bar{z}_0)} + \frac{(\bar{z} - \bar{z}_0)e^{i2\theta}}{(z - z_0)^2}\right], \tag{11}$$

$$\frac{\partial(u + iv)^{(2)}}{\partial\theta} = \frac{-izX}{4\pi G_2(1 + n_2)}\left\{\frac{n_2}{(z - z_0)} - \frac{1}{(\bar{z} - \bar{z}_0)} - \frac{\bar{z}}{z}\left[\frac{n_2}{(\bar{z} - \bar{z}_0)} + \frac{(z - z_0)}{(\bar{z} - \bar{z}_0)^2}\right]\right\} . \tag{12}$$

The stress combinations and differentiated displacement field due to the concentrated reaction moment M are

$$(\widehat{\rho\rho} + \widehat{\theta\theta})^{(3)} = 0 \,, \tag{13}$$

$$(\widehat{\rho\rho} - i\widehat{\rho\theta})^{(3)} = \frac{iMe^{i2\theta}}{2\pi(z - z_0)^2} \,, \tag{14}$$

$$\frac{\partial(u + iv)^{(3)}}{\partial\theta} = \frac{iz}{2G_2}\left[\frac{iMe^{-i2\theta}}{2\pi(\bar{z} - \bar{z}_0)^2}\right] . \tag{15}$$

The forms in Eqs. (10)–(15) are easily found from results given in ref [*1*].[1] Since the sums of these three stress and displacement fields do not satisfy the required boundary conditions, a solution of the equilibrium and compatibility equations corresponding to zero body force must be superimposed to obtain the solution of the problem under consideration. It is well known that such a solution is obtainable in the form

$$(\widehat{\rho\rho} + \widehat{\theta\theta})^{(4)} = 2[\Omega(z) + \bar{\Omega}(\bar{z})] \,, \tag{16}$$

$$(\widehat{\rho\rho} - i\widehat{\rho\theta})^{(4)} = \Omega(z) + \bar{\Omega}(\bar{z}) - e^{i2\theta}[\bar{z}\Omega'(z) + \chi(z)] \,, \tag{17}$$

$$\frac{\partial(u + iv)^{(4)}}{\partial\theta} = \frac{iz}{2G_2}\left[n_2\Omega(z) - \bar{\Omega}(\bar{z}) + \bar{z}\bar{\Omega}'(z) + \frac{\bar{z}}{z}\bar{\chi}(\bar{z})\right] , \tag{18}$$

where $\Omega(z)$ and $\chi(z)$ are functions analytic in the region of the insert. The total stress combination $\widehat{\rho\rho} - i\widehat{\rho\theta}$ in the insert is given by the sum of Eqs. (8), (11), (14), and (17) as

$$\begin{aligned}\widehat{\rho\rho} - i\widehat{\rho\theta} &= \Omega(z) + \bar{\Omega}(\bar{z}) - e^{i2\theta}[\bar{z}\Omega'(z) + \chi(z)] + \frac{X(z_0 - \bar{z}_0)e^{i2\theta}}{4\pi(z - z_0)^2} \\ &\quad - \frac{(1 + \nu_2)q_2\omega^2}{4}(z - p)(\bar{z} - p) - \frac{(1 - \nu_2)q_2\omega^2 e^{i2\theta}}{8}(\bar{z} - p)^2 \\ &\quad - \frac{X}{2\pi(1 + n_2)}\left[\frac{1 + n_2 e^{i2\theta}}{(z - z_0)} + \frac{1}{(\bar{z} - \bar{z}_0)} + \frac{(\bar{z} - \bar{z}_0)e^{i2\theta}}{(z - z_0)^2}\right] ,\end{aligned} \tag{19}$$

while the total differentiated displacement field in the insert is given by the sum of Eqs. (9), (12), (15), and (18) as

$$\begin{aligned}\frac{2}{iz}\frac{\partial(u + iv)}{\partial\theta} &= \frac{1}{G_2}\left[n_2\Omega(z) - \bar{\Omega}(\bar{z}) + \bar{z}\bar{\Omega}'(\bar{z}) + \frac{\bar{z}}{z}\bar{\chi}(\bar{z})\right] \\ &\quad - \frac{(1 - \nu_2)q_2\omega^2}{8G_2}\left[2(z - p)(\bar{z} - p) - \frac{\bar{z}}{z}(z - p)^2\right] + \frac{X(z_0 - \bar{z}_0)e^{-i2\theta}}{4\pi G_2(\bar{z} - \bar{z}_0)^2}\end{aligned}$$

---

[1] Reference [*1*], p 208.

$$-\frac{X}{2\pi G_2(1+n_2)}\left\{\frac{n_2}{(z-z_0)}-\frac{1}{(\bar{z}-\bar{z}_0)}-e^{-i2\theta}\left[\frac{n_2}{(\bar{z}-\bar{z}_0)}+\frac{(z-z_0)}{(\bar{z}-\bar{z}_0)^2}\right]\right\}. \tag{20}$$

The Ring

The stress and displacement fields due to the inertia forces can be obtained directly from those of the insert by replacing q_2, ν_2, and G_2 by q_1, ν_1, and G_1 respectively. Accordingly,

$$(\widehat{\rho\rho}+\widehat{\theta\theta})^{(5)}=-\frac{(1+\nu_1)q_1\omega^2}{2}(z-p)(\bar{z}-p), \tag{21}$$

$$(\widehat{\rho\rho}-i\widehat{\rho\theta})^{(5)}=-\frac{(1+\nu_1)q_1\omega^2}{4}(z-p)(\bar{z}-p)-\frac{(1-\nu_1)q_1\omega^2e^{i2\theta}}{8}(\bar{z}-p)^2, \tag{22}$$

$$\frac{\partial(u+iv)^{(5)}}{\partial\theta}=-\frac{iz(1-\nu_1)q_1\omega^2}{16G_1}\left[2(z-p)(\bar{z}-p)-\frac{\bar{z}}{z}(z-p)^2\right]. \tag{23}$$

Because of the presence of the insert, there is a resultant force F given by Eq. (3) on the inside edge of the ring. The stress and displacement fields corresponding to this force are

$$(\widehat{\rho\rho}+\widehat{\theta\theta})^{(6)}=\frac{-F}{\pi(1+n_1)}\left[\frac{1}{z}+\frac{1}{\bar{z}}\right], \tag{24}$$

$$(\widehat{\rho\rho}-i\widehat{\rho\theta})^{(6)}=\frac{-F}{2\pi(1+n_1)}\left[\frac{2}{z}+\frac{(1+n_1)}{\bar{z}}\right], \tag{25}$$

$$\frac{\partial(u+iv)^{(6)}}{\partial\theta}=\frac{izF}{2\pi G_1(1+n_1)\bar{z}}, \tag{26}$$

in which the required singularity is taken for simplicity to be at $z = 0$. To satisfy the boundary conditions, fields of the following form must be added to the two presented in Eqs. (21)–(26):

$$(\widehat{\rho\rho}+\widehat{\theta\theta})^{(7)}=2[\Phi(z)+\bar{\Phi}(\bar{z})], \tag{27}$$

$$(\widehat{\rho\rho}-i\widehat{\rho\theta})^{(7)}=\Phi(z)+\bar{\Phi}(\bar{z})-e^{i2\theta}[\bar{z}\Phi'(z)+\Psi(z)], \tag{28}$$

$$\frac{\partial(u+iv)^{(7)}}{\partial\theta}=\frac{iz}{2G_1}\left[n_1\Phi(z)-\bar{\Phi}(\bar{z})+\bar{z}\bar{\Phi}'(\bar{z})+\frac{\bar{z}}{z}\bar{\Psi}(\bar{z})\right], \tag{29}$$

where $\Phi(z)$ and $\Psi(z)$ are functions analytic in the region of the ring.

The total stress combination $\widehat{\rho\rho}-i\widehat{\rho\theta}$ in the ring is given by the sum of Eqs. (22), (25), and (28) to be

$$\widehat{\rho\rho} - i\widehat{\rho\theta} = \Phi(z) + \bar{\Phi}(\bar{z}) - e^{i2\theta}[\bar{z}\Phi'(z) + \Psi(z)] - \frac{(1+\nu_1)q_1\omega^2}{4}(z-p)(\bar{z}-p)$$

$$- \frac{(1-\nu_1)q_1\omega^2 e^{i2\theta}}{8}(\bar{z}-p)^2 - \frac{F}{2\pi(1+n_1)}\left[\frac{2}{z} + \frac{(1+n_1)}{\bar{z}}\right], \quad (30)$$

while the total differentiated displacement field in the ring is given by the sum of Eqs. (23), (26), and (29) to be

$$\frac{2}{iz}\frac{\partial(u+iv)}{\partial\theta} = \frac{1}{G_1}\left[n_1\Phi(z) - \bar{\Phi}(\bar{z}) + \bar{z}\bar{\Phi}'(\bar{z}) + \frac{\bar{z}}{z}\bar{\Psi}(\bar{z})\right]$$

$$+ \frac{F}{\pi G_1(1+n_1)\bar{z}} - \frac{(1-\nu_1)q_1\omega^2}{8G_1}\left[2(z-p)(\bar{z}-p) - \frac{\bar{z}}{z}(z-p)^2\right]. \quad (31)$$

Establishment of Boundary Equations

The boundary conditions for determining the four analytic functions $\Omega(z)$, $\chi(z)$, $\Phi(z)$, and $\Psi(z)$ can now be explicitly set up. The condition that the outside edge of the ring be traction free is equivalent to the condition that the stress combination $\widehat{\rho\rho} - i\widehat{\rho\theta}$ must vanish on that edge. This condition is obtained by setting the right side of Eq. (30) equal to zero on $z = R_1e^{i\theta}$, which gives

$$\{\Phi(z) + \bar{\Phi}(\bar{z}) - e^{i2\theta}[\bar{z}\Phi'(z) + \Psi(z)]\}_{z=R_1e^{i\theta}} = \sum_{-\infty}^{\infty} A_m e^{im\theta}, \quad (32)$$

where

$$\sum_{-\infty}^{\infty} A_m e^{im\theta} = \frac{(1+\nu_1)q_1\omega^2}{4}(R_1e^{i\theta} - p)(R_1e^{-i\theta} - p)$$

$$+ \frac{(1-\nu_1)q_1\omega^2}{8}(R_1 - pe^{i\theta})^2 + \frac{F}{2\pi(1+n_1)R_1}[2e^{-i\theta} + (1+n_1)e^{i\theta}]. \quad (33)$$

The condition of compatibility of normal and shear stresses at the interface is obtained by equating the right sides of Eqs. (19) and (30) on the interface circle $z = R_2e^{i\theta}$, which leads to

$$\{\Phi(z) + \bar{\Phi}(\bar{z}) - \Omega(z) - \bar{\Omega}(z) - e^{i2\theta}[\bar{z}\Phi'(z) + \Psi(z)$$

$$- \bar{z}\Omega'(z) - \chi(z)]\}_{z=R_2e^{i\theta}} = \sum_{-\infty}^{\infty} B_m e^{im\theta}, \quad (34)$$

where

$$\sum_{-\infty}^{\infty} B_m e^{im\theta} = \frac{X(z_0 - \bar{z}_0)e^{i2\theta}}{4\pi(R_2e^{i\theta} - z_0)^2} + \frac{[(1-\nu_1)q_1 - (1-\nu_2)q_2]\omega^2}{8}(R_2 - pe^{i\theta})^2$$

$$+ \frac{[(1+\nu_1)q_1 - (1+\nu_2)q_2]\omega^2}{4}(R_2e^{i\theta} - p)(R_2e^{-i\theta} - p)$$

$$- \frac{X}{2\pi(1+n_2)}\left[\frac{1+n_2e^{i2\theta}}{(R_2e^{i\theta} - z_0)} + \frac{1}{(R_2e^{-i\theta} - \bar{z}_0)} + \frac{(R_2e^{-i\theta} - \bar{z}_0)e^{i2\theta}}{(R_2e^{i\theta} - z_0)^2}\right]$$

$$+ \frac{F}{2\pi(1+n_1)R_2}[2e^{-i\theta} + (1+n_1)e^{i\theta}] . \quad (35)$$

The condition of compatibility of displacements at the interface is obtained by equating the right sides of Eqs. (20) and (31) on the interface circle $z = R_2e^{i\theta}$. The condition is

$$\left[n_1\Phi(z) - \bar{\Phi}(\bar{z}) + \bar{z}\bar{\Phi}'(\bar{z}) + \frac{\bar{z}}{z}\bar{\Psi}(\bar{z}) - n_2N\Omega(z) + N\bar{\Omega}(\bar{z})\right.$$

$$\left. - N\bar{z}\bar{\Omega}'(\bar{z}) - N\frac{\bar{z}}{z}\bar{\chi}(\bar{z})\right]_{z=R_2e^{i\theta}} = \sum_{-\infty}^{\infty} C_m e^{im\theta} , \quad (36)$$

where

$$\sum_{-\infty}^{\infty} C_m e^{im\theta} = \frac{-Fe^{i\theta}}{\pi(1+n_1)R_2} + \frac{XN(z_0 - \bar{z}_0)e^{-i2\theta}}{4\pi(R_2e^{-i\theta} - \bar{z}_0)^2}$$

$$+ \frac{[(1-\nu_1)q_1 - (1-\nu_2)q_2N]\omega^2}{8}[2(R_2e^{i\theta} - p)(R_2e^{-i\theta} - p) - (R_2 - pe^{-i\theta})^2]$$

$$- \frac{XN}{2\pi(1+n_2)}\left\{\frac{n_2}{(R_2e^{i\theta} - z_0)} - \frac{1}{(R_2e^{-i\theta} - \bar{z}_0)}\right.$$

$$\left. - e^{-i2\theta}\left[\frac{n_2}{(R_2e^{-i\theta} - \bar{z}_0)} + \frac{(R_2e^{i\theta} - z_0)}{(R_2e^{-i\theta} - \bar{z}_0)^2}\right]\right\} , \quad (37)$$

$$N = G_1/G_2 .$$

The A_m, B_m, and C_m evaluated directly from Eqs. (33), (35), and (37) are presented as follows:

$$A_m = 0 , \qquad m \geq 3 ,$$

$$A_2 = \frac{(1-\nu_1)q_1\omega^2p^2}{8} ,$$

$$A_1 = -\frac{q_1\omega^2 p R_2^2}{2R_1},$$

$$A_0 = \frac{q_1\omega^2}{8}[(3+\nu_1)R_1^2 + 2(1+\nu_1)p^2],$$

$$A_{-1} = -\frac{(1+\nu_1)q_1\omega^2 R_2^2 p}{4R_1},$$

$$A_m = 0, \qquad m \leqq -2,$$

$$B_m = \frac{-X}{2\pi(1+n_2)R_2}\left(\frac{\bar{z}_0}{R_2}\right)^{m-1}, \qquad m \geqq 3,$$

$$B_2 = \frac{[(1-\nu_1)q_1 - (1-\nu_2)q_2]\omega^2 p^2}{8} - \frac{X\bar{z}_0}{2\pi(1+n_2)R_2^2},$$

$$B_1 = -\frac{q_1\omega^2 p R_2}{2},$$

$$B_0 = \frac{[(3+\nu_1)q_1 - (3+\nu_2)q_2]\omega^2 R_2^2}{8} + \frac{[(1+\nu_1)q_1 - (1+\nu_2)q_2]\omega^2 p^2}{4} + \frac{X(1-n_2)(z_0+\bar{z}_0)}{4\pi(1+n_2)R_2^2},$$

$$B_{-1} = \frac{Xz_0(z_0+\bar{z}_0-n_2\bar{z}_0)}{2\pi(1+n_2)R_2^3} + \frac{q_1\omega^2 p}{4R_2}[(\nu_1-\nu_2)R_1^2 - (1+2\nu_1-\nu_2)R_2^2],$$

$$B_m = \frac{-X(z_0/R_2)^{-m-1}}{4\pi(1+n_2)R_2^3}\bar{Y}_{-m}, \qquad m \leqq -2,$$

$$C_m = \frac{XN(\bar{z}_0/R_2)^{m-1}}{4\pi(1+n_2)R_2^3} Y_m, \qquad m \geqq 2,$$

$$C_1 = \frac{XN}{\pi(1+n_2)R_2} - \frac{F}{\pi(1+n_1)R_2} - \frac{[(1-\nu_1)q_1 - (1-\nu_2)q_2 N]R_2 p\omega^2}{4} - \frac{XN\bar{z}_0(\bar{z}_0+z_0-n_2 z_0)}{2\pi(1+n_2)R_2^3},$$

$$C_0 = \frac{XN(n_2-1)(z_0+\bar{z}_0)}{4\pi(1+n_2)R_2^2} + \frac{[(1-\nu_1)q_1 - (1-\nu_2)q_2 N](R_2^2+2p^2)\omega^2}{8},$$

$$C_{-1} = 0,$$

$$C_{-2} = \frac{-[(1-\nu_1)q_1 - (1-\nu_2)q_2 N]\omega^2 p^2}{8} - \frac{XNn_2 z_0}{2\pi(1+n_2)R_2^2},$$

$$C_m = -\frac{XNn_2}{2\pi(1+n_2)R_2}\left(\frac{z_0}{R_2}\right)^{-m-1}, \qquad m \leqq -3,$$

where

$$Y_m = 2(m+1)R_2^2 + (m+1)n_2 z_0 \bar{z}_0 - (m-1)n_2 \bar{z}_0^2 - (m+1)\bar{z}_0(z_0 + \bar{z}_0). \tag{38}$$

The three boundary equations, (32), (34), and (36), completely determine the two analytic functions $\chi(z)$ and $\Psi(z)$, and determine the functions $\Omega(z)$ and $\Phi(z)$ to within one arbitrary imaginary constant which corresponds to a rigid-body rotation of the entire ring and insert assembly. This arbitrary rigid-body rotation arises because the boundary condition on the displacements is a relative one, the actual displacements not being prescribed. The rigid-body rotation of the ring is taken to be zero.

Since the four unknown functions are analytic in their respective regions in the z plane, they may be expanded in Laurent series of the form:

$$\begin{aligned}
\Omega(z) &= \sum_0^\infty a_m z^m, \\
\chi(z) &= \sum_0^\infty b_m z^m, \\
\Phi(z) &= \sum_{-\infty}^\infty c_m z^m, \\
\Psi(z) &= \sum_{-\infty}^\infty d_m z^m.
\end{aligned} \tag{39}$$

Substitution of the expansions (39) into the boundary equation (32) and a comparison of coefficients of $e^{im\theta}$ gives

$$(1-m)c_m R_1^m + \bar{c}_{-m} R_1^{-m} - d_{m-2} R_1^{m-2} = A_m, \tag{40}$$

which holds for all m.

Substitution of Eq. (39) into the boundary equation (34) and a comparison of coefficients of $e^{im\theta}$ gives,

for $m \geqq 2$,

$$(1-m)c_m R_2^m - (1-m)a_m R_2^m + \bar{c}_{-m} R_2^{-m} - d_{m-2} R_2^{m-2} + b_{m-2} R_2^{m-2} = B_m, \tag{41}$$

$$\bar{c}_{-1} - d_{-1} = B_1 R_2, \tag{42}$$

$$c_0 + \bar{c}_0 - (a_0 + \bar{a}_0) - d_{-2} R_2^{-2} = B_0; \tag{43}$$

for $m \leqq -1$,

$$(1-m)c_m R_2^m + \bar{c}_{-m} R_2^{-m} - \bar{a}_{-m} R_2^{-m} - d_{m-2} R_2^{m-2} = B_m \,. \tag{44}$$

Similarly, from Eqs. (39) and (36), after setting $m = -m$ and taking the complex conjugate, one finds,

for $m \geqq 2$,

$$n_1 \bar{c}_{-m} R_2^{-m} - (1-m)c_m R_2^m + N(1-m)a_m R_2^m + d_{m-2} R_2^{m-2} - N b_{m-2} R_2^{m-2} = \bar{C}_{-m} \,, \tag{45}$$

$$n_1 c_{-1} + \bar{d}_{-1} = C_{-1} R_2 \,, \tag{46}$$

$$n_1 c_0 - \bar{c}_0 - N(n_2 a_0 - \bar{a}_0) + \bar{d}_{-2} R_2^{-2} = C_0 \,; \tag{47}$$

for $m \leqq -1$,

$$n_1 \bar{c}_{-m} R_2^{-m} - (1-m)c_m R_2^m + d_{m-2} R_2^{m-2} - n_2 N \bar{a}_{-m} R_2^{-m} = \bar{C}_{-m} \,. \tag{48}$$

Solution of Boundary Equations

The boundary Eqs. (41)–(48), together with the previously mentioned condition on the rigid-body rotation of the ring, uniquely determine the coefficients in the Laurent expansion (39) and thus provide the solution to the problem. The algebra of solving the equations for these coefficients is straightforward but tedious. Only the results are given here. The coefficients are:

$$c_m = \frac{XNk^{m-1}(\bar{z}_0/R_2)^{m-1}}{4\pi R_1^{m+3} Z_m} \{2(1+m)(1-k^2)(1+n_2N)R_1^2 + [(n_1+N)+(1-N)k^{2m-2}]Y_m\} \,, \qquad m \geqq 2 \,,$$

$$c_1 = \frac{\left\{\begin{aligned}&\pi\omega^2 pk^2 R_1^4 \{q_1 N(1-k^2)[(1+\nu_2) - n_2(\nu_1-\nu_2)] - (1-\nu_1)q_1 k^2 \\ &\quad + 2q_2 N k^2 - (1+\nu_1)q_1\} - 2XN\bar{z}_0(z_0+\bar{z}_0 - n_2 z_0)\end{aligned}\right\}}{4\pi R_1^4[1+n_2N+(n_1-n_2N)k^4]} \,,$$

$$c_0 = \frac{\left\{\begin{aligned}&2XN(n_2-1)(z_0+\bar{z}_0) + \pi\omega^2\{2q_1 R_1^2[(3+\nu_1)R_1^2 + 2(1+\nu_1)p^2] \\ &\quad + 2(1-\nu_1)q_1 R_2^2(R_2^2+2p^2) + N(n_2-1)[(3+\nu_1)q_1(R_1^4 - R_2^4) \\ &\quad + 2(1+\nu_1)q_1 p^2(R_1^2 - R_2^2) + 2q_2 R_2^4]\}\end{aligned}\right\}}{16\pi R_1^2[2+(n_1-1)k^2 - N(1-n_2)(1-k^2)]} \,,$$

$$c_{-1} = \frac{-(1+\nu_1)q_1\omega^2 pR_2^2}{8} \,,$$

$$c_m = \frac{-XN(z_0/R_2)^{-m-1}}{4\pi R_2^{m+3} Z_{-m}} \{2R_2^2[(1+n_2N)+(n_1-n_2N)k^{-2m+2}] + (1+m)(1-N)k^{-2m}(1-k^2)\bar{Y}_{-m}\} \,, \qquad m \leqq -2 \,,$$

$$a_0 = \frac{\left\{\begin{array}{l}(1-\nu_2)X(z_0+\bar{z}_0)[2+(n_1-1)k^2-2N(1-k^2)] \\ +\pi\omega^2R_2^2[(1+n_1)q_1(3+\nu_1)R_1^2+(1-n_1)2q_1k^4R_1^2+2(3+\nu_2)q_2k^2R_1^2 \\ -(1-n_1)(3+\nu_2)q_2k^4R_1^2-2(1+\nu_2)q_2p^2k^2(1-n_1) \\ +2(1-\nu_2)q_2N(R_2^2+2p^2)(1-k^2)+4(1+\nu_2)q_2p^2-8q_1k^2R_1^2]\end{array}\right\}}{16\pi R_2^2[2+(n_1-1)k^2-N(1-n_2)(1-k^2)]},$$

$$a_1 = \frac{\left\{\begin{array}{l}\pi\omega^2(1+n_2)pk^2R_1^4\{[(1+\nu_2)+(1-\nu_2)k^2]q_1(1+n_1k^4) \\ +(1-\nu_2)q_1Nn_2k^2(1-k^4)-[(1+\nu_1)+(1-\nu_1)k^2]q_1k^4(1+n_1) \\ +(1+\nu_2)q_1N(1-k^4)[n_2-(1-k^2)(1+n_2N)] \\ -2q_2Nk^2(1-k^4)(1+n_2N)-(1+n_1)k^4q_1N(1-k^2)n_2(\nu_1-\nu_2)\} \\ -2X\bar{z}_0(z_0+\bar{z}_0-n_2z_0)[1+n_1k^4+N(1+n_1n_2)k^4-N(1-n_2) \\ -n_2N^2(1-k^4)]\end{array}\right\}}{4\pi R_2^4(1+n_2)(1+n_2N)[1+n_2N+(n_1-n_2N)k^4]},$$

$$a_m = \frac{X(\bar{z}_0/R_2)^{m-1}}{4\pi(1+n_2)R_2^{m+3}(1+n_2N)Z_m}\Big((1-N)Y_mZ_m+(1+n_1)(1+n_2)Nk^{2m+2}$$
$$\times\{2(1+m)(1-k^2)(1+n_2N)R_1^2+[(n_1+N)+(1-N)k^{2m-2}]Y_m\}\Big), \qquad m \geqq 2,$$

$$d_{m-2} = \frac{-XNk^{m-1}(\bar{z}_0/R_2)^{m-1}}{4\pi R_1^{m+1}Z_m}\{2R_1^2[(1+n_2N)(m^2+k^2-m^2k^2)$$
$$+(n_1-n_2N)k^{2m+2}]+(m-1)[(n_1+N)+(1-N)k^{2m}]Y_m\}, \qquad m \geqq 3,$$

$$d_0 = -\frac{(1-\nu_1)q_1\omega^2p^2}{8} - \frac{XN\bar{z}_0}{4\pi R_1^4Z_2}\{[(n_1+N)+(1-N)k^4]Y_2$$
$$+2R_1^2[(4-3k^2)(1+n_2N)+(n_1-n_2N)k^6]\},$$

$$d_{-1} = \frac{(3-\nu_1)q_1\omega^2pR_2^2}{8},$$

$$d_{-2} = \frac{\left\{\begin{array}{l}2XN(n_2-1)(z_0+\bar{z}_0)+\pi\omega^2R_2^2\{2(1-\nu_1)q_1R_2^2-(n_1-1)q_1(3+\nu_1)R_1^2 \\ +N(n_2-1)[(3+\nu_1)q_1(R_1^2-R_2^2)+2q_2R_2^2]\}\end{array}\right\}}{8\pi[2+(n_1-1)k^2-N(1-n_2)(1-k^2)]},$$

$$d_{-3}=\frac{\left\{\begin{array}{l}\pi\omega^2 pk^2R_1^4\{q_1N(1-k^2)[(1+\nu_2)-n_2(\nu_1-\nu_2)]-(1-\nu_1)q_1k^2\\ \qquad +2q_2Nk^2-(1+\nu_1)q_1\}-2XNz_0(z_0+\bar{z}_0-n_2\bar{z}_0)\end{array}\right\}}{4\pi[1+n_2N+(n_1-n_2N)k^4]},$$

$$d_{m-2}=\frac{-XN(z_0/R_2)^{-m-1}}{4\pi R_2^{m+1}Z_{-m}}\{2(1-m)R_2^2[(1+n_2N)+(n_1-n_2N)k^{-2m}]$$

$$-[(n_1+N)+(1-N)k^{-2m-2}(k^2+m^2-m^2k^2)]\bar{Y}_{-m}\},\qquad m\leqq -2,$$

$$b_0=-\frac{(1-\nu_2)q_2\omega^2p^2}{8}$$

$$-\frac{\left\{\begin{array}{l}X(\bar{z}_0/R_2)\Big((1-N)^2Y_2Z_2+2R_2^2(1+n_2N)^2Z_2\\ +(1-N)(1+n_1)(1+n_2)N\{k^4(1-k^2)(1+n_2N)Y_2\\ +k^6[(n_1+N)+(1-N)k^2]Y_2+6R_1^2k^6(1-k^2)(1+n_2N)]\\ -2R_2^2N(1+n_2N)(1+n_2)(1+n_1)[(1+n_2N)+(n_1-n_2N)k^6]\Big)\end{array}\right\}}{4\pi R_2^3(1+n_2)(1-N)(1+n_2N)Z_2},\ \text{for } m\geqq 3,$$

$$b_{m-2}=\frac{-X(\bar{z}_0/R_2)^{m-1}}{4\pi R_2^{m+1}(1+n_2)(1-N)(1+n_2N)Z_m}\Big((m-1)(1-N)^2Y_mZ_m$$

$$+2R_2^2(1+n_2N)^2Z_m+(m-1)(1-N)(1+n_1)(1+n_2)Nk^{2m}$$

$$\times\{(1-k^2)(1+n_2N)Y_m+k^2[(n_1+N)+(1-N)k^{2m-2}]Y_m$$

$$+2R_2^2(1+m)(1-k^2)(1+n_2N)\}-2R_2^2N(1+n_2N)(1+n_2)(1+n_1)$$

$$\times[(1+n_2N)+(n_1-n_2N)k^{2m+2}]\Big).$$

In these expressions, $k=R_2/R_1$ and

$$Z_m=(m^2-1)(1+n_2N)(1-N)k^{2m-2}(1-k^2)^2$$

$$+[(1+n_2N)+(n_1-n_2N)k^{2m+2}][(n_1+N)+(1-N)k^{2m-2}]. \quad (49)$$

Some Limiting Cases

As has been previously mentioned, the analysis presented contains many limiting configurations of interest. Attention is confined here to just three of them.

Elastic Ring with a Rigid Insert Rotating About an Arbitrary Point. – The stresses in a thin ring with a rigid insert rotating about an arbitrary point can be obtained from the preceding analysis by making the rigidity of the insert G_2 equal to infinity, which gives $N = 0$. For this case the only nonzero coefficients are c_1, c_0, c_{-1}, and $d_0, d_{-1}, d_{-2}, d_{-3}$. One finds that

$$\Phi(z) = \frac{q_1\omega^2}{8}\left\{\frac{[(3+\nu_1)R_1^2 + 2(1+\nu_1)p^2 + (1-\nu_1)k^2R_2^2 + 2(1-\nu_1)k^2p^2]}{[2+(n_1-1)k^2]} - \frac{2pk^2[(1+\nu_1)+(1-\nu_1)k^2]z}{(1+n_1k^4)} - \frac{(1+\nu_1)pR_2^2}{z}\right\}, \quad (50)$$

$$\Psi(z) = \frac{-q_1\omega^2}{8}\left\{(1-\nu_1)p^2 + \frac{2(1-\nu_1)R_2^2[(3+\nu_1)R_1^2 - (1+\nu_1)R_2^2]}{(1+\nu_1)[2+(n_1-1)k^2]z^2} - (3-\nu_1)\frac{pR_2^2}{z} + \frac{2pk^2R_1^4[(1+\nu_1)+(1-\nu_1)k^2]}{(1+n_1k^4)z^3}\right\}, \quad (51)$$

and that the stress combinations

$$\widehat{\rho\rho} + \widehat{\theta\theta} = \frac{q_1\omega^2}{2}\left\{\frac{R_1^2[(3+\nu_1)+(1-\nu_1)k^4]}{[2+(n_1-1)k^2]} - \frac{(1+\nu_1)pR_1^2}{2}\left[\frac{1}{z}+\frac{1}{\bar{z}}\right] - (1+\nu_1)z\bar{z} + \frac{p[(1+\nu_1)(1-k^2)+2k^4](z+\bar{z})}{(1+n_1k^4)}\right\}, \quad (52)$$

$$\widehat{\theta\theta} - \widehat{\rho\rho} + 2i\widehat{\rho\theta} = \frac{q_1\omega^2}{2}\left\{\frac{(1-\nu_1)}{2}z\bar{z} + \frac{(1+\nu_1)pR_1^2}{2}\left[\frac{1}{z}+\frac{n_1}{\bar{z}}\right] - (1-\nu_1)pz - \frac{pk^2[(1+\nu_1)+(1-\nu_1)k^2]}{(1+n_1k^4)}\left[z+\frac{R_1^4}{z^2\bar{z}}\right] - \frac{(1-\nu_1)R_2^2[(3+\nu_1)R_1^2-(1+\nu_1)R_2^2]}{(1+\nu_1)[2+(n_1-1)k^2]z\bar{z}}\right\}. \quad (53)$$

If in Eqs. (52) and (53), p is set equal to zero, the solution for the stresses in an elastic ring rotating concentrically about a rigid insert is obtained. If, in addition, R_2 is set equal to zero, the well-known formulas for the stresses in a disk rotating about its center are obtained.

Circular Disk Rotating About an Arbitrary Point. – The stresses in a thin circular disk rotating about an arbitrary point may be obtained by setting $G_1 = q_1 = N = 0$. The resulting series for $\Omega(z)$ and $\chi(z)$ may be summed to give:

$$\Omega(z) = \frac{q_2\omega^2}{16}\left[(3+\nu_2)R_2^2 + 2(1+\nu_2)p^2 - (1-\nu_2)p(z_0+\bar{z}_0) - 2\frac{(1+\nu_2)pR_2^2}{\bar{z}_0} - 4(1+\nu_2)pz + \frac{2(1+\nu_2)pR_2^6}{\bar{z}_0(R_2^2-\bar{z}_0z)^2} + \frac{2(1-\nu_2)pz_0R_2^4}{(R_2^2-\bar{z}_0z)^2} - \frac{4p\bar{z}_0R_2^4}{(R_2^2-\bar{z}_0z)^2} + \frac{2(3-\nu_2)p\bar{z}_0R_2^2}{(R_2^2-\bar{z}_0z)}\right], \quad (54)$$

$$\chi(z) = -\frac{(1-\nu_2)q_2\omega^2p^2}{8} - \frac{q_2\omega^2p\bar{z}_0}{8}\left[\frac{(1+\nu_2)R_2^2}{(R_2^2-\bar{z}_0z)} + \frac{(1+\nu_2)R_2^4}{(R_2^2-\bar{z}_0z)^2}\right.$$

$$+ \frac{(1-\nu_2)R_2^2\bar{z}_0(z_0+\bar{z}_0)}{(R_2^2-\bar{z}_0z)^2} + \frac{2(1+\nu_2)R_2^6}{(R_2^2-\bar{z}_0z)^3} - \frac{4\bar{z}_0^2R_2^4}{(R_2^2-\bar{z}_0z)^3}$$

$$\left. + \frac{2(1-\nu_2)z_0\bar{z}_0R_2^4}{(R_2^2-\bar{z}_0z)^3}\right], \tag{55}$$

and the complex stress combinations in the disk are given by

$$\widehat{\rho\rho} + \widehat{\theta\theta} = \frac{q_2\omega^2}{4}\left\{(3+\nu_2)R_2^2 - 2(1+\nu_2)z\bar{z} - (1-\nu_2)p(z_0+\bar{z}_0)\right.$$

$$- (1+\nu_2)pR_2^2\left[\frac{1}{z_0} + \frac{1}{\bar{z}_0}\right] + (1+\nu_2)pR_2^6\left[\frac{1}{\bar{z}_0(R_2^2-\bar{z}_0z)^2} + \frac{1}{z_0(R_2^2-z_0\bar{z})^2}\right]$$

$$+ (1-\nu_2)pR_2^4\left[\frac{z_0}{(R_2^2-\bar{z}_0z)^2} + \frac{\bar{z}_0}{(R_2^2-z_0\bar{z})^2}\right] - 2pR_2^4\left[\frac{\bar{z}_0}{(R_2^2-\bar{z}_0z)^2}\right.$$

$$\left. + \frac{z_0}{(R_2^2-z_0\bar{z})^2}\right] + (3-\nu_2)pR_2^2\left[\frac{\bar{z}_0}{(R_2^2-\bar{z}_0z)} + \frac{z_0}{(R_2^2-z_0\bar{z})}\right]$$

$$\left. - (1+\nu_2)pR_2^2\left[\frac{1}{z-z_0} + \frac{1}{\bar{z}-\bar{z}_0}\right]\right\}, \tag{56}$$

$$\widehat{\theta\theta} - \widehat{\rho\rho} + 2i\widehat{\rho\theta} = \frac{q_2\omega^2e^{i2\theta}}{4}\left\{(1-\nu_2)\bar{z}^2 - 4p\bar{z} - \frac{2pR_2^2(z_0-\bar{z}_0)}{(z-z_0)^2}\right.$$

$$+ (1+\nu_2)pR_2^2\left[\frac{(\bar{z}-\bar{z}_0)}{(z-z_0)^2} + \frac{n_2}{(z-z_0)}\right] + \frac{2(1+\nu_2)pR_2^6(\bar{z}-\bar{z}_0)}{(R_2^2-\bar{z}_0z)^3}$$

$$+ \frac{2(1-\nu_2)pz_0\bar{z}_0R_2^4(\bar{z}-\bar{z}_0)}{(R_2^2-\bar{z}_0z)^3} - \frac{4p\bar{z}_0^2R_2^4(\bar{z}-\bar{z}_0)}{(R_2^2-\bar{z}_0z)^3} + \frac{(3-\nu_2)pR_2^2\bar{z}_0^2\bar{z}}{(R_2^2-\bar{z}_0z)^2}$$

$$\left. - \frac{(1+\nu_2)R_2^2p\bar{z}_0}{(R_2^2-\bar{z}_0z)} - \frac{(1+\nu_2)R_2^4p\bar{z}_0}{(R_2^2-\bar{z}_0z)^2} - \frac{(1-\nu_2)R_2^2\bar{z}_0^2(z_0+\bar{z}_0)p}{(R_2^2-\bar{z}_0z)^2}\right\}. \tag{57}$$

If in Eqs. (56) and (57), $p = z_0 = \bar{z}_0$, the stresses in a circular disk rotating about a point in the disk are obtained. This result agrees with that obtained in ref [3]. Again, if $p = z_0 = \bar{z}_0 = 0$, the stresses in a circular disk rotating about its center are obtained.

Rotating Disk Mounted Eccentrically Inside a Constraining Rigid Ring. – The stresses in a thin circular disk mounted in a constraining rigid ring which is rotating about an arbitrary point are obtained by setting $G_1 = N = \infty$.

The resulting series for $\Omega(z)$ and $\chi(z)$ may be summed to give:

$$\Omega(z) = \frac{(1+\nu_2)q_2\omega^2(R_2^2+2p^2)}{16} + \frac{(1+\nu_2)[H+\pi q_2\omega^2 pR_2^2][2R_2^2-\bar{z}_0(z_0+\bar{z}_0)]}{16\pi n_2\bar{z}_0R_2^2}$$

$$-\frac{(1-\nu_2)q_2\omega^2pz}{4n_2} - \frac{(H+\pi q_2\omega^2pR_2^2)}{4\pi(1+n_2)n_2\bar{z}_0}\left[\frac{2R_2^4}{(R_2^2-\bar{z}_0z)^2} + \frac{n_2R_2^2z_0\bar{z}_0}{(R_2^2-\bar{z}_0z)^2}\right.$$

$$\left.-\frac{n_2\bar{z}_0^2R_2^2}{(R_2^2-\bar{z}_0z)^2} - \frac{\bar{z}_0(z_0+\bar{z}_0)R_2^2}{(R_2^2-\bar{z}_0z)^2} + \frac{2n_2\bar{z}_0^2}{(R_2^2-\bar{z}_0z)}\right], \quad (58)$$

$$\chi(z) = \frac{-(1-\nu_2)q_2\omega^2p^2}{8} + \frac{(H+\pi q_2\omega^2pR_2^2)\bar{z}_0}{4\pi n_2(1+n_2)}\left[\frac{2n_2^2}{(R_2^2-\bar{z}_0z)} + \frac{2R_2^2}{(R_2^2-\bar{z}_0z)^2}\right.$$

$$+\frac{(n_2-1)\bar{z}_0(z_0+\bar{z}_0)}{(R_2^2-\bar{z}_0z)^2} + \frac{2n_2R_2^2\bar{z}_0(z_0-\bar{z}_0)}{(R_2^2-\bar{z}_0z)^3} + \frac{4R_2^4}{(R_2^2-\bar{z}_0z)^3}$$

$$\left.-\frac{2R_2^2\bar{z}_0(z_0+\bar{z}_0)}{(R_2^2-\bar{z}_0z)^3}\right], \quad (59)$$

in which H denotes the total mass of the rigid ring whose center of mass is at $z = 0$. Hence

$$\widehat{\rho\rho} + \widehat{\theta\theta} = \frac{(1+\nu_2)q_2\omega^2R_2^2}{4} - \frac{(1+\nu_2)q_2\omega^2z\bar{z}}{2} + \frac{q_2\omega^2p(z+\bar{z})}{n_2}$$

$$+\frac{(1+\nu_2)[H+\pi q_2\omega^2pR_2^2][2R_2^2-\bar{z}_0(z_0+\bar{z}_0)]}{4\pi n_2\bar{z}_0R_2^2} - \frac{(H+\pi q_2\omega^2pR_2^2)}{\pi(1+n_2)}$$

$$\times\left[\frac{1}{(z-z_0)} + \frac{1}{(\bar{z}-\bar{z}_0)}\right] - \frac{(1+\nu_2)(H+\pi q_2\omega^2pR_2^2)}{8\pi n_2\bar{z}_0}\left\{\frac{2n_2\bar{z}_0^2}{(R_2^2-\bar{z}_0z)}\right.$$

$$+\frac{2n_2z_0^2}{(R_2^2-z_0\bar{z})} + \frac{R_2^2[2R_2^2+n_2\bar{z}_0(z_0-\bar{z}_0)-\bar{z}_0(z_0+\bar{z}_0)]}{(R_2^2-\bar{z}_0z)^2}$$

$$\left.+\frac{R_2^2[2R_2^2-n_2z_0(z_0-\bar{z}_0)-z_0(z_0+\bar{z}_0)]}{(R_2^2-z_0\bar{z})^2}\right\}, \quad (60)$$

$$\widehat{\theta\theta} - \widehat{\rho\rho} + 2i\widehat{\rho\theta} = \frac{(1-\nu_2)q_2\omega^2z\bar{z}}{4} - \frac{2(1-\nu_2)q_2\omega^2pz}{(3-\nu_2)}$$

$$+\frac{(H+\pi q_2\omega^2pR_2^2)e^{i2\theta}}{\pi(1+n_2)}\left[\frac{\bar{z}-\bar{z}_0}{(z-z_0)^2} + \frac{n_2}{(z-z_0)}\right] - \frac{(H+\pi q_2\omega^2pR_2^2)e^{i2\theta}(z_0-\bar{z}_0)}{2\pi(z-z_0)^2}$$

$$-\frac{(H+\pi q_2\omega^2pR_2^2)e^{i2\theta}}{2\pi(1+n_2)n_2}\left[\frac{2n_2R_2^2\bar{z}_0(z_0-\bar{z}_0)(\bar{z}-\bar{z}_0)}{(R_2^2-\bar{z}_0z)^3} + \frac{4R_2^4(\bar{z}-\bar{z}_0)}{(R_2^2-\bar{z}_0z)^3}\right.$$

$$- \frac{2R_2^2\bar{z}_0(z_0+\bar{z}_0)(\bar{z}-\bar{z}_0)}{(R_2^2-\bar{z}_0 z)^3} + \frac{2n_2\bar{z}_0^2\bar{z}}{(R_2^2-\bar{z}_0 z)^2} - \frac{(n_2-1)\bar{z}_0^2(z_0+\bar{z}_0)}{(R_2^2-\bar{z}_0 z)^2}$$

$$- \frac{2R_2^2\bar{z}_0}{(R_2^2-\bar{z}_0 z)^2} - \frac{2n_2^2\bar{z}_0}{(R_2^2-\bar{z}_0 z)}\Bigg] . \quad (61)$$

CASE 2: ASSEMBLY FASTENED IN RING

The analysis required in this case, shown in Fig. 2, proceeds along lines similar to those given above. The total stress and differentiated displacement fields in the insert are found to be

$$\widehat{\rho\rho} + \widehat{\theta\theta} = 2[\Omega(z) + \bar{\Omega}(\bar{z})] - \frac{(1+\nu_2)q_2\omega^2}{2}(z-p)(\bar{z}-p) , \quad (62)$$

$$\widehat{\rho\rho} - i\widehat{\rho\theta} = \Omega(z) + \bar{\Omega}(\bar{z}) - e^{i2\theta}[\bar{z}\Omega'(z) + \chi(z)]$$

$$- \frac{(1+\nu_2)q_2\omega^2}{4}(z-p)(\bar{z}-p) - \frac{(1-\nu_2)q_2\omega^2 e^{i2\theta}}{8}(\bar{z}-p)^2 , \quad (63)$$

$$\frac{\partial(u+iv)}{\partial\theta} = \frac{iz}{2G_2}\left[n_2\Omega(z) - \bar{\Omega}(\bar{z}) + \bar{z}\bar{\Omega}'(\bar{z}) + \frac{\bar{z}}{z}\bar{\chi}(\bar{z})\right]$$

$$- \frac{iz(1-\nu_2)q_2\omega^2}{16G_2}\left[2(z-p)(\bar{z}-p) - \frac{\bar{z}}{z}(z-p)^2\right] , \quad (64)$$

and the total stress and differentiated displacement fields in the ring are

$$\widehat{\rho\rho} + \widehat{\theta\theta} = 2[\Phi(z) + \bar{\Phi}(\bar{z})] - \frac{(1+\nu_1)q_1\omega^2}{2}(z-p)(\bar{z}-p)$$

$$- \frac{X}{\pi(1+n_1)}\left[\frac{1}{(z-z_0)} + \frac{1}{(\bar{z}-\bar{z}_0)}\right] - \frac{F'}{\pi(1+n_1)}\left[\frac{1}{z} + \frac{1}{\bar{z}}\right] , \quad (65)$$

$$\widehat{\rho\rho} - i\widehat{\rho\theta} = \Phi(z) + \bar{\Phi}(\bar{z}) - e^{i2\theta}[\bar{z}\Phi'(z) + \Psi(z)] + \frac{X(z_0-\bar{z}_0)e^{i2\theta}}{4\pi(z-z_0)^2}$$

$$- \frac{(1+\nu_1)q_1\omega^2}{4}(z-p)(\bar{z}-p) - \frac{(1-\nu_1)q_1\omega^2 e^{i2\theta}}{8}(\bar{z}-p)^2$$

$$- \frac{X}{2\pi(1+n_1)}\left[\frac{1+n_1 e^{i2\theta}}{(z-z_0)} + \frac{1}{(\bar{z}-\bar{z}_0)} + \frac{(\bar{z}-\bar{z}_0)e^{i2\theta}}{(z-z_0)^2}\right]$$

$$- \frac{F'}{2\pi(1+n_1)}\left[\frac{2}{z} + \frac{(1+n_1)}{\bar{z}}\right] , \quad (66)$$

$$\frac{\partial(u+iv)}{\partial\theta} = \frac{iz}{2G_1}\quad n_1\Phi(z) - \bar{\Phi}(\bar{z}) + \bar{z}\bar{\Phi}'(\bar{z}) + \frac{\bar{z}}{z}\bar{\Psi}(\bar{z})$$

$$- \frac{iz(1-\nu_1)q_1\omega^2}{16G_1}\left[2(z-p)(\bar{z}-p) - \frac{\bar{z}}{z}(z-p)^2\right] + \frac{izF'}{2\pi G_1(1+n_1)\bar{z}}$$

$$- \frac{izX}{4\pi G_1(1+n_1)}\left\{\frac{n_1}{(z-z_0)} - \frac{1}{(\bar{z}-\bar{z}_0)} - e^{-i2\theta}\left[\frac{n_1}{(\bar{z}-\bar{z}_0)}\right.\right.$$

$$\left.\left. + \frac{(z-z_0)}{(\bar{z}-\bar{z}_0)^2}\right]\right\} + \frac{iz}{2G_1}\left[\frac{X(z_0-\bar{z}_0)e^{-i2\theta}}{4\pi(\bar{z}-\bar{z}_0)^2}\right], \qquad (67)$$

where F' is the resultant force on the inside edge of the ring due to the presence of the insert and is given by

$$F' = -\pi q_2\omega^2 pR_2^2\,. \qquad (68)$$

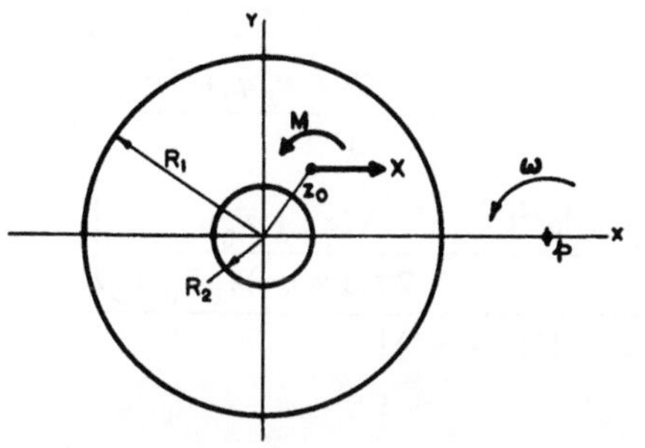

Fig. 2. Rotating Elastic Ring with Circular Elastic Insert. Assembly mounted in ring.

The coefficients in the Laurent expansions of the four complex potential functions $\Phi(z)$, $\Psi(z)$, $\Omega(z)$, and $\chi(z)$ are found from the boundary conditions to be:

$$c_m = \frac{X}{4\pi(1+n_1)R_1^m Z_m}\left([(n_1+N)+(1-N)k^{2m-2}]\left\{\frac{(1+n_2N)(\bar{z}_0/R_1)^{m-1}}{R_1^3}\right.\right.$$

$$\left.\times[2(1+m)R_1^2+Y_m'] - \frac{2(n_1-n_2N)}{z_0}\left(\frac{R_2}{z_0}\right)^m k^{m+2}\right\}$$

$$-(1+m)(1+n_2N)(1-N)(1-k^2)k^m\left\{\frac{2(\bar{z}_0/R_1)^{m-1}k^{m-2}}{R_1}\right.$$

$$\left.\left. + \frac{(R_2/z_0)^{m-2}}{z_0^3}[2(1-m)R_2^2+\bar{Y}'_{-m}]\right\}\right), \qquad m \geq 2\,,$$

$$c_m = \frac{X(1-N)R_2^{-m}}{4\pi(1+n_1)Z_{-m}}\left([(1+n_2N)+(n_1-n_2N)k^{-2m+2}]\left\{\frac{2(z_0/R_1)^{-m-1}k^{-m-2}}{R_1}\right.\right.$$

$$+ \frac{(R_2/\bar{z}_0)^{-m-2}}{\bar{z}_0^3}[2(1+m)R_2^2 + Y'_m]\Bigg\} + (1+m)(1-k^2)k^{-m-2}$$

$$\times \left\{\frac{2(n_1 - n_2 N)}{\bar{z}_0}\left(\frac{R_2}{\bar{z}_0}\right)^{-m} k^{-m+2} - \frac{(1+n_2 N)}{R_1^3}\left(\frac{z_0}{R_1}\right)^{-m-1}\right.$$

$$\left.\times [2(1-m)R_1^2 + \bar{Y}'_{-m}]\right\}\Bigg), \qquad m \leqq -2 ,$$

$$c_{-1} = \frac{-q_1 \omega^2 p R_2^2}{2(1+n_1)},$$

$$c_0 = \frac{\left\{\begin{aligned} &[2 + N(n_2 - 1)]\left[\frac{(3+\nu_1)q_1\omega^2 R_1^2}{8} + \frac{(1+\nu_1)q_1\omega^2 p^2}{4}\right. \\ &\left. + \frac{(n_1 - 1)X(z_0 + \bar{z}_0)}{4\pi(1+n_1)R_1^2}\right] + \frac{[(1-\nu_1)q_1 - (1-\nu_2)q_2 N]\omega^2 k^2 (R_2^2 + 2p^2)}{4} \\ &- \frac{2Xk^2(n_1 - 1)(z_0 + \bar{z}_0)}{4\pi(1+n_1)z_0\bar{z}_0} - k^2 N(n_2 - 1)\left\{\frac{[(3+\nu_1)q_1 - (3+\nu_2)q_2]\omega^2 R_2^2}{8}\right. \\ &\left. + \frac{[(1+\nu_1)q_1 - (1+\nu_2)q_2]\omega^2 p^2}{4} - \frac{X(z_0 + \bar{z}_0)}{2\pi(1+n_1)z_0\bar{z}_0}\right\} \end{aligned}\right\}}{2[2 + (n_1 - 1)k^2 + N(n_2 - 1)(1 - k^2)]},$$

$$c_1 = \frac{\left\{\begin{aligned} &\pi\omega^2 p R_2^4[(1+\nu_1)q_2 - 2q_1 + q_2 N(1 - \nu_2 + n_2\nu_1 - n_2\nu_2)] \\ &- \frac{2X}{(1+n_1)z_0^2}[(n_1 - n_2 N)R_2^4 + (1+n_2 N)z_0^2\bar{z}_0(z_0 + \bar{z}_0 - n_1 z_0)] \end{aligned}\right\}}{4\pi R_1^4[(1+n_2 N) + (n_1 - n_2 N)k^4]},$$

$$d_{m-2} = \frac{X}{4\pi(1+n_1)R_1^{m-1}Z_m}\Bigg((1-m)[(n_1 + N) + (1-N)k^{2m}]$$

$$\times \left\{\frac{(1+n_2 N)}{R_1^2}\left(\frac{\bar{z}_0}{R_1}\right)^{m-1}[2(1+m)R_1^2 + Y'_m] - 2(n_1 - n_2 N)\left(\frac{R_2}{z_0}\right)^{m+1} k^{m+1}\right\}$$

$$+ k^{m-1}(1-N)[(1+n_2 N)(k^2 + m^2 - k^2 m^2) + (n_1 - n_2 N)k^{2m+2}]$$

$$\times \left\{2\left(\frac{\bar{z}_0}{R_1}\right)^{m-1} k^{m-1} + \frac{(R_2/z_0)^{m-1}}{z_0^2}[2(1-m)R_2^2 + \bar{Y}'_{-m}]\right\}$$

$$- 2\left(\frac{\bar{z}_0}{R_1}\right)^{m-1} Z_m\Bigg), \qquad m \geqq 3 ,$$

$$d_{m-2}=\frac{XR_2^{-m}}{4\pi(1+n_1)R_1Z_{-m}}\Bigg((1-m)(1-N)[(1+n_2N)+(n_1-n_2N)k^{-2m}]$$

$$\times\left\{2\left(\frac{z_0}{R_1}\right)^{-m-1}k^{-m-2}R_2^2+\frac{R_1}{\bar{z}_0}\left(\frac{R_2}{\bar{z}_0}\right)^{-m}[2(1+m)R_2^2+Y'_m]\right\}$$

$$+(1-m)(1+m)(1-N)(1-k^2)k^{-m-2}\left\{2(n_1-n_2N)\left(\frac{R_2}{\bar{z}_0}\right)^{-m+1}k^{-m+1}R_1^2\right.$$

$$\left.-k^2(1+n_2N)\left(\frac{z_0}{R_1}\right)^{-m-1}[2(1-m)R_1^2+\bar{Y}'_{-m}]\right\}-[(n_1+N)$$

$$+(1-N)k^{-2m-2}](n_1-n_2N)\left\{2\left(\frac{R_2}{\bar{z}_0}\right)^{-m+1}kR_1^2+(z_0/R_1)^{-m-1}k^{-m+2}\right.$$

$$\left.\times[2(1-m)R_1^2+\bar{Y}'_{-m}]\right\}\Bigg),\qquad m\leqq -2,$$

$$d_0=-\frac{(1-\nu_1)q_1\omega^2p^2}{8}-\frac{X}{4\pi(1+n_1)R_1Z_2}\Bigg([(n_1+N)+(1-N)k^4]$$

$$\times\left\{\frac{(1+n_2N)\bar{z}_0}{R_1^3}[6R_1^2+Y'_2]-2\,\frac{(n_1-n_2N)R_2^3k^3}{z_0^3}\right\}+\frac{2\bar{z}_0Z_2}{R_1}$$

$$-k(1-N)[(1+n_2N)(4-3k^2)+(n_1-n_2N)k^6]\left[\left(\frac{2\bar{z}_0}{R_1}\right)k\right.$$

$$\left.+\frac{R_2}{z_0^3}(-2R_2^2+\bar{Y}'_{-2})\right]\Bigg),$$

$$d_{-1}=\frac{(3-\nu_1)q_1\omega^2pR_2^2}{8},$$

$$d_{-2}=\frac{\left\{\begin{array}{l}R_2^2\left\{N(n_2-1)\left[\dfrac{q_2\omega^2R_2^2}{4}+\dfrac{(3+\nu_1)q_1\omega^2R_1^2(1-k^2)}{8}\right.\right.\\[2ex] \left.+\dfrac{X(z_0+\bar{z}_0)}{2\pi(1+n_1)z_0\bar{z}_0}+\dfrac{(n_1-1)X(z_0+\bar{z}_0)}{4\pi(1+n_1)R_1^2}\right]-\dfrac{(1-\nu_1)q_1\omega^2R_1^2(1-k^2)}{4}\\[2ex] \left.-\dfrac{(n_1-1)q_1\omega^2R_1^2}{4}-\dfrac{2X(n_1-1)(z_0+\bar{z}_0)}{4\pi(1+n_1)z_0\bar{z}_0}-\dfrac{(n_1-1)^2X(z_0+\bar{z}_0)}{4\pi(1+n_1)R_1^2}\right\}\end{array}\right\}}{[2+(n_1-1)k^2+N(n_2-1)(1-k^2)]},$$

$$d_{-3} = \frac{\left\{\begin{array}{l}\pi\omega^2 pR_2^4[(1+\nu_1)q_2 - 2q_1 + q_2N(1-\nu_2+n_2\nu_1-n_2\nu_2)] \\ \qquad + \dfrac{2X(n_1-n_2N)}{(1+n_1)\bar{z}_0^2}[k^4\bar{z}_0^2 z_0(z_0+\bar{z}_0-n_1\bar{z}_0) - R_2^4]\end{array}\right\}}{4\pi[(1+n_2N)+(n_1-n_2N)k^4]},$$

$$\begin{aligned} a_m = \frac{X}{4\pi R_2^m Z_m} & \left\{k^m[(n_1+N)+(1-N)k^{2m-2}]\frac{1}{R_1^3}\left(\frac{\bar{z}_0}{R_1}\right)^{m-1}[2(m+1)R_1^2 + Y_m']\right. \\ & -(m+1)(1-N)(1-k^2)k^{2m}\left[\frac{2(\bar{z}_0/R_1)^{m-1}k^{m-2}}{R_1} + \frac{(R_2/z_0)^{m-2}}{z_0^3}\bar{Y}_{-m}\right] \\ & + \frac{2}{z_0}\left(\frac{R_2}{z_0}\right)^m [k^{2m-2}(m^2-1)(1-N)(1-k^2) + (n_1+N) \\ & \left. + (1-N)k^{2m-2}]\right\}, \quad m \geqq 2, \end{aligned}$$

$$a_1 = \frac{\left\{\begin{array}{l}\dfrac{2X}{z_0^2}[R_1^4 - z_0^2\bar{z}_0(z_0+\bar{z}_0-n_1z_0)] + \pi\omega^2 pR_1^4\Big[2q_1(1-k^4) \\ \qquad - (1-\nu_2)q_2N(1-k^4) - (1+\nu_2)q_2 - \nu_2q_2n_1k^4 \\ \qquad\qquad + \dfrac{1}{4}(1+n_1)(1+5\nu_1)q_2k^4\Big]\end{array}\right\}}{4\pi R_1^4[(1+n_2N)+(n_1-n_2N)k^4]},$$

$$a_0 = -\frac{X(z_0-\bar{z}_0)}{4\pi z_0\bar{z}_0N(n_2-1)}$$

$$+\frac{\left\{\begin{array}{l}\dfrac{(1+n_1)(3+\nu_1)q_1\omega^2R_1^2}{8} + \dfrac{(n_1-1)X(z_0+\bar{z}_0)}{4\pi R_1^2} + \dfrac{X(z_0+\bar{z}_0)}{2\pi z_0\bar{z}_0} \\ - \dfrac{[4q_1-(3+\nu_2)q_2]\omega^2R_2^2}{4} + \dfrac{(1+\nu_2)q_2\omega^2p^2}{2} \\ + \dfrac{(n_1-1)(1+\nu_2)q_2\omega^2p^2k^2}{4} - \dfrac{(n_1-1)[2q_1-(3+\nu_2)q_2\omega^2k^2R_2^2}{8} \\ \qquad + \dfrac{(1-\nu_2)q_2N\omega^2(1-k^2)(R_2^2+2p^2)}{4}\end{array}\right\}}{2[2+(n_1-1)k^2+N(n_2-1)(1-k^2)]},$$

$$b_{m-2} = \frac{-X}{4\pi R_2^{m-2}Z_m}\left\{\left[\frac{2(\bar{z}_0/R_1)^{m-1}k^{m-2}}{R_1} + \frac{(R_2/z_0)^{m-2}}{z_0^3}\bar{Y}_{-m}'\right][(1+n_2N)\right.$$

$$+ (n_1 - n_2 N)k^{2m+2} - (m+1)(m-1)(1-N)(1-k^2)k^{2m}]$$

$$+ (m-1)k^{m-2} \frac{(\bar{z}_0/R_1)^{m-1}}{R_1^3} [2(1+m)R_1^2 + Y'_m]\{(1+n_2 N)(1-k^2)$$

$$+ k^2[(n_1 + N) + (1-N)k^{2m-2}]\} + (m-1)\frac{2}{z_0}\left(\frac{R_2}{z_0}\right)^m$$

$$\times [(m+1)(m-1)(1-N)(1-k^2)k^{2m-2} + (n_1+N) + (1-N)k^{2m-2}$$

$$- (n_1 - n_2 N)k^{2m} - (1 + n_2 N)]\Big\}, \qquad m \geqq 3,$$

$$b_0 = -\frac{(1-\nu_2)q_2\omega^2 p^2}{8} - \frac{X}{4\pi Z_2}\left\{\left[\frac{2\bar{z}_0}{R_1^2} + \frac{1}{z_0^3}\bar{Y}'_{-2}\right][(1+n_2 N) + (n_1 - n_2 N)k^6\right.$$

$$- 3(1-N)(1-k^2)k^4] + \frac{\bar{z}_0}{R_1^4}[6R_1^2 + Y'_2][1 + n_2 N)(1-k^2) + (n_1 + N)k^2$$

$$+ (1-N)k^4] + \frac{2R_2^2}{z_0^3}[3k^2(1-N)(1-k^2) + (n_1+N) + (1-N)k^2$$

$$\left. - (n_1 - n_2 N)k^4 - (1 + n_2 N)]\right\}.$$

In these expressions, the notation of case 1 has been used, and in addition,

$$Y'_m = \bar{z}_0[(1+m)n_1 z_0 + (1-m)n_1 \bar{z}_0 - (1+m)(z_0 + \bar{z}_0)]. \tag{69}$$

Some Limiting Cases

As in case 1, a number of configurations of interest are readily attained by choosing special values for some of the arbitrary parameters.

Elastic Ring Rotating About an Arbitrary Point. – This case is obtained by setting $q_2 = 0$ and $N = \infty$, which corresponds to $G_2 = 0$ in the expressions for c_m and d_m.

Elastic Ring with Rigid Insert Rotating About an Arbitrary Point. – This is obtained by setting $N = 0$, which corresponds to $G_2 = \infty$ in the expressions for c_m and d_m.

Elastic Insert in a Rigid Ring Rotating About an Arbitrary Point. – The stress field in an elastic circular insert in a rigid ring rotating about an arbitrary point may be obtained from the preceding analysis by setting $G_1 = N = \infty$ in the expressions for a_m and b_m. The only nonzero terms are found to be a_1, a_0, and b_0, giving

$$\Omega(z) = \frac{(1+\nu_2)q_2\omega^2(R_2^2 + 2p^2)}{16} - \frac{(1-\nu_2)q_2\omega^2 pz}{4n_2}, \tag{70}$$

$$\chi(z) = -\frac{(1-\nu_2)q_2\omega^2 p^2}{8}, \tag{71}$$

from which

$$\widehat{\rho\rho} + \widehat{\theta\theta} = \frac{(1+\nu_2)q_2\omega^2}{4}\left[R_2^2 + \frac{4}{(3-\nu_2)}\, p(z+\bar{z}) - 2z\bar{z}\right], \tag{72}$$

$$\widehat{\theta\theta} - \widehat{\rho\rho} + 2i\widehat{\rho\theta} = \frac{(1-\nu_2)q_2\omega^2 z}{4}\left[\bar{z} - \frac{8p}{(3-\nu_2)}\right]. \tag{73}$$

REFERENCES

[1] N. I. Muskhelishvili, *Some Basic Problems in the Mathematical Theory of Elasticity*, 3d ed., Noordhoff, Groningen, Netherlands, 1953.

[2] A. C. Stevenson, "Complex Potential in Two Dimensional Elasticity," *Proc. Roy. Soc. (London)* **A184**, 129 (1945).

[3] B. Sen, "Boundary Value Problems of Circular Discs Under Body Forces. Part II," *Bull. Calcutta Math. Soc.* **36**, 83 (1944).

PERTURBATION SOLUTION OF A CLASS OF AXISYMMETRICAL SHELLS SUBJECTED TO NORMAL LOADING[1]

Kin N. Tong

Syracuse University

This paper presents a method for finding the stresses and displacements in a shell of revolution which does not differ much in shape from a spherical dome. By using as the geometrical parameter the ratio of the two local radii of curvature and examining its difference from unity, a perturbation solution is derived in which the recurrence step is the solution of the spherical problem.

INTRODUCTION

The problem investigated in this paper is that of finding the stress distribution in an axisymmetrical shell which is reasonably close to being a sphere or a hemisphere and is loaded by axisymmetrically distributed normal force.

This problem is studied with the purpose of finding a theory to assess the effects of imperfection in a spherical shell subjected to uniform pressure, and to form the basis of a stepwise integration procedure in the study of linear-creep deformation of initially spherical shells.

BASIC EQUATIONS

According to the notation shown in Fig. 1, the fundamental equations of the problem are as follows:

Equations of Equilibrium. –

$$2N_\phi \sin\phi + 2Q\cos\phi + \bar{p}r_0 = 0\,, \tag{1}$$

$$N_\phi r_0 + N_\theta r_1 \sin\phi + \frac{d}{d\phi}(Qr_0) + pr_1 r_0 = 0\,, \tag{2}$$

$$\frac{d}{d\phi}(M_\phi r_0) - M_\theta r_1 \cos\phi - Qr_1 r_0 = 0\,, \tag{3}$$

[1]Sponsored by Oak Ridge National Laboratory, Union Carbide Nuclear Co., a Division of Union Carbide Corp., as a part of subcontract No. 913 under contract No. W-7405-eng-26.

NOTATION

r_0	Distance to axis of rotation
r_1, r_2	Principal radii of curvature (Fig. 1)
N_ϕ, N_θ	Normal-stress resultants (Fig. 1)
M_ϕ, M_θ	Bending moments (Fig. 1)
Q	Shear-stress resultant (Fig. 1)
p	Normal surface load (Fig. 1)
$\bar{p}$	Average projected pressure (Eq. 3*a*)
V	Rotation of meridional element (Eq. 4)
v, w	Meridional and normal displacements
χ_ϕ, χ_θ	Curvature changes
$\epsilon_\phi, \epsilon_\theta$	Midsurface strains
h	Shell thickness
E	Young's modulus
ν	Poisson's ratio
D	$Eh^3/12(1-\nu^2)$
R	Radius of curvature at apex
U	Stress function $r_2 Q$
ρ	r_2/r_1
P	Loading function (Eq. 15)
x, y, z, f	Meissner's variables (Eq. 18)
A	$EhR/4$
B	$R/4D$
$\mathbf{L}, \mathbf{M}$	Linear differential operators (Eq. 21)
ϵ	Perturbation parameter
$\delta(x)$	$\rho(x) - 1$
$q_n(x)$	Normalized Jacobi polynomials
k_n	Normalizing factor
$W(x)$	Weighting function
λ_n	Eigenvalues of $\mathbf{L}$

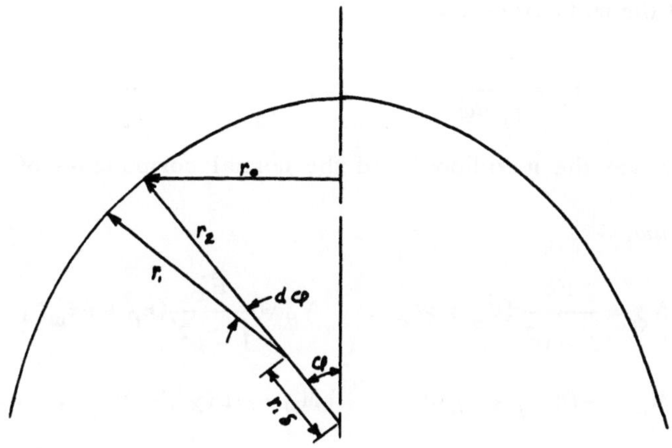

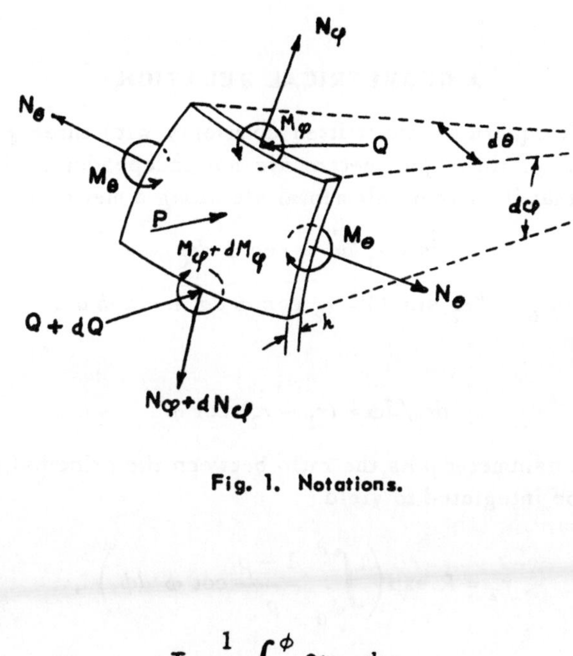

Fig. 1. Notations.

where

$$\bar{p} = \frac{1}{r_0^2} \int_0^{\phi} 2pr_0 \, dr_0 \, . \tag{3a}$$

Displacements and Curvature Changes. – The rotation of a meridional element is given by

$$V = \frac{1}{r_1}\left(v + \frac{dw}{d\phi}\right). \tag{4}$$

The curvature changes are

$$\chi_\phi = \frac{1}{r_1}\frac{dV}{d\phi}, \quad \chi_\theta = \frac{1}{r_2} V \cot \phi \, . \tag{5}$$

The strains of the midsurface are

$$\epsilon_\phi = \frac{1}{r_1}\frac{dv}{d\phi} - w \,, \qquad \epsilon_\theta = \frac{v}{r_2}\cot\phi - \frac{w}{r_2} \,, \tag{6}$$

where v and w are the meridional and the normal components of the displacement.

Hooke's Law. –

$$N_\phi = \frac{Eh}{1-\nu^2}(\epsilon_\phi + \nu\epsilon_\theta) \,, \qquad N_\theta = \frac{Eh}{1-\nu^2}(\epsilon_\theta + \nu\epsilon_\phi) \,, \tag{7}$$

$$M_\phi = -D(\chi_\phi + \nu\chi_\theta) \,, \qquad M_\theta = -D(\chi_\theta + \nu\chi_\phi) \,, \tag{8}$$

where $D = Eh^3/12(1-\nu^2)$.

A GEOMETRICAL RELATION

Although shell equations are written commonly with three geometric parameters, r_1, r_2, and ϕ, these parameters are not independent. It can be easily seen from Fig. 1 that if ds is an elemental arc along a meridian, one has

$$ds = r_1\, d\phi = \sec\phi\, dr_0 \,,$$

$$dr_0 = d(r_2 \sin\phi) = r_2\cos\phi\, d\phi + \sin\phi\, dr_2 \,.$$

Hence,

$$dr_2/d\phi = (r_1 - r_2)\cot\phi \,. \tag{9}$$

Upon defining the parameter ρ as the ratio between the principal radii, $\rho = r_2/r_1$, Eq. (9) can then be integrated to yield

$$r_2 = R\exp\left(\int_{\phi_0}^{\phi} \frac{1-\rho}{\rho}\cot\phi\, d\phi\right) \,, \tag{10}$$

in which R is the value of r_2 at $\phi = \phi_0$. For shells having no cusp or hole at the apex, one may set $\phi_0 = 0$ and

$$R = r_2(0) = r_1(0) \,, \quad \text{or} \quad \rho(0) = 1 \,.$$

Thus, if the parameter ρ as a function of the angle ϕ together with the integration constant R is specified, all the other geometric parameters are mathematically determined.

REDUCTION TO MEISSNER'S EQUATIONS

The introduction of the function $U = r_2 Q$ converts Eqs. (1) and (2) into the form

$$N_\theta = -\frac{1}{r_1}\frac{dU}{d\phi} + \frac{r_2(\rho\bar{p} - 2p)}{2},$$
$$N_\phi = -\frac{1}{r_2} U \cot\phi - \frac{\bar{p} r_2}{2}. \tag{11}$$

In combining Eqs. (4), (6), (7), and (11), one obtains

$$Ebr_1 V = \rho \frac{d^2 U}{d\phi^2} + \left(\frac{d\rho}{d\phi} + \rho \cot\phi\right)\frac{dU}{d\phi} - \left(\frac{1}{\rho}\cot^2\phi - \nu\right) U + P, \tag{12}$$

where

$$2P = -r_1 r_2 \cot\phi[\bar{p}(\rho^2 + 2\rho\nu + 1) - 2p(\rho + \nu)] + \frac{d}{d\phi}\{r^2[2p - \bar{p}(\rho + \nu)]\}. \tag{13}$$

Now the definition of $\bar{p}$, according to Eq. (3*a*), can be written as

$$\bar{p} = \frac{1}{r_2^2 \sin^2\phi}\int_0^\phi \frac{p}{\rho} r_2^2 \sin 2\phi \, d\phi,$$

or

$$\frac{d\bar{p}}{d\phi} = \frac{2}{\rho}\cot\phi(p - \bar{p}). \tag{14}$$

The substitution of the relations (9) and (14) into Eq. (13) finally yields

$$2P = -r_1 r_2 \left\{[(\rho - 1)4p - (\nu - 1)\bar{p}] \cot\phi + \rho\bar{p}\frac{d\rho}{d\phi} - 2\rho\frac{dp}{d\phi}\right\}. \tag{15}$$

For the special case of uniform pressure,

$$p = \bar{p} = \text{a constant},$$

$$2P = -pr_1 r_2 \cot\phi(\rho - 1)(\rho - 1 + 2\nu) + 2pr_2\frac{dr_2}{d\phi}(2 - \rho - \nu) - pr_2^2\frac{d\rho}{d\phi}. \tag{16}$$

Equation (12) thus represents the equilibrium between the surface load and the stress resultants in a shell element. Similarly, by combining Eqs. (3), (4), (5), and (8), one obtains

$$-\frac{r_1 U}{D} = \rho\frac{d^2 V}{d\phi^2} + \left(\frac{d\rho}{d\phi} + \rho\cot\phi\right)\frac{dV}{d\phi} - \left(\frac{1}{\rho}\cot^2\phi + \nu\right) V, \tag{17}$$

which represents the equilibrium between the bending moments.

Let the following transformation of variables now be introduced:

$$x = \sin^2 \phi , \qquad y = \frac{U}{\sin \phi}, \qquad z = \frac{V}{\sin \phi}, \qquad f = \frac{P}{4 \sin \phi}. \tag{18}$$

This change of variables transforms Eqs. (12) and (17) into

$$\frac{Ar_1 z}{R} = \rho \mathbf{L}(y) + \left[\left(\rho - \frac{1}{\rho} \right) \frac{(1-x)}{x} + \nu \right] \frac{y}{4} + \frac{d\rho}{dx} \mathbf{M}(y) + f(x) , \tag{19}$$

$$-\frac{Br_1 y}{R} = \rho \mathbf{L}(z) + \left[\left(\rho - \frac{1}{\rho} \right) \frac{(1-x)}{x} - \nu \right] \frac{z}{4} + \frac{d\rho}{dx} \mathbf{M}(z) , \tag{20}$$

where $A = EbR/4$, $B = R/4D$, and $\mathbf{L}$ and $\mathbf{M}$ are two linear differential operators defined by

$$\mathbf{L} = x(1-x) \frac{d^2}{dx^2} + \left(2 - \frac{5}{2} x \right) \frac{d}{dx} - \frac{1}{4},$$
$$\mathbf{M} = x(1-x) \frac{d}{dx} + \frac{(1-x)}{2}. \tag{21}$$

The nonhomogeneous term $f(x)$ in Eq. (19) represents the loading. It is found from Eqs. (15) and (18) to be

$$f(x) = -\frac{r_2^2}{8} \sqrt{1-x} \left\{ \frac{\rho - 1}{\rho x} [4p - \overline{p}(\rho + 1)] + 2\overline{p} \frac{d\rho}{dx} - 4 \frac{dp}{dx} \right\} . \tag{22}$$

In terms of x, the two radii of curvature are given by

$$r_2 = R \exp \left[\int_0^x \frac{(1-\rho)}{2\rho x} dx \right] ,$$
$$r_1 = \frac{R}{\rho} \exp \left[\int_0^x \frac{(1-\rho)}{2\rho x} dx \right] . \tag{23}$$

PERTURBATION EQUATIONS

The only parameter specifying the geometry of the shell in the two differential Eqs. (19) and (20) is the ratio between the two radii of curvature ρ. This parameter as a function of ϕ and, hence, also of x is supposed to be known. The constant R representing the radius of curvature at the apex is equivalent to a scale factor. Now, if ρ is identically equal to unity, the shell is spherical. Otherwise, it differs from unity by an amount, say, $\delta(x)$:

$$\rho(x) = 1 + \delta(x) .$$

Let it now be assumed that $|\delta(x)| < 1$ everywhere. Introduce also a perturbation parameter ϵ, $|\epsilon| \leqq 1$, and set, in general,

$$\rho = 1 + \epsilon\delta(x) . \tag{24}$$

Let the solution of the problem be now represented in powers of ϵ as

$$\begin{aligned} y(x) &= y_0(x) + \epsilon y_1(x) + \epsilon^2 y_2(x) + \dots , \\ z(x) &= z_0(x) + \epsilon z_1(x) + \epsilon^2 z_2(x) + \dots . \end{aligned} \tag{25}$$

The two differential Eqs. (19) and (20) can then be expanded also in powers of ϵ. By equating the coefficients of like powers of ϵ, the two original equations are now converted into an infinite set of equations. The steps are as follows.

First, r_2 and r_1 are represented as

$$r_2 = R(1 + \alpha_1 \epsilon + \alpha_2 \epsilon^2 + \alpha_3 \epsilon^3 + \dots) , \tag{26}$$

$$r_1 = R(1 + \beta_1 \epsilon + \beta_2 \epsilon^2 + \beta_3 \epsilon^3 + \dots) , \tag{27}$$

where

$$\alpha_1(x) = \frac{1}{R}\left(\frac{\partial r_2}{\partial \epsilon}\right)_0 = \left[\frac{\partial}{\partial \epsilon} \exp - \int_0^x \frac{\epsilon\delta}{2(1 + \epsilon\delta)x}\, dx\right]_{\epsilon=0} = - \int_0^x \frac{\delta}{2x}\, dx .$$

Similarly,

$$\begin{aligned} \alpha_2(x) &= \frac{1}{2}\alpha_1^2 + \int_0^x \frac{\delta^2\, dx}{2x} , \\ \alpha_3(x) &= \frac{-1}{3}\alpha_1^3 + \alpha_1\alpha_2 - \int_0^x \frac{\delta^3\, dx}{2x} . \end{aligned} \tag{28}$$

The β's can be obtained through the relation

$$r_1 = r_2\rho^{-1} = r_2(1 + \epsilon\delta)^{-1} .$$

Thus,

$$\begin{aligned} \beta_1 &= \alpha_1 - \delta , \\ \beta_2 &= \alpha_2 - \alpha_1\delta + \delta^2 , \\ \beta_3 &= \alpha_3 - \alpha_2\delta + \alpha_1\delta^2 - \delta^3 . \end{aligned} \tag{29}$$

Now write Eq. (22) as

$$f(x) = \frac{\sqrt{1-x}}{8x} R^2(\gamma_0 + \gamma_1 \epsilon + \gamma_2 \epsilon^2 + \gamma_3 \epsilon^3 + \dots) . \tag{30}$$

The substitution of Eqs. (26), (27), and (29) into Eq. (22) yields

$$\gamma_0 = 4x\frac{dp}{dx},$$

$$\gamma_1 = 2\bar{p}(\delta - x\delta') - 4p\delta - 2\alpha_1\gamma_0,$$

$$\gamma_2 = \bar{p}\delta(\delta - 2x\delta') - (\gamma_1 - 2\alpha_1\gamma_0)(2\alpha_1 - \delta) - \gamma_0(2\alpha_2 + \alpha_1^2),$$

$$\gamma_3 = 2\bar{p}\delta(\delta - 2x\delta')\alpha_1 - 2\gamma_0(\alpha_3 + \alpha_1\alpha_2) - (\delta\gamma_0 + \gamma_1 - 2\alpha_1\gamma_0)(2\alpha_2 + \alpha_1^2) + \delta\gamma_2 .$$

Also,

$$\begin{aligned}\rho\mathbf{L}(y) &= (1 + \epsilon\delta)\mathbf{L}(y_0 + \epsilon y_1 + \epsilon^2 y_2 + \ldots) \\ &= \mathbf{L}(y_0) + \epsilon[\mathbf{L}(y_1) + \delta\mathbf{L}(y_0)] + \ldots,\end{aligned} \tag{31}$$

$$\begin{aligned}\left(\rho - \frac{1}{\rho}\right)y &= [1 + \epsilon\delta - (1 + \epsilon\delta)^{-1}](y_0 + \epsilon y_1 + \epsilon^2 y_2 + \ldots) \\ &= 2\delta\epsilon y_0 + (2\delta y_1 - \delta^2 y_0)\epsilon^2 + \ldots,\end{aligned} \tag{32}$$

$$\frac{d\rho}{dx}\mathbf{M}(y) = \epsilon\delta'\mathbf{M}(y_0) + \epsilon^2\delta'\mathbf{M}(y_1) + \ldots . \tag{33}$$

Similar equations can be written for z.

Substituting the relations (24)–(33) into the differential Eqs. (19) and (20) and equating the coefficients of like powers of ϵ, one finally arrives at the following set of equations:

$$\begin{aligned} Az_0 &= \mathbf{L}(y_0) + \frac{\nu y_0}{4} + \frac{R^2}{8x}\sqrt{1-x}\,\gamma_0, \\ -By_0 &= \mathbf{L}(z_0) - \frac{\nu z_0}{4},\end{aligned} \tag{34}$$

$$\begin{aligned} Az_1 &= \mathbf{L}(y_1) + \frac{\nu y_1}{4} - A\beta_1 z_0 + \delta\mathbf{L}(y_0) + \frac{(1-x)}{4x}2\delta y_0 + \delta'\mathbf{M}(y_0) + \frac{R^2}{8x}\sqrt{1-x}\,\gamma_1 \\ &= \mathbf{L}(y_1) + \frac{\nu y_1}{4} + A(2\delta - \alpha_1)z_0 + \left[\frac{(1-x)}{2x} - \frac{\nu}{4}\right]\delta y_0 \\ &\quad + \delta'\mathbf{M}(y_0) + \frac{R^2}{8x}\sqrt{1-x}\,(\gamma_1 - \delta\gamma_0),\end{aligned} \tag{35}$$

$$-By_1 = \mathbf{L}(z_1) - \frac{\nu z_1}{4} - B(2\delta - \alpha_1)y_0 + \left(\frac{1-x}{2x} + \frac{\nu}{4}\right)\delta z_0 + \delta'\mathbf{M}(z_0),$$

$$Az_2 = \mathbf{L}(y_2) + \frac{\nu y_2}{4} - A(\beta_1 z_1 + \beta_2 z_0) + \delta \mathbf{L}(y_1)$$

$$+ \frac{(1-x)}{4x}(2\delta y_1 - \delta^2 y_0) + \delta' \mathbf{M}(y_1) + \frac{R^2}{8x}\sqrt{1-x}\, y_2 \,,$$

(36)

$$-By_2 = \mathbf{L}(z_2) - \frac{\nu z_2}{4} + B(\beta_1 y_1 + \beta_2 y_0) + \delta \mathbf{L}(z_1)$$

$$+ \frac{(1-x)}{4x}(2\delta z_1 - \delta^2 z_0) + \delta' \mathbf{M}(z_1) \,.$$

In the last two equations, $\mathbf{L}(y_1)$ and $\mathbf{L}(z_1)$ may be eliminated by the previous set of equations. In general, one has the recurring situation that

$$Az_i = \mathbf{L}(y_i) + \frac{\nu y_i}{4} + S_i \,,$$

(37)

$$-By_i = \mathbf{L}(z_i) - \frac{\nu z_i}{4} + T_i \,,$$

where S_i and T_i are two known functions involving the previous result and the shell geometry. Their general expressions are

$$S_i = \delta \mathbf{L}(y_{i-1}) + \delta' \mathbf{M}(y_{i-1}) - A(\beta_i z_0 + \beta_{i-1} z_1 + \ldots + \beta_1 z_{i-1})$$

$$+ \frac{(1-x)}{4x}[2\delta y_{i-1} - \delta^2 y_{i-2} + \ldots - (-\delta)^i y_0] - \frac{R^2}{8x}\sqrt{1-x}\, y_i \,,$$

(37*a*)

$$T_i = \delta \mathbf{L}(z_{i-1}) + \delta' \mathbf{M}(z_{i-1}) + B(\beta_i y_0 + \beta_{i-1} y_1 + \ldots + \beta_1 y_{i-1})$$

$$+ \frac{(1-x)}{4x}[2\delta z_{i-1} - \delta^2 z_{i-2} + \ldots - (-\delta)^i z_0] \,.$$

REDUCTION TO SPHERICAL SHELL PROBLEM

The recurring set of Eqs. (37) is the nonhomogeneous version of the well-known Love-Meissner equation for spherical thin shells. The nonhomogeneous term S_i is equivalent to a distribution of surface load or body force, and the term T_i is equivalent to a distributed moment. The solution of the perturbation problem thus consists of the solution of successive problems of spherical shells under various distributions of loads. Drop the subscript i for the time being and write Eqs. (37) as

$$Az = \mathbf{L}(y) + my + S \,,$$

(38)

$$-By = \mathbf{L}(z) - my + T \,,$$

where $A = EbR/4$, $B = R/4D$, and $m = \nu/4$ are known constants, and S and T are two known functions of x. A certain set of boundary conditions are also to be prescribed.

As usual, the solution to Eqs. (38) can be decomposed into two parts: the complementary solution and the particular integral. Various methods are available for finding the complementary solution (see refs [1]–[4]). In a previous paper [5], the writer presented a method for finding the complementary solution of Eqs. (38) (see also ref [6]). This solution is in the form of a Jacobi series; the result is as follows:

Consider the set of Jacobi polynomials $q_n(x)$ defined by

$$q_n(x) = k_n F\left(-n\,,\, n + \frac{3}{2},\, 2; x\right)\,, \tag{39}$$

where $k_n = \sqrt{(4n + 3)(2n + 1)(n + 1)/2}$ and $n = 0, 1, 2, 3, \ldots$. The symbol F is the sign for hypergeometric function. Because the first argument, $-n$, is a nonpositive integer, F becomes a polynomial instead of an infinite power series for each n. This set of polynomials is orthonormal with respect to the weighting function

$$W(x) = \frac{x}{\sqrt{1 - x}}$$

in the interval $0 < x < 1$. Furthermore, the set is the eigenfunction of the differential equation

$$\mathsf{L}(q_n) - \lambda_n q_n = 0$$

under the boundary condition that q should be bounded in $0 < x < 1$. The corresponding eigenvalue is

$$\lambda_n = -n\left(n + \frac{3}{2}\right) - \frac{1}{4}\,.$$

The functions S and T can thus be expanded into two series of these polynomials:

$$S(x) = \sum_0^\infty s_n q_n(x)\,,$$
$$T(x) = \sum_0^\infty t_n q_n(x)\,, \tag{40}$$

where

$$s_n = \int_0^1 \frac{x}{\sqrt{1 - x}}\, S(x) q_n(x)\, dx\,,$$
$$t_n = \int_0^1 \frac{x}{\sqrt{1 - x}}\, T(x) q_n(x)\, dx\,. \tag{41}$$

Take the case where the shell is not a hemisphere or where $x = \sin^2 \phi \leqq \xi < 1$; to obtain s_n and t_n, any appropriate values, say zero, may be assigned to S and T in the integrands of Eqs. (41) for $\xi < x < 1$.

Denote by $Q_n(x)$ the set of functions defined by

$$Q_n(x) = \frac{1}{(\lambda_n^2 - m^2 + AB)} q_n(x) = \frac{4}{4n^4 + 12n^3 + 11n^2 + 3n + \beta^4} q_n(x) ,$$

with

$$\beta^4 = \frac{1 - \nu^2}{4} \left[12 \left(\frac{R}{h} \right)^2 + 1 \right] .$$

Now define six functions of x with β as a parameter, G, H, I, J, K, and N, by the following six convergent Jacobi series:

$$\begin{aligned}
G(x, \beta) &= \sum_0^\infty q_n(1) Q_n(x)/4 , \\
H(x, \beta) &= \sum_0^\infty \lambda_n q_n(1) Q_n(x) , \\
I(x, \beta) &= \sum_0^\infty s_n Q_n(x)/4 , \\
J(x, \beta) &= \sum_0^\infty \lambda_n s_n Q_n(x) , \\
K(x, \beta) &= \sum_0^\infty t_n Q_n(x)/4 , \\
N(x, \beta) &= \sum_0^\infty \lambda_n t_n Q_n(x) .
\end{aligned} \tag{42}$$

The solution to differential Eqs. (38), under the boundary condition that it be bounded at $x = 0$, can now be expressed as

$$\begin{aligned}
y(x) &= C_1(\nu G - H) - C_2(EhR)G - (EhR)K - J + \nu I , \\
z(x) &= C_1(R/D)G - C_2(\nu G + H) + (R/D)I - N - \nu K ,
\end{aligned} \tag{43}$$

where C_1 and C_2 are two integration constants whose values are to be determined by conditions at the boundary. If the shell is a hemisphere, in other words if $\phi = \pi/2$ and $x = 1$ at the edge, one has the additional knowledge that

$$\begin{aligned}
C_1 &= \frac{1}{2} \left(\frac{dy}{d\phi} \right)_{\pi/2} = \lim_{x \to 1} \sqrt{1 - x} \frac{dy}{d\phi} , \\
C_2 &= \frac{1}{2} \left(\frac{dz}{d\phi} \right)_{\pi/2} = \lim_{x \to 1} \sqrt{1 - x} \frac{dz}{dx} .
\end{aligned} \tag{44}$$

Also, it is clear that the complementary part of solutions (43) may be replaced by other expressions obtained by other methods both "exact" or approximate so long as the necessary integration constants are included for adjustment of the solutions to meet the boundary conditions.

The set of Eqs. (34)–(37), etc., do not contain ϵ. Upon setting $\epsilon = 1$, one has

$$y = y_0 + y_1 + y_2 + \dots = \sum_0^\infty y_i \,,$$

$$z = z_1 + z_2 + z_3 + \dots = \sum_0^\infty z_i \,,$$

with y_i and z_i given by appending the subscript i to the appropriate terms at the right-hand side of Eqs. (43).

BOUNDARY CONDITIONS

The shell being studied is assumed to have no hole or cusps at the apex and a continuously turning meridional tangent. One of the boundary points is thus at $\phi = 0$ or $x = 0$, and the other boundary is assumed to be at $\phi = \phi_0 \leqq \pi/2$ or $x = \xi \leqq 1$. Because of the rotational symmetry and continuity assumed, the shear Q and rotation V must vanish at $x = 0$. As

$$y \sin \phi = r_2 Q \,,$$

$$z \sin \phi = V \,,$$

the vanishing of Q and V at $\phi = 0$ implies that y and z are finite at $x = 0$. This implication has already been taken into account when one assumes that y and z possess infinite series expansions in Jacobi polynomials. As a result, there are only two integration constants in the solution of each set of perturbation equations, represented by C_1 and C_2 in Eqs. (43). At the other boundary two homogeneous or nonhomogeneous boundary conditions exist on y and z or their first derivatives. The corresponding conditions on the individual y_i and z_i in Eqs. (25) can, however, be imposed with a certain degree of arbitrariness.

A convenient way of handling the boundary conditions in many cases is as follows. Let the constants C_1 and C_2 be zero in solutions (43) for all y_i and z_i except the first pair y_0 and z_0. The constants associated with this pair are then adjusted to fit the boundary conditions imposed on y and z. The final solution then consists of a part that does not depend on the imposed boundary condition and another that does. Oftentimes the second part can be obtained by another existing method which is approximate but much simpler than the method presented. As an example, consider the following case.

Suppose that one is studying the stress in a shell subjected to a uniform pressure and of a shape which is essentially a hemisphere except for a small region around the apex where the curvature is not constant. Let this shell be simply supported along its equator. The boundary conditions are therefore

$$M_\phi = 0\,, \qquad \epsilon_\theta = 0\,,$$

at

$$\phi = \pi/2 \qquad \text{or} \qquad x = 1\,.$$

The first condition implies through Eqs. (5) and (8) that

$$(dV/d\phi)_{\pi/2} = 0\,.$$

Because of the relations (18) and (44) one has

$$(dy/d\phi) = 0 \qquad \text{at} \qquad x = 1\,.$$

Hence this boundary condition will be satisfied by setting

$$dz_i/d\phi = 0\,, \qquad\qquad i = 0, 1, 2, \dots,$$

or $C_2 = 0$ in solutions (43) for all i's.

The second boundary condition leads to

$$Eh\epsilon_\theta = N_\theta - \nu N_\phi = 0\,.$$

Upon letting $\phi = \pi/2$ and $p = \bar{p}$ in Eqs. (11) and substituting the results into the equation above, one arrives at

$$\frac{dU}{d\phi} = \frac{pr_1 r_2}{2}(1 - \nu - \delta)\,,$$

$$\phi = \pi/2\,.$$

Thus the second boundary condition specifies the value of $dU/d\phi$ and hence $dy/d\phi$ at $\phi = \pi/2$. Now if this given boundary condition is replaced by $dy/d\phi = 0$, the resulting ϵ_θ will not be zero at the boundary. The solution obtained with this set of boundary conditions can be made to fit the original boundary conditions by superposing another solution which corresponds to that produced by an edge force causing a radial displacement or strain at the equator to cancel the strain ϵ_θ resulting from letting $dy/d\phi = 0$. Because in this example the shell is essentially spherical in the region near the boundary, this solution can be obtained by other classical methods.

The coefficients s_n and t_n are, in general, to be computed by numerical integration according to formulas (41). To avoid treating singularities caused by $\sqrt{1 - x}$, the variable of integration is changed back to ϕ:

$$s_{i,n} = \int_0^{\pi/2} 2S_i q_n(\sin^2\phi)\sin^3\phi\, d\phi\,,$$

$$t_{i,n} = \int_0^{\pi/2} 2T_i q_n(\sin^2\phi)\sin^3\phi\, d\phi\,.$$

The functions S_i and T_i are as given by Eqs. (37*a*). These two functions are made up of a number of terms. Among them are terms involving the differential operators **L** and **M**. To avoid the use of numerical differentiation, the terms $\delta\mathbf{L}(y_{i-1})$ and $\delta\mathbf{L}(z_{i-1})$ should be replaced by terms involving y_{i-1}, z_{i-1}, S_{i-1}, and T_{i-1}. The first two are already known; the second two consist of either known functions or can be computed from the knowledge of $s_{i-1,n}$ and $t_{i-1,n}$. The terms $\delta'\mathbf{M}(y_{i-1})$ and $\delta'\mathbf{M}(z_{i-1})$ can be handled this way. Utilizing the definition of **M** as given by Eqs. (21), one can see that

$$\int_0^1 W(x)\mathbf{M}(y)\delta'(x)q_n(x)\,dx = \int_0^1 y(x)\,\frac{d}{dx}\left[x^2\sqrt{1-x}\,\delta'(x)q_n(x)\right]dx$$

$$+\int_0^1 \frac{x\sqrt{1-x}}{2}\,y(x)\delta'(x)q_n(x)\,dx\,.$$

Inasmuch as $\delta'(x)$ and $q_n(x)$ are both assumed to be differentiable, the above integrals can be evaluated without complications.

CONVERGENCE OF THE METHOD

A detailed analysis of the necessary and sufficient conditions for the convergence of the method would be too involved to warrant the effort, since in reality it is seldom practical to carry out the perturbation solution to beyond the first few terms. In the following discussion, therefore, only an estimate of the rapidity of convergence is attempted.

It was pointed out previously that one way of treating the integration constants in carrying out the perturbation procedure was to assign the value zero to all the integration constants, and then add to the solution obtained another solution such that the combined results satisfy the boundary conditions given. If this procedure is adopted, evidently the convergence of this solution hinges upon the manner with which S_i and T_i approach zero as i approaches infinity. Consider these two functions as points in Hilbert space. A comparison of the series expansions (40) and (42) reveals that the four functions I_i, J_i, K_i, and N_i are closer to the origin than the functions S_i and T_i that generate them. Now examine the expressions (37*a*). The makeup of S_i and T_i is essentially $\delta(x)$ times the previous perturbates plus terms originating from the expansion of the loading term. Hence if $\delta(x)$ is uniformly bounded by ± 1, and $\delta'(x)$ is uniformly bounded by some finite bounds, the procedure should converge. The rapidity of convergence will mostly depend on the rapidity of convergence of the expressions (26), (27), and (30), as well as the magnitude of $\delta(x)$.

REFERENCES

[1] S. Timoshenko and Woinowsky-Krieger, *Theory of Plates and Shells*, 2d ed., McGraw-Hill, New York, 1959.

[2] W. Flügge, *Stresses in Shells*, Springer-Verlag, Berlin, 1960.

[3] A. L. Luré, "Statics of Thin-Walled Elastic Shells," AEC Translation 3798 from Russian published by State Publishing House of Technical and Theoretical Literature, Moscow and Leningrad, 1947.

[4] F. A. Leckie, *Nuclear Reactor Containment Building and Pressure Vessels, Section III,* The Royal College of Science and Technology, Glassgow, Butterworths, London, 1960.

[5] K. N. Tong, *Perturbation Solution of A Class of Axisymmetrical Shells Subjected to A Uniform Pressure,* S.U.R.I. report ME790-617A.

[6] K. N. Tong, *Jacobi-Series Solution of Thin Spherical Shells Subjected to Axisymmetrical Loading,* S.U.R.I. report ME895-6112T.

SHELL SECTORS HAVING DOUBLE CURVATURE[1]

J. T. S. Wang
Georgia Institute of Technology
and
H. Bahiman
Purdue University

The work presented in this paper is a portion of an unpublished study of shell sectors in which the general membrane and bending theory of thin shells is discussed. Approximate solutions for truncated hyperbolic paraboloidal shell sectors are obtained. When the shell boundaries do not all coincide with coordinate lines, satisfactory solutions are obtained which satisfy the boundary conditions approximately.

GENERAL THEORY[2]

Let $\mathbf{R}$ represent the position vector of a point on the middle surface of a thin, homogeneous, isotropic, and linearly elastic shell. The base vector $\mathbf{a}_i$ is defined as

$$\mathbf{a}_i = \mathbf{R},_i \, , \tag{1}$$

where[3] the index $i = 1$ and 2 represents two parametric coordinate lines, t^1 and t^2, on the surface, and, in Cartesian coordinates,

$$\mathbf{R} = x_\alpha \mathbf{i}_\alpha \quad (\alpha = 1,\ 2,\ \text{and}\ 3)\ . \tag{2}$$

The metric surface tensor, a_{ij}, is defined as

$$a_{ij} = \mathbf{a}_i \cdot \mathbf{a}_j = (x_{\alpha},_i) \cdot (x_{\beta},_j)\ . \tag{3}$$

---

[1]Sponsored by the Dow Chemical Company. This paper presents a portion of an unpublished study, "Investigation of Shell Sectors Having Double Curvature," by J. T. S. Wang (Ph.D. thesis, Purdue University, 1961).

[2]Detailed information on the geometry of surfaces can be found in almost any textbook on differential geometry or tensor analysis.

[3]All Latin indices take on the values 1 and 2, and the repetition of an index denotes summation convention unless noted otherwise. No summation is needed when indices are underlined. A comma followed by an index denotes partial differentiation.

NOTATION

For convenience of presentation, tensor notation will be used. All symbols are defined in the text where they first appear, and some of the major symbols are listed below:

$\mathbf{a}_i$	Base vectors
a_{ij}, b_{ij}	Surface tensors
a^*	Determinant of a_{ij}
b^*	Determinant of b_{ij}
D	Plate rigidity
E	Young's modulus
e_{ij}	Strain tensor
F	Stress function
h	Thickness of shell
l_{ia}	Direction cosines
n_{ij}/h	Membrane stress tensor
$\mathbf{n}$	Unit normal vector
$\mathbf{R}$	Position vector
t^i, u, v	Parametric coordinates
w	Vertical or normal displacement
x_a	Cartesian coordinates
δ_{ij}	Kronecker delta
ϵ_{ij}	Permutation symbol
λ_k	Direction cosines
λ, μ	Lamé constants
ν	Poisson's ratio

The basic line elements along coordinate lines are $\sqrt{a_{\underline{ii}}}\,dti$. The unit normal vector $\mathbf{n}$ becomes

$$\mathbf{n} = \frac{\mathbf{a}_1 \times \mathbf{a}_2}{|\mathbf{a}_1 \times \mathbf{a}_2|} . \tag{4}$$

The vectors $\mathbf{a}_1$, $\mathbf{a}_2$, and $\mathbf{n}$ form a reference system on a surface. Let $d\beta_i$ be the projections of $d\mathbf{n}$ on the $\mathbf{a}_i$ directions. They also equal the magnitude of the rotation of the normal projected on these two directions. Thus

$$\frac{1}{\sqrt{a_{\underline{ii}}}}\beta_{i,j} = \frac{1}{\sqrt{a_{\underline{ii}}a_{\underline{jj}}}}\mathbf{R},_i\cdot\mathbf{n},_j = -\frac{b_{ij}}{\sqrt{a_{\underline{ii}}a_{\underline{jj}}}} , \tag{5}$$

where

$$b_{ij} = -\mathbf{R},_i\cdot\mathbf{n},_j = -\mathbf{a}_i\cdot\mathbf{n},_j = \mathbf{n}\cdot\mathbf{a}_{i},_j = (\mathbf{a}_{i},_j\mathbf{a}_1\mathbf{a}_2)/\sqrt{a^*} , \tag{6}$$

$$a^* = |a_{ij}| .$$

The Cartesian coordinate system (x_i, z) is often a natural system to be used in the analysis of shells. Consider the equation of the middle surface of a shell to be written as

$$z = z(x_i) .$$

By summing forces in the x_1, x_2, and z directions and moments in x_1 and x_2 directions, the following equilibrium equations result:

$$n_{ij},_j + p_i = 0 , \tag{7}$$

$$v_{i},_i + (z,_j n_{ij}),_i + p_z = 0 , \tag{8}$$

$$v_i = m_{ij},_j . \tag{9}$$

Here n_{ij}/h denotes the stress tensor, h is the thickness of the shell, v_i is the transverse shear, m_{ij} denotes the stress couple, p_i and p_z are the components of tractions on the surface of the shell in the x_i directions and in the z direction respectively. For membrane theory, v_i and m_{ij} will be neglected, and Eq. (7) may be substituted into Eq. (8), yielding

$$z,_{ij}n_{ij} - z,_j p_i + p_z = 0 . \tag{10}$$

The stress tensor may be expressed in terms of a stress function F in the following manner:

$$n_{ij} = (\nabla^2 F - \int p_i\, dx_j)\delta_{ij} - F,_{ij} . \tag{11}$$

Substitution of Eq. (11) into Eq. (10) yields

$$z,_{ij}[(\nabla^2 F - \int p_i\, dx_j)\delta_{ij} - F,_{ij}] - z,_j p_j + p_z = 0 ,$$

which may be written in the following compact form [5]:

$$\epsilon_{im}\epsilon_{jn}z,_{ij}F,_{mn} = q , \tag{12}$$

$$q = \int p_i\, dx_i + z,_j p_j - p_z , \tag{13}$$

where ϵ_{ij} is the permutation symbol. Equations (12) and (13) therefore govern the behavior of a membrane of arbitrary shape. Let $\mathbf{u}$ be the displacement vector of any point on the middle surface of the shell and $\mathbf{R}$ be the position vector of any point P on the middle surface of a shell:

$$\mathbf{R} = x_j\mathbf{i}_j + z\mathbf{k} , \tag{14}$$

where $z = z(x_j)$ represents the equation of the middle surface of the shell in Cartesian coordinates. A line element at P on the middle surface of a shell before deformation may be expressed by the following vector:

$$\frac{\partial \mathbf{R}}{\partial x_j}\, dx_j = (\mathbf{i}_j + z,_j \mathbf{k})\, dx_j \,. \tag{15}$$

The corresponding element after defomation can be expressed as

$$\frac{\partial \mathbf{R}^*}{\partial x_j}\, dx_j = \frac{\partial}{\partial x_j}\,(\mathbf{R} + \Delta\mathbf{R})\, dx_j \,, \tag{16}$$

where

$$\Delta\mathbf{R} = u_j \mathbf{i}_j + w\mathbf{k} \,, \tag{17}$$

in which u_j represents the displacement of any point P on the middle surface of the shell in the x_j direction and w is the displacement of P in the z direction. Then Eq. (16) may be written as

$$\frac{\partial \mathbf{R}^*}{\partial x_j}\, dx_j = [(\delta_{ij} + u_{i},_{j})\mathbf{i}_i + (z + w),_j \mathbf{k}]\, dx_j \,. \tag{18}$$

Let ds be the length of a line element on the middle surface of a shell, which may be expressed as follows:

$$(ds)^2 = \left(\frac{\partial \mathbf{R}^*}{\partial x_i}\, dx_i\right) \cdot \left(\frac{\partial \mathbf{R}}{\partial x_j}\, dx_j\right) dx_i\, dx_j \,. \tag{19}$$

Upon defomation, the line element ds changes to ds^*, given by

$$\begin{aligned}(ds^*)^2 &= \left(\frac{\partial \mathbf{R}^*}{\partial x_i}\, dx_i\right) \cdot \left(\frac{\partial \mathbf{R}^*}{\partial x_j}\, dx_j\right) \\ &= (ds + \mathcal{E} ds)^2 \\ &= \delta_{ij} + z,_i z,_j + u_i,_j + u_j,_i \\ &= z,_j w,_i + z,_i w,_j + w,_j w,_i + u_k,_i u_k,_j \,,\end{aligned} \tag{20}$$

where $\mathcal{E}$ is the unit elongation of the element ds. If $\mathcal{E}^2 (\mathcal{E} \ll 1)$ is neglected in the expression

$$(ds^*)^2 = (ds + \mathcal{E} ds)^2 = (1 + 2\mathcal{E} + \mathcal{E}^2)(ds)^2 \,, \tag{21}$$

one obtains

$$2\mathcal{E}(ds)^2 = (ds^*)^2 - (ds)^2 \,; \tag{22}$$

thus

$$2\mathcal{E}(ds)^2 = (u_i,_j + u_j,_i + z,_i w,_j + z,_j w,_i + u_k,_j u_k,_i + w,_i w,_j)\, dx_i\, dx_j \,. \tag{23}$$

Let λ_k be the direction cosine of the line element of the surface, given by

$$\lambda_k = \frac{dx_k}{ds} . \tag{24}$$

Substitution of Eq. (24) into Eq. (23) yields

$$\mathcal{E} = \frac{1}{2} (u_{i,j} + u_{j,i} + z_{,i} w_{,j} + z_{,j} w_{,i} + u_{k,j} u_{k,i} + w_{,j} w_{,i}) \lambda_i \lambda_j . \tag{25}$$

According to the theory of elasticity,

$$\mathcal{E} = e_{ij} \lambda_i \lambda_j . \tag{26}$$

After substitution of Eq. (26) into Eq. (25) and elimination of the higher-order terms (e.g., $w_{,i} w_{,j}$), the Cartesian strain tensor of the thin elastic shell becomes

$$e_{ij} = \frac{1}{2} (u_{i,j} + u_{j,i} + z_{,i} w_{,j} + z_{,j} w_{,i}) . \tag{27}$$

From the theory of elasticity, the stress-strain relation for the case of plane stress is known to be

$$e_{ij} = \frac{1 + \nu}{Eh} n_{ij} - \frac{\nu}{Eh} \Theta \delta_{ij} , \tag{28}$$

or

$$\frac{1}{h} n_{ij} = \frac{E}{1 - \nu^2} \theta \delta_{ij} + 2\mu e_{ij} , \tag{29}$$

where $\Theta = n_{ii}$ and $\theta = e_{ii}$ are, respectively, the stress and strain invariants, λ and μ are Lamé constants, and ν is Poisson's ratio. The moment-displacement relationship may be written as follows:

$$m_{ij} = - \frac{1}{2} D[2\nu w_{,kk} \delta_{ij} + (1 - \nu)(w_{,ij} + w_{,ji})] . \tag{30}$$

From Eqs. (27) and (29), the following expression results:

$$n_{ij} = \frac{Eh}{1 - \nu^2} \left[\nu(u_{k,k} + z_{,k} w_{,k}) \delta_{ij} + \frac{1}{2} (1 - \nu) \times (u_{i,j} + u_{j,i} + z_{,i} w_{,j} + z_{,j} w_{,i}) \right] . \tag{31}$$

Substitution of Eq. (31) into Eq. (7) yields

$$\frac{Eh}{2(1 - \nu^2)} [(1 + \nu)(u_{k,ki} + z_{,ki} w_{,k} + z_{,k} w_{,ki}) + (1 - \nu)(U_{i,jj} + z_{,i} w_{,jj} + z_{,jj} w_{,i})] + p_i = 0 . \tag{32}$$

Substitution of Eq. (19) into Eq. (18) by using the relations shown in Eqs. (30) and (31) results in the following expression:

$$w_{,iijj} - \frac{Eh}{D(1-\nu^2)} \Big[\nu z_{,ii}(u_{k,k} + z_{,k} w_{,k})$$

$$+ \frac{1}{2}(1-\nu) z_{,ij}(u_{i,j} + u_{j,i} + z_{,i} w_{,j} + z_{,j} w_{,i}) \Big] = p_z - z_{,j} p_j . \tag{33}$$

Equations (32) and (33) govern the behavior of a shell in terms of displacements, u_i and w. Alternatively, the governing differential equations for the bending theory of shells may be expressed in terms of the vertical displacement w and the stress function F. For the two-dimensional problem, only the following equation needs to be considered in order to satisfy compatibility conditions [6]:

$$e_{11,22} + e_{22,11} - 2e_{12,12} = -(z_{,11} w_{,22} + z_{,22} w_{,11} - 2z_{,12} w_{,12}) . \tag{34}$$

Substitution of Eq. (28) into Eq. (34) yields

$$\frac{1+\nu}{Eh}(n_{11,22} + n_{22,11} - 2n_{12,12}) - \frac{\nu}{Eh}(\Theta_{,11} + \Theta_{,22})$$

$$= -(z_{,11} w_{,22} + z_{,22} w_{,11} - 2z_{,12} w_{,12}) . \tag{35}$$

By use of Eq. (11) and relation $n_{ij,ji} = p_{i,i}$, Eq. (35) may be reduced to the following form:

$$\nabla^4 F = (\textstyle\int p_1\, dx_1)_{,22} + (\int p_2\, dx_2)_{,11} - Eh \epsilon_{ij} \epsilon_{ab} z_{,ij} w_{,ab} - \nu(p_{i,i}) . \tag{36}$$

Substitution of Eqs. (9), (11), and (30) into Eq. (8) gives

$$-D\nabla^4 w + z_{,ji}(\nabla^2 F \delta_{ij} - F_{,ij}) = z_{,ji} \textstyle\int p_k\, dx_k + z_{,j} p_{,j} - p_z . \tag{37}$$

Equations (36) and (37) become the governing differential equations for the bending theory of shells. The two equations may be combined into an eighth-order differential equation by the assumption of a new function V, which is related to w and F. The form of V depends upon the geometry of the shell [8].

MEMBRANE ANALYSIS OF HYPERBOLIC PARABOLOIDAL SHELL SECTORS

First consider the hyperbolic paraboloidal shell sector shown in the plan view in Fig. 1. The region of the shell is bounded by four vertical planes $x = a$, $y = a$, $y = x \tan\theta$, and $x = y \tan\theta$. The equation of the middle surface of the shell is

$$z = \frac{c}{a^2} xy , \tag{38}$$

where c is the difference of elevation between point O and point A. The governing equations for membrane analysis are

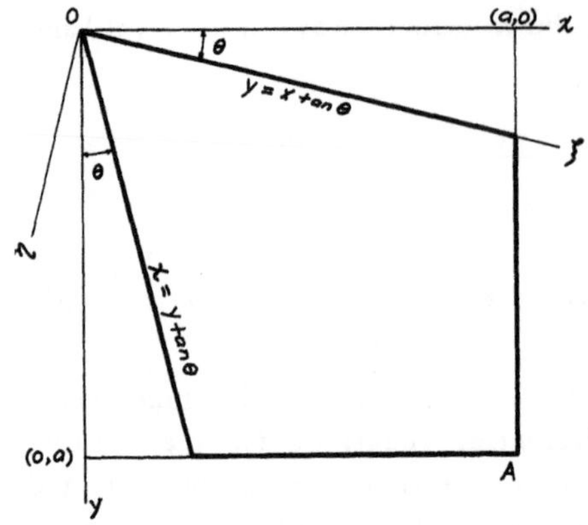

Fig. 1. Plan View of a Shell Sector.

$$\epsilon_{im}\,\epsilon_{jn}\,z,_{ij}\,F,_{mn} = q\;, \tag{39}$$

$$q = \int p_j\,dx_j + z,_j p_j - p_z\;, \tag{40}$$

where $x_1 = x$ and $x_2 = y$. These equations are the same as Eqs. (12) and (13). Substitution of Eqs. (38) and (40) into (39) and double integration yields

$$F = -\iint \frac{qa^2}{2c}\,xy\,dx\,dy + f_1(y) + g_1(x)\;. \tag{41}$$

Consequently, the membrane stresses n_{ij} become

$$n_{xy} = \frac{qa^2}{2c}\,, \tag{42a}$$

$$n_{xx} = F,_{yy} - \int p_1\,dx = -\left(\frac{qa^2}{2c}\,dx\right),_y - \int p_1\,dx + f(y)\;, \tag{42b}$$

$$n_{yy} = F,_{xx} - \int p_2\,dy = -\left(\frac{qa^2}{2c}\,dy\right),_x - \int p_2\,dy + g(x)\;. \tag{42c}$$

This is the general solution for membrane stresses of a hyperbolic paraboloidal shell. Since the two boundaries along $y = x \tan\theta$ and $x = y\tan\theta$ do not coincide with the coordinate lines, some difficulty arises in determining the arbitrary functions $f(y)$ and $g(x)$. To surmount this difficulty, let ξ be the coordinate in the tangential direction along the boundary $y = x\tan\theta$ and η the coordinate in the direction normal to this boundary. Consider the edge along ξ to be simply supported; this leads to the following boundary condition:

$$n_{\eta\eta} = 0\;. \tag{43}$$

The transformation of stresses is given by

$$n_{ab} = n_{ij}\,l_{ia}\,l_{jb}\;, \tag{44}$$

where n_{ab} represents a set of stresses in the new orthogonal coordinates (ξ, η), n_{ij} represents the set of stresses associated with the Cartesian coordinates (x, y), and l_{ia} and l_{jb} are the direction cosines between the two sets of coordinates. For simplicity, consider the loading conditions to be symmetrical about the diagonal OA. Then the boundary conditions along either $y = x \tan \theta$ or $x = y \tan \theta$ are to be satisfied. By use of Eq. (44), one finds

$$n_{\eta\eta} = f(y) \sin^2 \theta + g(x) \cos^2 \theta - \int P_1 \, dx \sin^2 \theta - \int P_2 \, dy \cos^2 \theta$$
$$- \sin^2 \theta \left(\int \frac{qa^2}{2c} dx \right)_{,y} - \cos^2 \theta \left(\int \frac{qa^2}{2c} dy \right)_{,x} - \frac{qa^2}{2c} \sin 2\theta = 0 . \quad (45)$$

The functions $f(y)$ and $g(x)$ are assumed to be in the form of the following series:

$$g(x) = \sum_{m=0}^{\infty} A_m (x/a)^m , \quad (46a)$$

$$f(y) = \sum_{m=0}^{\infty} B_m (y/a)^m . \quad (46b)$$

Due to the symmetry in geometry and loading about the diagonal, one may conclude that $A_m = B_m$. No negative term is included, because the stresses are bounded in the region including the point $x = y = 0$. Assume the loading function expressed in polynomial form. Substitution of Eqs. (46*a*) and (46*b*) into Eq. (45) and comparison of the terms leads to the evaluation of A_m. Exactly the same procedure could be applied to the nonsymmetrical case. Coefficients A_m and B_m will be obtained from the solution of two simultaneous equations for each m.

For illustrative purposes, consider a shell sector under its own weight. Thus

$$A_m = B_m ,$$
$$q = -P_z = -p_0[1 - (cx/a^2)^2 + (cy/a^2)^2]^{1/2} \quad (47)$$
$$= -p_0\{1 - \tfrac{1}{2}[(cx/a^2)^2 - (cy/a^2)^2] - \tfrac{1}{8}[(cx/a^2)^2 - (cy/a^2)^2]^2 + \dots \} ,$$

where p_0 is the weight per unit surface area of the shell. Then, from Eq. (45),

$$n_{\eta\eta} = \sum_{m=0}^{\infty} A_m (x/a)^m (\tan^m \theta \sin^2 \theta + \cos^2 \theta)$$
$$+ \tfrac{1}{2} p_0 c (x/a)^2 \tan \theta \{1 - \tfrac{1}{2}(c/a)^2 (x/a)^2 [\tan^2 \theta (1 - \tfrac{2}{3} \cos^2 \theta)$$
$$+ (1 - \tfrac{2}{3} \sin^2 \theta)] + \dots \} + \tfrac{1}{2} p_0 a (a/c) \{1 - \tfrac{1}{2}(c/a)^2 (x/a)^2 \sec^2 \theta$$
$$- \tfrac{1}{8}(c/a)^4 (x/a)^4 \sec^4 \theta + \dots \} \sin 2\theta = 0 . \quad (48)$$

Finally,

$$A_m (x/a)^m (\tan^m \theta \sin^2 \theta + \cos^2 \theta)$$
$$= -\tfrac{1}{2} p_0 c \tan \theta (x/a)^2 \{1 - \tfrac{1}{2}(c/a)^2 (x/a)^2 [\sec^2 \theta - \tfrac{4}{3} \sin^2 \theta] + \dots \}$$

$$- \tfrac{1}{2}p_0(a/c)^2 \sin^2\theta\{1 + \tfrac{1}{2}\sec^2\theta(c/a)^2(x/a)^2 - \tfrac{1}{8}\sec^4\theta(c/a)^4(x/a)^4 + \dots\}$$

$$= -\tfrac{1}{2}p_0\, a\{(a/c)\sin 2\theta - 2(c/a)\tan\theta(x/a)^2 - (c/a)^3\tan\theta(\tfrac{3}{4}\sec^2\theta - \tfrac{2}{3}\sin^2\theta)(x/a)^4 + \dots\}. \qquad (49)$$

The coefficients of the series given in Eq. (49) are

$$\begin{aligned} A_0 &= -\frac{p_0 a^2}{2c}\sin 2\theta\,, \\ A_1 &= A_3 = 0\,, \\ A_2 &= -\frac{p_0 c\sin 2\theta}{2(\sin^4\theta + \cos^4\theta)}\,, \\ A_4 &= \tfrac{1}{2}p_0 c\left(\frac{c}{a}\right)^2 \frac{\tan\theta(\tfrac{3}{4}\sec\theta - \tfrac{2}{3}\sin^2\theta)}{\tan^4\theta\sin^2\theta + \cos^2\theta}\,. \end{aligned} \qquad (50)$$

When (c/a) is small, the terms on the right side of Eq. (49) converge rapidly, and consideration of only a few terms will give good results. The results are summarized as follows:

$$\begin{aligned} n_{xx} &= -\frac{p_0 a^2}{2c}\sin 2\theta - \frac{p_0 c\sin 2\theta}{2(\sin^4\theta + \cos^4\theta)}\left(\frac{y}{a}\right)^2 + \dots \\ &\quad + \tfrac{1}{2}p_0 c(x/a)(y/a)\{1 - \tfrac{1}{2}(c/a)^2(y/a)^2 - \tfrac{1}{6}(c/a)^2(x/a)^2 + \dots\}\,, \\ n_{yy} &= -\frac{p_0 a^2}{2c}\sin 2\theta - \frac{p_0 c\sin^2\theta}{2(\sin^4\theta + \cos^4\theta)}\left(\frac{y}{a}\right)^2 + \dots \\ &\quad + \tfrac{1}{2}p_0 c(x/a)(y/a)\{1 - \tfrac{1}{2}(c/a)^2(x/a)^2 - \tfrac{1}{6}(c/a)^2(y/a)^2 + \dots\}\,, \\ n_{xy} &= -\frac{p_0 a^2}{2c}\{1 + \tfrac{1}{2}(c/a)^2[(x/a)^2 + (y/a)^2] - \tfrac{1}{8}(c/a)^4[(x/a)^2 \\ &\quad + (y/a)^2]^2 + \dots\}\,. \end{aligned} \qquad (51)$$

Consider next a hyperbolic paraboloidal shell sector whose plan view is shown in Fig. 2. The equation of the middle surface of the shell in Cartesian coordinates is

$$\frac{z}{c} = \frac{x^2}{a^2} - \frac{y^2}{a^2}\,. \qquad (52)$$

Let $t^1 = u$ and $t^2 = v$ be a set of surface coordinates which match the boundary (see Fig. 2). The coordinates u and v are related to the Cartesian coordinates in the following manner:

$$\begin{aligned} x &= au\cosh v\,, \\ y &= au\sinh v\,, \\ z &= cu^2\,. \end{aligned} \qquad (53)$$

Fig. 2. Plan View of a Shell Sector.

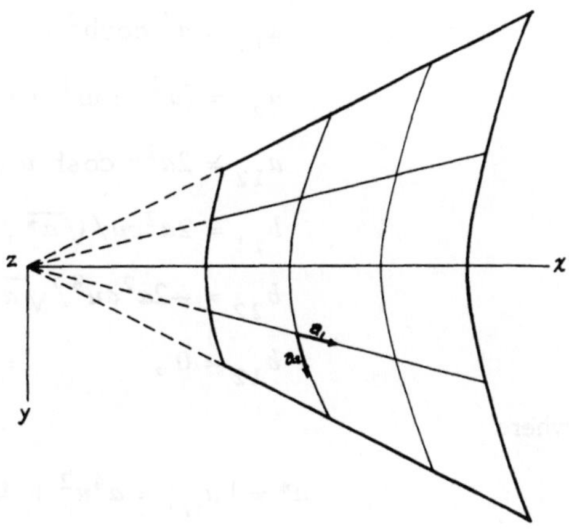

The line u = a constant represents a hyperbola on the surface, and the line v = a constant represents the intersecting curve between the surface and a vertical plane passing through the origin. The base vectors of this coordinate system vary from point to point; however, for certain geometrical configurations, this variation is small. If $\mathbf{a}_i$ and a_{ij} are considered as constants in differentiation [4], the equilibrium equations become:

$$\frac{\partial n_{11}}{\sqrt{a_{11}}\,\partial u} + \frac{\partial n_{12}}{\sqrt{a_{22}}\,\partial v} + \sqrt{\frac{a^*}{a_{11}a_{22}}}\,p_1 = 0\,,$$

$$\frac{\partial n_{12}}{\sqrt{a_{11}}\,\partial u} + \frac{\partial n_{22}}{\sqrt{a_{22}}\,\partial v} + \sqrt{\frac{a^*}{a_{11}a_{22}}}\,p_2 = 0\,, \tag{54}$$

$$n_{11}\frac{b_{11}}{a_{11}} + 2n_{12}\frac{b_{12}}{a_{12}} + n_{22}\frac{b_{22}}{a_{22}} + p_3 = 0\,.$$

Let F be the stress function, such that

$$n_{11} = a_{11}F,_{22} - \int\sqrt{a^*}\,p_1\,du\,,$$
$$n_{22} = a_{22}F,_{11} - \int\sqrt{a^*}\,p_2\,dv\,, \tag{55}$$
$$n_{12} = -\sqrt{a_{11}a_{22}}\,F,_{12}\,.$$

Equations (54) will be satisfied if

$$b_{11}F,_{22} - 2b_{12}F,_{12} + b_{22}F,_{11} = q\,, \tag{56}$$

$$q = -p_3 + \frac{b_{11}}{a_{11}}\int\sqrt{\frac{a^*}{a_{11}a_{22}}}\,p_1\,du + \frac{b_{22}}{a_{22}}\int\sqrt{\frac{a^*}{a_{11}a_{22}}}\,p_2\,dv\,. \tag{57}$$

The surface tensors a_{ij} and b_{ij} are found to be

$$
\begin{aligned}
a_{11} &= a^2 \cosh^2 v + a^2 \sinh^2 v + 4c^2u^2 , \\
a_{22} &= (a^2 \sinh^2 v + a^2 \cosh^2 v)u^2 , \\
a_{12} &= 2a^2u \cosh v \sinh v , \\
b_{11} &= 2a^2cu/\sqrt{a^*} , \\
b_{22} &= -2a^2cu^3/\sqrt{a^*} , \\
b_{12} &= 0 ,
\end{aligned} \tag{58}
$$

where

$$a^* = |a_{ij}| = a^4u^2 + 4a^2c^2u^4(1 + 2 \sinh^2 v) .$$

Therefore, for this case Eq. (56) reduces to

$$u^2F,_{11} - F,_{22} = -\frac{\sqrt{a^*}}{2a^2cu} q . \tag{59}$$

The complementary solution of Eq. (59) is assumed to be

$$F_c = X(u) Y(v) . \tag{60}$$

Substitution of Eq. (60) into Eq. (59) leads to

$$u^2X'' + \alpha^2X = 0 , \tag{61a}$$

$$Y'' + \alpha^2Y = 0 , \tag{61b}$$

from which the following solution is obtained:

$$Y = A_1 \cos \alpha v + B_1 \sin \alpha v . \tag{62}$$

Let

$$t = \log u , \tag{63}$$

and substitute Eq. (63) into Eq. (61a); this leads to

$$\frac{d^2X}{dt^2} + \frac{dX}{dt} + \alpha^2X = 0 . \tag{64}$$

The solution of Eq. (64) is

$$
\begin{aligned}
X &= e^{t/2}(C_1 \cos \sqrt{4\alpha^2 - 1}\, t + D_1 \sin \sqrt{4\alpha^2 - 1}\, t) \\
&= u^{1/2}[C_1 \cos (\sqrt{4\alpha^2 - 1} \log u) + D_1 \sin (\sqrt{4\alpha^2 - 1} \log u)] .
\end{aligned} \tag{65}
$$

Consider the loading condition to be uniform on the horizontal projection of the shell and enforce the following boundary conditions: along $v = \pm v_0$,

$$n_{22} = a_{22}F,_{11} = 0 ;$$

then

$$B_1 = 0\,, \qquad \alpha = m\pi/2t_0\,.$$

The total solution is given as

$$F = \sum_{m=1,3}^{\infty} u^{1/2}\left\{A_m \cos\left[\sqrt{\left(\frac{m}{2v_0}\right)^2 - 1}\,\log u\right] + B_m \sin\left(\sqrt{\left(\frac{m}{2v_0}\right)^2 - 1}\,\log u\right)\right\}\cos\frac{m\pi v}{2v_0} + F_p\,. \tag{66}$$

The integration constants A_m and B_m may be determined by the boundary conditions along the other two boundaries.

BENDING ANALYSIS OF A HYPERBOLIC PARABOLOIDAL SHELL SECTOR

The behavior of the shell is governed by the following differential equations in terms of vertical displacement w and the stress function F:

$$-D\nabla^4 w + z,_{ij}(\nabla^2 F\delta_{ij} - F,_{ij}) = z,_{ii}\int p_{\underline{i}}\,dx_{\underline{i}} + z,_j p_j - p_3\,, \tag{67}$$

$$\nabla^4 F = \left(\int p_1\,dx_1\right),_{22} + \left(\int p_2\,dx_2\right),_{11} - Eh\epsilon_{ia}\epsilon_{jb}z,_{ij}w,_{ab} - \nu(p_i,_i)\,. \tag{68}$$

Consider a hyperbolic paraboloidal shell sector which is described by the equation

$$z = \frac{c}{a^2}xy\,, \tag{69}$$

where $x = x_1$ and $y = x_2$. Substitution of Eq. (69) into Eqs. (67) and (68), with a new function V given by

$$w = \nabla^4 v\,, \tag{70}$$

$$F = Eh\frac{2c}{a^2}V,_{xy}\,, \tag{71}$$

leads to the following eighth-order differential equation in terms of this new function V:

$$D\nabla^8 V + Eh(2c/a^2)^2 V,_{xxyy} = -z,_{ij}\int p_{\underline{i}}\,dx_{\underline{i}} - z,_j p_j + p_3\,. \tag{72}$$

Consider first a shell sector which has the boundaries $x = a$, $y = a$, $y = x\tan\theta$, and $x = y\tan\theta$ (see Fig. 1), and then let the boundaries be simply supported. The complementary solution is assumed to be in the form of the following series:

$$V_c = \sum_{n=1}^{\infty}\sum_{m=1}^{\infty} A_{mn}(x - a)^m(y - a)^n\,. \tag{73}$$

Substitution of Eq. (73) into Eq. (72) results in the following general solution:

$$
\begin{aligned}
V_c = {} & A_{11}(x-a)(y-a) + A_{21}(x-a)^2(y-a) + A_{12}(x-a)(y-a)^2 \\
& + A_{31}(x-a)^3(y-a) + A_{13}(x-a)(y-a)^3 + A_{51}(x-a)^5(y-a) \\
& + A_{15}(x-a)(y-a)^5 + A_{61}(x-a)^6(y-a) + A_{16}(x-a)(y-a)^6 \\
& + A_{71}(x-a)^7(y-a) + A_{17}(x-a)(y-a)^7 \\
& + A_{81}\left\{(x-a)^8(y-a) - \frac{8}{12}a^4(x-a)^3(y-a)^2\right. \\
& \left. + \frac{3360}{\gamma}a^4(x-a)^4(y-a)\right\} + A_{18}\left\{(x-a)(y-a)^8 - \frac{8}{12}a^4(x-a)^2(y-a)^3\right. \\
& \left. + \frac{3360}{\gamma}a^4(x-a)(y-a)^4\right\} + A_{91}\left\{(x-a)^9(y-a) - \frac{8!}{4\gamma}(a^4)(x-a)^3(y-a)^3\right\} \\
& + A_{19}\left\{(x-a)(y-a)^9 - \frac{8!}{4\gamma}(a^4)(x-a)^3(y-a)^3\right\} \\
& + A_{(10)1}\left\{(x-a)^{10}(y-a) - \frac{50400}{\gamma}a^4(x-a)^4(y-a)^3\right\} \\
& + A_{1(10)}\left\{(x-a)(y-a)^{10} - \frac{50400}{\gamma}a^4(x-a)^3(y-a)^4\right\} \\
& + A_{(11)1}\left\{(x-a)^{11}(y-a) - \frac{55440}{\gamma}a^4(x-a)^5(y-a)^3\right\} \\
& + A_{1(11)}\left\{(x-a)(y-a)^{11} - \frac{55440}{\gamma}a^4(x-a)^3(y-a)^5\right\} , \qquad (74)
\end{aligned}
$$

where

$$
\gamma = \frac{Eh(2c/a^2)^2}{D} , \qquad a^4 = 48(1-\nu^2)(c/h)^2 . \qquad (75)
$$

From this,

$$
\begin{aligned}
w = {} & \nabla^4 V \\
= {} & 240(x-a)(y-a)A_{51} + 360(x-a)^2(y-a)A_{61} + (x-a)(y-a)^2A_{16} \\
& + 840\{(x-a)^3(y-a)A_{71} + (x-a)(y-a)^3A_{17}\} + 1680\{(x-a)^4(y-a)A_{81} \\
& + (x-a)(y-a)^4A_{18}\} + 3024\left\{\left[(x-a)^5(y-a) - \frac{2}{\gamma}(5!)a^4(x-a)(y-a)\right]A_{91}\right.
\end{aligned}
$$

$$+\left[(x-a)(y-a)^5-\frac{2}{\gamma}(5!)a^4(x-a)(y-a)\right]A_{19}\Bigg\}$$

$$+\,5040\left\{\left[(x-a)^6(y-a)-\frac{240}{\gamma}a^4(y-a)^3-\frac{1440}{\gamma}a^4(x-a)^2(y-a)\right]A_{(10)1}\right.$$

$$\left.+\left[(x-a)(y-a)^6-\frac{240}{\gamma}a^4(x-a)^3-\frac{1440}{\gamma}a^4(x-a)(y-a)^2\right]A_{1(10)}\right\}$$

$$+\,7920\left\{\left[(x-a)^7(y-a)-\frac{2520}{\gamma}a^4(x-a)^3(y-a)\right]A_{(11)1}\right.$$

$$\left.+\left[(x-a)(y-a)^7-\frac{2520}{\gamma}a^4(x-a)(y-a)^3\right]A_{1(11)}\right\}+\nabla^4V_p\,, \qquad (76)$$

$$F=Eb\frac{2c}{a^2}\{A_{11}+2[(x-a)A_{21}+(y-a)A_{12}]$$

$$+\,3[(x-a)^2A_{31}+(y-a)^2A_{13}]+5[(x-a)^4A_{51}+(y-a)^4A_{15}]$$

$$+\,6[(x-a)^5A_{61}+(y-a)^5A_{16}]+7[(x-a)^6A_{71}+(y-a)^6A_{17}]$$

$$+\left[8(x-a)^7-\frac{20168}{\gamma}a^4(x-a)(y-a)^2+\frac{13440}{\gamma}a^4(x-a)^3\right]A_{81}$$

$$+\left[8(y-a)^7-\frac{20168}{\gamma}a^4(x-a)(y-a)^2+\frac{13440}{\gamma}a^4(y-a)^3\right]A_{18}$$

$$+\left[9(x-a)^8-9\left(\frac{10080}{\gamma}\right)a^4(x-a)^2(y-a)^2\right]A_{91}$$

$$+\left[9(y-a)^8-9\left(\frac{10080}{\gamma}\right)a^4(x-a)^2(y-a)^2\right]A_{19}$$

$$+\left[10(x-a)^9-\frac{604800}{\gamma}a^4(x-a)^3(y-a)^2\right]A_{(10)1}$$

$$+\left[10(y-a)^9-\frac{604800}{\gamma}a^4(x-a)^2(y-a)^3\right]A_{1(10)}$$

$$+\left[11(x-a)^{10}-\frac{831600}{\gamma}a^4(x-a)^4(y-a)^2\right]A_{(11)1}$$

$$+\left[11(y-a)^{10}-\frac{831600}{\gamma}a^4(x-a)^2(y-a)^4\right]A_{1(11)}+V_{p,12}\Bigg\}\,, \qquad (77)$$

where V_p is the particular solution.

UNIFORMLY LOADED SHELL

It is seen that if the loading condition is symmetrical about the diagonal plane, the coefficients appearing in the above general results will be symmetrical as well; that is,

$$A_{mn} = A_{nm} \,. \tag{78}$$

Furthermore, observe that the coefficients A_{11}, A_{12}, and A_{21} can be perfectly arbitrary; that is, any value can be assigned to these coefficients without voiding the solution. Therefore they may be assumed to be zero for the sake of convenience. There are 16 constants left in the general solution which will be determined by the boundary conditions. With a careful examination, it is seen that the solution for a simply supported square hyperbolic paraboloidal shell with uniform vertical load will be $w = 0$ and $F = -p_0(a^2/2c)xy$. The result coincides exactly with the result discussed by Reissner [10].

For the particular shell sector under consideration, the general solution cannot satisfy all the boundary conditions exactly. Therefore the solution is made to satisfy the mean value of some of the boundary conditions. For simplicity, a uniform loading condition is assumed. The shell is assumed to be simply supported along the boundaries. Due to the symmetry of the loading, $A_{mn} = A_{nm}$. Apply the following boundary conditions: along $x = a$ (where $a \tan\theta \leqq y \leqq a$),

$$w = 0 \,, \qquad m_{11} = m_{xx} = 0 \qquad (\text{or } w,_{xx} = 0) \,;$$

this yields

$$A_{61} = A_{(10)1} = 0 \,. \tag{79}$$

The rest of the boundary conditions are as follows: along $y = x \tan\theta$,

$$\begin{aligned}
&\frac{1}{\cos\theta}\int_{a\tan\theta}^{a} n_{\eta\eta}\, dx = 0 \,, \\
&\frac{1}{\cos\theta}\int_{a\tan\theta}^{a} e_{\xi\xi}\, dx = \frac{1}{\cos\theta}\int_{a\tan\theta}^{a} n_{\xi\xi}\, dx = 0 \,, \\
&\frac{1}{\cos\theta}\int_{a\tan\theta}^{a} w\, dx = 0 \,, \\
&\frac{1}{\cos\theta}\int_{a\tan\theta}^{a} m_{\eta\eta}\, dx = \frac{1}{\cos\theta}\int_{a\tan\theta}^{a} w,_{\eta\eta}\, dx = 0 \,;
\end{aligned} \tag{80}$$

at point 0,

$$\begin{aligned} x = y = 0 \,, \\ w = 0 \,; \end{aligned} \tag{81}$$

along $x = a$,

$$\int_{a \tan \theta}^{a} F_{,yy}\, dy = -\frac{q}{Eb}\left(\frac{a^2}{2c}\right)^2 a(1 - \tan\theta)\,. \tag{82}$$

The following general equation will be used for the transformation of stresses and stress couples:

$$X_{ab} = X_{ij}\, l_{ia}\, l_{jb}\,, \tag{83}$$

where the indices a and b refer to ξ and η coordinates, and i and j refer to x and y coordinates. These boundary conditions give the following six simultaneous algebraic equations for the six unknown coefficients:[4]

$$[C_{ij}][A_j] = [B_i]\,, \tag{84}$$

where i and j range from 1 to 6, and

$$\begin{aligned} &A_1 = A_{31}\,, && A_2 = A_{51}a^2\,, && A_3 = A_{71}a^4\,,\\ &A_4 = A_{81}a^5\,, && A_5 = A_{91}a^6\,, && A_6 = A_{(11)1}a^8\,. \end{aligned} \tag{85}$$

For an example, consider a uniformly loaded shell with the following geometrical dimensions:

$$\begin{aligned} &a = 10\text{ ft}\,, && b = 1.5\text{ in.} = 0.125\text{ ft}\,, && c = 2.5\text{ ft}\,,\\ &\nu = 0\,, && \theta = 15^\circ\,, && \gamma = 192 \times 10^2\,. \end{aligned}$$

Equation (84) is solved by use of a Datatron 205 computer and yields the following results:

$$\begin{aligned} &A_1 = -0.679623680\,, && A_2 = 1.195136600\,,\\ &A_3 = -1.548234500\,, && A_4 = -0.928761270\,,\\ &A_5 = -0.095344736\,, && A_6 = -0.017071517\,. \end{aligned}$$

The matrices $[C_{ij}]$ and $[B_j]$ for this case are shown below:

$[C_{ij}] =$

6	21.699016	47.479730	−57.981736	60.791069	−178.92054
6	43.66065	118.31442	−175.01588	228.49508	305.79052
0	1	4.76267185	−8.373097	12.257564	24.924254
0	1	30.564311	−45.466558	114.76573	481.57471
0	1	7	−14	24.885000	57.33745
4.3923	10.717940	12.061804	−11.517126	11.08098	6.6414128

[4]Refer to the dissertation cited in footnote 1 for the detailed expression of $[C_{ij}]$.

$$[B_i] = \begin{bmatrix} -0.5 \\ 0.5 \\ 0 \\ 0 \\ 0 \\ -0.366025 \end{bmatrix} \frac{p_0}{Eh} \left(\frac{a^2}{2c}\right)^2 .$$

Based upon the above results, the deflections along the diagonal section ($x = y$) were calculated by use of an LGP 30 computer and are shown in Table 1. The

Table 1. Deflection Along Diagonal

x/a	$\frac{wEh}{p_0}\left(\frac{2c}{a^2}\right)^2$	x/a	$\frac{wEh}{p_0}\left(\frac{2c}{a^2}\right)^2$
0.000	0.0000000	0.6	10.789605
0.1	−25.107997	0.7	12.829329
0.2	−36.544719	0.8	8.6180712
0.3	−30.806757	0.9	2.7144999
0.4	−15.400443	1.0	0.0000000
0.5	0.6290338		

maximum deflection shown in the table is at $x/a = 0.2$. Consider $E = 432 \times 10^6$ psf for concrete and $p_0 = 20$ psf; the magnitude of the deflection is found to be

$$w_{(x/a=0.2)} = -\frac{36.544719}{(432 \times 10^6)(0.125)} (100/5)^2(20)(12) = -0.0648 \text{ in.}$$

For the second example, a model was made of thin plastic sheet and has the following characteristics:

$$a = 10 \text{ in.}, \quad c = 2.5 \text{ in.}, \quad \nu = 0.3, \quad h = 0.05 \text{ in.},$$

$$\theta = 15^\circ \quad \gamma = 10.92 \times 10^4, \quad E = 380{,}000 \text{ psi}.$$

The plan view of the model has boundaries along $x = a$, $y = a$, $y = x \tan\theta - a(1 - \tan\theta)$, and $x = y \tan\theta - a(1 - \tan\theta)$; see Fig. 3. For this case, the elements of $[C_{ij}]$ are generated by use of the LGP 30 electronic computer. The matrices $[C_{ij}]$ and $[B_i]$ are found as follows:

$$[C_{ij}] = \begin{bmatrix} 12 & 176.42410 & 1310.3547 & -4076.5159 & 10288.999 & 145160.77 \\ 12 & 388.72824 & 3771.0558 & -10825.043 & 29157.290 & 296552.54 \\ 0 & 192.01153 & 2536.8815 & -4427.598 & 13939.597 & 114950.61 \\ 0 & 240 & 51300.634 & -296387.68 & 1300743.2 & 22575938 \\ 0 & 240 & 6720 & -26880 & 95437.44 & 999139.7 \\ 15.2154 & 326.15674 & 4404.6343 & -14926.058 & 48473.341 & 471715.75 \end{bmatrix} ,$$

$$[B_j] = \begin{bmatrix} -1.0 \\ 1.0 \\ 0 \\ 0 \\ 0 \\ -0.73205 \end{bmatrix} \frac{p_0}{Eh} \left(\frac{a^2}{2c} \right)^2 .$$

Equation (84) was then solved by use of the Datatron 205, giving the following results:

$$A_1 = -0.35468349 , \qquad A_2 = 0.04975161 ,$$
$$A_3 = -0.00501126 , \qquad A_4 = 0.00073711 ,$$
$$A_5 = 0.00055533 , \qquad A_6 = -0.00001146 .$$

Based upon the above results, the deflections along the diagonal section ($x = y$) were calculated by use of the LGP 30 and are shown in Table 2. The maximum

Table 2. Deflection Along Diagonal

x/a	$\frac{wEh}{p_0}\left(\frac{2c}{a^2}\right)^2$	x/a	$\frac{wEh}{p_0}\left(\frac{2c}{a^2}\right)^2$
0.0	4.1899660		
0.1	4.3600691	−0.1	3.6659299
0.2	4.2018019	−0.2	2.8027800
0.3	3.7717105	−0.3	1.6638548
0.4	3.1473637	−0.4	0.36712182
0.5	2.4174488	−0.5	0.91158223
0.6	1.6724741	−0.6	−1.9390987
0.7	0.99655437	−0.7	−2.4302601
0.8	0.46068094	−0.8	−2.0590490
0.9	0.11780644	−0.9	−0.47670984
1.0	0.00000000	−1.0	0.00000000

deflection shown in the table is at $x/a = 0.1$. Consider $E = 380{,}000$ psi (information supplied by Dow Chemical Company for Styron 475) and $p_0 = 0.3$ psi (or 43.2 psf); the magnitude of the deflection is found to be

$$w_{(x/a=0.1)} = + \frac{4.3600691}{(380{,}000)(0.05)} (100/5)^2(0.3) = +0.02754 \text{ in.}$$

The calculated and the experimental deflection curves are shown in Fig. 3. The deflection modes agree satisfactorily. However, the magnitudes of the

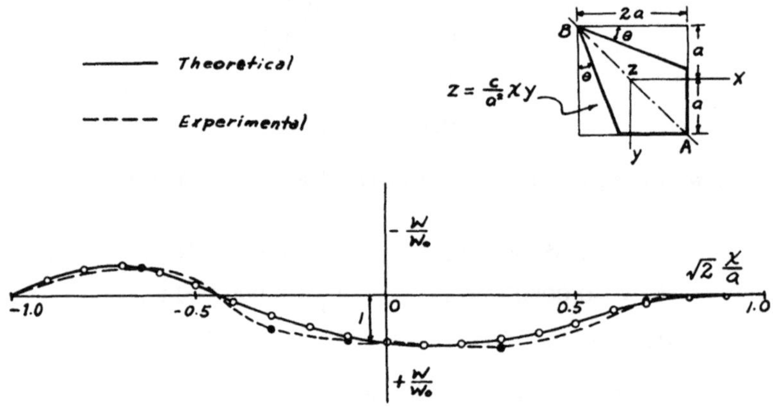

Fig. 3. Deflection Curves Along Diagonal *AB*.

measured deflections were larger than those calculated by the method derived in this study. For example, at the point $x = 0$, $y = 0$ (Fig. 3), the calculated deflection was 0.0264 in., whereas the experimentally determined deflection was 0.0440 in. The experimental discrepancies are probably due to the following reasons:

1. uncertainty of Young's modulus,
2. nonhomogeneity of material,
3. variation in the thickness of the model due to forming techniques.

REFERENCES

[1] Aimond, F., "Etude Statique des Voiles Minces en Paraboloide Hyperbolique Travaillant sans Flexion," *Intern. Assoc. Bridge and Structural Eng. Publics.* **4**, 1–112 (1936).

[2] Chien, W. Z., "The Intrinsic Theory of Thin Shells and Plates," *Quart. Appl. Math.* **1**(4), 297–327 (1944); **2**(1), 43–49 (1944); **2**(2) (1944).

[3] Gerard, F. A., "The Analysis of Hyperbolic Paraboloidal Shell Roofs," *Trans. Eng. Inst. of Canada* **3**(1) (1959).

[4] Goldenveizer, A. L., "Several Devices for Integrating the Equations of the Theory of Shells," *Prikl. Matem. i Mekh.* **10**(3), 396 (1946).

[5] Green, A. E. and W. Zerna, *Theoretical Elasticity*, Clarendon, Oxford, 1954.

[6] Libai, Avinoam, *Theory of Doubly Curved Shells*, Ph.D. thesis, Purdue University, 1959.

[7] Margenau, H., and G. M. Murphy, *The Mathematics of Physics and Chemistry*, 2d ed., Van Nostrand, New York, 1956.

[8] Mircea Mihailescu, "On the Hyperbolic Paraboloids – Bending Theory," *Stud. Cerc. Mech. Aplic.* (April–June 1956).

[9] Novoshilov, V. V., "On an Error in a Hypothesis of the Theory of Shells," *Doklady Akad. Nauk S. S. S. R.* **38**, 160–64 (1943) (in English).

[10] Reissner, Eric, "On Some Aspects of the Theory of Thin Elastic Shells," *J. Boston Soc. Civil Engrs.* **42**, 100–133 (April 1955).

[11] Sanders, J. Lyell, Jr., "An Improved First-Approximation Theory for Thin Shells," *NASA* TR-R-24 (1959).

[12] Sokolnikoff, I. S., *Mathematical Theory of Elasticity*, McGraw-Hill, New York, 1956.

[13] Sokolnikoff, I. A., *Tensor Analysis, Theory and Applications*, Wiley, New York, 1951.

[14] Struik, D. J., *Lectures on Classical Differential Geometry*, Addison-Wesley, Cambridge, 1950.

[15] Timoshenko, S., and S. Woinowsky-Krieger, *Theory of Plates and Shells*, 2d ed., McGraw-Hill, New York, 1959.

[16] Vlasov, V. S., "Basic Differential Equations in General Theory of Elastic Shells," *NACA* TM-1241 (February 1951).

[17] Wang, C. T., *Applied Elasticity*, McGraw-Hill, New York, 1953.

[18] Wilson, B. S., *A Linear Theory of Elastic Shells*, Ph.D. thesis, University of California, 1957.

STRESSES DUE TO ROTARY INERTIA FORCES IN A HOLLOW CIRCULAR CYLINDER[1]

H. B. Wilson, Jr.
University of Illinois

A method for solving plane-elasticity problems for annular regions by application of complex variables and the problem of linear relationship is considered, and invalidities are indicated in the analysis of another author. Appropriate corrections are given, and the method is employed to solve a plane-strain problem for a hollow cylinder which rotates about an eccentric axis. The assumed boundary conditions are zero internal stress and zero external displacement.

INTRODUCTION

This paper presents an analysis of the stresses caused in a hollow circular cylinder due to rotation with uniform angular velocity about an axis parallel to the geometrical axis of the cylinder. The cylinder is assumed to be linearly elastic and in a state of plane strain. The boundary conditions are zero stress at the internal boundary and zero displacement at the external boundary. The solution is based on complex-variable techniques and a simple application of the problem of linear relationship [*1*]. Numerical calculations of the stresses on the boundaries are given for the case where Poisson's ratio equals $\frac{1}{2}$.

REDUCTION TO A PROBLEM OF ZERO BODY FORCE

In Fig. 1, a homogeneous cylinder of mass ρ per unit volume rotates with an angular velocity ω about an axis O. The inner boundary Γ_b is free from stress, and the displacement on the outer boundary Γ_a is zero relative to a set of axes which rotate with the cylinder. It is required to determine the stresses and displacements caused in the cylinder by rotary inertia body forces.

The body forces due to rotation are

$$X = \rho\omega^2(x + h) , \qquad Y = \rho\omega^2 y . \tag{1}$$

---

[1] Performed under a research project of the Department of Theoretical and Applied Mechanics, University of Illinois, sponsored by the Department of the Army under the technical supervision of Redstone Arsenal, Alabama. Contract No. DA-11-022-506-ORD-3488, ORD Project No. TW-205.

NOTATION

$\widehat{xx}, \widehat{yy}, \widehat{xy}$	Stresses in rectangular coordinates
$\widehat{rr}, \widehat{\theta\theta}, \widehat{r\theta}$	Stresses in polar coordinates
ω	Angular velocity
ρ	Mass per unit volume
u, v	Horizontal and vertical components of displacement
μ	Shear modulus
ν	Poisson's ratio
b, a	Internal and external radii of cylinder
Γ_b, Γ_a	Internal and external boundaries of cylinder
h	Distance from axis of rotation to longitudinal axis of cylinder
ξ	a/b
η	h/b

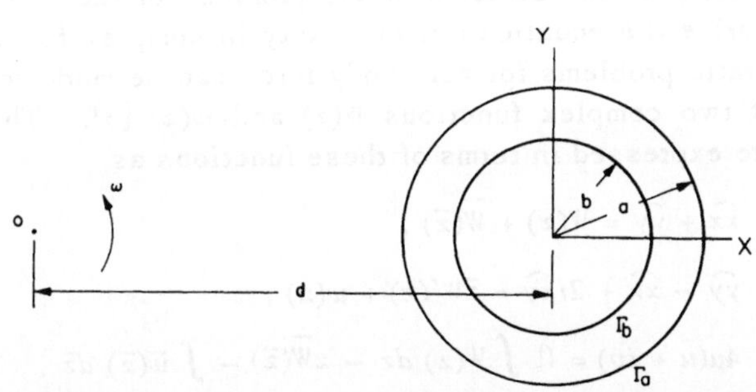

Fig. 1. Rotation of a Hollow Cylinder About an Eccentric Axis.

To obtain a solution to the problem it is most convenient to deal with modified stresses and displacements which satisfy the equations of elasticity for zero body force. For this purpose the actual stresses and displacements can be taken equal to the sum of a particular set of stresses and displacements which satisfy the equations of elasticity with body force included plus a modified set of stresses and displacements which satisfy the equations for zero body force. The boundary conditions for the modified problem are the stresses or displacements in the actual problem minus the stresses or displacements given by the particular solution.

It can be verified by substitution that a particular solution of the equations of plane elasticity [2] for the body forces given by Eqs. (1) is

$$\widehat{xx}^{(p)} = -\alpha(x+b)^2 - \beta y^2 , \qquad u^{(p)} = -\frac{(\alpha-\beta)}{4\mu}[(x+b)^2+y^2](x+b) ,$$

$$\widehat{yy}^{(p)} = -\beta(x+b)^2 - \alpha y^2 , \qquad v^{(p)} = -\frac{(\alpha-\beta)}{4\mu}[(x+b)^2+y^2]y , \tag{2}$$

$$\widehat{xy}^{(p)} = -(\alpha-\beta)(x+b)y , \qquad \alpha = \frac{(3-2\nu)\rho\omega^2}{8(1-\nu)} , \qquad \beta = \frac{(1+2\nu)\rho\omega^2}{8(1-\nu)} .$$

In Eqs. (2), ν is Poisson's ratio, μ is the shear modulus, and the superscript (p) denotes a particular solution.

The boundary conditions for the modified stress problem for the annulus become

$$\begin{aligned} \widehat{rr} &= -\widehat{rr}^{(p)} , & \widehat{r\theta} &= -\widehat{r\theta}^{(p)} , & &\text{on } \Gamma_b , \\ u &= -u^{(p)} , & v &= -v^{(p)} , & &\text{on } \Gamma_a . \end{aligned} \tag{3}$$

After the solution of the modified problem is obtained, the actual solution is simply the sum of the modified solution and the particular solution from Eqs. (2).

SUMMARY OF COMPLEX-VARIABLE FORMULAS

Before proceeding to the solution of the problem for the annulus, it is convenient to summarize the equations of elasticity in complex form. The solution of plane elastostatic problems for zero body force can be made to depend on the determination of two complex functions $W(z)$ and $w(z)$ [3]. The stresses and displacements are expressed in terms of these functions as

$$\begin{aligned} &\widehat{xx} + \widehat{yy} = W(z) + \overline{W}(\bar{z}) , \\ &\widehat{yy} - \widehat{xx} + 2i\widehat{xy} = \bar{z}W'(z) + w(z) , \\ &4\mu(u+iv) = \eta \int W(z)\,dz - z\overline{W}(\bar{z}) - \int \overline{w}(\bar{z})\,d\bar{z} , \\ &2(\widehat{rr} + i\widehat{r\theta}) = \widehat{xx} + \widehat{yy} + (\widehat{xx} - \widehat{yy} + 2i\widehat{xy})e^{-2i\theta} . \end{aligned} \tag{4}$$

In the above displacement equation, μ is the shear modulus, and $\eta = 3 - 4\nu$ for plane strain, where ν is Poisson's ratio. If $z = re^{i\theta}$, partial differentiation of the third equation of Eqs. (4) yields another useful formula,

$$4\mu\left(\frac{\partial u}{\partial\theta} + i\frac{\partial v}{\partial\theta}\right) = iz[\eta W(z) - \overline{W}(\bar{z}) + \bar{z}\overline{W}'(\bar{z}) + \frac{\bar{z}}{z}\overline{w}(\bar{z})] . \tag{5}$$

Other relations which will be required are the Plemelj formulas [1]. Let C denote the union of a finite number of smooth nonintersecting arcs in the complex plane, and let $f(\xi)$ be a continuous function defined on C. The arcs which compose C may be open or closed. The function

$$F(z) = \frac{1}{2\pi i}\int_C \frac{f(\xi)\,d\xi}{\xi - z} \tag{6}$$

is analytic for z not on C. Let z tend to a point t on C, where t is not an end of C. Then $F(z)$ has different limits, depending on whether t is approached from the left or from the right of C. The two limiting expressions are

$$\begin{aligned} F^L(t) &= \frac{1}{2} f(t) + \frac{1}{2\pi i} \int_C \frac{f(\xi)\, d\xi}{\xi - t}, \\ F^R(t) &= -\frac{1}{2} f(t) + \frac{1}{2\pi i} \int_C \frac{f(\xi)\, d\xi}{\xi - t}, \end{aligned} \tag{7}$$

where the singular integrals are to be taken as Cauchy principal-value integrals. Equations (7) are the Plemelj formulas, which may be written in an equivalent form,

$$\begin{aligned} F^L(t) - F^R(t) &= f(t), \\ F^L(t) + F^R(t) &= \frac{1}{\pi i} \int_C \frac{f(\xi)\, d\xi}{\xi - t}. \end{aligned} \tag{8}$$

SOLUTION OF THE MODIFIED-STRESS PROBLEM

The solution of the modified-stress problem will now be considered. By use of the last of Eqs. (4), the stress and displacement components frce Eqs. (2) may be written as

$$\begin{aligned} 2[\widehat{rr}^{(p)} + i\widehat{r\theta}^{(p)}] &= -[(\alpha - \beta)h^2 e^{-2i\theta} + (3\alpha - \beta)hre^{-i\theta} \\ &\qquad + (\alpha + \beta)h^2 + 2\alpha r^2 + (\alpha + \beta)hre^{i\theta}], \\ 4\mu[u^{(p)} + iv^{(p)}] &= -(\alpha - \beta)[b^2 re^{-i\theta} + (b^3 + 2br^2) \\ &\qquad + (2b^2 r + r^3)e^{i\theta} + br^2 e^{2i\theta}]. \end{aligned} \tag{9}$$

With the use of these relations, the boundary conditions for the modified-stress problem become

$$\begin{aligned} 2(\widehat{rr} + i\widehat{r\theta}) &= B_{-2}t^{-2} + B_{-1}t^{-1} + B_0 + B_1 t, & t &= be^{i\theta} \text{ on } \Gamma_b, \\ 4\mu(u + iv) &= \eta(A_{-1}t^{-1} + A_0 + A_1 t + A_2 t^2), & t &= ae^{i\theta} \text{ on } \Gamma_a. \end{aligned} \tag{10}$$

The constants $B_{-2} \ldots B_1$ and $A_{-1} \ldots A_2$ are given by

$$\begin{aligned} B_{-2} &= (\alpha - \beta)h^2 b^2, & B_{-1} &= (3\alpha - \beta)hb^2, \\ B_0 &= (\alpha + \beta)h^2 + 2\alpha b^2, & B_1 &= (\alpha + \beta)h, \\ A_{-1} &= \frac{(\alpha - \beta)b^2 a^2}{\eta}, & A_0 &= \frac{(\alpha - \beta)(b^3 + 2ba^2)}{\eta}, \\ A_1 &= \frac{(\alpha - \beta)(2b^2 + a^2)}{\eta}, & A_2 &= \frac{(\alpha - \beta)b}{\eta}. \end{aligned} \tag{11}$$

The solution of the modified-stress problem depends on the determination of two analytic functions $W(z)$ and $w(z)$ for the region $b < |z| < a$. Although the solution can be obtained by conventional series techniques, an alternate method based on the problem of linear relationship will be employed. This leads to a formulation in terms of a single-stress function. The technique also has the advantage of being applicable to mixed boundary value problems.

A method for formulating plane elasticity problems for an annulus in terms of the problem of linear relationship was presented by Milne-Thomson [3]. However, the general solutions given there for an annulus are invalid due to certain assumptions concerning orders of singularities of the stress-potential function and values of Cauchy integrals involving boundary stresses or displacements. These assumptions are discussed below.

It is convenient to extend the definition of $W(z)$ into the regions $(b^2/a) < |z| < b$ and $a < |z| < (a^2/b)$ by the following relationships:

$$W(z) = -\overline{W}\left(\frac{a^2}{z}\right) + \frac{a^2}{z}\,\overline{W}'\left(\frac{a^2}{z}\right) + \frac{a^2}{z^2}\,\overline{w}\left(\frac{a^2}{z}\right), \qquad b < \left|\frac{a^2}{z}\right| < a\,,$$
$$W(z) = -\overline{W}\left(\frac{b^2}{z}\right) + \frac{b^2}{z}\,\overline{W}'\left(\frac{b^2}{z}\right) + \frac{b^2}{z^2}\,\overline{w}\left(\frac{b^2}{z}\right), \qquad b < \left|\frac{b^2}{z}\right| < a\,. \tag{12}$$

Equations (12) give two definitions of $w(z)$ for $b < |z| < a$. These definitions are readily seen to be

$$w(z) = \frac{a^2}{z^2}\left[\overline{W}\left(\frac{a^2}{z}\right) + W(z) - zW'(z)\right],$$
$$w(z) = \frac{b^2}{z^2}\left[\overline{W}\left(\frac{b^2}{z}\right) + W(z) - zW'(z)\right], \quad \text{for } b < |z| < a\,. \tag{13}$$

Since both definitions of $w(z)$ must be identical, there results the identity

$$(a^2 - b^2)[W(z) - zW'(z)] \equiv b^2\overline{W}\left(\frac{b^2}{z}\right) - a^2\overline{W}\left(\frac{a^2}{z}\right). \tag{14}$$

By use of Eqs. (13), the formulas for stress and displacement can be expressed in terms of the single function $W(z)$, in the form

$$2(\widehat{rr} + i\widehat{r\theta}) = W(z) - \frac{a^2}{z\bar{z}}\,W\left(\frac{a^2}{\bar{z}}\right) + \left(1 - \frac{a^2}{z\bar{z}}\right)[\overline{W}(\bar{z}) - \bar{z}\overline{W}'(\bar{z})]\,,$$
$$4\mu\left(\frac{\partial u}{\partial\theta} + i\,\frac{\partial v}{\partial\theta}\right) = iz\left\{\kappa W(z) + \frac{a^2}{z\bar{z}}\,W\left(\frac{a^2}{\bar{z}}\right) - \left(1 - \frac{a^2}{z\bar{z}}\right)[\overline{W}(\bar{z}) - \bar{z}\overline{W}'(\bar{z})]\right\}. \tag{15}$$

Equations (15) are also valid if a is replaced by b.

Now let t denote the complex variable on either contour of the annulus, and let the positive direction on Γ_a and Γ_b be counterclockwise and clockwise respectively. As $z \to t$, Eqs. (15) have the limiting forms

$$
\begin{aligned}
2(\widehat{rr} + i\widehat{r\theta}) &= W^L(t) - W^R(t) \,, \\
4\mu \frac{d}{dt}(u + iv) &= \eta W^L(t) + W^R(t) \,,
\end{aligned} \tag{16}
$$

where L and R denote the left and right limits of $W(z)$ as z approaches the contour. Therefore, the solution of the modified-stress problem requires determination of a function $W(z)$ defined for $(b^2/a) < |z| < (a^2/b)$ except on Γ_b and Γ_a, where it has discontinuities:

$$
\begin{aligned}
W^L(t) - W^R(t) &= B_{-2}t^{-2} + B_{-1}t^{-1} + B_0 + B_1 t \,, && \text{on } \Gamma_b \,, \\
W^L(t) + \frac{1}{\eta} W^R(t) &= -A_{-1}t^{-2} + A_1 + 2A_2 t \,, && \text{on } \Gamma_a \,.
\end{aligned} \tag{17}
$$

Now introduce the function $F(z)$ defined by

$$
\begin{aligned}
F(z) &= W(z) \,, && \text{for } \frac{b^2}{a} < |z| < a \,, \\
F(z) &= -\frac{1}{\eta} W(z) \,, && \text{for } a < |z| < \frac{a^2}{b} \,.
\end{aligned} \tag{18}
$$

With these definitions, the boundary value Eqs. (17) become

$$
\begin{aligned}
F^L(t) - F^R(t) &= B_{-2}t^{-2} + B_{-1}t^{-1} + B_0 + B_1 t \,, && \text{on } \Gamma_b \,, \\
F^L(t) - F^R(t) &= -A_{-1}t^{-2} + A_1 + 2A_2 t \,, && \text{on } \Gamma_a \,.
\end{aligned} \tag{19}
$$

The solution of Eqs. (19) follows immediately from Eqs. (7) and (8). Therefore,

$$
F(z) = -\frac{1}{2\pi i} \oint_{\Gamma_b} \frac{(B_{-2}t^{-2} + B_{-1}t^{-1} + B_0 + B_1 t)\, dt}{t - z} + \frac{1}{2\pi i} \oint_{\Gamma_a} \frac{(-A_{-1}t^{-2} + A_1 + 2A_2 t)\, dt}{t - z} + f(z) \,, \tag{20}
$$

where $f(z)$ is a function which is continuous across Γ_a and Γ_b. Integration of Eq. (20) and use of Eqs. (18) yields

$$
\begin{aligned}
W(z) &= (A_1 - B_0) + (2A_2 - B_1)z + f(z) \,, && \text{for } (b^2/a) < |z| < b \,, \\
W(z) &= B_{-2}z^{-2} + B_{-1}z^{-1} + A_1 + 2A_2 z + f(z) \,, && \text{for } b < |z| < a \,, \\
W(z) &= -\eta(A_{-1} + B_{-2})z^{-2} - \eta B_{-1}z^{-1} - \eta f(z) \,, && \text{for } a < |z| < (a^2/b) \,.
\end{aligned} \tag{21}
$$

The function $f(z)$ must be determined such that identity (14) is satisfied.

Before proceeding with the solution of the above problem, certain remarks should be made concerning the analysis of the first and second fundamental problems for an annular region given in ref [3]. Milne-Thomson argues that, in general, the function corresponding to $f(z)$ in Eqs. (21) can be taken in the form

$$\frac{A}{z^2} + \frac{B}{z} + C \, .$$

However, Eq. (14) cannot always be satisfied with this form of $f(z)$. For example, if

$$\widehat{rr} + i\widehat{r\theta} = kt^3 \, , \qquad \text{on } \Gamma_b \, ,$$

$$\widehat{rr} + i\widehat{r\theta} = 0 \, , \qquad \text{on } \Gamma_a \, ,$$

it can be shown that to satisfy Eq. (14) and also have single-valued displacements, $f(z)$ must be taken as

$$C_{-3} z^{-3} + C_3 z^3 \, .$$

Since $W(z)$ will, in general, be analytic for $b < |z| < a$, $f(z)$ can be taken as a Laurent series about $z = 0$.

Furthermore, in articles 5.51 and 5.6 of ref [3], statements occur which are not completely general concerning Cauchy integrals which involve boundary stresses or displacements. For instance, it is assumed that

$$\frac{1}{2\pi i} \oint_{\Gamma_a} \frac{(\widehat{rr} + i\widehat{r\theta})\, dt}{t - z} = 0 \, , \qquad \text{for } |z| > a \, .$$

This would be true if $\widehat{rr} + i\widehat{r\theta}$ were the boundary value of a function holomorphic inside Γ_a. However, if the boundary stresses have a finite number of ordinary discontinuities, they may be expanded in a Fourier series of the form

$$\widehat{rr} + i\widehat{r\theta} = \sum_{n=-\infty}^{n=\infty} C_n e^{in\theta} = \sum_{n=-\infty}^{n=\infty} \frac{C_n}{a^n} t^n \, , \qquad |t| = a \, .$$

Then it follows that

$$\frac{1}{2\pi i} \oint_{\Gamma_a} \frac{(\widehat{rr} + i\widehat{r\theta})\, dt}{t - z} = - \sum_{n=-\infty}^{n=-1} C_n \left(\frac{z}{a} \right)^n , \qquad \text{for } |z| > a \, ,$$

instead of zero.

Return now to the solution of the original problem. Assuming that $f(z)$ has a Laurent expansion about $z = 0$ of the form

$$f(z) = \sum_{n=-\infty}^{n=\infty} \alpha_n z^n \, ,$$

substitution of Eqs. (21) into Eq. (14) gives

$$3B_{-2}(a^2 - b^2)z^{-2} + 2B_{-1}(a^2 - b^2)z^{-1} + A_1(a^2 - b^2)$$

$$+ \sum_{n=-\infty}^{n=\infty} \alpha_n(a^2 - b^2)(1 - n)z^n \equiv (2A_2 - B_1)b^4z^{-1} + (A_1 - B_0)b^2$$

$$+ \kappa B_{-1}z + \kappa(A_{-1} + B_{-2})a^{-2}z^2 + \sum_{n=-\infty}^{n=\infty} \bar{\alpha}_{-n}(b^{-2n+2} + \kappa a^{-2n+2})z^n . \qquad (22)$$

By equating the coefficients of equal powers of z on both sides of Eq. (22), the following system of simultaneous equations is obtained:

$$\begin{aligned}
A_1(a^2 - b^2) + (a^2 - b^2)\alpha_0 &= (A_1 - B_0)b^2 + \bar{\alpha}_0(b^2 + \kappa a^2) , \\
2B_{-1}(a^2 - b^2) + 2(a^2 - b^2)\alpha_{-1} &= (2A_2 - B_1)b^4 + \bar{\alpha}_1(b^4 + \kappa a^4) , \\
0 &= \kappa B_{-1} + \bar{\alpha}_{-1}(1 + \kappa) , \\
3B_{-2}(a^2 - b^2) + 3(a^2 - b^2)\alpha_{-2} &= \bar{\alpha}_2(b^6 + \kappa a^6) , \\
-(a^2 - b^2)\alpha_2 &= \kappa(A_{-1} + B_{-2})a^{-2} + \bar{\alpha}_{-2}(b^{-2} + \kappa a^{-2}) , \\
(a^2 - b^2)(1 - n)\alpha_n &= \bar{\alpha}_{-n}(b^{-2n+2} + \kappa a^{-2n+2}) , \qquad \text{for } |n| \geqq 3 .
\end{aligned} \qquad (23)$$

The solution of the system of Eqs. (23) is

$$\begin{aligned}
\alpha_0 &= \frac{(-a^2 + 2b^2)A_1 - b^2B_0}{a^2(1 - \kappa) - 2b^2} , \\
\alpha_{-1} &= -\frac{\kappa B_{-1}}{1 + \kappa} , \\
\alpha_1 &= \frac{2(a^2 - b^2)B_{-1} - (1 + \kappa)b^4(2A_2 - B_1)}{(1 + \kappa)(b^4 + \kappa a^4)} , \\
\alpha_{-2} &= \frac{-3(a^2 - b^2)^2B_{-2} - \kappa a^{-2}(b^6 + \kappa a^6)(A_{-1} + B_{-2})}{(b^6 + \kappa a^6)(b^{-2} + \kappa a^{-2}) + 3(a^2 - b^2)^2} , \\
\alpha_2 &= \frac{3(a^2 - b^2)(b^{-2}B_{-2} - \kappa a^{-2}A_{-1})}{(b^6 + \kappa a^6)(b^{-2} + \kappa a^{-2}) + 3(a^2 - b^2)^2} , \\
\alpha_n &= 0 , \qquad \text{for } |n| \geqq 3 .
\end{aligned} \qquad (24)$$

Therefore, $W(z)$ is completely determined, and $w(z)$ can be evaluated with the aid of either of Eqs. (13). After some algebraic reduction, it follows from Eqs. (11), (13), (21), and (24) that for $b < |z| < a$,

$$\begin{aligned}
W(z) &= K_{-1}z^{-1} + K_0 + K_1 z , \\
w(z) &= L_{-3}z^{-3} + L_{-2}z^{-2} + \alpha_{-1}z^{-1} + \alpha_{-2}b^{-2} ,
\end{aligned} \qquad (25)$$

where

$$K_{-1} = \frac{\rho\omega^2 h b^2}{1 + \eta},$$

$$K_0 = -\frac{(\alpha - \beta)a^4 + 2\alpha b^4 + 2(\alpha - \beta)h^2 a^2 + (\alpha + \beta)h^2 b^2}{a^2(1 - \eta) - 2b^2},$$

$$K_1 = \frac{2(a^2 - b^2)\rho\omega^2 h b^2 + 2(\alpha - \beta)(1 + \eta)h a^4 + (1 + \eta)(\alpha + \beta)h b^4}{(1 + \eta)(b^4 + \eta a^4)},$$

$$L_{-3} = \frac{\rho\omega^2[-(1 - \eta)a^4 + 2a^2 b^2]}{(1 + \eta)(b^4 + \eta a^4)},$$

$$L_{-2} = -\frac{2(\alpha - \beta)a^4 b^2 + 2\alpha(1 - \eta)a^2 b^4}{a^2(1 - \eta) - 2b^2},$$

$$\alpha_{-1} = -\frac{\eta\rho\omega^2 h b^2}{1 + \eta},$$

$$\alpha_{-2} = -(\alpha - \beta)h^2 b^2 .$$

When the stresses determined by $W(z)$ and $w(z)$ are combined with the stresses from the particular solution, it follows that for plane strain

$$\widehat{rr}^{(A)} + \widehat{\theta\theta}^{(A)} = -\frac{\rho\omega^2 r^2}{2(1 - \nu)} + \frac{\rho\omega^2[(1 - 2\nu)a^4 + (3 - 2\nu)b^4]}{4(1 - \nu)[(1 - 2\nu)a^2 + b^2]} + \left\{-\frac{\rho\omega^2 h[2(1 - \nu)a^4 + b^4 - a^2 b^2]r}{(1 - \nu)[b^4 + (3 - 4\nu)a^4]} + \frac{\rho\omega^2 h b^2 r^{-1}}{2(1 - \nu)}\right\}\cos\theta\,, \qquad (26)$$

$$\begin{aligned}
-\widehat{rr}^{(A)} + \widehat{\theta\theta}^{(A)} + 2i\widehat{r\theta}^{(A)} &= \frac{\rho\omega^2(1 - 2\nu)r^2}{4(1 - \nu)} + \frac{\rho\omega^2 a^2(1 - 2\nu)[a^2 - (3 - 2\nu)b^2]r^{-2}}{4(1 - \nu)[(1 - 2\nu)a^2 + b^2]} \\
&\quad + \left\{\frac{\rho\omega^2 h[(1 - 2\nu)b^4 + 4(1 - 2\nu)(1 - \nu)a^4 + a^2 b^2]r}{2(1 - \nu)[b^4 + (3 - 4\nu)a^4]}\right. \\
&\quad \left. - \rho\omega^2 h b^2 r^{-1} + \frac{\rho\omega^2 h b^4[(1 - 2\nu)a^4 + a^2 b^2]r^{-3}}{2(1 - \nu)[b^4 + (3 - 4\nu)a^4]}\right\}\cos\theta \\
&\quad + i\left\{\frac{\rho\omega^2 h[(1 - 2\nu)b^4 + 4(1 - 2\nu)(1 - \nu)a^4 + a^2 b^2]r}{2(1 - \nu)[b^4 + (3 - 4\nu)a^4]}\right. \\
&\quad \left. - \frac{\rho\omega^2 h b^2(1 - 2\nu)r^{-1}}{2(1 - \nu)}\right\}\sin\theta - i\,\frac{\rho\omega^2 h b^4[(1 - 2\nu)a^4 + a^2 b^2]r^{-3}}{2(1 - \nu)[b^4 + (3 - 4\nu)a^4]}\sin\theta\,. \qquad (27)
\end{aligned}$$

It is of interest to note that the corresponding gravity loading solution can be readily deduced from Eqs. (26) and (27) by taking $h\omega^2 = g$ and letting h tend to infinity.

In the case of an incompressible material ($\nu = \frac{1}{2}$), the stress relations simplify considerably. The stresses which occur on the boundaries are of particular interest. These are found to be given by

$$\widehat{\theta\theta}^{(A)} = \widehat{rr}^{(A)} = -\frac{\rho\omega^2 b^2}{2}(\xi^2 - 1)\left[1 + \frac{(2\xi^4 - \xi^2 + 1)\eta}{\xi(1+\xi^4)}\cos\theta\right],$$

$$\widehat{r\theta}^{(A)} = \frac{\rho\omega^2 hb}{2}\left[\frac{\xi^4 - 1}{\xi(\xi^4 + 1)}\right]\sin\theta\,, \qquad \text{on } \Gamma_a\,, \tag{28}$$

where $\xi = a/b$ and $\eta = h/b$. The only nonvanishing stress components on Γ_b are the circumferential stresses which are given by

$$\widehat{\theta\theta}^{(A)} = -\rho\omega^2 hb\,\frac{(\xi^2 - 1)^2}{(\xi^4 + 1)}\cos\theta\,. \tag{29}$$

In Figs. 2–5 are shown dimensionless plots of maximum boundary stresses in terms of ξ and η for $\nu = \frac{1}{2}$.

CONCLUSION

The above analysis considered a special case of the general-stress problem for an annulus with arbitrary stresses on one boundary and arbitrary displacements on the other boundary. If the boundary stresses and displacements are expandable in Fourier series, the assumption that $f(z)$ can be represented as a Laurent series about $z = 0$ readily leads to the general solution of this problem. The same technique may also be employed to solve the first and second fundamental problems for the annulus which are considered in ref [2].

ACKNOWLEDGMENT

The author wishes to thank I. C. Wang, Research Associate, Theoretical and Applied Mechanics, University of Illinois, for his assistance in checking portions of the computations in this analysis.

REFERENCES

[1] N. I. Muskhelishvili, *Singular Integral Equations,* P. Noordhoff, Groningen, Holland, 1952.

[2] N. I. Muskhelishvili, *Some Basic Problems of the Mathematical Theory of Elasticity,* P. Noordhoff, Groningen, Holland, 1953.

[3] Milne-Thomson, *Plane Elastic Systems,* Springer-Verlag, Berlin, 1960.

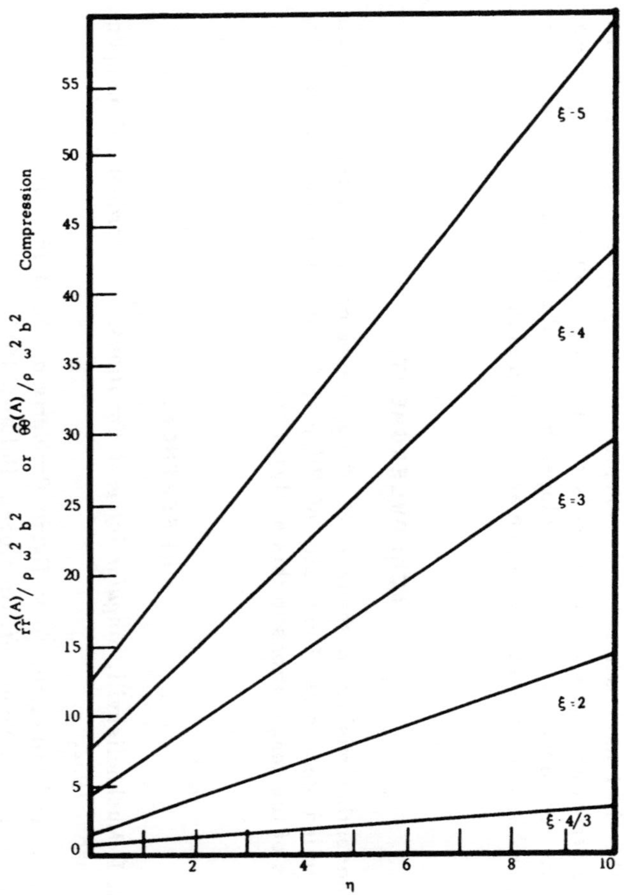

Fig. 2. Normal and Circumferential Stresses on Γ_a for $\theta = 0$ and $\nu = \frac{1}{2}$.

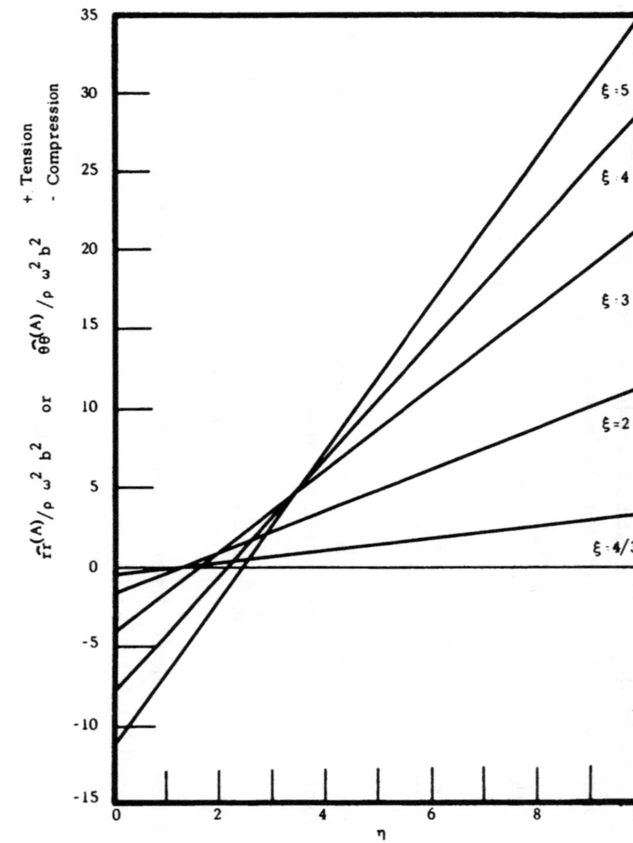

Fig. 3. Normal and Circumferential Stresses on Γ_a for $\theta = \pi$ and $\nu = \frac{1}{2}$.

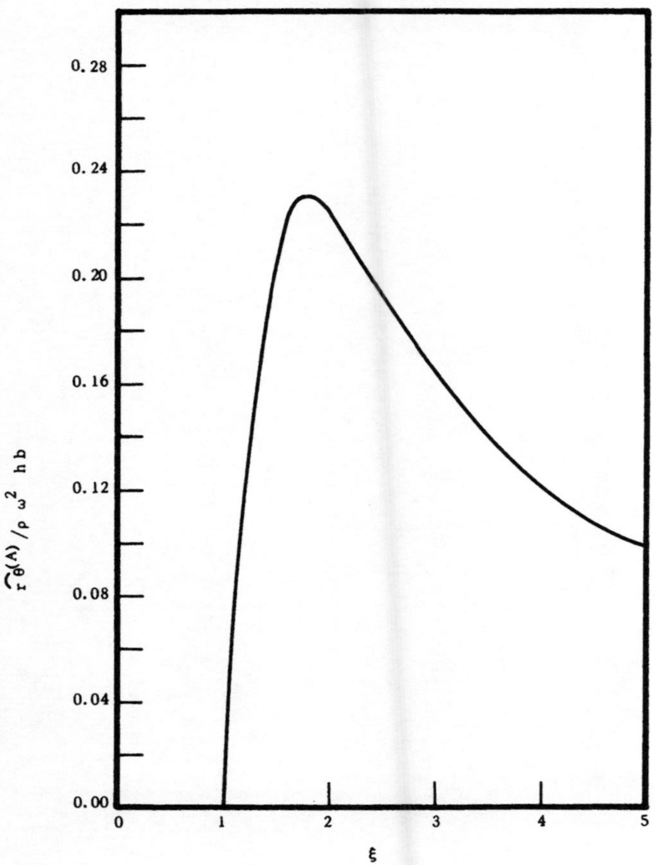

Fig. 4. Shear Stress on Γ_a for $\theta = \pi/2$ and $\nu = \frac{1}{2}$.

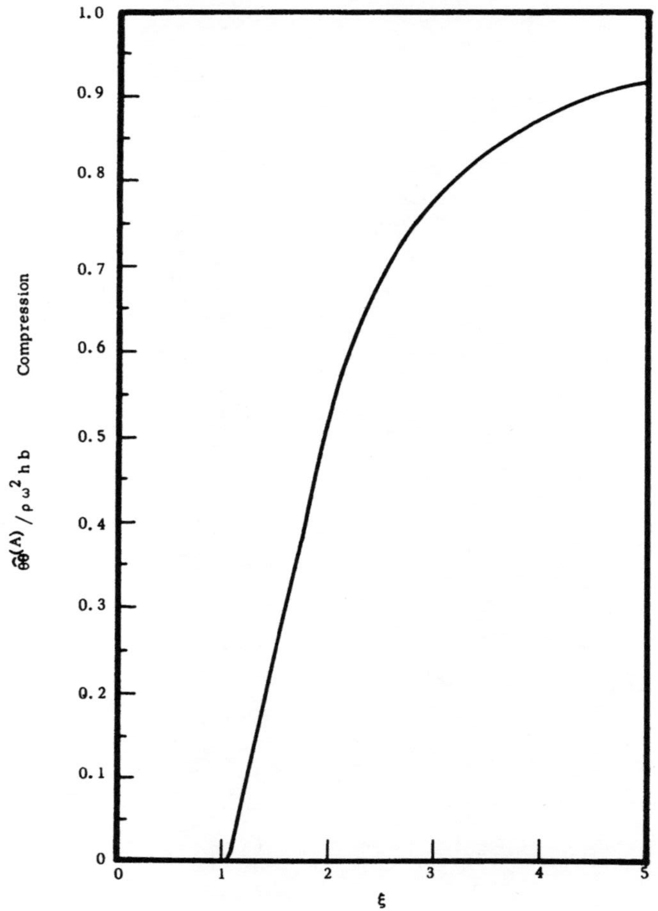

Fig. 5. Circumferential Stress on Γ_b for $\theta = 0$ and $\nu = \frac{1}{2}$.

Fluid Mechanics

DETERMINATION OF THE LIMITING VISCOSITY OF FLOCCULATED SUSPENSIONS[1]

D. M. Eissenberg
Oak Ridge National Laboratory[2]

An evaluation of the limiting viscosity of a shear-thinning suspension at high rates of shear is useful in characterizing a slurry. In order to measure a limiting viscosity for an aqueous thoria slurry, it was found necessary to pass the slurry through a small-bore capillary tube.

It was determined experimentally for the suspensions tested that the limiting viscosity is equal to the chemically dispersed Newtonian suspension viscosity, which can be obtained in low-shear viscometers. A model that accounts for this result was developed.

INTRODUCTION

Studies of the flow behavior and heat transfer characteristics of flocculated aqueous suspensions of small particles have been in progress for several years at the Oak Ridge National Laboratory. The objective is to develop means for predicting heat and momentum transfer of aqueous thoria and thoria-urania suspensions when pumped at elevated temperatures in a nuclear reactor system. Such suspensions have been found to exhibit shear-thinning (non-Newtonian) behavior under most conditions.

Particular suspensions are characterized by means of their laminar flow properties as measured in capillary-tube viscometers [*1*] by use of a rheological model containing two arbitrary constants. The numerical constants evaluated from the laminar-flow data are then used to calculate the viscosities needed to correlate and predict laminar- and turbulent-flow pressure drops and heat transfer coefficients [*2*].

All the presently used non-Newtonian rheological models are, at best, semi-empirical, and experimental evaluation of constants is required, even though the form of the equation may be dictated by theoretical considerations. This restricts the ability to predict rheological constants beyond the range of the experimental data. It is the purpose of this study to overcome this difficulty by

---

[1]Supported by the USAEC under contract No. W-7405-eng-26.

[2]Operated for USAEC by Union Carbide Corporation.

NOTATION

D	Inside diameter of pipe or capillary tube
E_s	Power per unit volume required to maintain a given rate of shear against the interparticle attractive forces
E_T	Total power per unit volume required to shear a flocculated suspension at a given shear rate
E_V	Power per unit volume required to overcome viscous dissipation at a given shear rate, including the effect of the presence of particulates
$-dv/dr$	Local rate of shear
$(-dv/dr)_w$	Wall shear rate
V_m	Mean velocity
η	Bingham coefficient of rigidity
μ_a	Viscosity defined by Newton's law
μ_{aw}	Newton's law viscosity evaluated at the wall
μ_D	Viscosity of dispersed (Newtonian) suspension
μ_e	Viscosity defined by Poiseuille's law
μ_{sp}	Specific viscosity (equal to μ_D/μ_w)
μ_w	Water viscosity
τ	Local shear stress
τ_w	Wall shear stress (equal to $D/4$ times pressure drop per length of pipe)
τ_y	Bingham yield stress

defining and measuring the limiting values of the viscosity of flocculated thoria suspensions which should be valid for conditions beyond the range of the experimental data. The measurements reported were made with aqueous thoria suspensions, but the results of the study are probably applicable to other non-Newtonian suspensions.

THEORY

The particular approach most commonly used to characterize flocculated thoria suspensions has been to assume that they behave as Bingham plastics; that is, they obey the Bingham plastic equation,

$$\tau - \tau_y = \frac{\eta}{g_c}\left(-\frac{dv}{dr}\right), \tag{1}$$

which relates the local shear stress τ to the local shear rate $-dv/dr$, by means of the rheological constants η (the plastic viscosity or coefficient of rigidity) and τ_y (the yield value or yield stress). In the absence of elastic effects, Eq. (1) can be combined with Newton's equation,

$$\mu_a = \frac{\tau}{\frac{1}{g_c}\left(-\frac{dv}{dr}\right)}, \tag{2}$$

to yield

$$\mu_a = \eta \frac{1}{1 - (\tau_y/\tau_w)} . \tag{3}$$

Equation (3) illustrates why the rheological constant η is often referred to as a viscosity. It represents the limiting Newton's law viscosity as the ratio τ_y/τ_w approaches zero, that is, as the shear stress τ_w becomes much greater than the yield stress τ_y.

Capillary-Tube Theory

The capillary-tube viscometer has been used to obtain the rheological constants of flocculated thoria suspensions because the effects of the settling rate of thoria agglomerates can be eliminated by the use of vertical capillaries. However, the data obtained with such viscometer-pressure drops and flow rates cannot be substituted directly into the Bingham plastic equation, since they are not readily convertible to shear rates. For this reason, a second viscosity is defined for capillary tubes based on Poiseuille's law,

$$\mu_e = \frac{\tau_w}{\frac{1}{g_c}\left(\frac{8V_m}{D}\right)} \tag{4}$$

The relationship between the true wall shear rate $(-dv/dr)_w$ and $8V_m/D$ is given by the Rabinowitch-Mooney equation [3],

$$\left(-\frac{dv}{dr}\right)_w = \frac{3}{4}\frac{8V_m}{D} + \frac{1}{4}\tau_w \frac{d(8V_m/D)}{d\tau_w} . \tag{5}$$

When the Bingham plastic equation [Eq. (1)] for the wall-shear condition is substituted into Eq. (5), the working equation for the Bingham plastic is obtained:

$$\frac{8V_m}{g_c D} = \frac{1}{\eta}\left(\tau_w - \frac{4}{3}\tau_y + \frac{1}{3}\frac{\tau_y^4}{\tau_w^3}\right). \tag{6}$$

When Eq. (6) is substituted into the Poiseuille's law equation, the resulting expression becomes

$$\mu_e = \frac{\eta}{1 - \dfrac{4}{3}\dfrac{\tau_y}{\tau_w} + \dfrac{1}{3}\left(\dfrac{\tau_y^4}{\tau_w}\right)} . \tag{7}$$

As the shear stress increases without limit, it is noted that the limiting viscosity is the plastic viscosity η in the case of Poiseuille's law, as it was for Newton's law.

Graphical Analysis

When the experimental data for a Bingham plastic are plotted as τ_w vs $8V_m/g_cD$, the rheological constants η and τ_y can be obtained directly from the graph by noting, from Eq. (6), that $1/\eta$ is the slope and that $4\tau_y/3$ is the intercept of the straight-line upper portion of the curve (where the term τ_y^4/τ_w^3 becomes negligible). The graphical evaluation, which is illustrated in Fig. 1, is referred to as a pseudoshear diagram. The curve shown is for a hypothetical Bingham plastic. It is noted that the calculated value of τ_y (equal to three-fourths of the intercept of the straight-line portion of the curve) coincides with the actual intercept, and that the curve becomes linear above about 2 to 3 times τ_y.

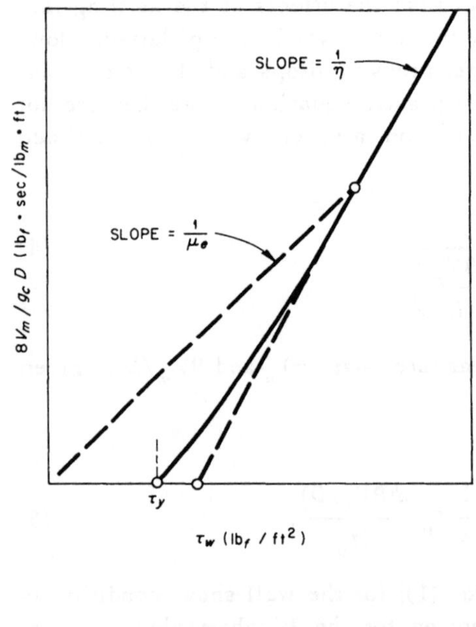

Fig. 1. Typical Bingham Plastic Pseudoshear Diagrams.

Application to Non-Bingham Plastics

The Bingham plastic equation, like any other rheological equation, can be applied to any shear-thinning suspension within a restricted range of data, so that the values of the rheological constants τ_y and η remain approximately fixed within the range. Flocculated thoria suspensions represent a class of suspensions which obey the Bingham plastic equation over a large range of shear stress. The characteristic pseudoshear diagram of thoria suspensions can be represented by the Bingham plastic equation at high shear stresses, but it has often been found to intersect the shear-stress axis at a value less than the yield stress calculated from the high shear-stress region [4]. In those cases, the viscosities previously defined will be valid if the Bingham plastic constants are not used to evaluate viscosities in the low shear-stress region.

EXPERIMENT

Experimental Approach

Evaluation of rheological constants at large rates of shear is made difficult by the fact that the viscometry must be carried out in laminar flow. As the shear rate is increased for a capillary tube of given size, the flow becomes unstable and turbulence occurs. This onset of turbulence is a function of the tube dimensions and occurs at higher shear rates as the size decreases. As a result, for flocculated suspensions which deviate from true Bingham plastic behavior, it is possible to obtain different values of the rheological constants η and τ_y by the use of too large a viscometer tube. This is illustrated by examining a pseudoshear diagram of a typical thoria suspension as obtained in viscometer tubes of different sizes. Figure 2 shows the curve for a 0.124-in.-diam viscometer tube. It is noted that the material is not a true Bingham plastic. This is seen in the sweeping curve at low shear stress, since the curve would have intercepted the shear-stress axis at 0.048 lbf/ft^2 if it were a true Bingham

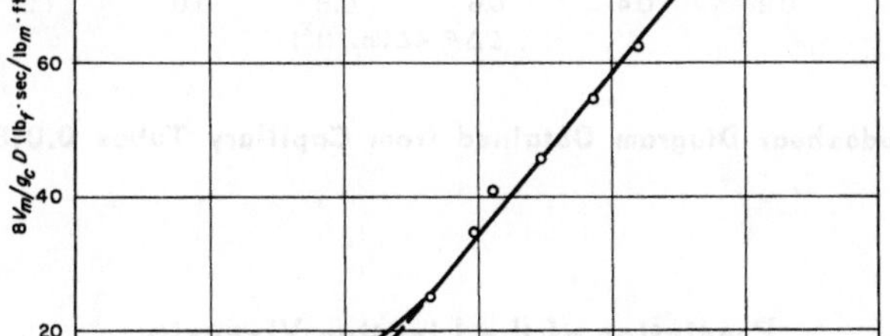

Fig. 2. Pseudoshear Diagram Obtained from a 0.124-in. Capillary Tube.

plastic. At the upper end of the curve is the indication of turbulence. To evaluate the Bingham constants, the technique has been [1] to use the linear portion of the curve and extend it to the shear-stress axis, as shown, to obtain the intercept (equal to $4\tau_y/3$).

Figure 3 shows the same data plotted on a different scale; data obtained with smaller viscometer tubes are included. A new pair of very different rheological constants is indicated, and again it can be seen that the material is not a true Bingham plastic.

Thus, in order to ensure that the rheological constants evaluated for high shear rates are not dependent on the viscometer tube size, some independent method of obtaining the limiting value of the viscosity is necessary.

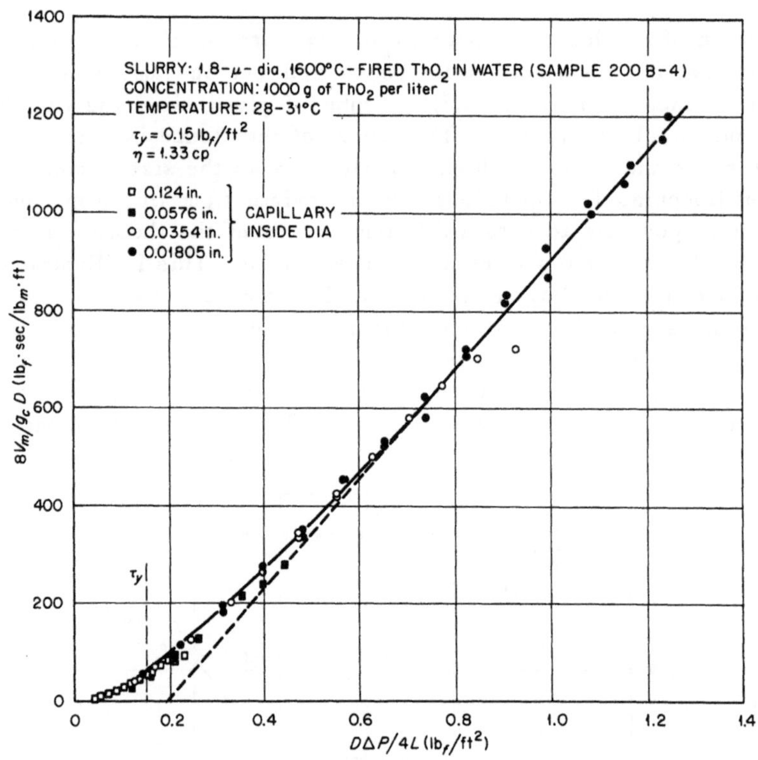

Fig. 3. Pseudoshear Diagram Obtained from Capillary Tubes 0.018 to 0.124 in. in Inside Diameter.

Prediction of the Limiting Viscosity

A first estimate of the value of the limiting viscosity of a shear-thinning suspension is that it is equal to or greater than that of the suspending medium. In the case of particulate suspensions, in which the volume occupied by the solids is significant, the limiting viscosity should be greater than that of the suspending medium, since there will be a flow disturbance because of the presence of particulates that will increase the viscosity. A better estimate of the limiting viscosity, then, would account for the presence of particulates in a manner similar to that used by Einstein [5] to calculate the viscosity of dilute suspensions of smooth, nonattracting, uniformly sized spheres and extended by many authors to include higher-volume concentrations. The difficulty with direct use of the Einstein equation is that the suspensions of interest are made up of particles of irregular shape and with a range of sizes; so the calculated limiting viscosity based on smooth uniform spheres does not apply. The use of the Einstein formula for calculating the limiting viscosity of flocculated clay suspensions [6], for example, rested on a synthetic volume concentration evaluated from sediment volume data in order to satisfy the experimental data. The approach used in this paper was to determine experimentally the noninteracting (Newtonian) viscosity by chemically dispersing the flocculated suspension. This dispersed viscosity was believed to represent the limiting viscosity of the original flocculated suspension.

Viscometer Description

The viscometer used in this experiment (Fig. 4) consisted of a pressurized reservoir, a vertical capillary tube, and a calibrated-volume sight glass. The reservoir contained a stirrer and could be electrically heated. Its capacity

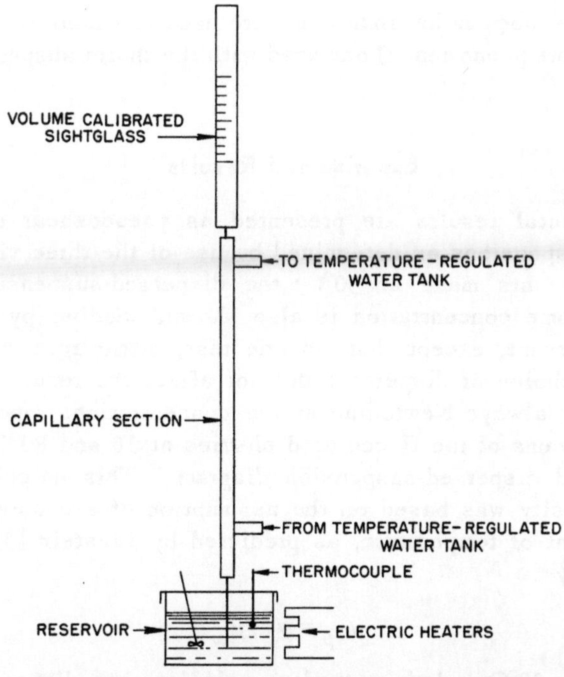

Fig. 4. Vertical Capillary-Tube Viscometer.

was 300 ml of suspension. Air pressure up to 35 psi was supplied to the reservoir and was measured by a mercury manometer. The reservoir could be maintained at temperatures between ambient and 80°C by means of an on-off temperature controller. For some of the runs, the temperature was regulated within ±2°C, whereas for the remaining runs the regulation was within ±0.1°C.

Three viscometer tubes were used. The tube sizes were 0.01805, 0.0354, and 0.0576 in., and the lengths were 18, 38, and 50 in. respectively. They were cut from stainless steel capillary tubing and brazed into $\frac{1}{4}$-in. stainless steel tubing to provide rigidity. The $\frac{1}{4}$-in. tubes were contained within $\frac{1}{2}$-in. copper-tubing jackets through which heated water was circulated at reservoir temperature. The tubes were interchangeable in the reservoir. They were mounted vertically to eliminate sedimentation effects. Flow was upward. The sight glass was coated with an antifilming compound, and the volume was calibrated with mercury.

The viscometer tube diameters were measured in calibration runs with a fluid of known viscosity, that is, distilled water. Corrections for static head and kinetic energy (entrance and exit losses) were included in water calibration and all suspension tests. Several tubes were rejected when the onset of turbulence was not a clear break in the laminar curve on the pseudoshear diagram. The tubes accepted for use showed a clear break into turbulence at Reynolds numbers of 2000 to 2200.

A cross calibration of the tubes, with a 50% glycerol–50% water mixture, was carried out. This consisted in filling the reservoir with the mixture and obtaining pseudoshear diagrams with the three tubes at the same temperature. With the water calibrations finally used, the three tubes gave identical laminar-flow lines. Three tubes rather than one were used to ensure that no appreciable wall dilution (sigma phenomenon) occurred with the thoria suspension.

Experimental Results

The experimental results are presented as pseudoshear diagrams of the shear-thinning suspensions as determined by use of the three viscometer tubes. For the measurements made at 30°C, the dispersed-suspension pseudoshear diagram of the same concentration is also shown. Sodium pyrophosphate was used for a dispersant, except that, in one test, nitric acid was also used to ensure that the choice of dispersant did not affect the result. The dispersed suspensions were always Newtonian at the concentrations reported here. The pseudoshear diagrams of the flocculated slurries at 50 and 80°C were compared with a calculated dispersed-suspension diagram. This calculated dispersed-suspension viscosity was based on the assumption of a constant specific viscosity independent of temperature, as predicted by Einstein [5]. The specific viscosity given by

$$\mu_{sp} = \mu_D/\mu_w \tag{8}$$

was evaluated at 30°C and then used to calculate the dispersed viscosity at higher temperatures. Experimental dispersed viscosities generally could not be obtained above room temperature because the suspension tended to re-flocculate.

Figures 5–7 show the dispersed- and flocculated-suspension pseudoshear diagrams for an unpumped, 650°C-fired, 1.8-μ mean-diameter thoria suspension at a thoria concentration of 500 g of ThO_2 per liter. The dispersed lines in Figs. 6 and 7 were obtained as explained above. It is seen in the plot that the limiting viscosity at high shear rates is approximately equal to the dispersed viscosity.

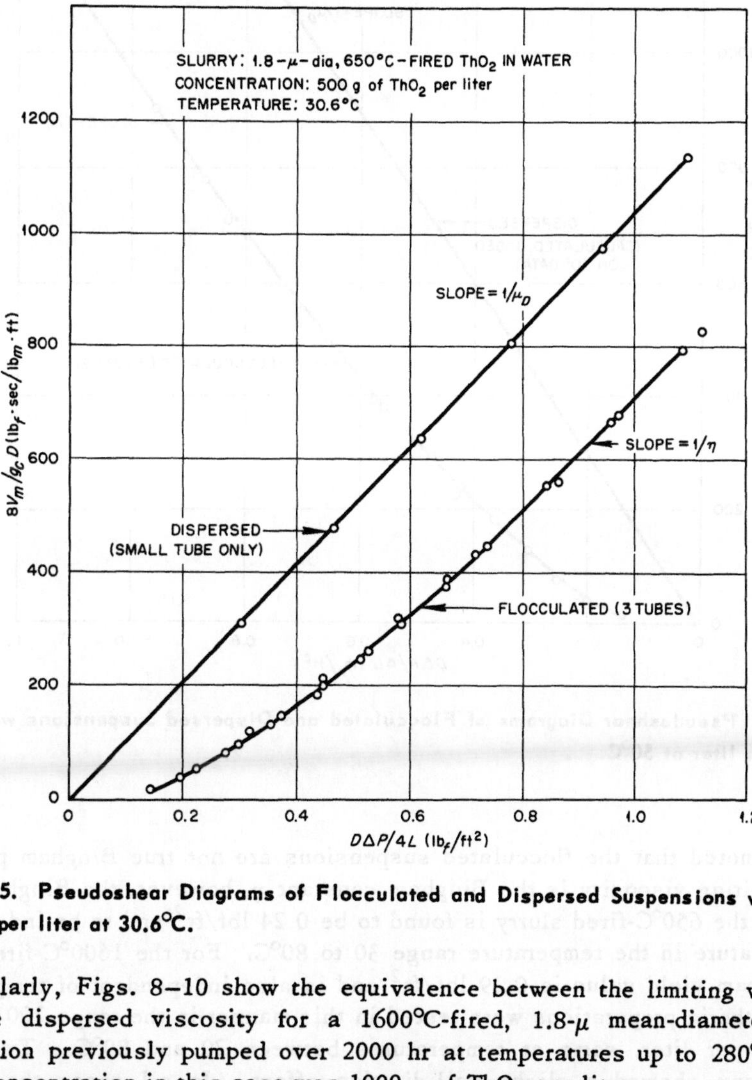

Fig. 5. Pseudoshear Diagrams of Flocculated and Dispersed Suspensions with 500 g of ThO_2 per liter at 30.6°C.

Similarly, Figs. 8–10 show the equivalence between the limiting viscosity and the dispersed viscosity for a 1600°C-fired, 1.8-μ mean-diameter thoria suspension previously pumped over 2000 hr at temperatures up to 280°C. The thoria concentration in this case was 1000 g of ThO_2 per liter.

The indicated deviation of some of the points from these curves is due to the onset of turbulence for the larger tubes. This was found to occur at a Reynolds number of close to 2100, based on the Poiseuille's law viscosity for that shear stress.

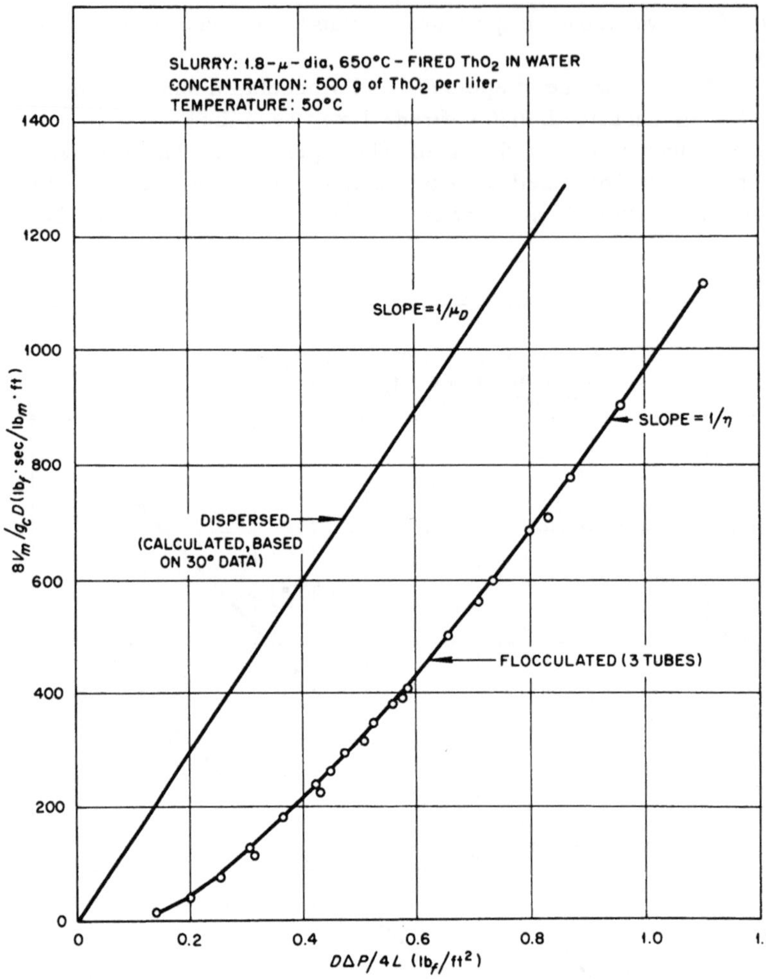

Fig. 6. Pseudoshear Diagrams of Flocculated and Dispersed Suspensions with 500 g of ThO_2 per liter at 50°C.

It is noted that the flocculated suspensions are not true Bingham plastics. If the limiting viscosity is the Bingham constant η, however, the Bingham yield value for the 650°C-fired slurry is found to be 0.24 lbf/ft^2 and to be independent of temperature in the temperature range 30 to 80°C. For the 1600°C-fired slurry the Bingham yield value is 0.19 lbf/ft^2 and is also independent of temperature.

Four thoria preparations were tested in this manner in the range 200 to 1400 g of ThO_2 per liter, some at temperatures between 20 and 80°C. Two of the suspensions showed a slight wall-dilution effect. In all cases the limiting flocculated viscosity was equal to the dispersed viscosity. In the cases where the limiting (flocculated) viscosity was not reached (the pseudoshear diagram had no linear segment), the dispersed viscosity was less than the minimum viscosity. The results are given in Table 1.

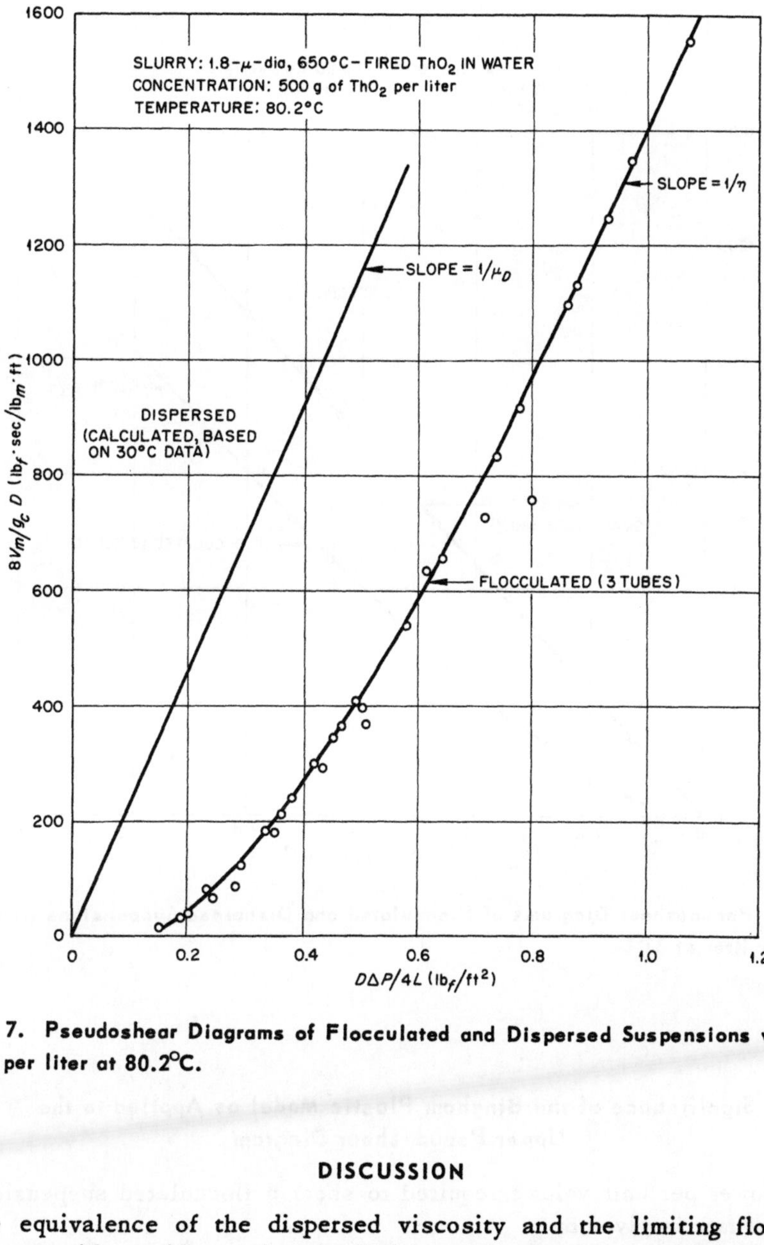

Fig. 7. Pseudoshear Diagrams of Flocculated and Dispersed Suspensions with 500 g of ThO_2 per liter at 80.2°C.

DISCUSSION

The equivalence of the dispersed viscosity and the limiting flocculated viscosity provides evidence that the limiting viscosity is a useful flow property of the flocculated suspension and as such may be useful in the correlation of turbulent-pressure drops and heat transfer behavior where high stresses may occur. Further, if the Bingham plastic model is used to define the pseudoshear diagram in the upper linear portion, it is possible to interpret physically the Bingham constants in the light of the results obtained, as outlined below.

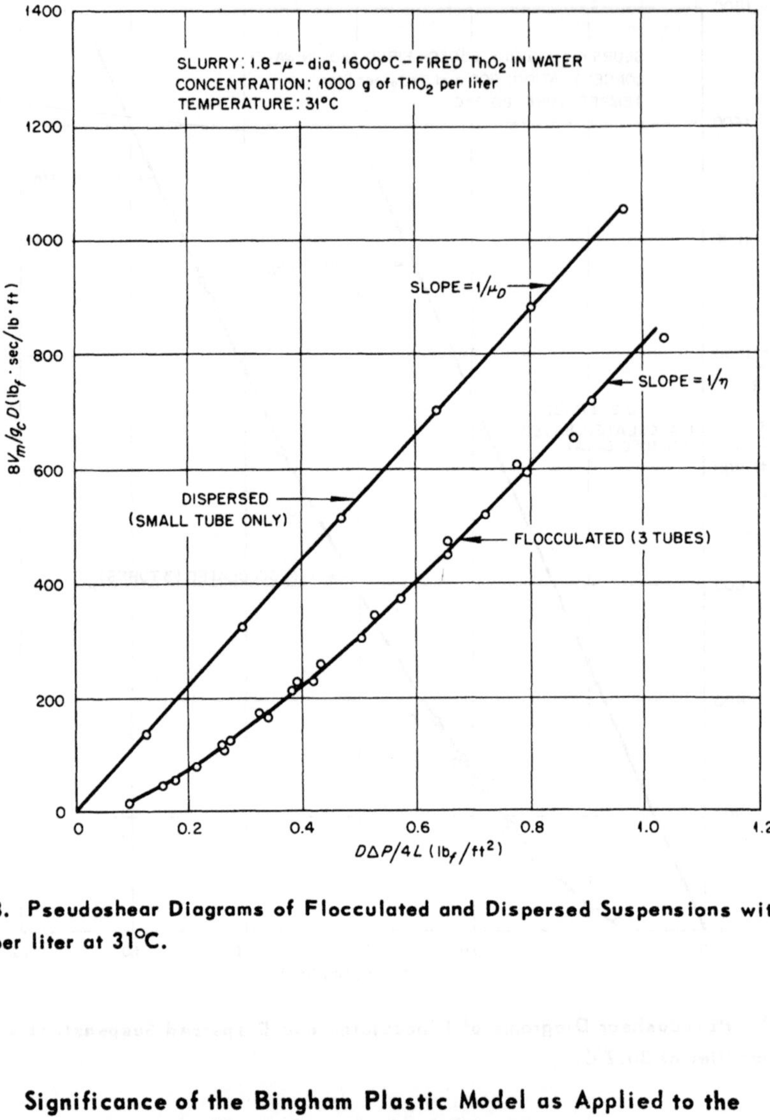

Fig. 8. Pseudoshear Diagrams of Flocculated and Dispersed Suspensions with 1000 g of ThO_2 per liter at 31°C.

Significance of the Bingham Plastic Model as Applied to the Upper Pseudoshear Diagram

The power per unit volume required to shear a flocculated suspension at a given shear rate is given by

$$E_T = \tau(-dv/dr) .$$

This power can be divided into two segments. Part of the power required is the same as would be required to shear a similar suspension of noninteracting (dispersed) particles and can be obtained from the dispersed suspension viscosity,

$$E_V = \mu_D(-dv/dr)^2 .$$

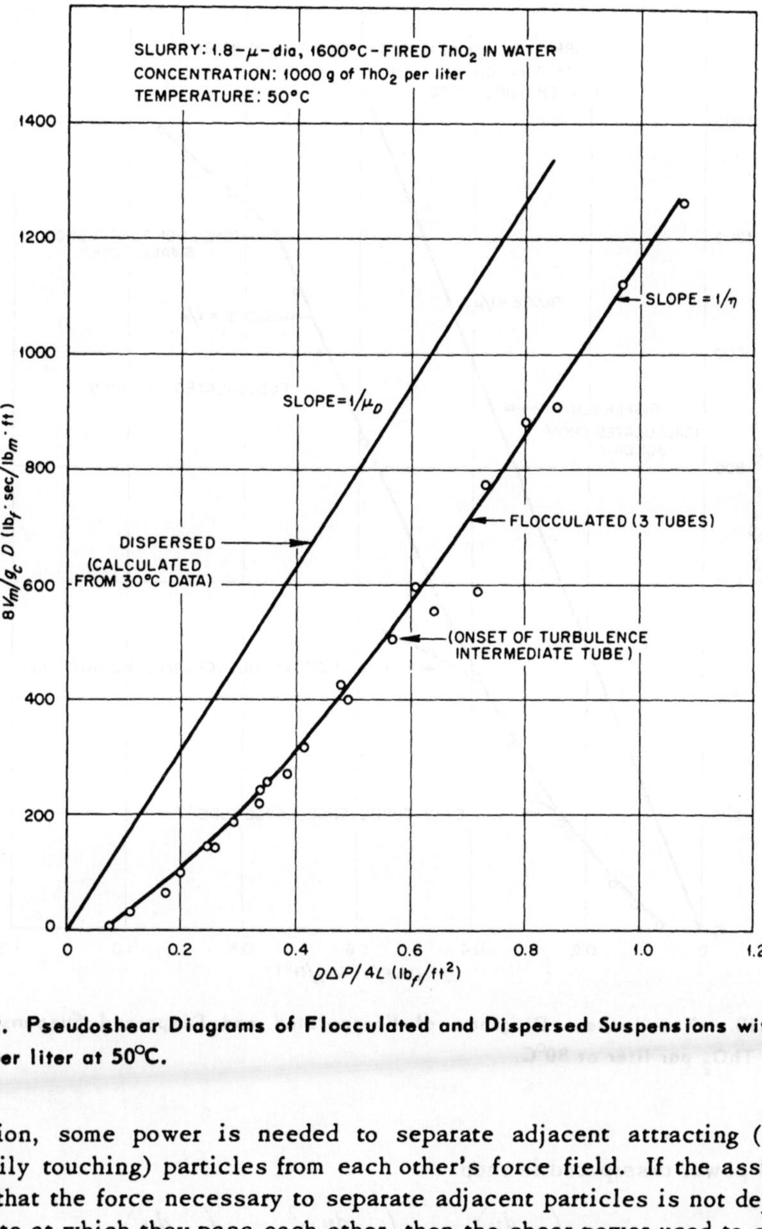

Fig. 9. Pseudoshear Diagrams of Flocculated and Dispersed Suspensions with 1000 g of ThO_2 per liter at 50°C.

In addition, some power is needed to separate adjacent attracting (but not necessarily touching) particles from each other's force field. If the assumption is made that the force necessary to separate adjacent particles is not dependent on the rate at which they pass each other, then the shear power used to separate the particles can be written

$$E_s = (\text{force to separate particle pair}) \times \left(\frac{\text{No. of particle pairs}}{\text{area}}\right) \times (\text{rate of passing of pairs}).$$

If we define the product of the first two terms as the dynamic Bingham yield stress and note that the last term is equal to the rate of shear, then

$$E_s = \tau_y(-dv/dr) .$$

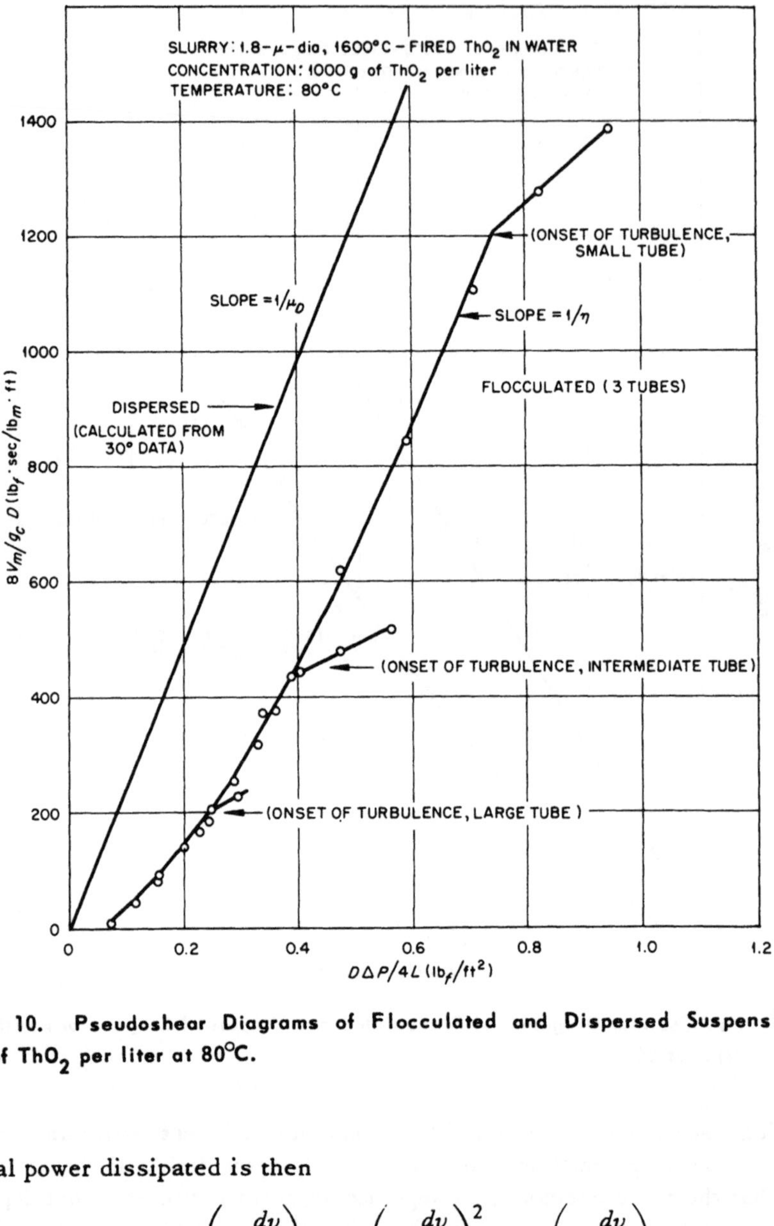

Fig. 10. Pseudoshear Diagrams of Flocculated and Dispersed Suspensions with 1000 g of ThO_2 per liter at 80°C.

The total power dissipated is then

$$E_T = \tau\left(-\frac{dv}{dr}\right) = \mu_D\left(-\frac{dv}{dr}\right)^2 + \tau_y\left(-\frac{dv}{dr}\right).$$

By simplifying, we obtain the Bingham plastic equation in which the Bingham constant η is replaced by the dispersed viscosity μ_D. This model thus predicts the experimental results obtained at high shear stress in this study.

A further observation can be made that if the shear rate is not high enough, some of the particles will not separate but will cluster into adhering flocs, even while flowing. This will result in a higher volume fraction occupied by flocs (because of immobilized water). Thus the equivalent dispersed viscosity would be increased. The dynamic Bingham yield stress, however, would be decreased, since the number of interacting discrete particle pairs would decrease. A curve of the form observed in the pseudoshear diagram of these flocculated suspensions would then be expected at low shear rates.

Table 1. Comparison of Specific Dispersed Viscosity and Specific Limiting Viscosity

Slurry	Slurry Concentration (g of ThO_2 per liter)	Temperature (°C)	μ_D/μ_w	η/μ_w
DT-4, unpumped, 650°C-fired,	500	30	1.8	1.8
1.8-μ thoria in water	350	30	1.5	1.5
DT-4, pumped, 650°C-fired,	600	30	1.6	1.6
1.8-μ thoria in water	500	30	1.5	1.5
	390–400	30	1.35	1.35
	290–300	30	1.27	1.27
	200	30	1.15	1.15
LO-41, unpumped, 1600°C-fired,	800	22–24	2.0	2.0
1.2-μ thoria in water				
DT-12, -13, pumped, 1600°C-fired,	550	30	1.3	1.3
1-μ thoria in water	1000	31	1.7	1.7

CONCLUSION

It is concluded from this study that, for flocculated thoria suspensions, the limiting viscosity (Bingham constant η) is equal to the dispersed viscosity.

REFERENCES

[1] D. G. Thomas, "Transport Characteristics of Suspensions: III. Laminar Flow Properties of Flocculated Suspensions," *A.I.Ch.E. Journal* **7**, 431–37 (1961).

[2] D. G. Thomas, "Heat and Momentum Transport Characteristics of Non-Newtonian Aqueous Thorium Oxide Suspensions," *A.I.Ch.E. Journal* **6**, 631–39 (1960).

[3] B. Rabinowitch, *Z. Physik. Chem. (Leipzig)* **145A**, 1 (1929).

[4] R. H. Nimmo and A. S. Kitzes, *A Collection of Pseudo-Shear Diagrams Obtained in a Low Temperature Capillary Tube Viscometer,* ORNL CF-56-12-124 (Dec. 7, 1956).

[5] A. Einstein, *Kolloid Z.* **27**, 137 (1920).

[6] J. C. Bolger, *Rheology of Kaolin Suspensions,* Ph.D. Thesis, MIT (August 1960).

CRITICAL REYNOLDS NUMBERS FOR INCOMPRESSIBLE PULSATING FLOW IN TUBES[1]

Donald A. Gilbrech and George D. Combs
University of Arkansas

A criterion for predicting laminar-turbulent transitions in the pulsating flow of a liquid through smooth tubes has been investigated, and experimental values for a critical Reynolds number have been determined. The results indicate a critical value that is higher under some conditions of flow than the steady-flow Reynolds number of 2220. A theoretical argument based upon the nonlinear generation and dissipation of turbulence is given to explain the behavior of the critical Reynolds number.

INTRODUCTION

There have been many investigations, since Professor Osborne Reynold's time, to determine a critical value for the Reynolds number which will differentiate stable and unstable flow in pipes with steady flow. On the other hand, pulsating flow has received relatively little attention despite the fact that such information would be extremely valuable to research in other aspects of pulsating flow by eliminating some of the guesswork in assuming either laminar or turbulent flow. This paper is a report and discussion of the experimental results obtained for the critical Reynolds numbers in pulsating flow of a liquid through smooth circular tubes. The pulsations considered were sinusoidal.

DEFINITIONS AND THEORY

The data presented in this report apply to the case of periodic flow in a long tube when the pressure gradient is of the form

$$-\frac{1}{\rho}\frac{\partial p}{\partial x} = a_0 + a_c \cos\theta \,, \tag{1}$$

where $\partial p/\partial x$ is the pressure gradient, ρ is the density of the fluid, a_0 is a measure of the steady component of the pressure gradient, and a_c is a measure

---

[1]Supported by the National Science Foundation.

NOTATION

a	Radius of the tube
a_0	Steady component of $-(1/\rho)(\partial p/\partial x)$
a_c	Amplitude of the periodic component of $-(1/\rho)(\partial p/\partial x)$
f	Frequency of the periodic component of the pressure gradient
k	Defined as $\sqrt{2\pi f/\nu}$
ka	Frequency parameter
p	Pressure
r	Distance from the center line of the tube radially
r'	Dimensionless radius r/a
t	Time
u	Instantaneous local velocity
$\overline{u}$	Steady component of the instantaneous local velocity
u'	Periodic component of the instantaneous local velocity
x	Distance along the axis of the tube
A	Function of ka and r/a in Eq. (5)
B	Function of ka and r/a in Eq. (5)
C	Constant in Eq. (12)
D, L	Diameter and length of the tube
G	Average growth rate of the turbulent plugs
G'	Instantaneous net growth rate of turbulent plugs
K	Parameter in Eq. (16)
L_C, L_D	Total length of N turbulent plugs on the recordings for photocells C and D respectively
N	Number of turbulent plugs observed
N_R	Reynolds number
N_R^c	Critical Reynolds number
N_R'	Instantaneous Reynolds number
S	Pulsator piston stroke
$\overline{V}$	Steady component of the cross-sectional mean velocity
V_A'	Amplitude of the periodic component of the cross-sectional mean velocity
V_L, V_T	Velocities of the leading and trailing edges, respectively, of a turbulent plug
Z	Instantaneous Reynolds number coefficient
θ	Time angle in Eq. (1) for the periodic component of the pressure gradient
ν	Kinematic viscosity
ρ	Mass density
σ_u	Coefficient of amplitude of the mean velocity

of the periodic component of the pressure gradient. The angle θ of the pressure gradient is defined by

$$\theta = 2\pi f t \ , \tag{2}$$

where t is the time and f is the frequency of the periodic component of the pressure gradient.

Uchida [1] solved the Navier-Stokes equations for the given conditions and found that the velocity is given by

$$\frac{u}{\overline{V}} = \frac{\overline{u}}{\overline{V}} + \frac{u'}{\overline{V}}, \tag{3}$$

where

$$\frac{\overline{u}}{\overline{V}} = 2\left[1 - \left(\frac{r}{a}\right)^2\right], \tag{4}$$

$$\frac{u'}{\overline{V}} = \frac{a_c}{a_0}\left[\frac{8B}{(ka)^2}\cos\theta + \frac{8(1-A)}{(ka)^2}\sin\theta\right], \tag{5}$$

$$k = \sqrt{2\pi f/\nu} \ . \tag{6}$$

In the preceding equations, ν is the kinematic viscosity, a is the tube radius, r is the radial distance from the center line of the tube, u is the instantaneous local velocity, $\overline{u}$ is the steady component of the instantaneous local velocity, and u' is the periodic component of the instantaneous local velocity. The quantities A and B are functions of the frequency parameter ka and the ratio r/a and are defined by

$$A = \frac{\text{Ber}\,(ka)\ \text{Ber}\,(kr) + \text{Bei}\,(ka)\ \text{Bei}\,(kr)}{\text{Ber}^2\,(ka) + \text{Bei}^2\,(ka)}, \tag{7}$$

$$B = \frac{\text{Bei}\,(ka)\ \text{Ber}\,(kr) - \text{Ber}\,(ka)\ \text{Bei}\,(kr)}{\text{Ber}^2\,(ka) + \text{Bei}^2\,(ka)} \ . \tag{8}$$

It should be observed that the dimensionless velocity profiles $u/\overline{V}$ are completely determined by the dimensionless groups ka and a_c/a_0.

Since the quantity a_c/a_0 is difficult to determine experimentally, the following relationship is used to convert from pressure gradients to velocity:

$$\frac{V'_A}{\overline{V}} = \sigma_u \frac{a_c}{a_0}, \tag{9}$$

where V'_A is the amplitude of the periodic component of the cross-sectional mean velocity, and $\overline{V}$ is the steady component of the cross-sectional mean velocity.

The coefficient of amplitude σ_u of the mean velocity is a function of the frequency parameter only. Thus, one concludes that the dimensionless velocity profiles are determined completely by the two parameters ka and $V_A'/\overline{V}$ in laminar flow. Both parameters can easily be controlled and measured in an experimental study. Figure 1 shows the velocity profiles for a particular value of these dimensionless parameters. Notice that at one point in the cycle the flow reverses near the wall.

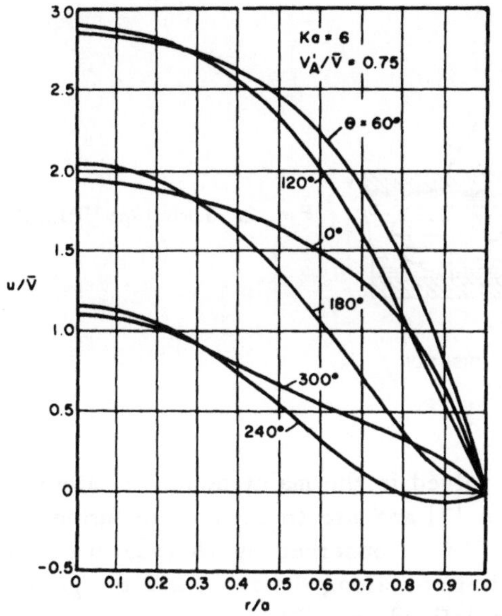

Fig. 1. Dimensionless Local Velocity vs Dimensionless Radius.

The question arises as to which velocity should be used in defining the Reynolds number. There are several velocities from which to choose. The authors chose to use the average of the cross-sectional mean velocity. Thus, the Reynolds number is defined by

$$N_R = \overline{V}D/\nu \,, \tag{10}$$

where D is the tube diameter. This definition allows a direct comparison of the results with the conventional steady-flow Reynolds number. An instantaneous Reynolds number is defined later, but it is not useful in correlating and presenting the data. It was assumed that the critical value of the Reynolds number should be determined by the shape of the laminar-velocity profiles and, therefore, by the parameters ka and $V_A'/\overline{V}$. The data were obtained and organized on this basis.

In addition to defining a Reynolds number it is necessary to define a *critical* value N_R^c for the Reynolds number. This has been done in several ways for steady flow with slightly different results. In this study a method used by Rotta [2] for steady flow has been adopted in which the critical value of the Reynolds number corresponds to the flow occurring when the growth rate of a disturbed portion of the fluid is zero. Rotta observed regions of spontaneous turbulence caused by entrance disturbances. However, spontaneous turbulence is not satisfactory for pulsating flow because of the tendency for the turbulence to occur at the same time in each cycle and the inherent difficulties involved in identifying "plugs" at consecutive observation stations. Thus, in the work reported herein, disturbances were created in otherwise laminar flow by means of a disturbance generator. These disturbances were introduced at different times throughout the cycles of pulsations and passed through the system as turbulent "plugs." A "plug," as shown in Fig. 2 for steady flow, is continually regenerated by the high center-line velocities at the rear while being dissipated

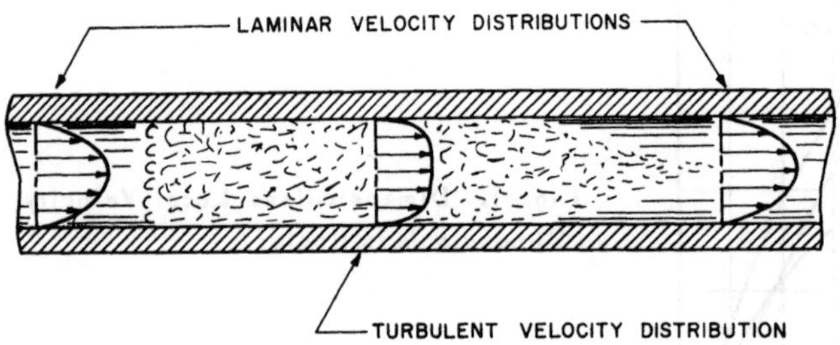

Fig. 2. Turbulent "Plug."

at the front. The detailed mechanisms involved in the maintenance of turbulent "plugs" have been discussed by Lindgren [3] and are to be studied further by the authors at a later date. The "plugs" were observed by photocells (to be described later) as they moved past two stations along the tube, and the growth rate was determined. The growth rate G is defined as

$$G = \frac{V_L - V_T}{\overline{V}}, \tag{11}$$

where V_L is the velocity of the leading edge of the "plug," V_T is the velocity of the trailing edge, and, as before, $\overline{V}$ is the time average of the cross-sectional mean velocity. Figure 3 illustrates one set of data in which the growth rate G

Fig. 3. Turbulence Growth Rate vs Reynolds Number.

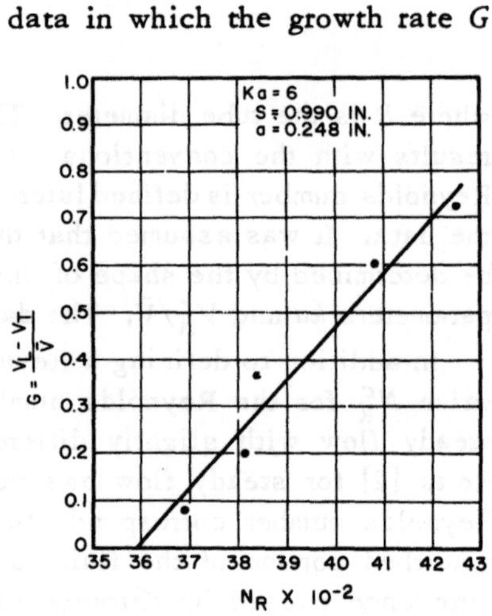

was determined as a function of the Reynolds number N_R. A least-squares straight line represents the data within the experimental accuracy near the critical value. Thus, the critical Reynolds number at which the growth rate was zero could be determined with a minimum of bias. For the conditions of Fig. 3 the critical value is 3590.

EXPERIMENTAL EQUIPMENT AND PROCEDURE

Several flow circuit designs were tried before the present system evolved. Two primary problems encountered in obtaining a workable system were: the spurious pressure and volume transients had to be limited to negligible values, and the flow controls (valves, switches, etc.) had to be either automatic or easily operated by one individual. The present system appears to answer the above requirements to a satisfactory degree and, in addition, allows a reasonable range of adjustment for the flow parameters.

Figure 4 illustrates schematically the overall flow system. The variable-speed positive-displacement pump produces a steady flow through the system, and the adjustable pulsator provides a superposed pulsating or oscillating flow in the system. The vertical column A serves as a check on the pulsating velocity component V_A' because the volume change in the column should equal the volume swept by the driving piston less any elastic volume changes in the system. The air reservoir minimizes these elastic deformations by maintaining a more constant pressure on the fluid even at large values of V_A' and ka. A disturbance generator in the form of a motor-driven piston is located at B. This device is used to insert turbulent "plugs" in the flow and is adjustable for amplitude as well as frequency. The test portion of the system is acrylic tubing carefully jointed and aligned. However, the maximum laminar-flow Reynolds numbers are limited by the pipe joints and the entrance during pulsating flow.

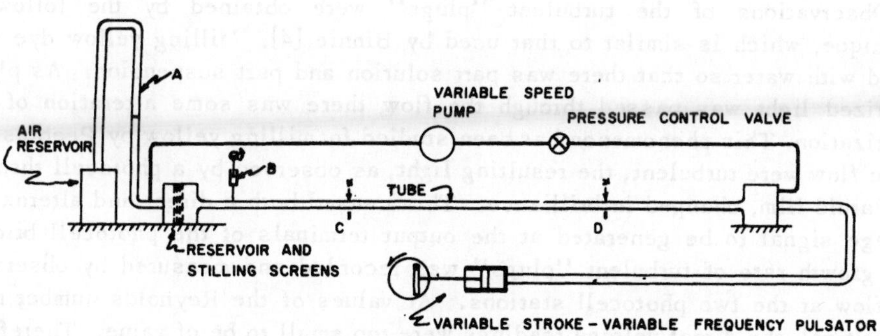

Fig. 4. Flow System Schematic.

This limit varies with the flow parameters but is easily detected by observing the output signals when the disturbance generator is not operating. The types of joints used for the 1.084-in.- and 0.496-in.-diam tubes are shown in Fig. 5. The joints in the larger tubing were hand-sanded to match, while those in the smaller tubes were standardized by pushing the heated ends on a mandrel tapered to 0.001 in./in. and allowing them to cool. This latter technique appears to work

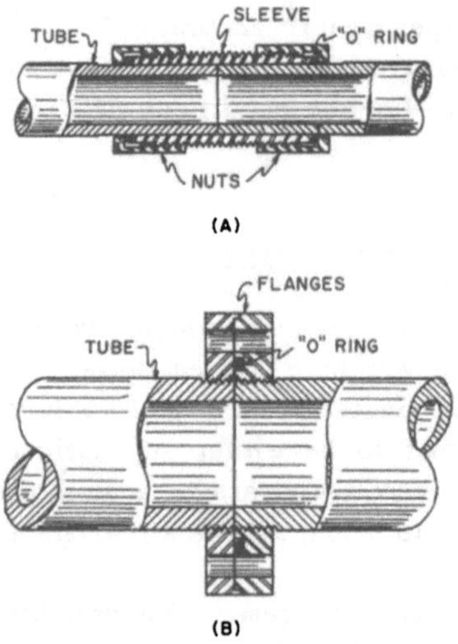

Fig. 5. Tube Couplings: *(A)* **0.496-in.,** *(B)* **1.084-in. (Diameters).**

quite well if the mandrel process is repeated one or two times to equalize any relaxation effects in the plastic. The dimensions of the flow system are:

Pipe inside diameter (in.)	1.084	0.496
Entrance to disturbance generator (L/D)	11	2
Disturbance generator to photocell C (L/D)	88	220
Photocell C to photocell D (L/D)	115	300
Photocell D to exit (L/D)	28	73

Observations of the turbulent "plugs" were obtained by the following technique, which is similar to that used by Binnie [4]. Milling yellow dye was mixed with water so that there was part solution and part suspension. As plane polarized light was passed through the flow there was some alteration of the polarization. This phenomenon has been studied for milling yellow by Peebles [5]. If the flow were turbulent, the resulting light, as observed by a photocell through a polaroid film, changed in brilliance. This caused both a direct and alternating voltage signal to be generated at the output terminals of the photocell bridge. The growth rate of turbulent "plugs" was recorded and measured by observing the flow at the two photocell stations. At values of the Reynolds number near the critical, the direct voltage readings were too small to be of value. Therefore, a capacitor was inserted in series with the bridge output to block any constant voltage and eliminate problems of signal drift. A capacitor was also placed across the bridge output to short undesirable 60-cycle ac pickup. Thus, a network was formed to pass only the desired signal oscillations caused by the turbulence. Figure 6 illustrates the mechanical arrangement of the photocell system and the electrical circuit employed. The output from each photocell station was recorded and analyzed to determine growth rates for the turbulent "plugs." Typical records for pulsating flow and for steady flow are shown on

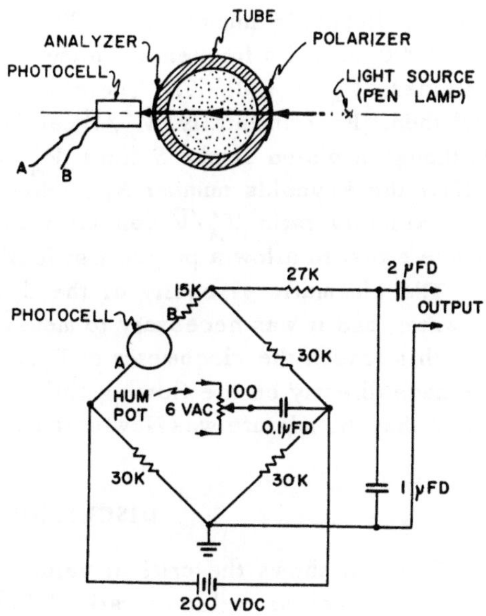

Fig. 6. Photocell Arrangement and Circuitry.

the left and right, respectively, of Fig. 7.

The growth rate G was calculated from the chart data by

$$G = \frac{C\bar{V}(L_D - L_C)}{N}, \tag{12}$$

where C is a constant depending upon the physical dimensions of the tube and the chart travel speed, L_C is the sum of the lengths of turbulent recordings from photocell C, L_D is the sum of the lengths of turbulent recordings from photocell D, and N is the number of "plugs" observed. Each growth rate on the graph of Fig. 3 is the result of averaging the growth of ten or twelve "plugs." The

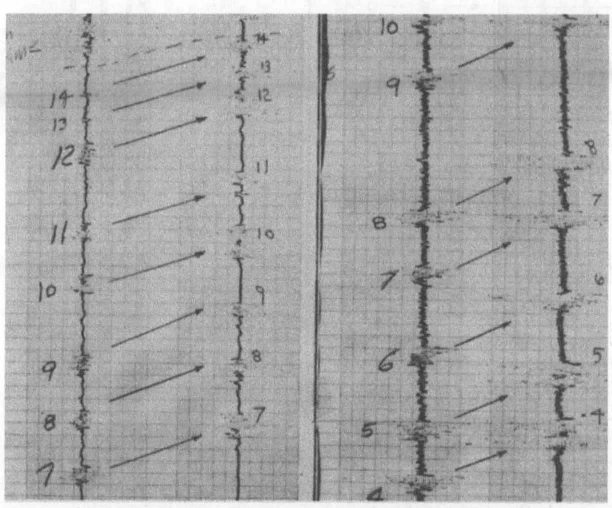

Fig. 7. Photocell Signal Recordings.

critical Reynolds number of 3590 corresponds to a pulsator piston stroke of $S = 0.990$ in. and a frequency parameter $ka = 6$. For the critical Reynolds number obtained, $V'_A/\overline{V} = 0.669$. Therefore, one point on the graph of Fig. 8 is determined. Experimentally, $V'_A/\overline{V}$ can be approximated only during a test because (although a piston stroke S fixes V'_A) the average velocity $\overline{V}$ must be varied to adjust the Reynolds number N_R. However, after some practice, the dimensionless velocity ratio $V'_A/\overline{V}$ can be estimated to a sufficient degree of accuracy before a test to allow a proper distribution of data.

The kinematic viscosity of the dye-water mixture differs slightly from that of water, and it was necessary to measure the viscosity at different temperatures. For this reason the viscometer of Fig. 9 was constructed so that a reading could be made directly on the flowing fluid during each run. Pressure-drop tests indicated that the mixture was Newtonian at the concentrations used.

DISCUSSION OF RESULTS

Figure 8 shows the critical values for the Reynolds number as a function of the dimensionless velocity ratio $V'_A/\overline{V}$ and the frequency parameter ka. An important feature of these results is the large increase in the critical value with $V'_A/\overline{V}$ within certain ranges. This is the reverse of what might be expected and what has generally been mentioned in the literature [6]. This increase in the critical value of the Reynolds number is difficult to explain except in general terms.

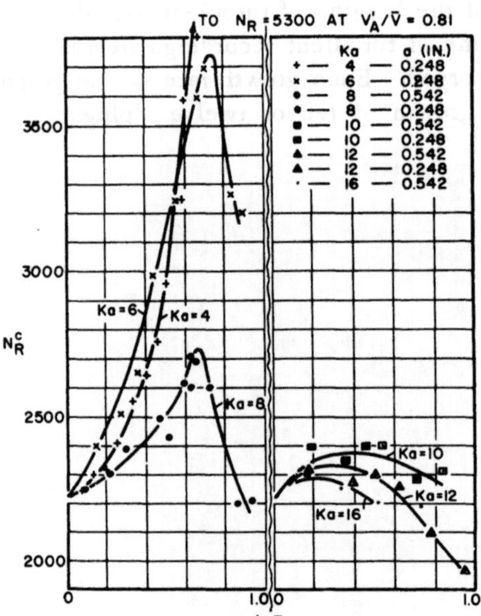

Fig. 8. Critical Reynolds Number vs Dimensionless Velocity.

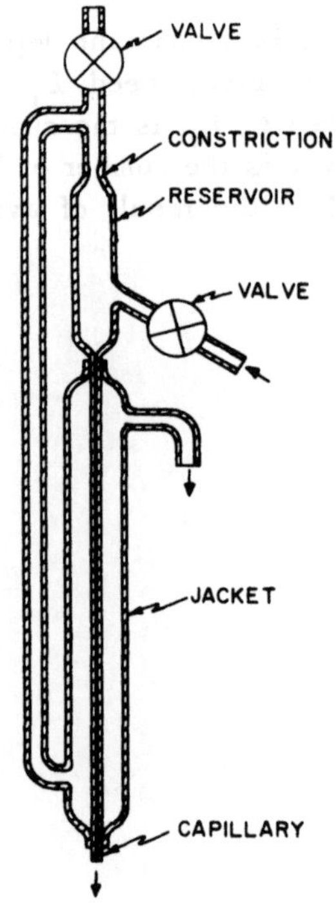

Fig. 9. Viscometer.

The relatively large amount of data used for calculating each value of the critical Reynolds number accounts in part for the small amount of scatter in the final results. Although the probable error is very difficult to analyze, the trend and reproducibility of the data indicate that the values of the critical Reynolds number N_R^c are within ±50 for $V_A'/\overline{V}$ less than 0.6. A more comprehensive discussion of the experimental techniques and errors will be given in an Experiment Station bulletin to be published locally.

It will be noted that at each value of *ka* there is a maximum in the curves of Fig. 8. The value of $V_A'/\overline{V}$ at which this maximum occurs decreases as *ka* increases. It is believed that this reversal in the trend of the Reynolds numbers is due to the occurrence of reverse flow during part of the pulsation cycle. A value of $V_A'/\overline{V}$ can be calculated for each *ka* above which a reversal in the direction of flow occurs in some part of the cycle. This reversal usually occurs close to the tube wall. The results of such calculations are shown in the following table, along with the values obtained by experiment:

ka	$V_A'/\overline{V}$ (for the experimental maximum)	$V_A'/\overline{V}$ (for reversal in the flow)
6	0.73	0.56
8	0.65	0.44
10	0.50	0.41

Notice that the trend is the same for both the experimental maximums and reversal in the flow. However, the experimental maximums occur at slightly larger values of $V_A'/\overline{V}$ in each case.

Experimental observations indicated that the turbulence generation was not constant throughout a cycle. This suggested that, although the Reynolds number based upon the average velocity $\overline{V}$ was adequate for correlating the data, an instantaneous Reynolds number was needed that would serve to indicate the flow conditions at any time in the entire cycle. It was assumed that the growth rate of the turbulent "plugs" at any instant would be a function of this instantaneous Reynolds number. This growth rate could then be integrated over a complete cycle to give the average growth rate. Thus,

$$G' = g(N_R') \, ,$$

$$G = f\int_0^{1/f} G' \, dt \, , \tag{13}$$

where N_R' is the instantaneous Reynolds number and G' is the instantaneous rate of growth for a turbulent "plug."

In order to test these ideas, the following instantaneous Reynolds number was defined for computation as

$$N_R' = \frac{3}{4}\frac{\overline{V}D}{\nu}\int_0^1 \left| \frac{\partial(u/\overline{V})}{\partial r'} \right| r' \, dr' \, , \tag{14}$$

or

$$N_R' = ZN_R ,$$

where

$$Z = \frac{3}{4}\int_0^1 \left| \frac{\partial(u/\overline{V})}{\partial r'} \right| r' \, dr' , \tag{15}$$

$$r' = \frac{r}{a} .$$

It should be noted that the quantity Z is a function of ka, $V_A'/\overline{V}$, and θ. Figure 10 presents a series of curves for calculated values of Z vs θ when $ka = 6$ and $V_A'/\overline{V} = 0.40$, 0.75, and 1.50.

An experimental G curve was determined as a function of N_R for values of N_R above 2220. It was assumed that this same curve could be used to give G' as a function of N_R'. For N_R' less than 2220 the function was assumed to be of the form

$$G' = -K \ln\left(\frac{4440 - N_R'}{2220}\right) . \tag{16}$$

The experimental and assumed curves are shown in Fig. 11 for the value of K used with $ka = 6$. In Eq. (16), K is a constant that was adjusted to fit the predicted values of the critical Reynolds numbers to the experimental curves. Thus, having chosen a value for K and a value for N_R at a given value of $V_A'/\overline{V}$, N_R' was calculated at various points in the cycle by using values of Z calculated from the laminar-velocity profiles. By using growth-rate functions as given in Fig. 11, G' was determined. Finally, the values of G' were averaged over a

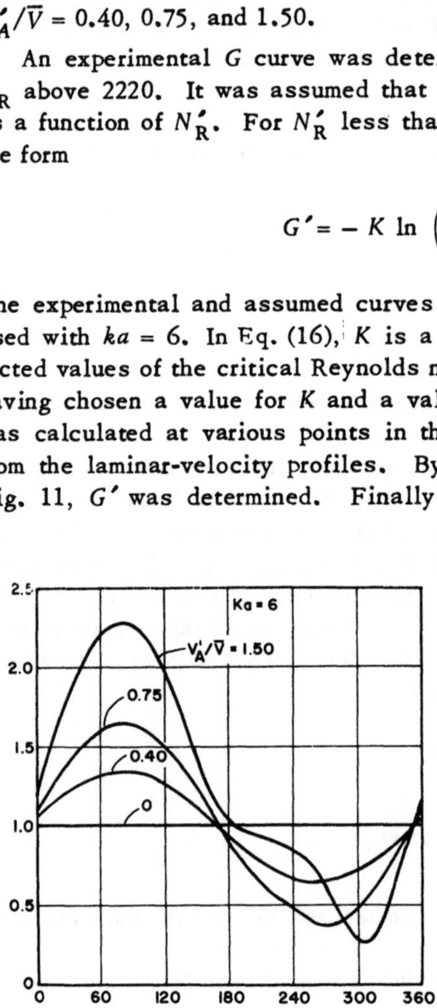

Fig. 10. Instantaneous Reynolds Number Coefficient vs Time.

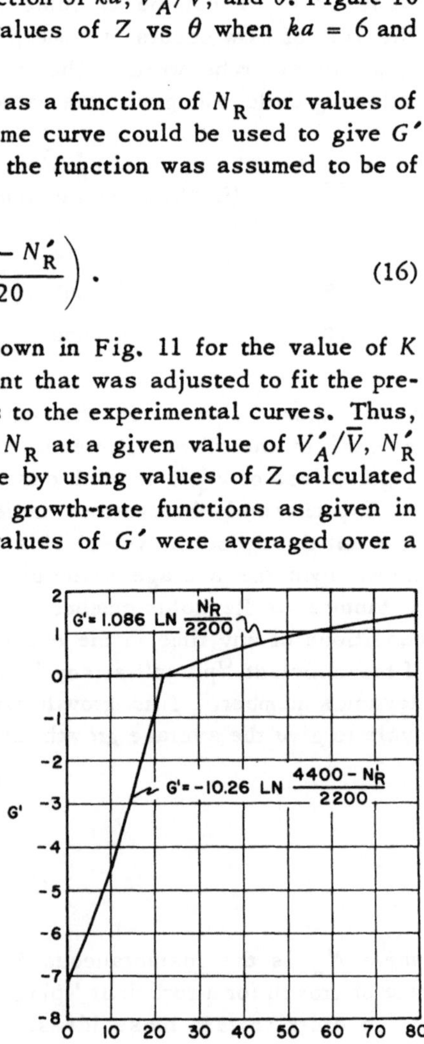

Fig. 11. Net Growth Rate vs Instantaneous Reynolds Number.

cycle to determine G. Adjusting N_R until G became zero gave a theoretical value for the critical Reynolds number. The theoretical values determined in this manner are shown in Figs. 12 and 13 along with the experimental values. The nonlinear nature of the growth-rate function causes the theoretical values of the critical Reynolds number to increase as $V'_A/\bar{V}$ is increased. The flow reversals eventually begin to affect the values of Z and cause the trend to reverse. The calculated curves agree very well with the experimental curves, considering the assumptions made. This is particularly true with respect to trends, and these trends were of prime interest. It should be pointed out that some doubt exists as to the validity of the positions of the experimental maximums because of the occurrence of spontaneous turbulence. The authors intend to investigate the effect of roughness and the attendant spontaneous turbulence on these maximums. If the correct definition of the instantaneous Reynolds number were used, the constant K in Eq. (16) would be the same for all ka. However, it was necessary to use different values for K when calculating the curves of Figs. 12 and 13.

Although the preceding definition for the instantaneous Reynolds number and assumptions for growth rate are speculative, they do offer one feasible explanation for the experimental data of Fig. 8.

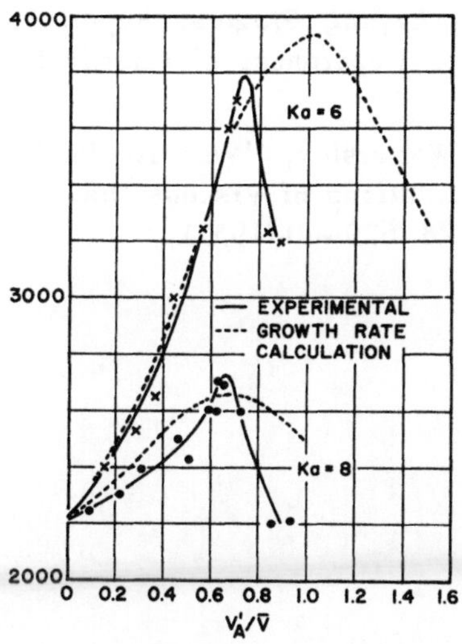

Fig. 12. Critical Reynolds Number vs Dimensionless Velocity.

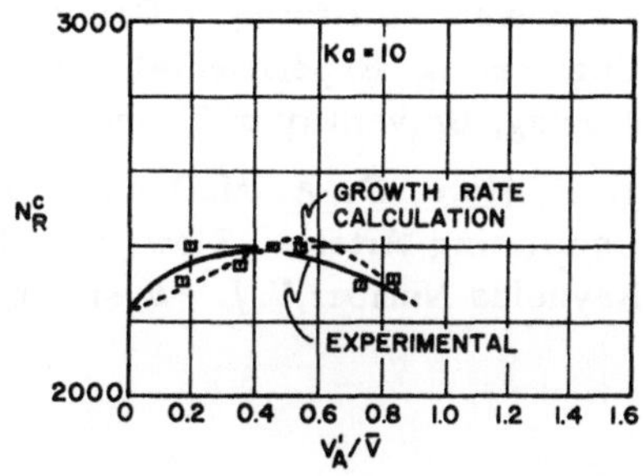

Fig. 13. Critical Reynolds Number vs Dimensionless Velocity.

SUMMARY

The experimental data indicate that critical values of the Reynolds number in pulsating flow through smooth tubes increase to a maximum and then decrease as the amplitude of the pulsations is increased from zero. The value of $V'_A/\bar{V}$ at which the maximum critical Reynolds number occurs appears to decrease with increasing values of the frequency parameter ka. The nonlinear growth rate for turbulent "plugs" offers one possible explanation for the experimentally determined trends. However, the ideas presented in this explanation need further investigation and application before a reasonable degree of confidence in their validity can be gained.

REFERENCES

[1] S. Uchida, "The Pulsating Viscous Flow Superposed on the Steady Laminar Motion of Incompressible Fluid in a Circular Pipe," *Z. angew. Math. u. Phys.* **7**(5), 404–22 (1956).

[2] J. Rotta, "Experimenteller Beitrag zur Entstehung Turbulenter Stroemung in Rohr," *Ingr.-Arch.* **24**(4), 258–81 (1956).

[3] E. R. Lindgren, "The Transition Process and Other Phenomena in Viscous Flow," *Arkiv Fysik* **12**(1), 1–169 (1957).

[4] A. M. Binnie, "A Double-Refraction Method of Detecting Turbulence in Liquids," *Proc. Phys. Soc. (London)* **57**, 390–402 (1945).

[5] F. N. Peebles, J. W. Prados, and E. H. Honeycutt, *A Study of Laminar Flow Phenomena Utilizing a Doubly Refracting Liquid*, Progress Report 1, The Engineering Experimental Station and The Department of Chemical Engineering, University of Tennessee, 1954.

[6] J. F. Hale, D. A. McDonald, and J. R. Womersley, "Velocity Profiles of Oscillating Arterial Flow, with Some Calculations of Viscous Drag and the Reynolds Number," *J. Physiol. (London)* **128**, 629–40 (1955).

APPROXIMATE ANALYSIS FOR THE TURBULENT-BOUNDARY-LAYER THICKNESS ON A CONE IN SUPERSONIC FLOW[1]

Henry F. Hrubecky
Tulane University

A simplified approximate analysis for the boundary-layer thickness, friction coefficient, and mass flow is presented for an initially turbulent boundary layer on a cone in a supersonic free stream. Considering the effect of curvature on the boundary-layer thickness a nonlinear differential equation is obtained whose simple form lends itself easily to a computer program. With the boundary-layer thickness small compared with the cone's radial dimensions, the nonlinear Bernoulli differential equation is obtained from the momentum integral equation, and hence is solvable by an appropriate change in variable. Allowing for various exponents in the Blasius friction relation and the velocity-profile relation, the method is applicable to a large range in Reynolds numbers. The simple mass-flow relation is readily applicable to those situations in which boundary-layer control may be desired.

INTRODUCTION

A nonlinear differential equation for the boundary-layer thickness is presented for a fully developed turbulent boundary layer on a cone. This equation is solved in a closed form when the boundary layer is assumed to be thin. The boundary layer is assumed to be turbulent starting with the leading edge. Furthermore, it is assumed that the boundary layer is subjected to a zero-pressure gradient and that its energy per unit mass is constant transverse to the direction of flow. Under these conditions, a constant relationship results between the cone and the flat-plate boundary-layer thickness for a given Mach number and a given Reynolds number. No such constant relationship exists when the effect of curvature is considered and the solution is of necessity a numerically iterative one. The upstream flow is assumed to be uniform with a Mach number designation of M_∞'. After passing through an assumed attached leading edge shock, the resultant flow is uniform and parallel to the cone surface with a Mach number designation of M_∞.

For the determination of the characteristic parameters of the boundary layer, a momentum integral analysis was utilized. In the momentum integral equation

[1]Paper was done while the author was at the University of Florida and extends to certain boundary-layer control studies for the Martin-Marietta Co. in Orlando, Florida.

NOTATION

x	Coordinate along and normal to the cone surface
r_0, r	Radial distances from axis of symmetry to wall and to a point in the boundary layer
U	Potential flow velocity
δ	Boundary layer thickness
θ	Momentum thickness
δ^*	Displacement thickness
M'	Upstream Mach number
M	Mach number along cone surface
τ	Shearing stress
ρ	Density
u	Velocity
γ	Cone half angle
$\theta_n', \theta_n, \theta_{2n}$	Dimensionless parameters
C_m	Constant in Blasius friction relation
β	Dimensionless parameter, $\overline{T}/T_C$
n	Exponent in velocity distribution law
N	$2n^2/(n+1)(2n+1)$
R	Reynolds number
m	Exponent in Blasius friction relation
μ	Viscosity
C_f	Local friction coefficient
G	Mass rate of flow
Δ_n, Δ_{2n}	Dimensionless parameters
T	Absolute temperature
λ	Coefficient of resistance
A	Area
ω	Exponent in viscosity law
E	Rate of energy transferred

Subscripts and superscripts:

0	Wall
∞	Free stream
c	Pipe center
–	Mean

used, the turbulent-velocity-fluctuation terms (u', v') are neglected; that is, the double and triple correlation terms are omitted from the shearing-stress formula. For convenience, bars over the mean-velocity terms are dropped. The present approximate analysis compares favorably with experiment for Mach numbers up to 3 and with the analysis of Van Driest [1]. The present analysis gives a much simpler form for the mass flow and energy flow in the boundary layer than the solution of Van Driest, and, consequently, offers a simpler form of estimating these parameters where boundary-layer-control application may be desired.

BOUNDARY-LAYER THICKNESS AND MOMENTUM EQUATIONS

We denote x as the current length measured along a meridian from the stagnation point as indicated in Fig. 1, while y denotes the coordinate at right angles to the cone surface. The contour of the cone is specified by the radii $r_0(x)$ of the sections at right angles to the axis of symmetry.

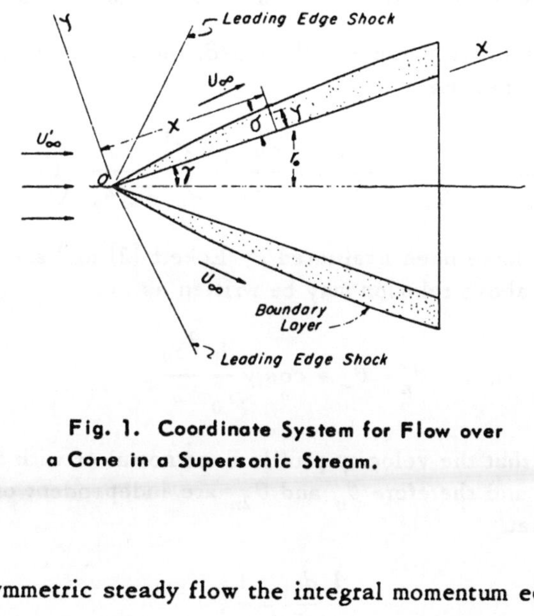

Fig. 1. Coordinate System for Flow over a Cone in a Supersonic Stream.

For axially symmetric steady flow the integral momentum equation is

$$U^2 \frac{d\theta}{dx} + (2\theta + \delta - M^2\theta)U \frac{dU}{dx} + U^2 \frac{\theta}{r_0}\frac{dr_0}{dx} = \frac{\tau_0}{\rho} .$$

If no detached shocks occur and essentially a zero-pressure gradient ($dU/dx = 0$) along x is assumed, the above equation becomes

$$\frac{\tau_0}{\rho u_\infty^2} = \frac{d\theta}{dx} + \frac{\theta}{r_0}\frac{dr_0}{dx} . \tag{1}$$

The definition of the momentum thickness θ for axisymmetric compressible flow is

$$\theta = \int_0^{\delta} \frac{\rho u}{\rho_\infty u_\infty}\left(1 - \frac{u}{u_\infty}\right)\frac{r}{r_0}\,dy\,. \tag{2}$$

From the geometry of the cone, $r = r_0 + y \cos \gamma$. If it is assumed that for a given wall radius r_0 the boundary thickness δ is small compared with r_0, then $r_0 \cong r$ and Eq. (2) reduces to

$$\theta = \int_0^{\delta} \frac{\rho u}{\rho_\infty u_\infty}\left(1 - \frac{u}{u_\infty}\right) dy\,. \tag{2a}$$

However, if no such assumption is made, the momentum thickness for the cone becomes

$$\theta = \int_0^{\delta} \frac{\rho u}{\rho_\infty u_\infty}\left(1 - \frac{u}{u_\infty}\right) dy + \frac{\cos\gamma}{r_0}\int_0^{\delta} \frac{\rho u}{\rho_\infty u_\infty}\left(1 - \frac{u}{u_\infty}\right) y\,dy\,. \tag{3}$$

Defining the dimensionless parameter $\theta_n' = \theta/\delta$, the above equation after a variable and limit change becomes

$$\theta_n' = \int_0^{1} \frac{\rho u}{\rho_\infty u_\infty}\left(1 - \frac{u}{u_\infty}\right) d\left(\frac{y}{\delta}\right) + \delta\,\frac{\cos\gamma}{r_0}\int_0^{1} \frac{\rho u}{\rho_\infty u_\infty}\left(1 - \frac{u}{u_\infty}\right)\frac{y}{\delta}\, d\left(\frac{y}{\delta}\right)\,. \tag{4}$$

The above integrals have been evaluated by Eckert [2] and are given in Table 1 as θ_n and θ_{2n}. The above relation may be written as

$$\theta_n' = \theta_n + \cos\gamma\,\frac{\delta}{r_0}\,\frac{\theta_{2n}}{2}\,. \tag{5}$$

If it is assumed that the velocity profile does not alter with the flow length x, the above integrals and therefore θ_n and θ_{2n} are independent of x. Noting from the cone geometry that

$$\frac{1}{r_0}\frac{dr_0}{dx} = \frac{1}{x}\,,$$

Eq. (5) may be substituted into Eq. (1) along with the above condition to give

$$\frac{\tau_0}{u^2} = \theta_n'\left(\frac{d\delta}{dx} + \frac{\delta}{x}\right) = \left(\theta_n + \frac{\cos\gamma}{2}\frac{\delta}{r}\theta_{2n}\right)\left(\frac{d\delta}{dx} + \frac{\delta}{x}\right)\,. \tag{6}$$

Furthermore, since r_0 varies with x, the above equation may be written as

$$\frac{\tau_0}{\rho_\infty u_\infty^2} = \left(\theta_n + \frac{\cot\gamma}{2}\,\theta_{2n}\,\frac{\delta}{x}\right)\left(\frac{d\delta}{dx} + \frac{\delta}{x}\right)\,. \tag{7}$$

Table 1. Dimensionless Parameters in Boundary-Layer Flow

M_∞	$\Delta_n = \delta^*/\delta$			β			$\theta_n = \theta/\delta$			Δ_{2n}/Δ_n		θ_{2n}/θ_n	
For n =	7	9	10	7	9	10	7	9	10	7	9	7	9
0	0.13	0.10	0.09	1.0	1.0	1.0	0.097	0.082	0.076	0.53	0.52	0.60	0.58
1	0.16	0.13	0.12	1.1	1.1	1.1	0.091	0.077	0.072	0.56	0.55	0.62	0.59
2	0.24	0.20	0.18	1.3	1.2	1.2	0.077	0.067	0.063	0.60	0.59	0.66	0.60
3	0.33	0.28	0.27	1.4	1.3	1.3	0.062	0.055	0.053	0.66	0.64	0.70	0.67
4	0.42	0.36	0.35	1.8	1.6	1.6	0.049	0.045	0.044	0.70	0.68	0.74	0.71
5	0.49	0.43	0.43	2.2	2.0	1.9	0.039	0.037	0.036	0.74	0.72	0.77	0.74
6	0.56	0.51	0.49	2.6	2.3	2.3	0.033	0.030	0.030	0.77	0.75	0.80	0.77
7	0.61	0.56	0.55	3.2	2.8	2.7	0.027	0.025	0.025	0.79	0.77	0.82	0.80
8	0.66	0.61	0.59	3.7	3.3	3.2	0.022	0.021	0.021	0.81	0.79	0.85	0.82
9	0.70	0.65	0.63	4.3	3.8	3.7	0.019	0.018	0.018	0.83	0.80	0.87	0.85
10	0.73	0.68	0.67	5.1	4.5	4.3	0.016	0.016	0.016	0.85	0.81	0.89	0.87

The preceding equation relates the boundary-layer thickness to the wall shearing stress for any point x. To determine the variation of the boundary-layer growth with the flow length, a wall shearing stress functional relation with x is needed.

Klebanoff [3] in his experiments found that the distribution of turbulent energy within the boundary-layer flow is similar to that for channel and pipe flow. Furthermore, although in fully developed pipe flow there is no free stream, the velocity profile for boundary-layer flow is quite similar to that for pipe flow. Frossel [4] investigated the turbulent flow of gases through pipes at high subsonic and supersonic velocities. He found the coefficient of resistance λ to be similar for both incompressible and compressible flow. In light of the above, Eckert [2] related the wall shearing stress variation over a flat plate to that through a pipe within a restricted range of Reynolds numbers and Mach numbers. Utilizing this pipe flow analogy, the Blasius law of friction may be written as

$$\tau_0 = \frac{\lambda}{8} \bar{\rho} \bar{u} ,$$

where $\lambda = C_m/R_x^m$, C_m is a constant, and the exponent m on the Reynolds number R_x may be assigned various values. If for this flow the power law of velocity distribution is assumed to be $\bar{u}/U = (y/\delta)\ 1/n$, then

$$N = \frac{\bar{u}}{U} = \frac{1}{AU} \int_0^A \bar{u}(dA) = \frac{2n^2}{(n+1)(2n+1)} .$$

The following relation, a generalized form of one first presented by Eckert [2], is utilized:

$$\frac{\tau_0}{\rho_\infty u_\infty^2} = C_2 \left(\frac{\mu_\infty}{u_\infty \rho_\infty} \right)^m \cdot (\delta)^{-m} \cdot \beta^p \cdot N^q , \tag{8}$$

where $C_2 = C_m/2^m \cdot 8$, $p = m(\omega + 1) - 1$, and $q = 2 - m$. The term ω is the exponent in the viscosity power law, and β is a temperature ratio and is presented in Table 1 as a function of the free-stream Mach number and the velocity profile exponent n.

Equating Eqs. (7) and (8), the following differential equation for the boundary-layer thickness $\delta(x)$ is obtained:

$$\frac{d\delta}{dx} + \frac{\delta}{x} = \frac{C_3 \delta^{-m}}{1 + \frac{\cot \gamma\, \theta'_{2n}}{2\theta'_n}\left(\frac{\delta}{x}\right)}, \tag{9}$$

where $C_3 = (C_2/\theta'_n)\beta^p \cdot N^q \cdot (\mu_\infty / u_\infty \rho_\infty)^m$ and is independent of x.

The above first-order nonlinear equation does not lead to a closed-fom solution. It is of a form, however, which can be easily programmed on a digital computer. If the δ/x term in the denominator on the right side of the equation is neglected, this is tantamount to physically restricting the solution to the case where the boundary-layer thickness is small compared with the wall radius $r_0(x)$. This in effect neglects the second integral in Eq. (3) for the momentum thickness θ. Therefore, placing this restriction upon the solution, the above equation may be simplified to (omitting the prime on θ'_n)

$$\frac{d\delta}{dx} + \frac{\delta}{x} = C_3 \delta^{-m}. \tag{10}$$

The above equation, though nonlinear, may be linearized by a change of variable. It is commonly referred to as the Bernoulli equation. By dividing through by δ^{-m} and letting $z = \delta^{m+1}$, the following linear equation is obtained:

$$\frac{dz}{dx} + \frac{(m+1)z}{x} = C_{3(m+1)} \cdot \tag{11}$$

For initial conditions, $x = 0$, $\delta = 0$, and $z = 0$, the general solution for the boundary-layer thickness is

$$\delta^{m+1} = C_3 \frac{m+1}{m+2} x . \tag{12}$$

For two-dimensional flow the momentum integral equation is

$$\frac{d\theta}{dx} + \frac{\theta}{U}\frac{dU}{dx}\left(2 + \frac{\delta}{\theta} - M^2\right) = \frac{\tau_0}{\rho_\infty u_\infty^2}. \tag{13}$$

For a zero pressure gradient this reduces to

$$\frac{\tau_0}{\rho_\infty u_\infty^2} = \frac{d\theta}{dx} = \theta_n \frac{d\delta}{dx}. \tag{14}$$

Utilizing the shearing stress relationship in Eq. (8), the resultant differential equation for the flat plate with U = a constant becomes

$$\frac{d\delta}{dx} = C_3 \delta^{-m} . \tag{15}$$

The above equation may be solved by simple quadrature to give

$$\frac{\delta^{m+1}}{m+1} = C_3 x . \tag{16}$$

For the same value of m and C_3, that is, the same Reynolds number, Mach number, and flow length x, a comparative ratio of cone to flat-plate boundary-layer thickness may be obtained. Denoting the flat-plate values by a prime, the following simple relation results:

$$\frac{\delta}{\delta'} = \left(\frac{1}{m+2}\right)^{1/(m+1)} \tag{17}$$

Furthermore, the local-friction coefficient may be found as

$$C_f = \frac{2\tau_0}{\rho_\infty u_\infty^2} = 2C_2 \left(\frac{\mu_\infty}{u_\infty \rho_\infty}\right)^m \beta^p N^q \delta^{-m} , \tag{18}$$

from which the following local-friction coefficient ratio may be obtained:

$$\frac{C_f}{C_f'} = \left(\frac{\delta'}{\delta}\right)^m = (m+2)^{m/m+1} . \tag{19}$$

Van Driest [1] showed that for the same cone Mach number and wall–to–free-stream temperature ratio, the cone solution for the local-friction and heat-transfer coefficients is the same as for the flat-plate solution when the cone Reynolds number is divided by 2. Using a value of $m = \frac{1}{4}$ in the Blasius equation gives $C_f/C_f' = 1.17$, as compared with the Van Driest value of 1.15.

The cone boundary-layer-thickness relation in terms of Mach number and Reynolds number may be given from Eq. (16) as

$$\frac{\delta^{m+1}}{x} = \frac{m+1}{m+2} \frac{C_2}{\theta_n} \frac{\beta^p N^q}{R_x^m} \tag{20}$$

or

$$\frac{\delta}{x} = \left(\frac{C_4 \beta^p N^q}{\theta_n R_x^m}\right)^{1/(m+1)} , \tag{20a}$$

where $C_4 = (m+1)/(m+2)C_2$. If $m = \frac{1}{4}$, $n = 7$, $C_m = 0.316$, and $\omega = 0.76$, then from definitions of p, q, N, and C_2 the above equation simplifies to

$$\frac{\delta}{x} = \frac{0.0308}{\theta_n^{0.8} \beta^{0.448} R_x^{0.2}} . \tag{21}$$

Values of θ_n and β as a function of Mach number are given in Table 1.

BOUNDARY-LAYER MASS FLOW RATE

The mass rate of flow G through the boundary layer for any x value can be represented as the flow through the annular ring, where cross-sectional profile is indicated as face $\overline{AB}$ in Fig. 2. This flow through face $\overline{AB}$ can be considered

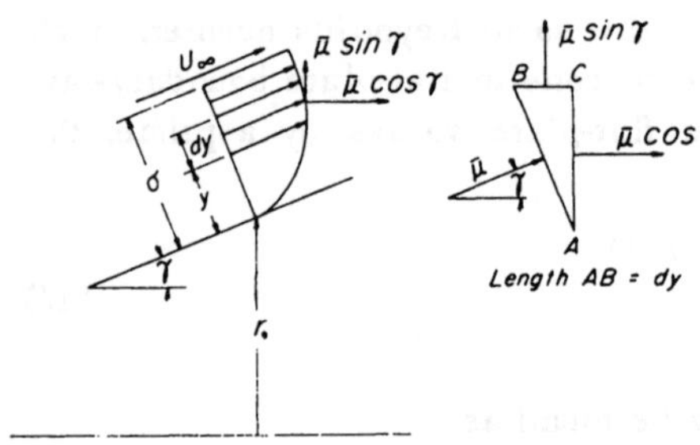

Fig. 2. Detailed View of Mean Flow in the Boundary Layer.

as the vector sum of the flows through faces $\overline{BC}$ and $\overline{AC}$. From geometry, this flow may be represented in view of the coordinate system used as

$$G = 2\pi \int_0^\delta \rho u r \, dy + 2\pi \int_0^\delta \rho u \cos \gamma \, y \, dy \, . \tag{22}$$

Noting that for any x, δ is independent of r and y, and dividing both sides of the equation by $2\pi \rho_\infty u_\infty \delta r$, the following expression results:

$$\frac{G}{2\pi \rho_\infty u_\infty \delta} = \left[\int_0^1 \frac{\rho u}{\rho_\infty u_\infty} \, d\left(\frac{y}{\delta}\right) \right] + \left[\frac{\delta \cos \gamma}{r} \int_0^1 \frac{\rho u}{\rho_\infty u_\infty} \frac{y}{\delta} \, d\left(\frac{y}{\delta}\right) \right] \, . \tag{23}$$

These integrals have been evaluated for various values of n and Mach number, and rewritten are

$$\begin{aligned} \frac{G}{2\pi \rho_\infty u_\infty r \delta} &= \left[1 - \left(\frac{\delta^*}{\delta}\right)_n \right] + \frac{1}{2} \, \frac{\delta \cos \gamma}{r} \left[1 - \left(\frac{\delta^*}{\delta}\right)_{2n} \right] \\ &= \left(1 - \Delta_n\right) + \frac{1}{2} \, \frac{\delta \cos \gamma}{r} \left(1 - \Delta_{2n}\right) \, . \end{aligned} \tag{24}$$

Values of Δ_n and Δ_{2n} may be found in Table 1. After the boundary-layer thickness δ has been determined, the above equation gives a relatively simple expression for the boundary-layer mass flow rate at any position x. If the energy per unit mass is constant across the boundary layer, the total rate of energy transferred at any boundary-layer section may be given as

$$E = c_p T_0 G \, . \tag{25}$$

DISCUSSION AND CONCLUSIONS

Relatively simple relations for the boundary-layer thickness, local-friction coefficient, and mass rate of flow are found for various values of m in the Blasius friction relation and for various values of n in the velocity-profile power relation. Experiments [5, 6, 7] indicate good agreement with $n = 1/7$ in the power relation for Mach numbers up to 2.0 and Reynolds numbers up to 20×10^6. At higher Reynolds numbers ($R_L = 50 \times 10^6$, $M = 2.0$), Wilson [5] obtained a velocity profile that followed close to a power law of $n = 9$. It is known that at high Reynolds numbers (due to a fuller velocity profile and consequent smaller effect of Reynolds number on the boundary-layer thickness), one may expect higher values of n and a deviation in the exponent of the Reynolds number. Correspondingly, to obtain better experimental agreement [8] at higher Reynolds numbers, smaller values of m in the Blasius equation, say 1/5 or 1/6 instead of 1/4, can be used. Accordingly, with the preceding statements, comparison of the present analysis, which enables adjustment of the m and n exponents to fit the Reynolds-number range, can be made with experiments on cones [9], where the boundary-layer thickness was determined. For the following conditions ($\alpha = 0.6°$, $R_L = 22.8 \times 10^6$, $M = 3.45$), a faired curve through the experimental points indicated a ratio of mean boundary-layer thickness to flow length (δ/x) of 0.010. The present analysis for $m = 1/5$ and $n = 9$ gives a value of $\delta/x = 0.011$, while for $m = 1/4$ and $n = 9$, $\delta/x = 0.0095$. This is, approximately, a 5 to 10% deviation from experiment.

In conclusion, the present approximate analysis allows for a simple evaluation of boundary-layer thickness, local-friction coefficient, and boundary-layer mass flow. This is especially significant in the mass flow, where boundary-layer-control applications may be considered, and an otherwise lengthy iterative solution can be substituted as an initial approximation.

REFERENCES

[1] E. R. Van Driest, "Turbulent Boundary Layer on a Cone in Supersonic Flow at Zero Angle of Attack," *J. Aeronaut. Sci.* **19**(1), 55–57 (1952).

[2] H. V. Eckert, "Characteristics of the Turbulent Boundary Layer on a Flat Plate in Compressible Flow from Measurements of Friction in Pipes," *J. Aeronaut. Sci.* **17**(9), 573–84 (1950).

[3] P. S. Klebanoff, "Characteristics of Turbulence in a Boundary Layer with Zero Pressure Gradient," *NACA* TR 1247 (1955).

[4] W. Frossel, "Stromung in Glatten, Geraden Rohren mit Uber und Unterschallgesctiwindigkeit," *Forsch. Gebiete Ingenieurw.* **7**, 75 (1936).

[5] R. E. Wilson, E. C. Young, and M. J. Thompson, *Second Interim Report on the Experimental Determination of Turbulent Boundary Layer Characteristics at Supersonic Speeds,* Defense Research Laboratory, University of Texas, Report CM501, ORL-198 (1949).

[6] M. W. Rubesin, R. C. Maydew, and S. A. Varga, "An Analytical and Experimental Investigation of the Skin Friction of the Turbulent Boundary Layer on a Flat Plate at Supersonic Speeds," *NACA* TN 2305 (1951).

[7] R. M. O'Donnell, "Experimental Investigation at a Mach Number of 2.41 of Average Skin Friction Coefficients and Velocity Profiles for Laminar and Turbulent Boundary Layers and an Assessment of Probe Effects," *NACA* TN 3122 (January 1954).

[8] K. Wieghardt, *Turbulente Grenzschichten,* part B5, Gottingen Monograph, 1946.

[9] A. Seiff and B. J. Short, "An Investigation of Supersonic Turbulent Boundary Layers on Slender Bodies of Revolution in Free Flight by Use of a Mach – Zehnder Interfermeter and Shadowgraphs," *NACA* TN 4364 (September 1958).

DIVISION OF ENERGY IN AXISYMMETRIC GRAVITY WAVES[1]

Lawrence R. Mack
The University of Texas

Expressions for the kinetic energy and the potential energy of periodic, first-mode, axisymmetric gravity waves are obtained. The difference between the mean kinetic and mean potential energies is shown to be proportional to the fourth power of the amplitude. It was found that the ratio of the difference between the mean kinetic energy and the mean potential energy to the total energy of the wave has, to the third approximation, the same variation with the depth of the liquid as the finite-amplitude correction to the square of the frequency. Thus there is a qualitative difference between deep water and shallow water in the time variation of the ratio of kinetic energy to total energy for finite-amplitude waves. The nature of this difference is illustrated and discussed.

INTRODUCTION

In a previous work [*1*] the author obtained a solution for periodic, axisymmetric, first-mode standing waves in a circular basin of uniform depth. By means of an iterative procedure, the velocity potential, the surface elevation, and the frequency of oscillation were found to the third approximation, the first approximation being the case of infinitesimal amplitude, and the wave of maximum amplitude was investigated. In the present paper the previous results will be applied to determine the division of the total energy of the wave into potential and kinetic portions, also the manner in which that division of energy varies with the depth of the liquid will be studied.

ANALYSIS

The principal difficulty in solving finite-amplitude gravity-wave problems lies in satisfying two nonlinear boundary conditions at the free surface. In order to meet this difficulty the velocity potential $\phi(r, z, t)$ and the surface elevation

[1]Portions of this work were supported by the Office of Naval Research, under Contract Nonr 1509(03) with the Iowa Institute of Hydraulic Research, and by the Bureau of Engineering Research of The University of Texas.

NOTATION

A_{11}	Amplitude parameter
A_{mn}	Coefficients
C	Curve formed by intersection of free surface with a plane containing the z axis
C_1, C_2, C_3, C_4	Functions of depth given by Eqs. (10–13)
dA	Element of area in a plane normal to the z axis
ds	Increment of length measured from the z axis along C
dS	Element of area in the boundaries of the fluid region
$d\tau$	Element of volume
g	Dimensional gravitational acceleration
G	Gravitational acceleration (serves as frequency variable)
G_c	Frequency correction factor
H	Mean depth
$J_{mn}, J_m(K_n r)$	Bessel function of first kind of order m and argument $K_n r$
K_n	Eigenvalues satisfying $J_1(K) = 0$
n	Distance measured into fluid region normal to dS
r	Radial coordinate
R	Dimensional radius of basin
t	Time
T	Kinetic energy
T_M	Temporal mean value of kinetic energy
th_n	Hyperbolic tangent of $K_n H$
V	Potential energy
V_M	Temporal mean value of potential energy
z	Vertical coordinate
$\alpha_n(F)$	nth coefficient in Dini series expansion of a function $F(r)$
Γ_n	Quantity defined by Eq. (5)
ζ_m	Functions of r
η	Surface elevation
ϕ	Velocity potential
ω	Dimensional frequency of oscillation

$\eta(r, t)$ were expressed as infinite series of eigenfunctions satisfying the linear governing equations

$$\phi = \sum_{m=1}^{\infty} \sum_{n=0}^{\infty} A_{mn} \frac{\cosh K_n(z+H)}{\cosh K_n H} J_0(K_n r) \sin mt\,, \tag{1}$$

$$\eta = \sum_{m=0}^{\infty} \zeta_m(r) \cos mt\,, \tag{2}$$

where H is the mean relative depth of the liquid, hereafter called merely depth, r and z are cylindrical coordinates as shown in Fig. 1, t is time, J_0 is a Bessel

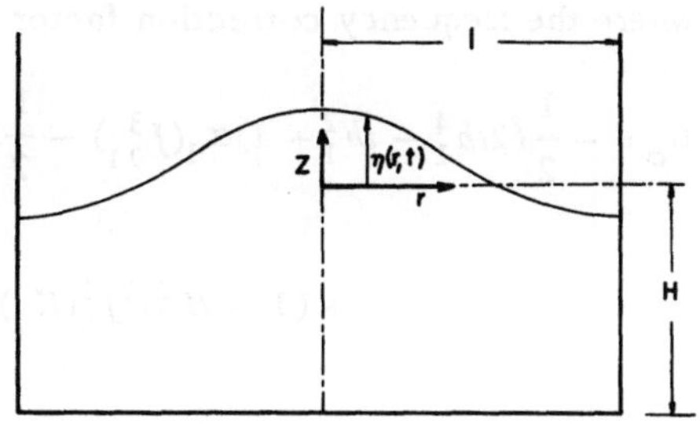

Fig. 1. Coordinate System and Geometric Configuration.

function of the first kind, ζ_m are functions to be determined, and the eigenvalues K_n are the consecutively numbered nonnegative roots of $J_1(K) = 0$ beginning with $K_0 = 0$. (Actually, cos mt terms in ϕ and sin mt terms in η were also assumed, but it was shown that their coefficients could be made zero by appropriate choice of the time origin.) Unless otherwise noted, all quantities used are dimensionless, being made so by division by appropriate combinations of the dimensional radius of the cylinder, frequency of oscillation, and density of the fluid.

To determine the particular choice of coefficients that would satisfy the nonlinear boundary conditions, an iterative procedure was followed which yielded all other A_{mn} in terms of the depth and an amplitude parameter A_{11} and yielded the functions ζ_m in terms of A_{11}, H, and r. This procedure required that the various functions of r generated by the substitution of ϕ and η into the free-surface boundary conditions be expanded in Dini series of Bessel functions [2, chap. 18]. That is,

$$F(r) = \sum_{n=0}^{\infty} a_n(F) J_{0n}\,,$$

where the constants a_n are given by

$$a_n(F) = \frac{\int_0^1 rF(r) J_{0n}\, dr}{\frac{1}{2} J_0^2(K_n)}$$

and the shortened notation J_{mn} is used for $J_m(K_n r)$.

The first, or linear, approximation gave the well-known solution for an infinitesimal-amplitude wave: $A_{11} = A_{11}$, $\zeta_1(r) = A_{11} K_1 th_1 J_{01}$, and $1/G = K_1 th_1$, where the shortened notation th_n is used for tanh $K_n H$ and the dimensionless gravitational acceleration G is a dependent variable giving the frequency. ($G = g/\omega^2 R$, where g, ω, and R are the dimensional gravitational acceleration, frequency of oscillation, and radius of tank respectively.) The second and third approximations introduced $A_{1n}(n = 0, n > 1)$, A_{2n}, A_{3n}, ζ_0, ζ_2, and ζ_3 terms as well as additions to ζ_1 and G. Of special interest to the present paper is the frequency equation obtained in the third approximation,

$$\frac{1}{G} = K_1 th_1 \left[1 + \frac{(A_{11} K_1^2)^2}{4} G_c\right], \tag{3}$$

where the frequency correction factor G_c is a function of H only given by

$$\begin{aligned} G_c = &-\frac{1}{2}(2th_1^4 - th_1^2 + 1)\alpha_1(J_{01}^3) - \frac{1}{2}(1 + th_1^2)\alpha_1(J_{01}J_{11}^2) + \alpha_1\left(\frac{J_{11}^3}{K_1 r}\right) \\ &+ (1 - th_1^2)^2 J_0^2(K_1) - \frac{1}{4}\sum_{n=1}^{\infty} \frac{K_n^2}{K_1^2}\Gamma_n \alpha_1(J_{01}J_{0n}) \\ &+ \frac{1}{8}(1 + 3th_1^2)\sum_{n=1}^{\infty} \frac{K_n th_n}{K_1 th_1}\Gamma_n \alpha_1(J_{01}J_{0n}) \\ &- \frac{1}{8}\sum_{n=1}^{\infty} \frac{K_n^2 th_n}{K_1^2 th_1}\Gamma_n \alpha_1(J_{11}J_{1n}), \end{aligned} \tag{4}$$

in which the quantity Γ_n is a function of H defined as

$$\Gamma_n = \frac{(3th_1^2 - 1)\alpha_n(J_{01}^2) + 2\alpha_n(J_{11}^2)}{1 - (K_n th_n / 4K_1 th_1)}. \tag{5}$$

As shown in Fig. 2, G_c is positive for small depths and negative for large depths. (Since G_c in the range $1 < H \lesssim \infty$ differs from its value at $H = 1$ by less than ¼%, only the range $0 \lesssim H \lesssim 1$ is shown in Fig. 2.) In fact, it was suggested that the depth at which G_c vanishes might be selected as the division between moderately shallow and moderately deep water.

Energy Equations

In a progressive wave of infinitesimal amplitude the total energy of the motion is one-half kinetic and one-half potential. Similarly, for an axisymmetric standing wave, the mean kinetic energy equals the mean potential energy to the first approximation. As will be shown, however, this is not the case when the amplitude is finite.

Fig. 2. Frequency-Correction Factor vs Depth.

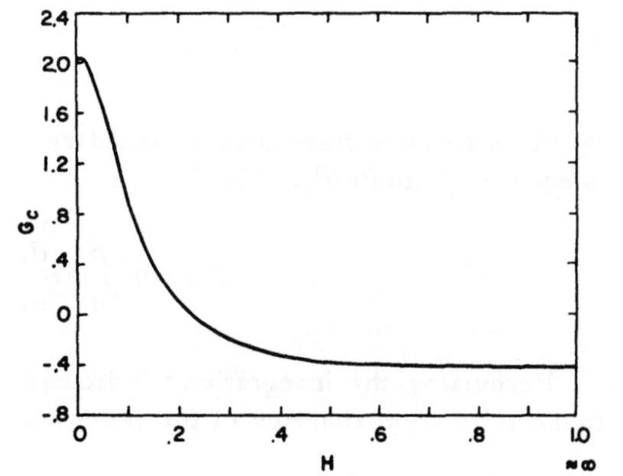

The potential energy V of the wave motion is

$$V = \frac{1}{2} G \iint_A \eta^2 \, dA \, ,$$

in which dA is an element of area in a plane normal to the z axis. Setting $dA = 2\pi r \, dr$ yields

$$V = \pi G \int_0^1 r\eta^2 \, dr \, . \tag{6}$$

The kinetic energy T may be expressed as

$$T = \frac{1}{2} \iiint_\tau \left[\left(-\frac{\partial \phi}{\partial r} \right)^2 + \left(-\frac{\partial \phi}{\partial z} \right)^2 \right] d\tau = -\frac{1}{2} \iint_S \phi \frac{\partial \phi}{\partial n} dS \, ,$$

where the latter form is obtained from the former by means of the Gauss theorem, $d\tau$ is an element of volume of the fluid, dS is an element of area in the boundaries of the fluid region, and n is distance measured into the fluid region normal to dS. Since $\partial\phi/\partial n$ is zero on the solid boundaries, only the integral over the free surface will contribute to T. Hence

$$T = -\pi \int_C r\phi \frac{\partial \phi}{\partial n} ds \, ,$$

where C is the curve formed by the intersection of the free surface with a plane containing the z axis, and ds is an increment of length measured along C away from the axis. Moreover,

$$\frac{\partial \phi}{\partial n} ds = \left(\frac{\partial \phi}{\partial r} \frac{\partial r}{\partial n} + \frac{\partial \phi}{\partial z} \frac{\partial z}{\partial n} \right) ds = \left[\frac{\partial \phi}{\partial r} \frac{\partial \eta}{\partial r} - \frac{\partial \phi}{\partial z} \right] dr$$

on C, since

$$\frac{\partial z}{\partial n}\,ds = -\,dr\,,\quad \frac{\partial r}{\partial n}\,ds = dz = \frac{\partial \eta}{\partial r}\,dr\,.$$

By the kinematic free-surface boundary condition, the quantity within brackets is equal on C to $\partial\eta/\partial t$. Finally,

$$T = -\pi \int_0^1 r\frac{\partial \eta}{\partial t}\,\phi\Big|_{z=\eta}\,dr\,. \tag{7}$$

Performing the integrations indicated in Eqs. (6) and (7) upon the solution to the third approximation of the previous work [1] gives

$$\frac{1}{\pi}V = \frac{A_{11}^2 K_1 th_1 J_0^2(K_1)}{4}\Bigg[(1+\cos 2t) + \frac{A_{11}K_1^2}{4}C_1(\cos t - \cos 3t) + \frac{(A_{11}K_1^2)^2}{16}(C_2 + C_3\cos 2t + C_4\cos 4t)\Bigg]\,, \tag{8}$$

$$\frac{1}{\pi}T = \frac{A_{11}^2 K_1 th_1 J_0^2(K_1)}{4}\Bigg\{[1-\cos 2t] + \frac{A_{11}K_1^2}{4}C_1[-\cos t + \cos 3t] + \frac{(A_{11}K_1^2)^2}{16}[(C_3 + C_4) - C_3\cos 2t - C_4\cos 4t]\Bigg\}, \tag{9}$$

where C_1, C_2, C_3, and C_4 are functions of H given by

$$C_1 = (2th_1^2 - 1)\alpha_1(J_{01}^2)\,, \tag{10}$$

$$C_2 = 2(9th_1^4 + 8th_1^2 - 5)\alpha_1(J_{01}J_{11}^2) + 6\alpha_1\left(\frac{J_{11}^3}{K_1 r}\right) + \frac{(1-9th_1^2)}{2}\sum_{n=1}^{\infty}\frac{th_n}{th_1}\Gamma_n\alpha_1(J_{11}J_{1n}) - 2\sum_{n=1}^{\infty}\frac{K_n}{K_1}\Gamma_n\alpha_1(J_{11}J_{1n}) + 2(1+3th_1^2)(1-th_1^2)J_0^2(K_1) + \frac{1}{4}\sum_{n=1}^{\infty}\frac{K_n th_n}{K_1 th_1}\Gamma_n^2\frac{J_0^2(K_n)}{J_0^2(K_1)}\,, \tag{11}$$

$$C_3 = 2(6th_1^4 + 19th_1^2 - 15)\alpha_1(J_{01}J_{11}^2) + 17\alpha_1\left(\frac{J_{11}^3}{K_1 r}\right) + \frac{(9-13th_1^2)}{4}\sum_{n=1}^{\infty}\frac{th_n}{th_1}\Gamma_n\alpha_1(J_{11}J_{1n}) - 5\sum_{n=1}^{\infty}\frac{K_n}{K_1}\Gamma_n\alpha_1(J_{11}J_{1n}) + 4(1-th_1^4)J_0^2(K_1)\,, \tag{12}$$

$$C_4 = 2(5 - 7th_1^2)\alpha_1(J_{01}J_{11}^2) - 5\alpha_1\left(\frac{J_{11}^3}{K_1 r}\right)$$

$$- \frac{(5 - th_1^2)}{4}\sum_{n=1}^{\infty}\frac{th_n}{th_1}\Gamma_n\alpha_1(J_{11}J_{1n}) + \sum_{n=1}^{\infty}\frac{K_n}{K_1}\Gamma_n\alpha_1(J_{11}J_{1n})$$

$$+ \frac{1}{4}\sum_{n=1}^{\infty}\frac{K_n th_n}{K_1 th_1}\Gamma_n^2\frac{J_0^2(K_n)}{J_0^2(K_1)}. \quad (13)$$

Because the system is conservative and no work is done at the boundaries, the total energy $T + V$ is a constant whose value is given by

$$\frac{1}{\pi}(T + V) = \frac{A_{11}^2 K_1 th_1 J_0^2(K_1)}{4}\left[2 + \frac{(A_{11}K_1^2)^2}{16}(C_2 + C_3 + C_4)\right]. \quad (14)$$

While the total energy of the wave motion is independent of time, both the kinetic and potential portions of that energy obviously are time-dependent. Letting the subscript M denote the temporal mean value over one oscillation, one sees by inspection from Eqs. (8) and (9) that

$$\frac{1}{\pi}V_M = \frac{A_{11}^2 K_1 th_1 J_0^2(K_1)}{4}\left[1 + \frac{C_2(A_{11}K_1^2)^2}{16}\right], \quad (15)$$

$$\frac{1}{\pi}T_M = \frac{A_{11}^2 K_1 th_1 J_0^2(K_1)}{4}\left[1 + \frac{(C_3 + C_4)(A_{11}K_1^2)^2}{16}\right]. \quad (16)$$

It is seen from Eqs. (15) and (16) that $T_M - V_M$ is of order A_{11}^4. This is in qualitative agreement with the results of Rayleigh [3], who first pointed out that the difference between the kinetic and potential energies of an oscillatory progressive wave of permanent form in water of infinite depth is of the fourth order, and of Phillips [4], who obtained several fairly general theorems for a random-wave field.

Partition of Energy

Platzman [5] in his investigation of the partition of energy in periodic progressive waves of finite amplitude and permanent form in water of infinite depth found that the kinetic energy exceeded the potential energy. For the wave of maximum amplitude he found $(T - V)/(T + V) \approx 1/16$, the first term in his expression yielding a value for $(T - V)/(T + V)$ of 4.97%. For the axisymmetric standing wave here considered, Eqs. (14–16) give

$$\frac{T_M - V_M}{T + V} = \frac{(A_{11}K_1^2)^2}{16}\left[\frac{C_3 + C_4 - C_2}{2}\right]$$

for the ratio of the difference between the mean energies to the total energy. Through the use of certain identities derived by Mack [6, Appendix II] involving integrals of products of Bessel functions (i.e., relating the coefficients a_n of the Dini expansions of two or more functions of r), it may be shown that $C_3 + C_4 - C_2 = 2G_c$, and hence that

$$\frac{T_M - V_M}{T + V} = \frac{(A_{11}K_1^2)^2}{16} G_c . \tag{17}$$

[Some of these identities have already been used in simplifying C_1, C_2, C_3, and C_4 as obtained directly by integration to the forms given in Eqs. (10–13).] This result means that, to the third approximation, the finite-amplitude correction to the frequency variable $1/G$ (i.e., to the square of the dimensional frequency) varies with depth in identically the same manner as does the mean energy difference (which, indeed, may be considered to be a finite-amplitude correction to the infinitesimal-amplitude value of zero for the difference in mean energies). The author can offer no physical explanation that would lead one to anticipate this rather startling result.

As may be seen from Fig. 2, the mean kinetic energy of axisymmetric standing waves exceeds the mean potential energy for small depths, whereas for large depths the mean potential energy is the greater. This finding for large depths is in contrast to that of Platzman [5] for progressive waves; however, as Starr [7] noted in his discussion of energy integrals, there appears to be no way of predicting the sign of $T - V$ (or $T_M - V_M$) from general considerations without making use of the detailed solution to the particular wave problem. For axisymmetric waves of approximately the maximum amplitude as determined previously [1], one finds from Eq. (17) that for very deep water V_M exceeds T_M by about 1.70% of the total energy, while for very shallow water T_M exceeds V_M by about 4.60% of the total energy.

For all depths the kinetic energy is identically zero twice each period, but the potential energy is always positive. Consequently, not only is the maximum value of the potential energy during a period greater than the maximum kinetic energy, but the minimum potential energy is also greater than the minimum kinetic energy. Thus it appears reasonable that $(T_M - V_M)/(T + V)$ be negative, as is indeed the case for deep water. That the mean kinetic energy is greater than the mean potential energy for small depths, however, implies a noticeable distortion of the kinetic-energy-to-total-energy ratio from its infinitesimal-amplitude value. To show this distortion, the quantity

$$\frac{T}{T + V} - \left.\frac{T}{T + V}\right|_{A_{11}=0} ,$$

which will be called the relative kinetic energy, has been plotted in Figs. 3 and 4 vs time over a half period for infinitely deep, moderately deep, moderately shallow, and very shallow depths for various amplitudes.

The second-order terms in the energy expressions [e.g., the C_1 terms in Eq. (9)] will cause a distortion of the above quantity which is skew-symmetric about $t = 0.5\pi$, is linearly proportional to A_{11}, and whose sign at a particular

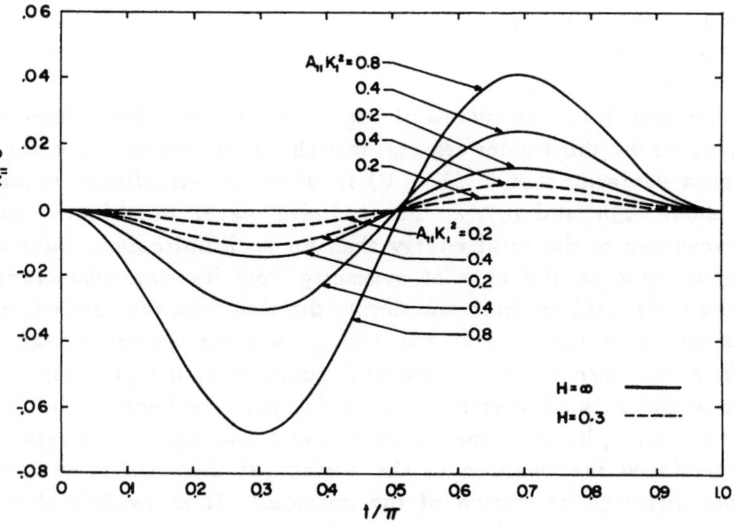

Fig. 3. Finite-Amplitude Kinetic-Energy Distortion in Deep Water.

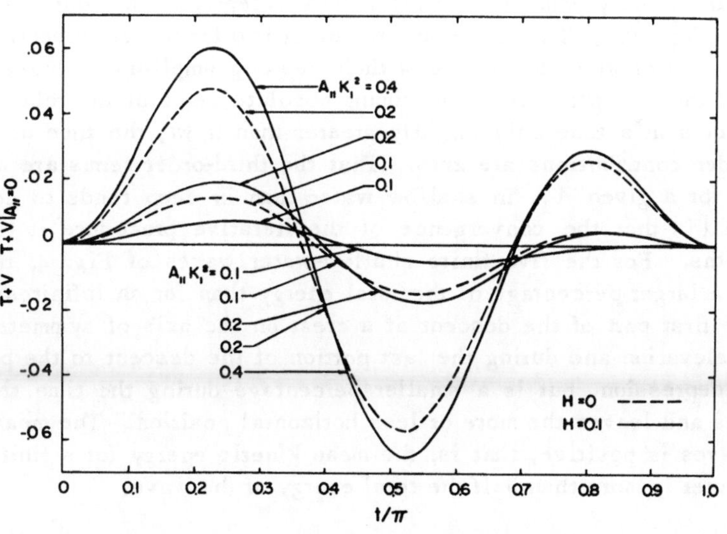

Fig. 4. Finite-Amplitude Kinetic-Energy Distortion in Shallow Water.

time depends on the sign of C_1. It may be seen from the region of low time values of Figs. 3 and 4 that C_1 has a sign for deep water opposite to that for shallow water (the zero of C_1 is at very nearly the same depth as the zero of G_c). The second-order terms do not contribute, however, to the difference in mean energies, whereas the third-order terms [e.g., the C_3 and C_4 terms in Eq. (9)] do. The third-order effects are evidenced in Figs. 3 and 4 by a shifting of the zero intercept away from $t = 0.5\pi$, by an inequality of the absolute values of the relative maxima, and by a tendency toward the introduction of a second zero in the range $0 < t < \pi$.

It may be seen from the curves of Fig. 3 that for infinitely deep and moderately deep water the third-order effects, though clearly present, are not dominant. The maximum deviations of $T/(T + V)$ from its zero-amplitude value occur at times of about 0.3π and 0.7π. For a finite wave, the kinetic energy is a smaller percentage of the total energy than for an infinitesimal wave during the descent of a crest on the axis of symmetry from its maximum elevation to a more or less horizontal position and during the rise from the latter position back to the maximum elevation; the kinetic energy is a greater percentage during the descent from the more or less horizontal position to the position of maximum depression at the axis of symmetry and during the rise back to the former position. (For any case, be the depth large or small, the sign of the relative kinetic energy depends on the position of the surface at the axis of symmetry rather than on the direction of motion of the surface.) It is evident that the mean ordinate for any curve in Fig. 3 is negative and hence that the mean kinetic energy for a finite wave in deep water is less than half the total energy of the wave.

A quite different picture is seen for shallow water in Fig. 4, the third-order effects being clearly dominant for all curves except possibly one, that for $H = 0.1$ and $A_{11}K_1^2 = 0.1$. This curve and the other two for the same depth illustrate the shift to third-order dominance with increasing amplitude. Indeed, for the greatest of these amplitudes the maximum absolute value of the relative kinetic energy occurs at a time only slightly greater than 0.5π, the time at which the second-order contributions are zero. That the third-order terms are much more important for a given A_{11} in shallow water than in deep tends to confirm the prediction [1] that the convergence of the iterative procedure is slower for small depths. For the five finite shallow-water waves of Fig. 4, the kinetic energy is a larger percentage of the total energy than for an infinitesimal wave during the first part of the descent of a crest on the axis of symmetry from its maximum elevation and during the last portion of the descent to the position of maximum depression, but is a smaller percentage during the time the surface approaches and leaves the more or less horizontal position. The mean ordinate for all curves is positive; that is, the mean kinetic energy for a finite wave in shallow water is more than half the total energy of the wave.

CONCLUSIONS

The author's earlier solution [1] for periodic axisymmetric gravity waves has been extended to obtain the kinetic energy, the potential energy, and hence the total energy of the wave motion. As expected, the difference between the mean kinetic and mean potential energies was found to be proportional to the fourth power of the amplitude. It was also shown, however, that the ratio of the difference of mean energies to the total energy varies with depth in precisely the same manner as does the ratio of finite-amplitude correction to the square of the (dimensional) frequency. For deep water the mean potential energy is more than half the total energy; for shallow water the mean kinetic energy is the greater. A qualitative difference between the manner in which the ratio of kinetic energy to total energy varies with time for deep water and for shallow water was noted and illustrated.

REFERENCES

[1] L. R. Mack, "Periodic, Finite-Amplitude, Axisymmetric Gravity Waves," *J. Geophys. Research* **67**, 829–43 (1962).

[2] G. N. Watson, *A Treatise on the Theory of Bessel Functions,* 2d ed., Cambridge University Press, London, 1944.

[3] Lord Rayleigh, "Hydrodynamical Notes – Periodic Waves in Deep Water Advancing Without Change of Type," *Phil. Mag.* **21**, 177–95 (1911).

[4] O. M. Phillips, "On the Dynamics of Unsteady Gravity Waves of Finite Amplitude; Part 2, Local Properties of a Random Wave Field," *J. Fluid Mech.* **11**, 143–55 (1961).

[5] G. W. Platzman, "The Partition of Energy in Periodic Irrotational Waves on the Surface of Deep Water," *J. Marine Research* (Sears Foundation) **6**, 194–202 (1947).

[6] L. R. Mack, *Periodic, Finite-Amplitude, Axisymmetric Gravity Waves,* Ph.D. Dissertation, University of Michigan, 1958.

[7] V. P. Starr, "Momentum and Energy Integrals for Gravity Waves of Finite Height," *J. Marine Research* (Sears Foundation) **6**, 175–93 (1947).

A STUDY OF THE MEAN MOTION AND TURBULENCE DOWNSTREAM OF A ROUGHNESS ELEMENT

Thomas J. Mueller and James M. Robertson
University of Illinois

The results of an experimental investigation of the flow downstream of a particular step-type roughness element in turbulent boundary-layer flow are reported. Three roughness heights were studied on a flat plate in air and one on a flat plate in water. The temporal mean velocity, turbulent fluctuations, and shear-stress variations were measured in the separated and redeveloping flow field for downstream distances as great as 40 times the roughness height. An essentially normal flat-plate turbulent boundary layer was achieved at this distance with a thickness about twice that obtained without the roughness element and about three times the roughness height. Information is presented on the manner in which the mean-flow and turbulence characteristics decay, and approach normal values, in the redeveloping region from the disturbed values produced in the separated flow.

INTRODUCTION

The flow past a single roughness element is of interest because of its application as a tripping and artificial thickening device for turbulent boundary layers. Intrinsically it forms a simple geometry in which to study separated or wake-flow regions. Thus, numerous investigators have studied incompressible turbulent boundary layers in the following situations: (a) behind rods [*1*], wires, and simple back steps [*2, 3*], (b) downstream of a jet issuing parallel, inclined, or normal to a flat plate [*4, 5*], and (c) behind the leading edge separation of blunt bodies [*6*]. These provide a considerable amount of experimental data on the flow characteristics up to the point of reattachment; however, there is a definite lack of information in the redeveloping flow following reattachment. The present study is more concerned with this latter region.

The work summarized herein has been concerned with the flow past a rounded wedge-shaped element with a blunt base mounted on a flat plate. The general features of the element and the nominally two-dimensional flow pattern are indicated in Fig. 1. The element, of height k, is placed some distance downstream from the plate leading edge. This distance is characterized by δ_0, the thickness of the turbulent boundary layer which would occur at the element location if the

NOTATION

C_f	Skin friction coefficient
C_p	Static pressure coefficient
d_1	Portion of the shear work transformed directly into heat
H	Shape factor defined as δ^*/θ
$\overline{H}$	Shape factor defined as δ_e/θ
k	Maximum height of roughness element
P	Static pressure
P_∞	Free-stream static pressure
R	Radius of curvature of roughness element
R_k	Element Reynolds number defined as $U_\infty k/\nu$
t_1	Portion of the shear work transformed to turbulence
U	Temporal mean velocity parallel to the plate
U_∞	Free-stream velocity
u'	The rms value of the turbulent velocity fluctuation in the x direction
v'	The rms value of the turbulent velocity fluctuation in the y direction
$\overline{uv}$	Average of the product of the turbulent fluctuations in velocity in the x and y directions
x	Distance along plate measured from the element
y	Distance measured perpendicular to the plate
δ	Boundary-layer thickness
δ_0	Boundary-layer thickness which would occur at the element location if the element were not present
δ^*	Boundary-layer displacement thickness
δ_e	Boundary-layer energy thickness
θ	Boundary-layer momentum thickness
ν	Fluid kinematic viscosity
ρ	Fluid density
τ	Shear stress

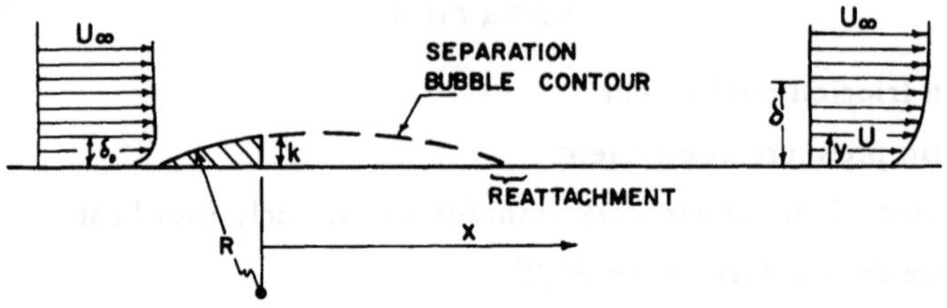

Fig. 1. General Features of Roughness Element Flow.

element were not present. The wake is closed by reattachment to form a separation bubble. Following reattachment, the boundary layer redevelops into a normal fully developed layer some distance downstream. Attention is focused on the location of reattachment and the change in the mean flow and turbulence characteristics in the redeveloping flow.

This study forms the basis for an integral analysis of the redeveloping boundary layer [7]. Such an analysis could be approached by either of two methods, that of Ross [8], which has been found suitable for boundary layers in adverse-pressure gradients and for redeveloping flow, or that of Truckenbrodt [9], for favorable and adverse-pressure-gradient boundary-layer flows. Application of the latter method to the results of the present study is indicated in a forthcoming paper by Mueller, Korst, and Chow [10]. As a verification and extension of the Truckenbrodt method to redeveloping flow, the present paper includes comparisons of the experimental results with some of his representations.

EXPERIMENTAL APPARATUS AND PROCEDURE

The experimental phase of this study was initiated on a flat plate in an open water channel as an attempt to artificially thicken the boundary layer. It soon became apparent that the mechanisms of separation, reattachment, and redevelopment of turbulent boundary layers behind single roughness elements needed to be studied in greater detail. Experiments in air were employed in order to obtain details on the fluctuating velocities and turbulent shear stress as well as the (temporal) mean flow.

The roughness elements used were of the type shown in Fig. 1, except for a preliminary study of a circular cylinder in water. All the other elements were cut from larger cylinders so as to have a gently curved surface facing the flow and a blunt base. With one exception, the curved surface had a radius *R* several times the step height *k*, as is apparent in Table 1. This table also includes values of the Reynolds number based on the element thickness; all of these values are larger than that (R_k = 826) necessary for transition to occur at the element [11]. Thus, in all the cases studied, the shear layer was turbulent downstream of the element.

Table 1. Experimental Flow Conditions and Geometry of Wedge-Type Roughnesses

k (in.)	R (in.)	R/k	δ_0 (in.)	U_∞ (fps)	Fluid	$R_k = \dfrac{kU_\infty}{\nu}$
0.2	0.2	1	0.08	2	Water	3.25×10^3
0.2	1.0	5	0.08	2	Water	3.25×10^3
0.25	2.0	8	0.25	97	Air	1.22×10^4
0.50	2.0	4	0.25	97	Air	2.44×10^4
0.75	2.0	2.7	0.25	97	Air	3.61×10^4

The water channel of 6-in. width was constructed on a water table, and the experiments were conducted with a water depth of 2 in. and a speed of about 2 fps. A flat plate, with a sharp nose, was placed on an edge near one side of the channel. The roughness elements were mounted on the opposite side of the plate. The mean-velocity profiles in the boundary layer following reattachment were obtained with a hypodermic pitot tube of 0.049-in. diameter. The impact pressures were referenced to wall piezometers and the differences obtained with an inverted U, water-oil (specific gravity of about 0.8) manometer. To avoid difficulties with low Reynolds-number effects on the pitot tube, the studies were limited to free-stream velocities of about 2 fps. Near the wall, corrections were made for velocity-gradient effects on the pitot tube. The two gently rounded roughness elements and the conditions under which they were studied are described in Table 1.

The roughness-element studies in air were carried out on a 20- by 10-in. flat plate fitted with side panels. This was placed in the 14-in. octagonal open jet of a low-speed air tunnel. The jet was preceded by a 20-in.-long contraction from a 30-in.-square section. A 60-mesh screen ahead of the contraction helped control the turbulence level so that it had a value between 0.3 and 0.8% at velocities above 70 fps. The advantage of the open test section was that the flow was effectively unlimited normal to the flat plate, thus providing an essentially constant-velocity flow outside the boundary layer. As indicated in Table 1, three roughness-element heights were studied in the air tunnel at a reference velocity of 97 fps.

The impact and static pressures, the longitudinal and transverse components of the fluctuating velocities, and the turbulent shear stress were measured in variation normal to the plate at discrete stations downstream of the roughness element. The pitot tubes were 0.028 and 0.049 in. in diameter, while the static pressures were found with a "screw-driver" type probe. The probe was made from 0.134-in. tubing with a 12° wedge tip, with 0.015-in.-diam holes through the side of the wedge. This type of probe has been shown [*12*] to be insensitive to angle of inclination with the flow for angles up to 10°, so that it did not have to be oriented with the flow in the separation bubble region. The mean-velocity profiles were obtained from the difference between the impact and static pressures.

The hot-wire anemometer used to measure the fluctuating velocities and the turbulent shear stress was of the constant-temperature type, developed at the State University of Iowa [13] and manufactured by the Hubbard-Ling Company. The hot wire was calibrated in the flow outside the boundary layer via reference to an adjacent pitot tube. A single wire was employed, so mounted that at any location it could be oriented at angles of 45, 90, and 135° with the flow direction. From the rms current readings in these orientations, it was possible to compute the rms values of the turbulence velocity fluctuations u', v' and the turbulent shear stress, $\rho\overline{uv}$. As a check on the turbulence measurements, results were obtained on the flat plate without the roughness element and compared with those of Klebanoff [14]. The agreement was good except that the intensities were slightly higher. This may be a result of the higher level of turbulence in the tunnel in the present study. The turbulence measurements with the roughness element in place included a calibration of the wire before, and a check after, each traverse. Whenever possible, all turbulence data for a given roughness were taken during a consecutive 8-hr period.

MEAN-FLOW RESULTS

Preliminary studies were conducted in water with an element of circular cross section 0.095 in. in diameter in the boundary layer of 0.080-in. thickness. One velocity profile found behind this element is included in Fig. 2. Since it was felt that with this type of roughness the point of separation could vary with the Reynolds number, the step-type element indicated in Fig. 1 was adopted.

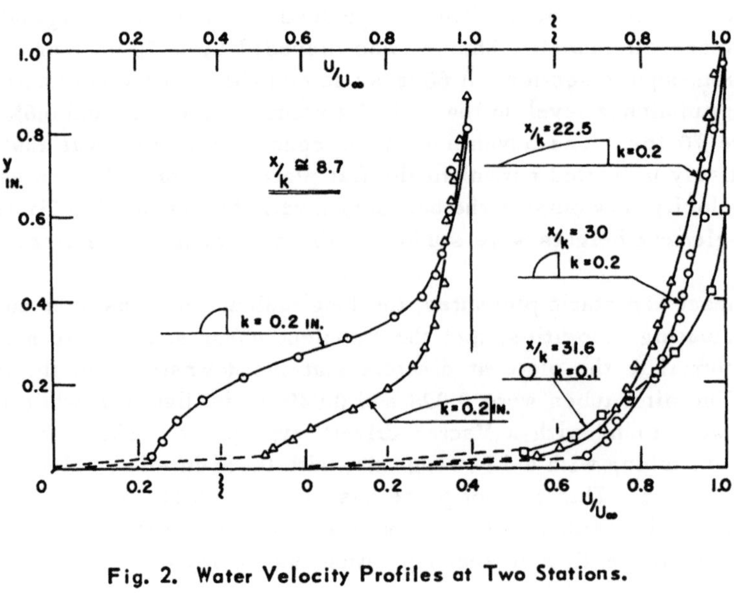

Fig. 2. Water Velocity Profiles at Two Stations.

This was first studied as a quarter round [$R = k$] and then with more gentle curvature, as indicated in Table 1. The first item of interest in these studies was the point of reattachment. For both wedge-type roughness elements in water, the point of reattachment, obtained from dye studies, was found to occur 5.62 roughness step heights downstream. This value corresponds to a mixing angle of 10.1° and agrees well with the results obtained with a simple back step [*15*]. For the air studies, reattachment was determined from the results of the wall shear stress determinations (see Fig. 3*a*) as the location where the zero value occurred. Velocity traverse taken at this location (see Fig. 4) served to verify this definition, although there was considerable fluctuation in the actual point. From the air studies, reattachment was thus found at downstream distances of 7.00, 6.80, and 6.87 times the step height. These values are probably more reliable than those found in water.

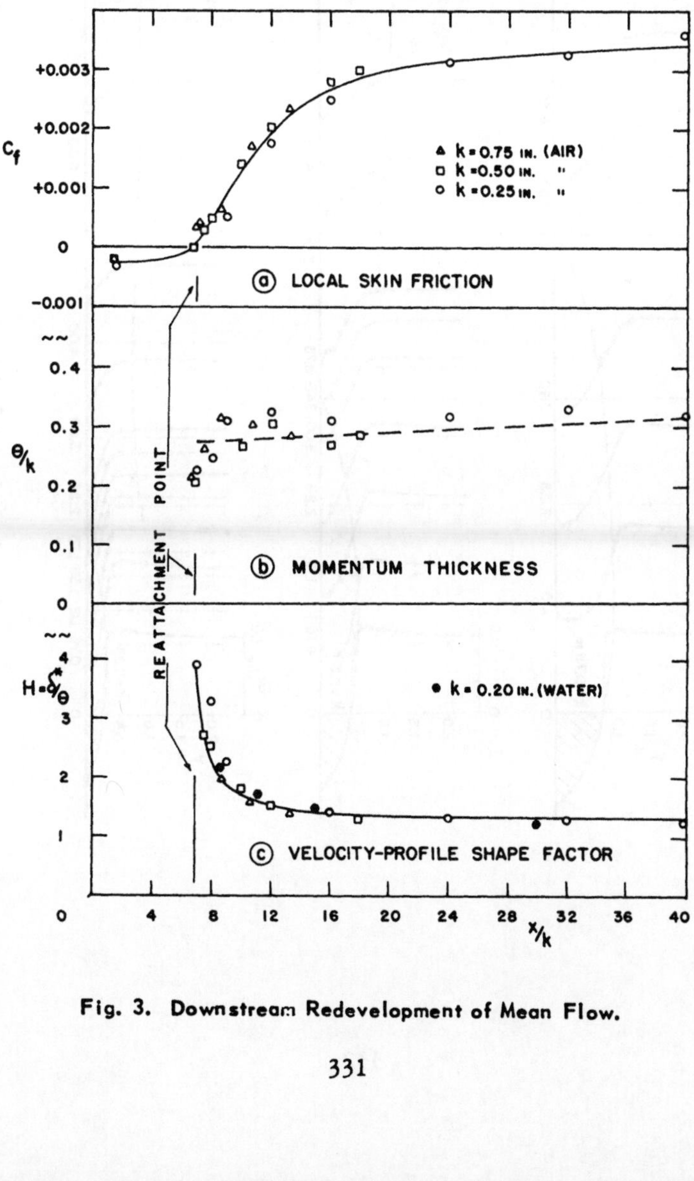

Fig. 3. Downstream Redevelopment of Mean Flow.

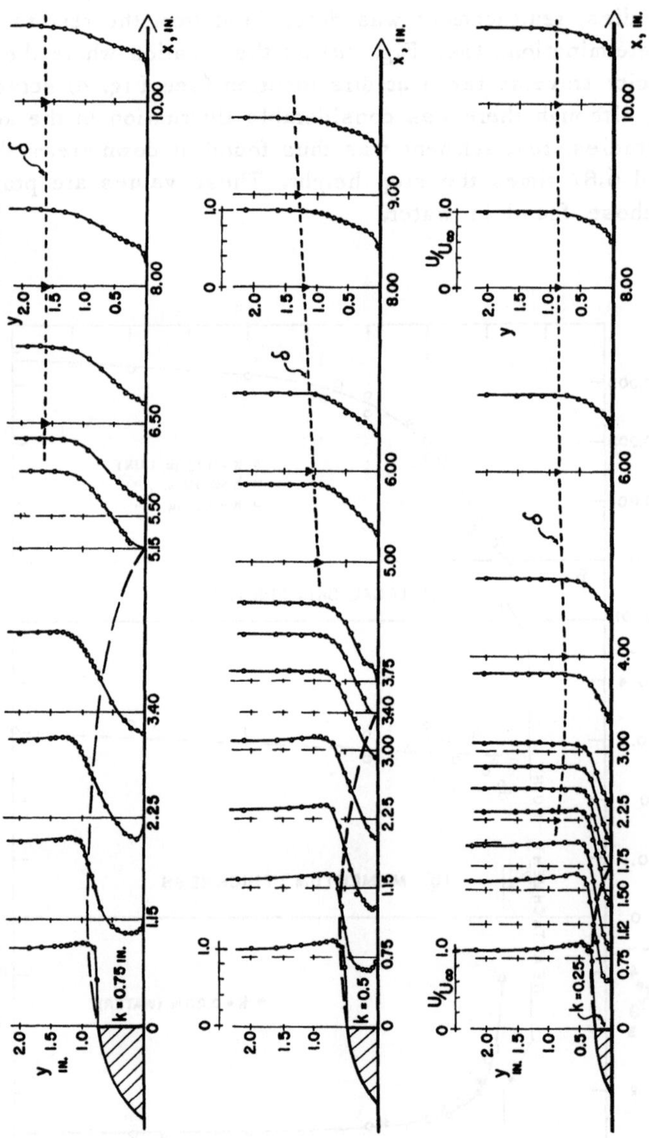

Fig. 4. Velocity Profiles in Air Flow Past Roughness Elements.

A more complete indication of the occurrences downstream from the step is obtained from the mean-velocity measurements. The family of velocity profiles found in the air studies behind the three roughness elements are indicated in Fig. 4. The greatest detail for the flow in the wake bubble and near reattachment is apparent for the largest roughness, while redevelopment of the flow to a normal boundary layer seems to have been achieved with only the smallest element. The associated variation in the pressures along the plate for these three experiments is indicated in Fig. 5. Actually, except for the initial point, these form a single curve when plotted vs the dimensionless distance x/k from the element. The pressure coefficient is approximately zero at reattachment.

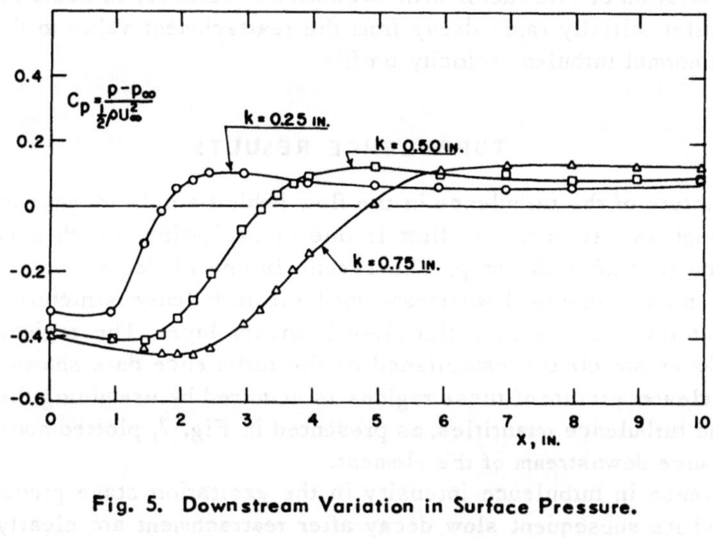

Fig. 5. Downstream Variation in Surface Pressure.

Typical velocity profiles from the water tests at two distances from the roughness element are shown in Fig. 2. The quarter-round element is seen to produce an initially larger disturbance than the gently curved element, but farther downstream they appear to produce about the same effect. Inspection of the profile behind the circular element at the same x/k of about 31 indicates that it produces a thinner but more disturbed boundary layer.

Local skin-friction coefficients C_f were obtained from the mean-velocity profiles near the wall by fitting them to the law of the wall expressed as

$$\frac{U}{U_\infty} = 5.6\sqrt{\frac{C_f}{2}}\left[1 + \log\left(\frac{yU_\infty}{\nu}\right) + \log\sqrt{\frac{C_f}{2}}\right]. \qquad (1)$$

The results of these determinations for the studies in air are presented in Fig. 3*a* as a function of the dimensionless distance downstream from the roughness element. Downstream of reattachment the skin friction is seen to increase rapidly toward the normal flat-plate value, essentially in reverse fashion to that in a boundary layer approaching separation. The variation of the skin-friction coefficient with the shape of the boundary-layer profile is similar [7, *10*] to that found by Truckenbrodt [9] for flows approaching separation.

Included in Fig. 4 is an indication of the boundary-layer thickness downstream from reattachment. This is seen to increase slowly with distance and to vary considerably with the height of the roughness element. Since the boundary-layer thickness – taken as the lateral distance to the point where the velocity has reached 99% of the stream velocity – is difficult to define, the momentum-loss thickness θ is more meaningful. Figure 3*b* indicates that following a sudden increase right after reattachment, θ increases slowly in a fashion similar to that for a normal-velocity profile. The form of a velocity profile is commonly characterized by the shape factor H, the ratio of the displacement to momentum thickness. The variation of this factor with downstream distance, indicated in Fig. 3*c*, shows a similar initially rapid decay from the reattachment value to that characteristic of a normal turbulent velocity profile.

TURBULENCE RESULTS

The structure of the turbulence in the flow behind single roughness elements passes through two stages. The first is one of excitation, which occurs in the mixing region just after the step. The second is one of decay, beginning at reattachment and continuing downstream until the turbulence structure reaches a state characteristic of a normal flat-plate boundary layer. The regions of excitation and decay are clearly established by the turbulence data shown in Fig. 6. However, a clearer picture of these regions is acquired by examining the maximum values of the turbulence quantities, as presented in Fig. 7, plotted nondimensionally vs distance downstream of the element.

The increase in turbulence intensity in the excitation stage preceding reattachment and its subsequent slow decay after reattachment are clearly apparent in the dimensionless plot of the maximum value of u' vs x presented in Fig. 7*a*. The data for all three roughness heights are seen to be in essential agreement, thus implying independence of upstream effects (i.e., δ_0/k) and the exact contour of the roughness element (i.e., R/k). The turbulence measurements obtained by Hsu [3] for a back step are also indicated and are seen to agree reasonably well with those of the present study. At large distances downstream, the maximum intensity is seen to approach an essentially constant value, which should be typical of a normal turbulent boundary layer. Comparison with the flat-plate results of Klebanoff [*14*] indicates that the asymptotic value is somewhat low. The maximum v' data yielded somewhat the same type of results, except that the actual values were several times larger. This result is questionable, and there is no apparent explanation for it.

The maximum turbulent shear-stress results are presented in dimensionless form as a function of distance in Fig. 7*b*. The shear is seen to have an initially high value in the excitation region and then to decay toward a value characteristic of the normal boundary layer, as indicated by Klebanoff's result. Two distinct curves of maximum shear appear in the excitation region before reattachment. The curve for the 0.75-in. element is characteristic of a fully developed

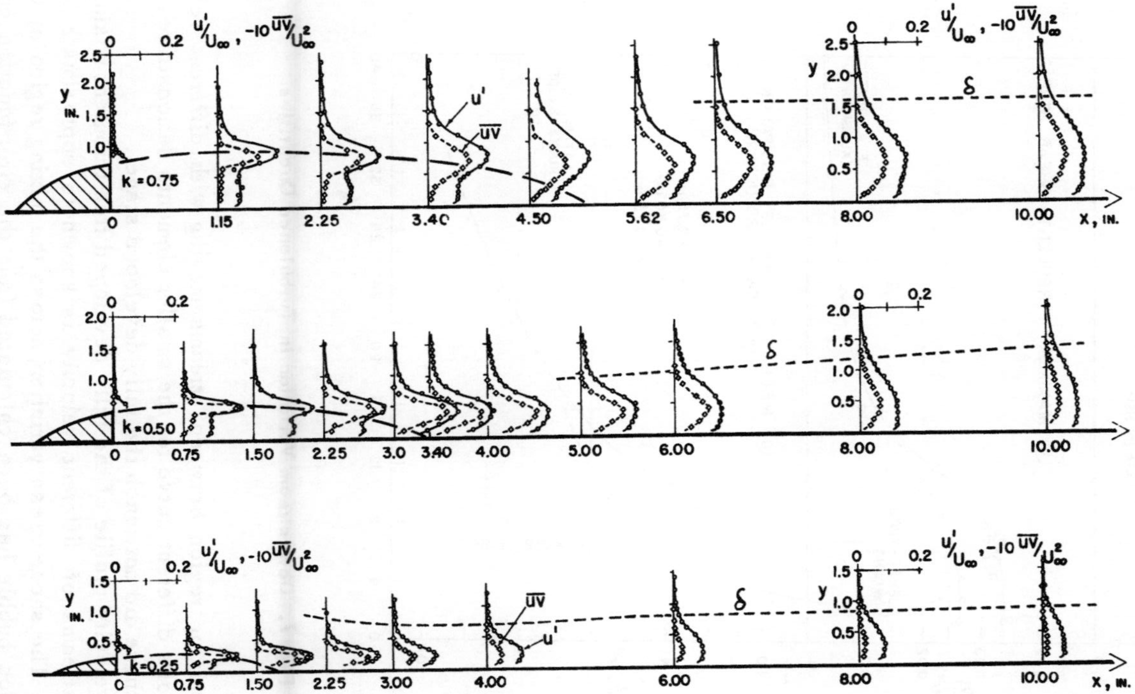

Fig. 6. Turbulence Profiles Behind Roughness Elements.

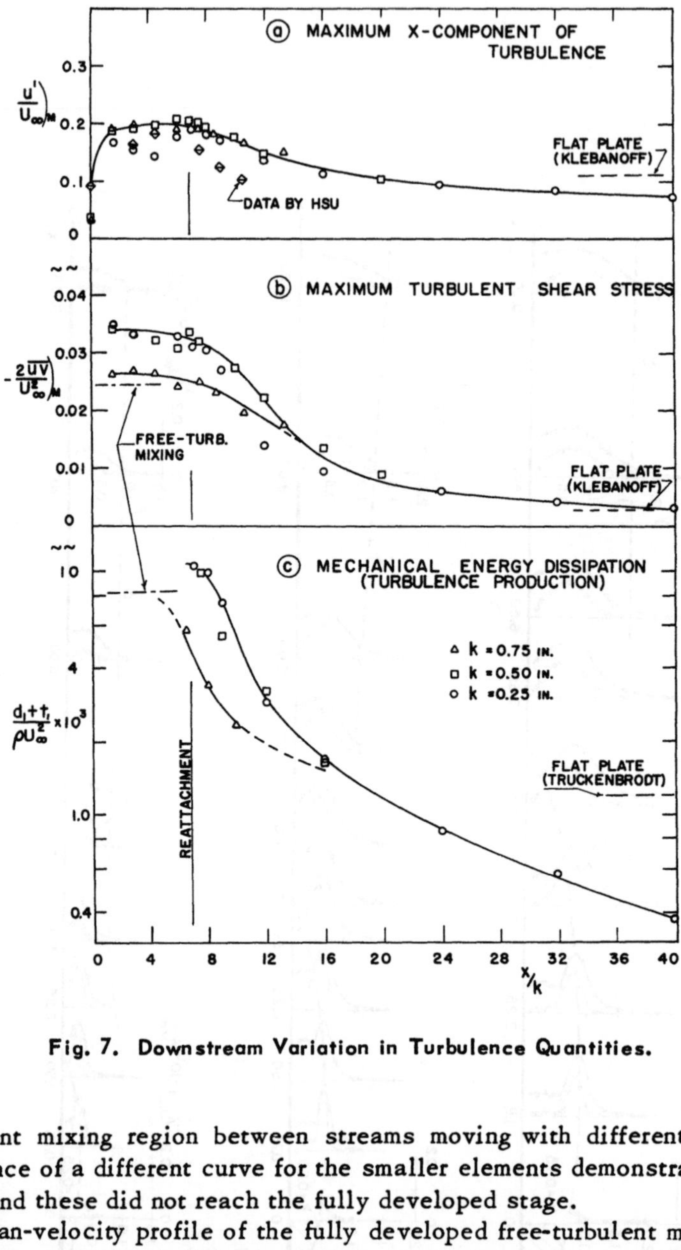

Fig. 7. Downstream Variation in Turbulence Quantities.

free-turbulent mixing region between streams moving with different velocities. The existence of a different curve for the smaller elements demonstrates that the mixing behind these did not reach the fully developed stage.

The mean-velocity profile of the fully developed free-turbulent mixing region between two streams of different velocity is known to approximate an error-function curve. The shear-stress variation across the mixing region at the edge of the separation bubble has been calculated from the error-function curve [7]. The peak value from this calculation is indicated in Fig. 7*b* and is seen to be in good agreement with the measurements for the largest step. The measured and calculated shear-stress profiles behind the largest element are compared in Fig. 8, in verification of the assumption that this was a fully developed free-turbulent mixing region. The width of this mixing region was found to increase linearly with distance from the roughness element.

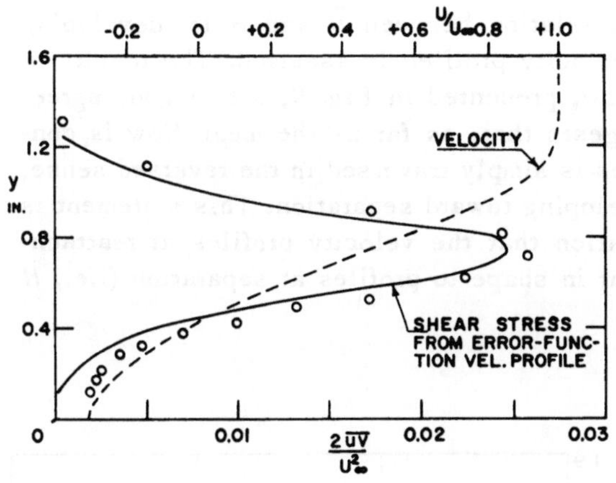

Fig. 8. Turbulent Shear-Stress Profile at $x/k = 4.5$ Behind 0.75-in. Element.

An important quantity associated with the energy processes in turbulent shear flows is the rate at which energy is being extracted from the mean flow through shearing action. This is calculated as the product of the shear stress times the lateral gradient of the mean velocity. When integrated across the flow it represents the rate of energy loss from the boundary-layer mean flow. This integral was evaluated from the expression

$$\frac{d_1 + t_1}{\rho U_\infty^3} = \int_0^\delta \frac{\tau}{\rho U_\infty^2} \frac{\partial}{\partial y} \left(\frac{U}{U_\infty} \right) dy \,, \tag{2}$$

where, following Truckenbrodt [9], d_1 represents the energy which goes directly into heat, and t_1 is the production of turbulent energy. The shear stress τ involves both turbulent and viscous contributions. In the type of flow studied, the viscous contribution is very small, and hence in the present calculations only the turbulent shear was used. This was taken from the $\overline{uv}$ measurements. The decrease in $d_1 + t_1$ in the redeveloping region downstream of reattachment is indicated in Fig. 7*c*. The value at reattachment for the 0.75-in. roughness element seems to be close to the value calculated from the error-function profile for the fully developed free-turbulent mixing region.

DISCUSSION OF RESULTS

The reverse similarity between redeveloping flows following reattachment and developing flows approaching separation in adverse pressure gradients has been noted in connection with the skin-friction data of Fig. 3*a*. It is also apparent in the shape-factor H results of Fig. 3*c*. A further verification of this similarity is available in terms of another shape factor involving the boundary-layer energy thickness δ_e,

$$\overline{H} = \frac{\delta_e}{\theta} \,. \tag{3}$$

Truckenbrodt [9] indicated a unique relation between $\bar{H}$ and H for developing flows if a one-parameter family of velocity profiles is assumed. The results of the present study of redeveloping flow, presented in Fig. 9, are in good agreement with his correlation. This suggests that, as far as the mean flow is concerned, the sequence of profile shapes is simply traversed in the reversed sense, as compared to a boundary layer developing toward separation. This statement is further supplemented by the observation that the velocity profiles at reattachment, as shown in Fig. 4, are similar in shape to profiles at separation (i.e., H is the order of 3).

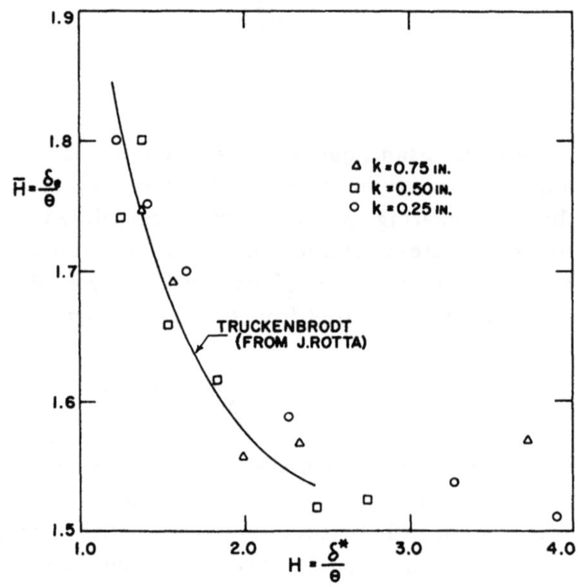

Fig. 9. Relation Between Boundary-Layer Shape Factors Compared with Result for Flows Approaching Separation.

The use of single roughness elements of the type considered here for thickening boundary layers on flat plates has been indicated. A fully developed turbulent boundary layer – with mean velocity free of any distortions introduced by the thickening process (as apparent in Figs. 4 and 3*c*) – was obtained at 40 step heights downstream of the roughness element. Except for the v' results, the turbulence data verify the occurrence of an essentially normal boundary layer at this distance. This layer was more than twice as thick as the boundary layer which would be obtained over the same length of plate without the element. By considering the momentum defect introduced by the 0.25-in. roughness element, it was found that a distance of about 75 times the height would be necessary with a flat plate to achieve the same boundary layer as was found with the element at 40 times the height.

Downstream of reattachment, the factors H, C_f, C_p, and θ/k, characterizing and specifying the mean flow, are independent of the specific roughness element height in terms of their relative downstream distance x/k. The turbulence parameters also suggest this independence. Thus it would appear that, in these experiments, the final boundary layer was independent of the upstream flow as specified by δ_0/k. Forty heights downstream of the element a normal boundary layer occurred which was some three times as thick as the element height, that is, of momentum thickness approximately one-third this height. The range of variables covered in these experiments limits this conclusion to undisturbed boundary-layer thicknesses (δ_0) equal to or less than the step height.

CONCLUSIONS

The following conclusions may be drawn from this investigation of the character of the mean motion and turbulence in the separation, reattachment, and redevelopment of the incompressible, turbulent, shear flow downstream of a single, curved, step-type, roughness element:

In the semi-infinite free-turbulent mixing region, which bounds the separation bubble, fully developed mixing was obtained with an element height as small as 0.75 in,, corresponding to a height Reynolds number of 3.6×10^4. The maximum shear stress in fully developed free-turbulent mixing was found to be about ten times that found in the normal boundary layer 40 heights downstream. The dimensionless rate of dissipation of mechanical energy $(d_1 + t_1)/\rho U_\infty^3$ was 8.3×10^{-3}, several times the value characteristic of the normal turbulent boundary layer.

The position of reattachment occurs at about 6.8 element heights downstream. Variations in the location of reattachment occur due to unsteadiness of the separation and wake flows.

In the redeveloping boundary layer following reattachment, the mean-velocity profiles appear to constitute a one-parameter family, as evidenced by the shape factors. The sequence of profile shapes traversed downstream of reattachment is much the same as that traversed in reverse sense in a boundary-layer flow developing toward separation.

At 40 roughness-element heights downstream, a turbulent boundary layer – with mean-velocity profile free of any distortions introduced by roughness – is obtained with momentum thickness about one-third the height of the element. Except for the v' fluctuations, which are somewhat in doubt, the turbulence characteristics also appear to have returned to normal ones at this distance. As long as the undisturbed boundary-layer thickness at the position of the roughness element is less than or equal to the element height, this result is independent of the upstream boundary layer.

ACKNOWLEDGEMENTS

The technical and financial assistance of the Caterpillar Tractor Company, Peoria, Illinois, is hereby acknowledged. Professor H. H. Korst contributed significantly to this study. The experiments were conducted on a water table of the Department of Mechanical and Industrial Engineering and in an air tunnel of the Department of Theoretical and Applied Mechanics.

REFERENCES

[*1*] P. S. Klebanoff and Z. W. Diehl, "Some Features of Artificially Thickened Fully Developed Turbulent Boundary Layers with Zero Pressure Gradient," *NACA* Rept. 1110 (1952).

[*2*] I. Tani, *Experimental Investigation on Flow Separation over a Step,* paper presented at Symposium on Boundary Layer Research, Freiburg, Germany, August 26–29, 1957.

[*3*] H. C. Hsu, *Characteristics of Mean Flow and Turbulence at an Abrupt Two-Dimensional Expansion,* Ph.D. Dissertation, State University of Iowa, 1950.

[*4*] C. Bourque and B. G. Newman, "Reattachment of a Two-Dimensional Incompressible Jet to an Adjacent Flat Plate," *Aeronaut. Quart.*, **XI** (3), 201 (1960).

[*5*] M. Arie and H. Rouse, "Experiments on Two Dimensional Flow Over a Normal Wall," *J. Fluid Mech.* **1**(2), 129–41 (1956).

[*6*] G. B. McCullough and D. E. Gault, *Boundary-Layer and Stalling Characteristics of the NACA 64 A OOL Airfoil Section, NACA* TN 1923 (August 1949).

[*7*] T. J. Mueller, *On Separation, Reattachment and Redevelopment of Turbulent Boundary Layers,* Ph.D. Dissertation, University of Illinois, 1961.

[*8*] D. Ross, "A Physical Approach to Turbulent-Boundary-Layer Problems," *Trans. Am. Soc. Civil Engrs.* **121**, 1219–54 (1956).

[*9*] E. Truckenbrodt, "A Method of Quadrature for Calculation of the Laminar and Turbulent Boundary Layer in Case of Plane and Rotationally Symmetrical Flow," *Ingr.-Arch.* **20**, 1952; *NACA* TM 1397 (May 1955).

[*10*] T. J. Mueller, H. H. Korst, and W. L. Chow, "On the Separation, Reattachment, and Redevelopment of Incompressible Turbulent Shear Flow," to be published in the *Trans. ASME.*

[*11*] J. C. Gibbings, *On Boundary-Layer Transitions Wires,* Brit. Min. of Supply, ARC 20593 (December 1958).

[*12*] H. Girerd and P. Guienne, "New Static Pressure Probes for Aerodynamic Measurements," *Compt. rend. acad. sci.* **228**, 651–53 (1949).

[*13*] P. G. Hubbard, *Operating Manual for the IIHR Hot Wire and Hot-Film Anemometers,* State University of Iowa Studies in Engineering, Bulletin 37, 1957.

[*14*] P. S. Klebanoff, *Characteristics of Turbulence in a Boundary Layer with Zero Pressure Gradient, NACA* Rept. 1247 (1955).

[*15*] R. A. Seban, A. Emery, and A. Levy, "Heat Transfer to Separated and Reattached Subsonic Turbulent Flows Obtained Downstream of a Surface Step," *J. Aero Space Sci.* **26**, (December 1959).

ANALYSIS OF VERY THICK LAMINAR BOUNDARY LAYERS IN AXISYMMETRIC HIGH-SPEED FLUID FLOW[1]

J. Leith Potter and J. A. Durand

Research Branch, von Kármán Gas Dynamics Facility, ARO, Inc.

This paper presents a new approximate method for calculation of laminar boundary-layer thicknesses in axisymmetric, high-speed flows with heat transfer and variable fluid properties. The particular point considered is the situation when boundary-layer thickness is approaching the duct radius so that neither thin nor fully developed boundary layers may be assumed.

The derivation combines an appropriate integral momentum equation with an approximate friction law which is based on local similarity. Continuum fluid-flow postulates are applied in developing the method. Either fluids in thermal and chemical equilibrium or fluids with these reactions frozen at a known state may be considered. Results are compared with experimental data, and generalized comparisons of the influences of several variables are drawn.

INTRODUCTION

Design of ducting for very low-density fluid flows at high Mach numbers may be critically affected by considerations of the thick boundary layer which will grow along the walls. Strong effects on the power requirements and the longitudinal and lateral distributions of velocity, pressure, temperature, and other properties may be expected when dealing with flow systems related to low-density wind tunnels, vacuum chambers, and certain low-thrust space propulsion engines. The low-density, hypervelocity wind-tunnel nozzle presents a good example of the general problem. The boundary-layer thickness may be of the same order as the nozzle radius, fluid properties may be variable, strong pressure gradients almost always are present, heat transfer may be a factor, and, in the face of all this, it is essential that the nozzle produce test-section flow uniform to a few percent in all quantities. This paper is directed particularly to the design of axisymmetric nozzles with thick laminar boundary layers which are typical of contemporary low-density wind tunnels used to study flight 40 miles and more from the earth.

---

[1]Sponsored by the Arnold Engineering Development Center, Air Force Systems Command, under contract No. AF 40(600)-800 S/A 24(61-73) with ARO, Inc.

NOTATION

a	Nozzle radius
C_f	Skin-friction coefficient
C_f'	Skin-friction coefficient based on T' or h'
d	Duct diameter
$G(x)$	Function defined in Eq. (25)
H	Form factor [Eq. (22)]
h	Enthalpy
h'	Reference enthalpy defined in Eq. (32)
$\dot{m}$	Mass-flow rate
p	Static pressure
p_0'	Experimental pitot pressure or impact pressure as measured in wind-tunnel nozzle
R	Gas constant
Re_s	Reynolds number based on length along nozzle
r	Coordinate measured from nozzle axis (Fig. 1)
s	Coordinate along nozzle wall, zero at throat (Fig. 1)
T	Temperature
T'	Reference temperature in boundary layer corresponding to h', the reference enthalpy [Eq. (32)]
u, v	Velocity in boundary layer in s direction and in y direction
$W(x)$	Function defined in Eq. (26)
x	Distance along nozzle axis measured from throat (Fig. 1)
y	Coordinate perpendicular to s direction (Fig. 1)
γ	Ratio of specific heats of fluid medium
δ	Boundary-layer thickness
δ^*	Boundary-layer displacement thickness
δ_1	Defined in Eqs. (11) and (12)
η	y/δ
θ	Boundary-layer momentum thickness
θ_1	Defined in Eqs. (13) and (14)
μ	Absolute viscosity
ρ	Mass density
τ	Shear stress

ω	Angle of inclination of boundary of flow relative to center line (Fig. 1)

Subscripts

0	Isentropic stagnation
1	Local value on edge of expansion core, except δ_1 and θ_1
$2d$	Two-dimensional
aw	Adiabatic wall
ec	Expansion core (Fig. 1)
w	Wall

Superscript

$( )^*$	Indicates sonic throat value, except δ^*

Methods for the design of axisymmetric nozzles for supersonic, low-density flows with low stagnation temperature have been developed [*1–4*] on the basis of definitions of displacement and momentum thicknesses which do not account for transverse curvature. Although this procedure is strictly applicable only when boundary-layer thickness is small compared with nozzle radius, satisfactory nozzles were built [*1–3*]. Another approach to the problem was believed worthwhile because of the higher Mach numbers and real gas effects to be taken into consideration. At the same time, terms related to transverse curvature were retained insofar as necessary.

When the boundary layer becomes very thick, the two-dimensional definitions of the displacement and momentum thicknesses of the boundary layer are no longer appropriate. Seban and Bond [*5*], Kelly [*6*], and Probstein and Elliott [*7*] have considered the characteristics of the thick boundary layer in axisymmetric flow. Kelly has given an appropriate definition of displacement thickness, and a similar definition of the momentum thickness can also be written. When these definitions are used with the other boundary-layer equations, a workable form of the integral momentum expression can be derived, and the boundary-layer thickness can be evaluated when a velocity distribution or skin-friction law is specified.

In high-temperature fluid flow a number of complications may arise. Variation of the ratio of specific heats and heat transfer to nozzle walls is taken into account in the development described. The possibility of thermal and chemical nonequilibrium and the location in the nozzle at which they may be manifest are questions outside the scope of this report. A considerable body of literature on this subject already exists [*8–10*], but the analysis of nonequilibrium nozzle flow is dependent on certain assumptions best dealt with for specific cases. Equilibrium fluids or frozen fluids acting as perfect gases may be easily dealt with in the calculation procedure.

ANALYSIS

Boundary-Layer Momentum Equation

Referring to the coordinate system shown in Fig. 1, it is assumed that a potential flow exists in the expansion core. The potential-flow properties may be obtained by the method of characteristics, the Foelsch method [*11*], or any other suitable procedure which takes into account the thermodynamic state of the flow, that is, perfect, thermally perfect, dissociated equilibrium, etc. This known flow in the expansion core, along with the boundary-layer correction δ^* "added" to the expansion-core radius at an axial station, will produce the desired coordinate of a point on the wall. Thus, the geometric nozzle contour is determined.[2]

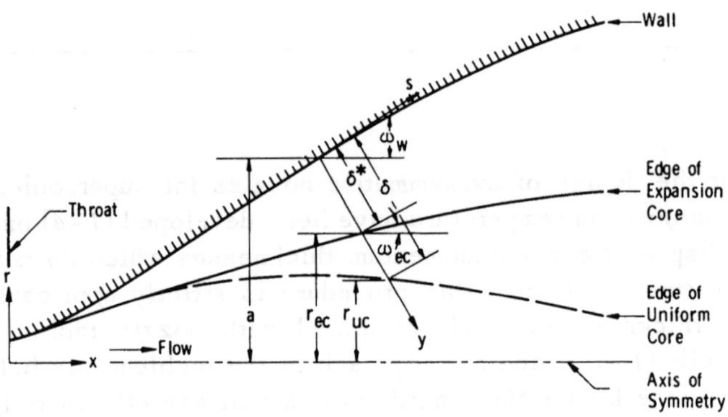

Fig. 1. Coordinate System.

It is assumed that the static pressure is constant across the boundary layer at any station along the nozzle and that similar velocity profiles exist in the boundary layer of the expanding flow.

The continuity and momentum equations for an axially symmetric boundary layer in compressible flow are written by Probstein and Elliott [7] as follows: Equation of continuity,

$$\frac{\partial}{\partial s}(\rho u r) + \frac{\partial}{\partial y}(\rho v r) = 0 \; ; \tag{1}$$

[2]It is realized that waves do not reflect from the boundary layer as though the displacement boundary layer were a solid wall. However, this usual assumption is perpetuated in the present work because of the resulting simplification in procedure. The resulting error may be greater than usual when very thick displacement layers are involved.

Momentum equation,

$$\rho u \frac{\partial u}{\partial s} + \rho v \frac{\partial u}{\partial y} = -\frac{dp}{ds} + \frac{\partial}{\partial y}\left(\mu \frac{\partial u}{\partial y}\right) + \frac{\mu}{r}\frac{\partial r}{\partial y}\frac{\partial u}{\partial y}. \qquad (2)$$

The coordinates and velocity components are shown in Fig. 1.

Multiplying both sides of Eq. (2) by r and integrating through the boundary layer from $y = 0$ to $y = \delta$ gives

$$\int_0^\delta \rho u r \frac{\partial u}{\partial s}\, dy + \int_0^\delta \rho v r \frac{\partial u}{\partial y}\, dy = -\int_0^\delta r \frac{dp}{ds}\, dy + \int_0^\delta r \frac{\partial}{\partial y}\left(\mu \frac{\partial u}{\partial y}\right) dy + \int_0^\delta \mu \frac{\partial u}{\partial y}\frac{\partial r}{\partial y}\, dy. \qquad (3)$$

Expressing the second term on the left side of Eq. (3) in terms of $\rho u r$ by using Eq. (1), it is found that

$$\int_0^\delta \rho u r \frac{\partial u}{\partial s}\, dy - \int_0^\delta \left[\frac{\partial u}{\partial y}\int \frac{\partial}{\partial s}(\rho u r)\, dy\right] dy = -\int_0^\delta r \frac{dp}{ds}\, dy + \int_0^\delta r \frac{\partial}{\partial y}\left(\mu \frac{\partial u}{\partial y}\right) dy + \int_0^\delta \mu \frac{\partial u}{\partial y}\frac{\partial r}{\partial y}\, dy. \qquad (4)$$

If the second term on both the left and right sides of Eq. (4) is integrated by parts, these terms can be combined to reduce Eq. (4) to

$$\frac{\partial}{\partial s}\int_0^\delta \rho r u^2\, dy - u_1 \frac{\partial}{\partial s}\int_0^\delta \rho u r\, dy = -\frac{dp}{ds}\int_0^\delta r\, dy - a\mu\left(\frac{\partial u}{\partial y}\right)_{y=0}, \qquad (5)$$

where a is the local radius of the nozzle.

When the relationship $-dp/ds = \rho_1 u_1(\partial u_1/\partial s)$ at the edge of the boundary layer is used in Eq. (5), one obtains

$$-\int_0^\delta \frac{\partial}{\partial s}(\rho r u^2)\, dy + u_1 \int_0^\delta \frac{\partial}{\partial s}(\rho u r)\, dy + \rho_1 u_1 \frac{\partial u_1}{\partial s}\int_0^\delta r\, dy = a\tau_w, \qquad (6)$$

where $\tau_w = \mu(\partial u/\partial y)_{y=0}$. Differentiation of the first two terms of Eq. (6) gives

$$\int_0^\delta (\rho u_1 u - \rho u^2)\frac{\partial r}{\partial s}\, dy - 2\int_0^\delta \rho r u \frac{\partial u}{\partial s}\, dy + \int_0^\delta r(u_1 u - u^2)\frac{\partial \rho}{\partial s}\, dy + \int_0^\delta \rho u_1 r \frac{\partial u}{\partial s}\, dy + \rho_1 u_1 \frac{\partial u_1}{\partial s}\int_0^\delta r\, dy = a\tau_w. \qquad (7)$$

If the term $\int_0^\delta \rho r u(\partial u_1/\partial s)\, dy$ is added to and subtracted from Eq. (7), the complete differential $(\partial/\partial s)[\rho r(u_1 u - u^2)]$ is realized:

$$\frac{\partial}{\partial s}\int_0^{\delta}\frac{r}{a}\left(1-\frac{u}{u_1}\right)\frac{\rho u}{\rho_1 u_1}\,dy+\frac{1}{u_1}\frac{\partial u_1}{\partial s}\int_0^{\delta}\frac{r}{a}\left(1-\frac{\rho u}{\rho_1 u_1}\right)dy$$

$$+\frac{1}{a\rho_1 u_1^2}\left[\int_0^{\delta}\frac{r\rho u}{a\rho_1 u_1}\left(1-\frac{u}{u_1}\right)dy\right]\frac{\partial}{\partial s}(a\rho_1 u_1^2)=\frac{\tau_w}{\rho_1 u_1^2}\,. \tag{8}$$

Boundary-Layer Thickness Ratios

The integrands in Eq. (8) can be expressed in terms of the boundary-layer thicknesses for axially symmetric, compressible flow. In order to take into account the transverse curvature effect of a thick boundary layer, the mass-flow defect and the momentum-flow defect must be defined in terms of nozzle geometry for axially symmetric flow. These defects are defined as (*a*) mass-flow defect,

$$\int_{y=0}^{y=\delta*} 2\pi r(\rho_1 u_1)\,dy=\int_{y=0}^{y=\delta} 2\pi r(\rho_1 u_1-\rho u)\,dy \tag{9}$$

(a similar definition for the mass-flow defect is given in [6] and [7]), and (*b*) momentum-flow defect,

$$\int_{y=0}^{y=\theta} 2\pi r(\rho_1 u_1^2)\,dy=\int_{y=0}^{y=\delta} 2\pi r\rho(u_1 u-u^2)\,dy\,. \tag{10}$$

At this point it is convenient to make the assumption that $\cos\omega_{ec}\approx\cos\omega_w$ (see Fig. 1). This results in errors on the order of 1% in calculated boundary-layer characteristics of interest in the following development, even when $\omega_w-\omega_{ec}=15°$.

Therefore, with $\omega_{ec}=\omega_w=\omega$, Eqs. (9) and (10) can be written

$$\delta_1=\int_0^{\delta}\frac{r}{a}\left(1-\frac{\rho u}{\rho_1 u_1}\right)dy\,, \tag{11}$$

where

$$\delta_1=\delta*-\frac{\delta*^2\cos\omega}{2a}\,, \tag{12}$$

$$\theta_1=\int_0^{\delta}\frac{r}{a}\left(1-\frac{u}{u_1}\right)\frac{\rho u}{\rho_1 u_1}\,dy\,, \tag{13}$$

$$\theta_1=\theta-\frac{\theta^2\cos\omega}{2a}\,. \tag{14}$$

When Eqs. (11) and (13) are substituted into Eq. (8), there is obtained the simple form of the integral momentum equation for axisymmetric flow,

$$\frac{d\theta_1}{ds} + \theta_1\left[\left(\frac{\delta_1/\theta_1}{u_1}\right)\frac{du_1}{ds} + \frac{1}{a\rho_1 u_1^2}\frac{d}{ds}(a\rho_1 u_1^2)\right] = \frac{C_f}{2}, \tag{15}$$

where $C_f/2 = \tau_w/\rho_1 u_1^2$. The ratio δ_1/θ_1 in Eq. (15) is, by the definitions of Eqs. (12) and (14), a function of the boundary-layer thicknesses δ^* and θ and the local radius of the nozzle. An expression for δ_1/θ_1 can be obtained in the following manner. With $\eta = y/\delta$ and, again, $r = a - y \cos\omega$, Eq. (11) can be written as

$$\delta_1 = \int_0^1 \frac{(a - y\cos\omega)}{a}\left(1 - \frac{\rho u}{\rho_1 u_1}\right)\delta\, d\eta, \tag{16}$$

which becomes

$$\frac{\delta_1}{\delta} = \int_0^1 \left(1 - \frac{\rho u}{\rho_1 u_1}\right) d\eta - \frac{\delta\cos\omega}{a}\int_0^1 \left(1 - \frac{\rho u}{\rho_1 u_1}\right)\eta\, d\eta. \tag{17}$$

Similarly, Eq. (13) can be written as

$$\theta_1 = \int_0^1 \left(\frac{a - y\cos\omega}{a}\right)\left(1 - \frac{u}{u_1}\right)\frac{\rho u}{\rho_1 u_1}\delta\, d\eta, \tag{18}$$

which becomes

$$\frac{\theta_1}{\delta} = \int_0^1 \left(1 - \frac{u}{u_1}\right)\frac{\rho u}{\rho_1 u_1} d\eta - \frac{\delta\cos\omega}{a}\int_0^1 \left(1 - \frac{u}{u_1}\right)\frac{\rho u}{\rho_1 u_1}\eta\, d\eta. \tag{19}$$

Division of Eq. (17) by Eq. (19) gives

$$\frac{\delta_1}{\theta_1} = \frac{\int_0^1 \left(1 - \frac{\rho u}{\rho_1 u_1}\right) d\eta - \frac{\delta\cos\omega}{a}\int_0^1 \left(1 - \frac{\rho u}{\rho_1 u_1}\right)\eta\, d\eta}{\int_0^1 \left(1 - \frac{u}{u_1}\right)\frac{\rho u}{\rho_1 u_1} d\eta - \frac{\delta\cos\omega}{a}\int_0^1 \left(1 - \frac{u}{u_1}\right)\frac{\rho u}{\rho_1 u_1}\eta\, d\eta}. \tag{20}$$

The ratio of the first term in the numerator to the first term in the denominator on the right side of Eq. (20) is the usual definition of δ^*/θ for two-dimensional, compressible flow. The second term in the numerator and denominator accounts for axially symmetric flow.

Velocity distributions which match closely the Blasius profile for laminar flow are the approximations

$$u/u_1 = 2\eta - \eta^2,$$
$$u/u_1 = \sin(\pi\eta/2).$$

The density may be obtained from a chart of gas properties such as a Mollier diagram, where it may be read off at intervals along the path of constant-pressure departure from a point representing the isentropic core of flow in the center of the nozzle. A number of variants in the details of this procedure obviously are possible. However, one must have a link between the velocity ratio u/u_1 and some other characteristic, such as enthalpy. Since static pressure is assumed constant at a given station, the density distribution in the boundary layer may be considered a function of the distribution of static enthalpy alone which may be computed by Crocco's relation for $P_r = 1$;[3]

$$h = h_w - (h_w - h_{aw})u/u_1 - (h_{aw} - h_1)(u/u_1)^2 \, . \tag{21}$$

Inasmuch as it is convenient to tabulate local conditions such as T_1 and ρ_1 during calculation of the isentropic core, it may be simpler to compute ρ from the equation of state applicable to the case in question. If conditions permit the assumption of constant specific heats, the temperature relation

$$T = T_w - (T_w - T_{aw})u/u_1 - (T_{aw} - T_1)(u/u_1)^2 \tag{21a}$$

may be used instead of Eq. (21).

Use of either of the velocity distributions in conjunction with a density distribution based on Eq. (21) permits evaluation of the integrands in Eq. (20) for a given $(\delta \cos \omega)/a$. The ratio δ_1/θ_1 is graphically presented in Fig. 2,

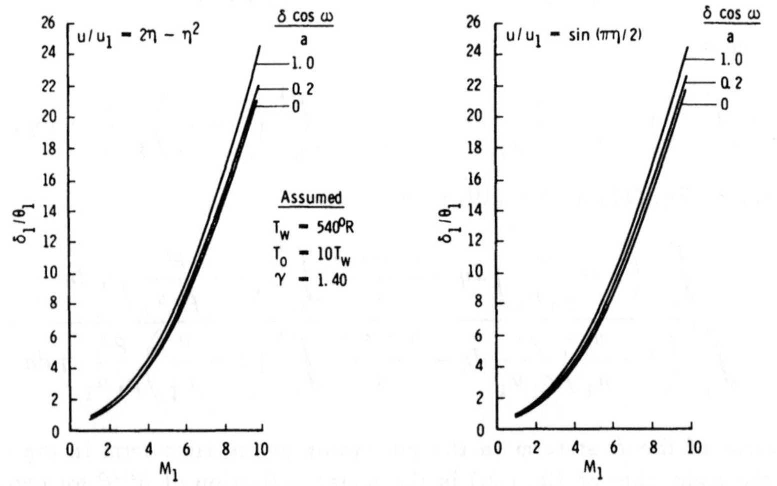

Fig. 2. Variation of δ_1/θ_1 with Mach Number and Velocity Distribution.

based on various values of $(\delta \cos \omega)/a$. In Fig. 3 are shown the corresponding values of the ratio δ^*/θ computed with the first integral in the numerator and denominator of the right side of Eq. (20). A comparison of the curves in Figs. 2 and 3 shows that both velocity profiles produce almost identical values of

[3] Note that $h_{aw} = h_0$ and $T_{aw} = T_0$ in these expressions, since they are derived on the basis that $P_r = 1$. However, it has become customary to insert values of h_{aw} or T_{aw} based on Prandtl numbers other than unity.

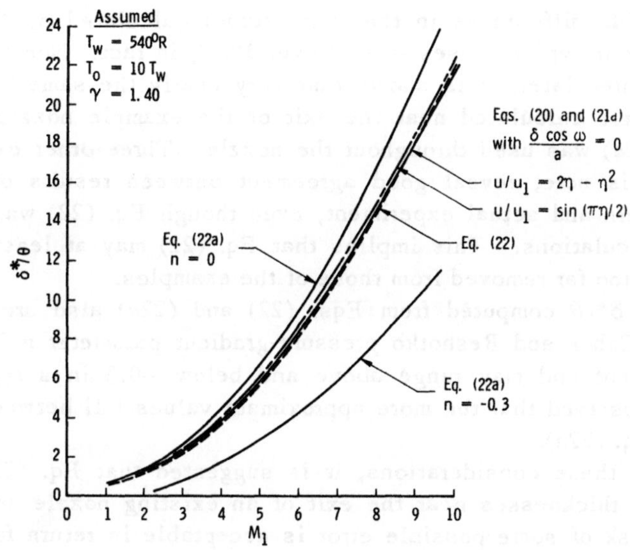

Fig. 3. Variation of δ^*/θ with Mach Number and Velocity Distribution.

δ_1/θ_1 and δ^*/θ. There too, it is seen that the contribution of the second term in the numerator and denominator of the right side of Eq. (20) causes the value of δ_1/θ_1 to change only moderately from the value of δ^*/θ.

Thus, it is assumed in this analysis that the value of δ_1/θ_1 may be approximated by the two-dimensional value of δ^*/θ. With this simplification then, instead of the integration of Eq. (20), recourse may be had to the results of previous work on the laminar boundary layer in two-dimensional flow. An expression which takes heat transfer into account has been derived [*12*] on the basis of the earlier, widely used, Illingworth-Stewartson transformation. It is

$$H = (\delta^*/\theta)_{2d} = H_i(T_w/T_1) + (T_{aw}/T_1) - 1\,, \tag{22}$$

where the temperatures correspond to h_w, h_1, and h_{aw}. For laminar flat-plate flow, $H_i = 2.605$. However, as mentioned [*4*], Eq. (22) does not reflect the influence of a pressure gradient when H_i is taken to be constant, and this may strongly influence the form factor H. Although the approximation $\delta_1/\theta_1 \approx (\delta^*/\theta)_{2d}$ is accepted here, it is not necessary to relax requirements to the extent of neglecting pressure gradients too. One may, for example, use an expression given by Cohen and Reshotko [*13*], namely,

$$H = H_{tr} + \frac{\gamma - 1}{2} M_1^2 (H_{tr} + 1)\,. \tag{22a}$$

Figure 7 of ref [*13*] gives values of H_{tr} as a function of wall enthalpy and pressure gradient. Unfortunately, determination of the pressure-gradient parameter used in ref [*13*] entails appreciable added labor, but it easily may be shown that considerable differences in the form factors calculated by Eqs. (22) and (22*a*) may exist in typical cases — well over 100%, in fact. Nonetheless, in an example presented later, it is shown that very nearly the same boundary-layer thicknesses were calculated near the exit of the example nozzle when either Eq. (22) or (22*a*) was used throughout the nozzle. Three other examples, also included in this note, reveal good agreement between results of the method developed herein and actual experiment, even though Eq. (22) was used in the theoretical calculations. This implies that Eq. (22) may at least be used for conditions not too far removed from those of the examples.

Values of δ^*/θ computed from Eqs. (22) and (22*a*) also are compared in Fig. 3. The Cohen and Reshotko pressure-gradient parameter n is zero for no pressure gradient and may range above and below -0.3 in a typical nozzle. It should be observed that the more approximate values fall between the curves representing Eq. (22*a*).

In view of these considerations, it is suggested that Eq. (22) be used if boundary-layer thicknesses near the exit of an existing nozzle are to be estimated, or if risk of some possible error is acceptable in return for easier calculation. Equation (22*a*) presumably is more reliable and may be used for design of complete, new nozzles if the added labor is justified. In the latter situation, however, it will generally be necessary to extrapolate the needed curves in ref [*13*] in order to obtain data for the pressure gradients typical of nozzles for low-density, hypersonic flow. This extrapolation appears easiest for problems involving highly cooled walls.

Solution of Integral Momentum Equation and Evaluation of Boundary-Layer Thickness

The momentum Eq. (15) for axisymmetric, compressible flow, with the assumption that $\delta_1/\theta_1 \approx (\delta^*/\theta)_{2d}$, can be written as

$$\frac{d\theta_1}{dx} + \theta_1 \left\{ \left[\frac{(\delta^*/\theta)_{2d}}{u_1} \frac{du_1}{dx} \right] + \frac{1}{a\rho_1 u_1^2} \frac{d}{dx} (a\rho_1 u_1^2) \right\} = \frac{C_f}{2} \sec \omega \ , \tag{23}$$

since $ds = dx \sec \omega$.

Equation (23) is a linear, first-order differential equation which has the solution

$$\theta_1 = \frac{\theta_1(0)}{G(x)} + \frac{1}{G(x)} \int_{x=0}^{x} W(x) G(x)\, dx \ , \tag{24}$$

where

$$G(x) = \exp \int_0^x \left[\frac{(\delta^*/\theta)_{2d}}{u_1} \frac{du_1}{dx} + \frac{1}{a\rho_1 u_1^2} \frac{d}{dx} (a\rho_1 u_1^2) \right] dx \ , \tag{25}$$

$$W(x) = \frac{C_f}{2} \sec \omega \ . \tag{26}$$

If it is assumed that the boundary-layer calculations will be started at the nozzle throat with $x = 0$, $M_1 = 1.0$, and $\theta_1 = 0$, then the momentum thickness along the nozzle may be obtained from the expression

$$\theta_1(x) = \frac{1}{G(x)} \int_{x=0}^{x} W(x)G(x)\, dx \,, \tag{27}$$

with the definitions for $G(x)$ and $W(x)$ from Eqs. (25) and (26). An iteration process must be applied in solving Eq. (25) because a is not known at the beginning. Setting $G(x) = 1$ is recommended as a good first approximation. The term $W(x)$ requires elaboration, which is deferred to the next section.

After the value of θ_1 is obtained for stations along the nozzle, δ_1 may be calculated from the local value of $\delta_1/\theta_1 \approx \delta^*/\theta = H$. With the value of δ_1 known, the displacement thickness δ^* may be obtained from the relationship between δ_1 and δ^* given in Eq. (12). With $a = r_{ec} + \delta^* \cos\omega$, the resulting expression for δ^* is

$$\delta^* = (\delta_1 - r_{ec} \sec\omega) + \sqrt{(r_{ec}\sec\omega - \delta_1)^2 + 2r_{ec}\sec\omega\,\delta_1} \,. \tag{28}$$

After δ^* is determined, a nozzle contour may be obtained by drawing a smooth curve tangent to circular arch of radius δ^* laid off from selected points along the expansion core.

In order to obtain the radius of the uniform core of flow, a knowledge of the total boundary-layer thickness δ is needed. From the known local values of δ^*, ω, and the nozzle radius, values of δ can be obtained from the definitions of Eqs. (11) and (12) and the integrations indicated in the numerator of the right side of Eq. (20). Equating Eqs. (11) and (12) gives

$$\delta^* - \frac{\delta^{*2}\cos\omega}{2a} = \int_0^{\delta} \frac{r}{a}\left(1 - \frac{\rho u}{\rho_1 u_1}\right) dy \,. \tag{29}$$

Then again with $\eta = y/\delta$ and $r = a - y\cos\omega$, Eq. (29) can be reduced to

$$\frac{\delta^*}{\delta} = \frac{\displaystyle\int_0^1 \left(1 - \frac{\rho u}{\rho_1 u_1}\right) d\eta - \frac{\delta\cos\omega}{a}\int_0^1 \left(1 - \frac{\rho u}{\rho_1 u_1}\right)\eta\, d\eta}{1 - \dfrac{\delta^*\cos\omega}{2a}} \,, \tag{30}$$

where $a = r_{ec} + \delta^*\cos\omega$.

After a choice of the form of the velocity and temperature variations in the boundary layer is made, the integrals in Eq. (30) can be evaluated. With δ^*, ω, and a known, Eq. (30) may be solved for the desired velocity boundary-layer thickness δ. In the present procedure, the velocity and temperature boundary-layer thicknesses are identical. This is not necessarily the case, particularly for nozzle flow characterized by highly favorable pressure gradients, but neglect of the difference probably is not inconsistent with the overall accuracy of this analysis.

To implement the procedure outlined above, it is necessary to define $W(x)$, or, specifically, the skin-friction law. Ideally, this law should be appropriate to axisymmetric flows with pressure gradients, compressibility, heat transfer, variable fluid properties, and large ratios of boundary-layer thickness to nozzle radius. Rather than attempt an elaborate derivation meeting these difficult requirements, attention has been concentrated on a usable approximation described in the next section.

Skin-Friction Coefficient

Several earlier analyses have made use of the well-known "reference enthalpy" approach in obtaining the local skin-friction coefficient for compressible flow with heat transfer in a wind-tunnel nozzle. This leads to the expression [4, 12]

$$C_f = C_f' \frac{\rho'}{\rho_1} , \tag{31}$$

with ρ' determined at the enthalpy h', where h' is E. R. G. Eckert's reference enthalpy of the boundary layer expressed in the form

$$h' = 0.5h_w + 0.22h_{aw} + 0.28h_1 . \tag{32}$$

The skin-friction coefficient C_f' is obtained by modifying the basic Blasius relationship for incompressible, laminar flow over a flat plate, and it is given by

$$C_f' = \frac{0.664}{\sqrt{\mathrm{Re}_s'}} , \tag{33}$$

where $\mathrm{Re}_s' = \rho' u_1 s/\mu'$ is the Reynolds number based on a length along the nozzle from the throat.[4] The density and viscosity are determined by the temperature corresponding to the reference enthalpy h'.

The approach exemplified by Eq. (31) relies heavily on local similarity, even to the extent that transformed axial and radial lengths are not used.[5] Local similarity appears to be a justifiable concept to invoke in the present case, because this study is specifically for situations where (1) pressure gradients are favorable, that is, $\partial p/\partial x < 1$, (2) walls are highly cooled, that is, $T_w \ll T_{aw}$, and (3) throughout the greater part of the nozzle the variables most influencing the boundary layer vary relatively slowly with x. It has been

---

[4] The distance s properly should be used in computing the Reynolds number, but in most applications $s(x)$ will not be known when the calculation is first made. Since it has been established by calculation of examples that little error is introduced, the substitution of x for s is recommended.

[5] Discussions of this and related approximations used by many investigators in attacking problems involving arbitrary pressure and fluid property distributions may be found in ref [14].

shown that these conditions are most compatible with the assumption of local similarity [*14, 15*].

If similar solutions were available for the case at hand, a correlation could be attempted, following Thwaites [*16*]. Or, as with the form factor, the results of Cohen and Reshotko [*13*] could be extrapolated to the extent necessary to extract friction coefficients from their report. Another alternative is the use of Blasius' relation between momentum thickness and friction coefficient for flat-plate flow. Although this would enable expressing C_f in terms of θ_1, it would represent the same type of approximation as Eq. (33). It is not immediately obvious that a form of friction law derived for two-dimensional flow with thin boundary layers is suitable for the present problem, wherein transverse curvature and very thick boundary layers are combined. However, it has been found that formulas for flat-plate flows may be applied, after suitable transformations, to axisymmetric flows with thin boundary layers [*17*] and to flows parallel to right circular cylinders with thick boundary layers [*5*]. With these results in mind and lacking a formulation clearly applicable to the case at hand, use is made of the simple approximation[6]

$$C_f = C_f'(\rho'/\rho_1) . \tag{34}$$

APPLICATION OF THE CALCULATION METHOD

Low-Density, Hypervelocity Wind-Tunnel Nozzle

The boundary-layer calculation method developed herein was applied to the flow in the 15° half-angle, water-cooled, conical nozzle of the von Kármán Gas Dynamics Facility (VKF) low-density tunnel described in ref [*18*].

Typical operating conditions with nitrogen as the working fluid and reservoir temperatures of 4000 and 5400°R were investigated as a test of the theory. These correspond to exit Mach numbers of 10.0 and 9.4 respectively. Numerous experimental axial and transverse pitot-pressure surveys well upstream from the nozzle exit provided the axial velocity distributions used in the boundary-layer calculations. Taking the experimentally determined axial-velocity distribution for use in the boundary-layer calculations causes the comparison to be based on identical theoretical and experimental core flows rather than identical nozzle geometry. As a result of starting with an "expansion core" which represents the nozzle giving the prescribed axial Mach numbers or velocities in an inviscid fluid, the calculation yields a theoretical nozzle wall contour which may be compared with the physical or real nozzle as a test of the computational method. This enables the calculation to be organized in the same manner as would be convenient in designing a new nozzle for a desired Mach number distribution. In that problem too, one would begin by establishing a potential-flow contour which, for an inviscid fluid, would produce the design

[6]This formulation allows the anomalous result, $\delta > a$. Therefore it must be understood that the method is applicable only when $\delta/a \leqq 1$.

Mach number.[7] Then it would be assumed that addition of displacement thicknesses to this potential contour would give the corresponding physical contour. Since, in reality, only part of the flow contained within the hypothetical potential core of a nozzle is uniform and isentropic, the term "potential core" has been replaced by "expansion core" in the present terminology. The radius of the uniform core of flow is denoted by r_{uc}.

The velocity and expansion core radius distributions used for the boundary-layer calculations are given in Fig. 4. The calculated boundary-layer displacement thicknesses are shown in Fig. 5, and the resulting theoretical nozzle

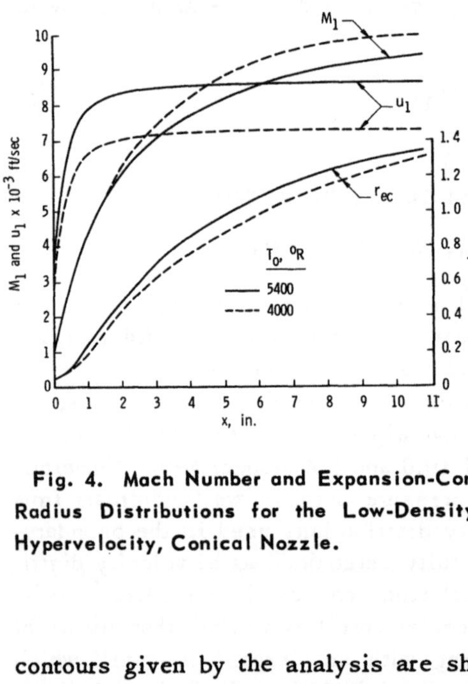

Fig. 4. Mach Number and Expansion-Core Radius Distributions for the Low-Density, Hypervelocity, Conical Nozzle.

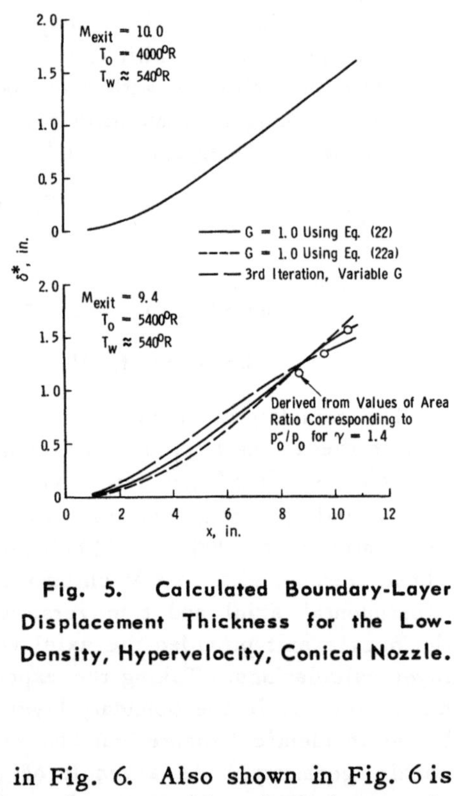

Fig. 5. Calculated Boundary-Layer Displacement Thickness for the Low-Density, Hypervelocity, Conical Nozzle.

contours given by the analysis are shown in Fig. 6. Also shown in Fig. 6 is the actual conical nozzle contour based on less careful calculations made during design of the tunnel. In principle, the degree of agreement between results of the computational method and the actual nozzle radii represents the test of the method.

It is interesting to see that the approximation $G(x) = 1$ leads to good results in these examples. Also apparent is the improvement effected through use of

[7] In the two examples relating to the conical nozzle, hypothetical potential-flow core radii were calculated from one-dimensional theory to match the given axial-velocity distributions. If the entire potential flow were calculated by, say, the Foelsch method, core flow properties and radii would come out together.

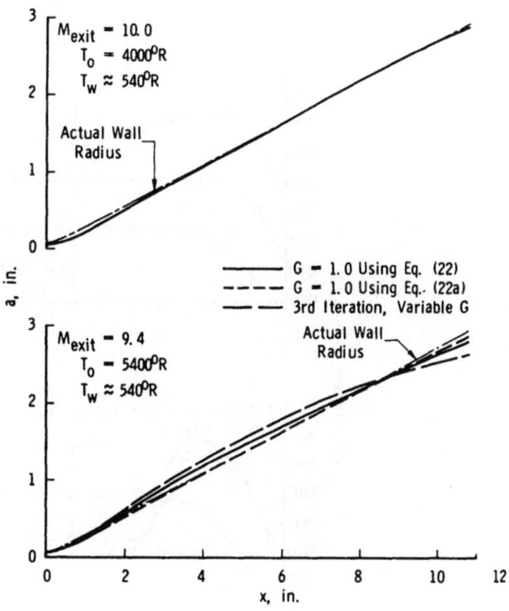

Fig. 6. Calculated and Actual Wall Radii for the Low-Density, Hypervelocity Wind-Tunnel Nozzle. Calculated for thermal equilibrium.

Eq. (22*a*) in the second example in Figs. 5 and 6. As remarked earlier, it was necessary to extrapolate curves given in ref [*13*] in order to apply Eq. (22*a*), so it is not certain if the improvement indicated is partly coincidental. However, Eq. (22*a*) clearly represents greater rigor of computational method.

In Fig. 7, experimental, transverse, pitot-pressure surveys at various stations along the nozzle axis are shown. The calculated free-stream core radii are in close agreement with the experimental values in all cases. The ratio δ^*/δ used to obtain δ was calculated by use of a density distribution based on Eq. (21) with $u/u_1 = \sin(\pi\eta/2)$.

University of California and University of Toronto Wind-Tunnel Nozzles

Boundary-layer calculations were also carried out for typical nozzles of the University of California and the University of Toronto low-density tunnels. Values of δ^*, δ, the wall radius, and the uniform core radius at the nozzle exit were obtained. The results of these calculations are compared with the data of refs [*2*] and [*3*] in Figs. 8 and 9, where calculated and actual nozzle contours are presented. Quite good agreement results, the discrepancies being about 4 and 6% for the California and Toronto examples respectively.

Effect of Wall Temperature

The calculation procedure is useful for comparative studies of various factors affecting growth of the boundary layer. It is interesting to see the influence of heat transfer at the nozzle wall as shown by the example presented

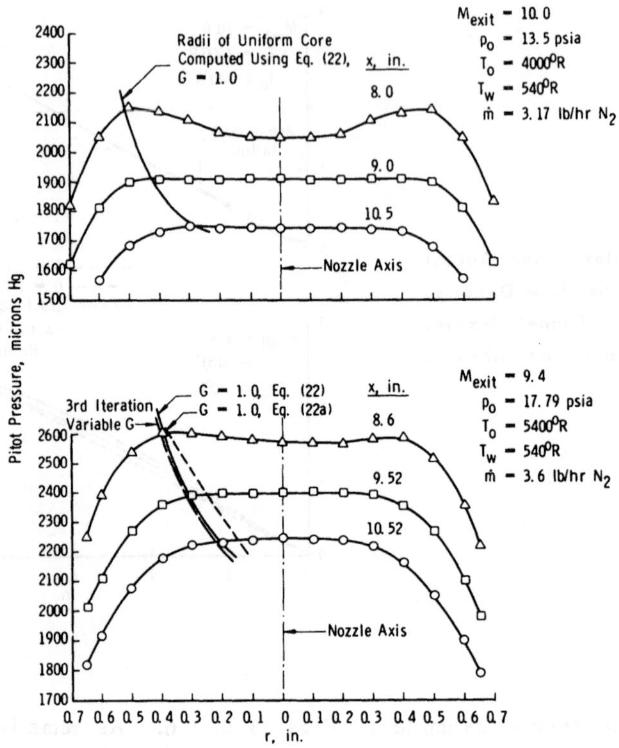

Fig. 7. Experimental and Calculated Uniform Core Radii.

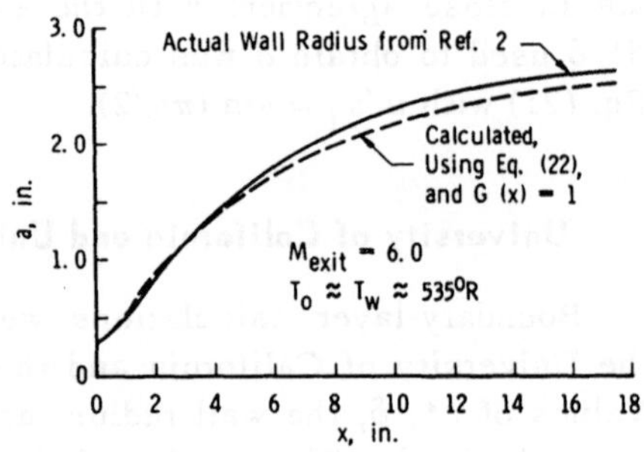

Fig. 8. Calculated and Actual Wall Radii for the University of California Wind-Tunnel Nozzle of Ref 2.

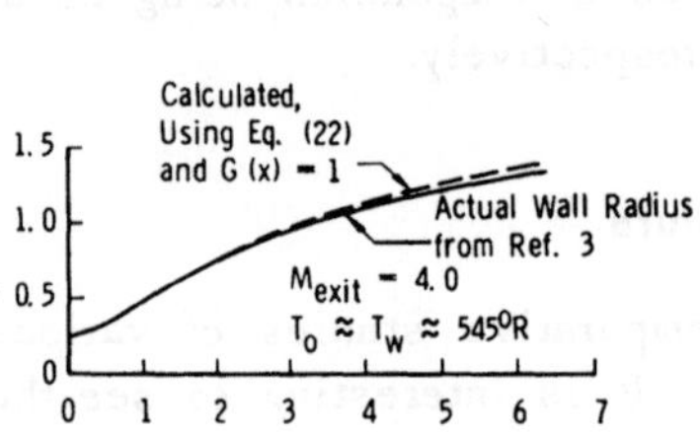

Fig. 9. Calculated and Actual Wall Radii for the University of Toronto Wind-Tunnel Nozzle of Ref 3.

in Fig. 10. The cases depicted in Fig. 10 are based on a rather long nozzle with an expansion core derived by the Foelsch method. Under the conditions assumed for the calculations, the nozzle exit in all cases is nearly filled with boundary layer.

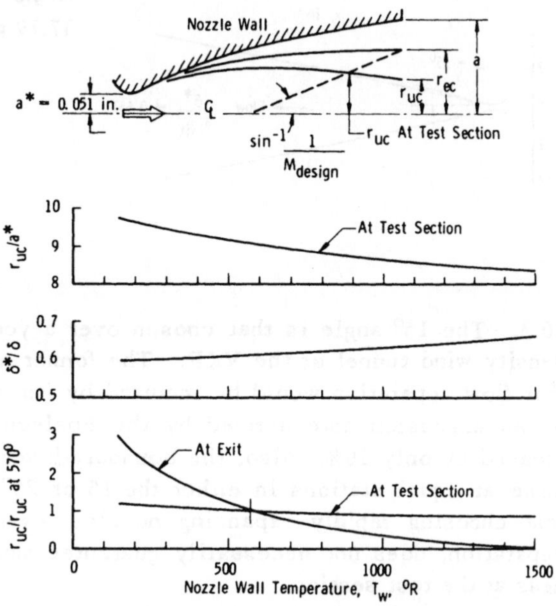

Fig. 10. Effect of Wall Temperature on the Boundary Layer ($M_{exit} = 9.5$, $T_0 = 5400°R$, $p_0 = 17.79$ psia, Molecular Vibration Assumed Frozen Downstream of Throat).

Indirect or back-side water cooling is adequate to maintain the nozzle wall temperature nearly constant at approximately 570°R in the case represented in Fig. 10. Cooling with liquid nitrogen might reduce the wall temperature to, say, 150°R. If this were done, it is predicted by the method of this paper that the radius of the uniform core at the test section would increase 6%. On the other hand, if the wall temperature were allowed to rise to, say, 1500°R, about one-half the melting point of copper, the radius of the uniform core at the test section would decrease 9%. At the same time, physical dimensions of the nozzle exit would have to be increased to enable the exit flow to attain a Mach number of 9.5. For the nozzle exit, as distinguished from the test section, greater changes are shown in Fig. 10.

Rapidly Expanding Nozzles

The tendency is to choose rapidly expanding nozzles for applications where thick boundary layers are expected, because, in this way, it would seem that the relative thickness of the boundary layer could be reduced. Two rates of expansion are represented in the results summarized in Fig. 11. The 25° half-angle expansion is near the maximum theoretical for the exit Mach number of

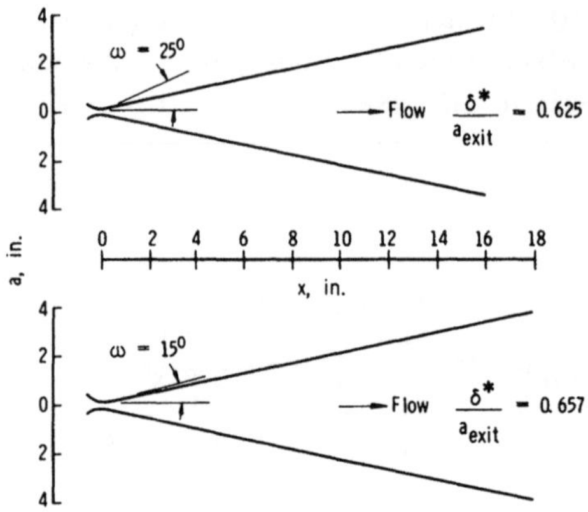

Fig. 11. Effect of Initial Expansion Angle (M_{exit} = 10.3, T_0 = 5400°R, p_0 = 17.79 psia).

10.3. The 15° angle is that chosen over a year ago for the nozzle of the low-density wind tunnel at the VKF. The former angle is so great that it is likely that flow separation would be incurred by its use. Yet, if the nozzle is based on an expansion core derived by the Foelsch method, the total length is decreased by only 10%. Also, the boundary-layer thickness is almost exactly the same at equal stations in either the 15 or 25° cases. This example indicates that choosing rapidly expanding nozzles, even aside from the danger of flow separation, does not necessarily guarantee much reduced boundary-layer thickness at the test section.

CONCLUSIONS

The difficult problem of calculating the thickness distribution of a thick laminar boundary layer in an axisymmetric flow of low-density, high-speed fluid has been attacked by a combination of integral momentum and local similarity concepts in this analysis. The outcome appears to represent a reasonably accurate method for general use when very thick boundary layers are encountered in axisymmetric nozzles.

Results of the application of the method to typical low-density wind-tunnel nozzles having Mach numbers up to approximately 10 with either unheated or heated gases up to 5400°R reservoir temperature are shown to agree with experimental data. The error in computed radii or displacement thicknesses found in these examples ranged from zero to something less than 10%.

ACKNOWLEDGMENT

The authors wish to express their gratitude to W. C. Armstrong for his preparation of the machine computation program in connection with this work.

REFERENCES

[1] J. M. Owen and F. S. Sherman, *Fluid Flow and Heat Transfer – Design and Testing of a Mach 4 Axially Symmetric Nozzle for Rarefied Gas Flows,* University of California Report HE-150-104, July 1952.

[2] G. J. Maslach and F. S. Sherman, *Design and Testing of an Axisymmetric Hypersonic Nozzle for a Low Density Wind Tunnel,* WADC-TR-56-341, University of California Report 150-134, August 1956.

[3] A. K. Sreekanth, *Performance of a Mach 4 Axially Symmetric Nozzle Designed to Operate at 40 Microns Hg in the UTIA Low Density Wind Tunnel,* University of Toronto Tech. Note 10, AFOSR-TN-57-94, September 1956.

[4] A. F. Johnson, *A Method of Calculating Boundary-Layer Thickness in Axisymmetric Nozzles with Laminar Hypersonic Flow,* Sandia Corporation Report SC-4370(RR), October 1959.

[5] R. A. Seban and R. Bond, "Skin-Friction and Heat-Transfer Characteristics of a Laminar Boundary Layer on a Cylinder in Axial Incompressible Flow," *J. Aeronaut. Sci.* **18**(10), 671–75 (1951).

[6] H. R. Kelly, "A Note on the Laminar Boundary Layer on a Circular Cylinder in Axial Incompressible Flow," *J. Aeronaut. Sci.* **21**(9), 634 (1954).

[7] R. F. Probstein and David Elliott, "The Transverse Curvature Effect in Compressible Axially Symmetric Laminar Boundary-Layer Flow," *J. Aeronaut. Sci.* **23**(3), 208–22 (1956).

[8] K. N. C. Bray, *Departure from Dissociation Equilibrium in a Hypersonic Nozzle,* University of Southampton, A. R. C. 19,983, March 1958.

[9] J. G. Hall and A. L. Russo, *Studies of Chemical Nonequilibrium in Hypersonic Nozzle Flows,* Cornell Aeronautical Lab., Inc., Report 1118-A-6, November 1959.

[10] P. P. Wegener, "Experiments on the Departure from Chemical Equilibrium in a Supersonic Flow," *Jet Propulsion* **30**(4), 322–29 (1960).

[11] Kuno Foelsch, "The Analytical Design of an Axially Symmetric Laval Nozzle for a Parallel and Uniform Jet," *J. Aeronaut. Sci.* **16**(3), 161–66, 188 (1949).

[12] J. C. Sivells and R. G. Payne, *A Method of Calculating Turbulent-Boundary-Layer Growth at Hypersonic Mach Numbers,* AEDC-TR-59-3, March 1959.

[13] C. B. Cohen and Eli Reshotko, "The Compressible Laminar Boundary Layer with Heat Transfer and Arbitrary Pressure Gradient," *NACA* TR 1294 (1956).

[14] W. D. Hayes and R. F. Probstein, *Hypersonic Flow Theory,* Academic Press, New York, 1959.

[15] Lester Lees, "Laminar Heat Transfer Over Blunt-Nosed Bodies at Hypersonic Flight Speeds," *Jet Propulsion,* **26**(4), 259–69, 274 (1956).

[16] B. Thwaites, "Approximate Calculation of the Laminar Boundary Layer," *Aeronaut. Quart.* **1**, 245–80 (1949).

[17] W. Mangler, "Zusammenhang Zwischen, ebenen und rotationssymmetrischen Grenzschichten in kompressiblen Flüssigkeiten," *Z. angew. Math. Mech.* **28**(4), 97 (1948).

[18] J. L. Potter *et al.*, *Description and Preliminary Calibration of a Low-Density, Hypervelocity Wind Tunnel*, AEDC TN-61-83, August 1961.

DETERMINATION OF THE SIX REYNOLDS STRESSES BY THE LINEARIZED-RESPONSE HOT-WIRE ANEMOMETER[1]

R. P. Wichner and F. N. Peebles
Oak Ridge National Laboratory[2]

This paper discusses one of the several statistical quantities which may be obtained by use of the hot-wire anemometer (HWA) to describe turbulent flows; these are the "one-point double-velocity correlations" which are proportional to the Reynolds stresses. All six of the Reynolds stresses may be determined by the described method. The treatment is one suggested by Traugott [*1*].

Experimental measurements of Reynolds stresses which make use of the analytical methods of this paper will be presented at a later date.

DETERMINATION OF THE REYNOLDS STRESSES

Before we can proceed with the development of the relation between the electrical signal and fluctuating velocity, some assumption must be made as to the manner in which the wire is cooled by the fluid. Almost universally it is assumed that only that component of the velocity which is perpendicular to the wire is effective in cooling it. Actually, this cannot be strictly true, for even when the perpendicular component is zero (i.e., flow is along the axis of the wire), some cooling takes place. Hinze [*2*] gives the following expression for the velocity *V*:

$$V^2_{\text{effective}} = V^2_{\text{perpendicular}} + C_1 V^2_{\text{parallel}} \, . \tag{1a}$$

The value of the coefficient C_1 is given as ranging between 0.09 and 0.01, decreasing with increasing velocity. Hinze goes on to state that it appears to be possible to ignore the parallel-velocity component when the axis of the wire is more than 20° from the direction of the velocity. Equation (1*a*) then reduces to

$$V_{\text{effective}} = V_{\text{perpendicular}} \tag{1b}$$

In extending this assumption to turbulent flow, we will place the axis of the wire outside a 20° cone about the *mean* velocity vector and state that only the component of the *instantaneous* velocity perpendicular to the wire is effective in cooling

---

[1]Supported by the USAEC under contract No. W-7405-eng-26.
[2]Operated for USAEC by Union Carbide Corporation.

NOTATION

$A, B, \ldots, F$	Constants in Eq. (10)
b	Scale factor, related to $\sqrt{\overline{e^2}}$ by $\sqrt{\overline{e^2}} = bI$
$\mathfrak{D}$	Determinant of the coefficients of Eq. (13), sixth order
$\mathfrak{D}_i$	Sixth-order determinant formed from $\mathfrak{D}$ by replacing the ith column with the six HWA root-mean-square voltage readings
$\mathfrak{D}_i^j$	Fifth-order determinant formed from $\mathfrak{D}$ by eliminating the ith column and jth row
E	HWA voltage, total
e	HWA voltage, fluctuating
I	Root-mean-square meter reading
k, m	Constants in Eq. (3)
r'	Distance from pipe wall
r_0	Pipe radius
V	Velocity, total
v	Velocity, fluctuating part
$\vec{w}_1$	Unit vector along wire
α, β	Angles defined by Fig. 1

Subscripts

x, y, z	Coordinate axes directions
x, θ, r	Cylindrical polar coordinates
p	Indicates perpendicular
1	Indicates unit vector

Superscripts

$\overline{}$	Indicates time average
$\rightarrow$	Indicates vector

it. Since the instantaneous velocity may often be closer than 20° to the wire axis, and hence out of the range of applicability of Eq. (1*b*), this assumption is not entirely satisfactory; the results so obtained will undoubtedly be somewhat warped. However, there seems to be no practical alternative.

In order to continue, a relationship between the instantaneous current $I(t)$ and the instantaneous perpendicular velocity $V_p(t)$ is required. The most common starting point is King's law,

$$\overline{I^2} = C_2 + C_3\sqrt{\overline{V}_p}\,, \tag{2a}$$

which strictly applies to time-averaged turbulent flow (and perhaps steady laminar flow) and extends its applicability to each turbulent instant, that is,

$$I^2(t) = C_2 + C_3\sqrt{V_p(t)}\,. \tag{2b}$$

If, on the contrary, we assume that the HWA circuit contains an element which linearizes the output signal with respect to the perpendicular velocity component, we have, instead of King's law,

$$k\overline{E} + m = \overline{|\vec{V}_p(t)|}\,, \tag{3a}$$

where

$\overline{E}$ = time-average output signal,

k, m = constants,

$\overline{|\vec{V}_p(t)|}$ = average value of the magnitude of the perpendicular velocity.

Equation (3a) is an experimental fact in an HWA system containing a linearizing element [3]. As before, we must extend the applicability of the experimentally observed time-average behavior to each instant, that is,

$$kE(t) + m = |\vec{V}_p(t)|\,. \tag{3b}$$

Now define $v_p(t)$ as follows:

$$v_p(t) \equiv |\vec{V}_p(t)| - \overline{|\vec{V}_p(t)|}\,. \tag{3c}$$

Squaring and combining Eqs. (3a) and (3b) yields, by using Eq. (3c),

$$k^2\overline{e^2} = \overline{v_p^2}\,. \tag{4}$$

The problem now becomes one of expressing the quality $\overline{v_p^2}$ in terms of velocity fluctuations in convenient directions, for example, the coordinate-axes directions. As shown in Fig. 1, $\vec{i}_1$, $\vec{j}_1$, and $\vec{k}_1$ are the unit vectors along the x, y, and z axes. The time-varying velocity $\vec{V}(t)$ is the vector sum of the time-averaged and fluctuating velocities $\vec{V}$ and $\vec{v}(t)$. The direction of the unit vector $\vec{w}_1$ along the axis of the wire is specified by the angles α and β.

It is seen, then, that the velocity perpendicular to the wire may be written as

$$\vec{V}_p(t) = \vec{V}(t) - \vec{w}_1[\vec{V}(t)\cdot\vec{w}_1]\,. \tag{5}$$

The second term on the right side is the time-varying velocity parallel to the wire, which, when subtracted from the total velocity, yields the perpendicular component $\vec{V}_p$.

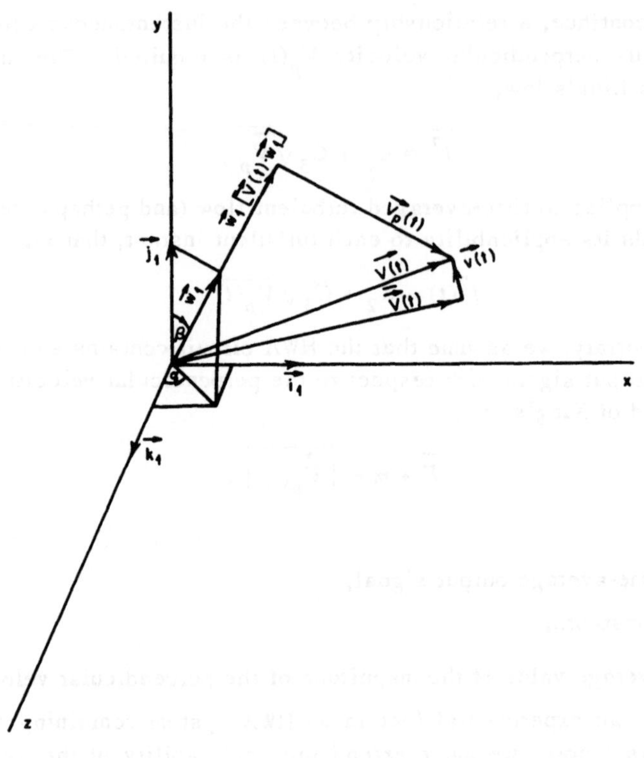

Fig. 1. Orientation of the Wire and Velocities Relative to the Coordinate Axes.

From Eq. (5), we obtain

$$|\vec{V}_p(t)|^2 = |\vec{V}(t)|^2 - [\vec{V}(t) \cdot \vec{w}_1]^2 \,, \tag{5a}$$

$$\overline{\vec{V}}_p = \overline{\vec{V}} - \vec{w}_1(\overline{\vec{V}} \cdot \vec{w}_1) \,, \tag{5b}$$

$$|\overline{\vec{V}}_p|^2 = |\overline{\vec{V}}|^2 - (\overline{\vec{V}} \cdot \vec{w}_1)^2 \,. \tag{5c}$$

Squaring and time-averaging Eq. (3c) yields

$$\overline{v_p^2} = \overline{|\vec{V}_p(t)|^2} - \overline{|\vec{V}_p(t)|}^2 \,. \tag{6}$$

We will now approximate $\overline{|\vec{V}_p(t)|}$ by $|\overline{\vec{V}}_p|$ and discuss the consequences of this approximation later. Therefore,

$$\overline{|\vec{V}_p(t)|} \approx |\overline{\vec{V}}_p| \,. \tag{7}$$

Thus, Eq. (6) may be rewritten as

$$\overline{v_p^2} = \overline{|\vec{V}_p|^2} - |\overline{\vec{V}_p}|^2 . \tag{8}$$

Since

$$\vec{V}(t) = (V_x + v_x)\vec{i}_1 + (V_y + v_y)\vec{j}_1 + (V_z + v_z)\vec{k}_1 , \tag{9}$$

$$\vec{w}_1 = (\sin\alpha \sin\beta)\vec{i}_1 + (\cos\beta)\vec{j}_1 + (\cos\alpha \sin\beta)\vec{k}_1 , \tag{10}$$

we obtain, by combining Eqs. (5*a*), (5*c*), (8), (9), and (10),

$$\overline{v_p^2} = A\overline{v_x^2} + B\overline{v_y^2} + C\overline{v_z^2} + D\overline{v_x v_y} + E\overline{v_x v_z} + F\overline{v_y v_z} , \tag{11}$$

where

$A \equiv 1 - \sin^2 A \sin^2\beta,$
$B \equiv \sin^2\beta,$
$C \equiv 1 \cos^2\alpha \sin^2\beta,$
$D \equiv -\sin\alpha \sin 2\beta,$
$E \equiv -\sin 2\alpha \sin^2\beta,$
$F \equiv -\cos\alpha \sin 2\beta.$

Equation (11) is the sought-after relation between the magnitude of the mean-square fluctuating perpendicular velocity component and the Reynolds stresses. Combining Eqs. (4) and (11) yields

$$k^2 \overline{e^2} = A\overline{v_x^2} + B\overline{v_y^2} + C\overline{v_z^2} + D\overline{v_x v_y} + E\overline{v_x v_z} + F\overline{v_y v_z} , \tag{12}$$

which describes the relation between the HWA signal and the six Reynolds stresses; note that, in general, the reading is a function of all six stresses.

When the wire axis is oriented parallel to the *x*, *y*, and *z* axes, Eq. (11) reduces to

$$\overline{v_p^2} = \overline{v_y^2} + \overline{v_z^2} , \tag{11a}$$

$$\overline{v_p^2} = \overline{v_x^2} + \overline{v_z^2} , \tag{11b}$$

$$\overline{v_p^2} = \overline{v_x^2} + \overline{v_y^2} , \tag{11c}$$

respectively. We therefore conclude that, in general, the HWA reading is equally sensitive to the cooling effects of the mean-square velocity fluctuations in two directions perpendicular to the wire, irrespective of the direction of the mean velocity.

To determine the Reynolds stresses experimentally, it is necessary to take a set of six readings ($\overline{e_j^2}$, $j = 1, 2, \ldots, 6$), each one taken at a different wire orientation $[(\alpha, \beta)_j, \; j = 1, \ldots, 6]$, and to solve the resulting six simultaneous linear equations:

$$k^2 \, \overline{e_j^2} = A_j \overline{v_x^2} + B_j \overline{v_y^2} + C_j \overline{v_z^2} + D_j \overline{v_x v_y} + E_j \overline{v_x v_z} + F_j \overline{v_y v_z} \, , \tag{13a}$$

where

$j = 1, 2, \ldots, 6$.

Let $\mathcal{D}$ be the determinant of the coefficients, and $\mathcal{D}_j$ be $\mathcal{D}$ with the jth column replaced by the six voltage readings, that is,

$$\mathcal{D} \equiv \begin{vmatrix} A_1 & B_1 & \ldots & F_1 \\ \cdot & & & \\ \cdot & & & \\ \cdot & & & \\ A_6 & & \ldots & F_6 \end{vmatrix} \, , \tag{14a}$$

$$\mathcal{D}_1 \equiv \begin{vmatrix} \overline{e_1^2} & B_6 & \ldots & F_6 \\ \cdot & & & \\ \cdot & & & \\ \cdot & & & \\ \overline{e_6^2} & B_6 & \ldots & F_6 \end{vmatrix} \, , \tag{14b}$$

etc. The solutions to Eq. (13a) may therefore be written as

$$\overline{v_x^2} = \frac{k^2 \mathcal{D}_1}{\mathcal{D}} \, , \qquad (a)$$

$$\overline{v_y^2} = \frac{k^2 \mathcal{D}_2}{\mathcal{D}} \, , \qquad (b)$$

$$\overline{v_z^2} = \frac{k^2 \mathcal{D}_3}{\mathcal{D}} \, , \qquad (c)$$

(15)

$$\overline{v_x v_y} = \frac{k^2 \mathcal{D}_4}{\mathcal{D}} \, , \qquad (d)$$

$$\overline{v_x v_z} = \frac{k^2 \mathcal{D}_5}{\mathcal{D}} \, , \qquad (e)$$

$$\overline{v_y v_z} = \frac{k^2 \mathcal{D}_6}{\mathcal{D}} \, . \qquad (f)$$

Equations (15) may be applied to yield the cylindrical polar Reynolds stresses simply by identifying the x, y, and z directions with the x, θ, and r directions and by rewriting Eqs. (15) with the change in subscripts.

It should be noted that the components of the time-averaged velocity can be calculated from three readings of the signal $\bar{E}$. Combine Eqs. (3a), (5b), and (10) by using approximation (7) to yield

$$(k\bar{E}_j + m)^2 = A_j V_x^2 + B_j V_y^2 + C_j V_z^2 + D_j V_x V_y + E_j V_x V_z + F_j V_y V_z . \tag{13b}$$

Thus, three readings of $\bar{E}$ fixes V_x, V_y, and V_z.

We now examine the consequences of the assumption involved in Eq. (7) by considering the special cases

$$\vec{V}(t) = (V_x + v_x)\vec{i}_1 + V_y \vec{j}_1 + v_z \vec{k}_1 ,$$

$$\vec{w}_1 = \vec{j}_1 .$$

From Eq. (5b), we find that $|\overline{\vec{V}_p}| = V_x$. Equation (5a) yields, for this case,

$$|\vec{V}_p(t)| = V_x \sqrt{\left(1 + \frac{v_x}{V_x}\right)^2 + \frac{v_z^2}{V_x^2}} .$$

By expanding the square root in a Maclaurin series and assuming v_x and v_z are much less than V_x, then

$$|\vec{V}_p(t)| = V_x + \frac{v_x}{V_x} + \text{higher-order terms},$$

so that $\overline{|\vec{V}_p(t)|} = V_x$ also. We therefore conclude that Eq. (7) is essentially a low-turbulence approximation.

ERROR ANALYSIS

A general discussion of the errors inherent in HWA determinations is given by Hinze [4]; here we will merely estimate the instrument error to determine if the lack of precision of available apparatus precludes utilization of the described method. Since the least precise reading is considered to be that of the root-mean-square voltmeter for $\overline{e^2}$, the maximum possible cumulative error in $\overline{v_x^2}$, due to the required six voltmeter readings, may be estimated by

$$\Delta\overline{v_x^2} \approx \sum_{j=1}^{6} \frac{\partial \overline{v_x^2}}{\partial I_j} \Delta I_j , \tag{16}$$

where

I_j = meter reading, related to $\overline{e_j^2}$ by $\sqrt{\overline{e_j^2}} = bI_j$,

b = scale factor ,

$\Delta\overline{v_x^2}$ = error in $\overline{v_x^2}$ caused by six errors, ΔI_j .

By utilizing Eq. (15*a*) and assuming $\Delta I_1 = \Delta I_2 = \ldots = \Delta I_6$, it can easily be shown that

$$\frac{\Delta\overline{v_x^2}}{\overline{v_x^2}} \approx \frac{2b^2\Delta I}{\mathfrak{D}_1}\sum_{j=1}^{6} I_j \mathfrak{D}_1^j , \tag{17a}$$

where $\mathfrak{D}_1^j$ is the fifth-order determinant formed by $\mathfrak{D}$ by eliminating the jth row and the first column. In Eqs. (17), the absolute value of each term is taken to make the sum. Similarly, for the other stresses, the relative errors are given by

$$\frac{\Delta\overline{v_\theta^2}}{\overline{v_\theta^2}} \approx \frac{2b^2\Delta I}{\mathfrak{D}_2}\sum_{j=1}^{6} I_j \mathfrak{D}_2^j , \tag{17b}$$

$$\frac{\Delta\overline{v_r^2}}{\overline{v_r^2}} \approx \frac{2b^2\Delta I}{\mathfrak{D}_3}\sum_{j=1}^{6} I_j \mathfrak{D}_3^j , \tag{17c}$$

$$\frac{\Delta\overline{v_x v_\theta}}{\overline{v_x v_\theta}} \approx \frac{2b^2\Delta I}{\mathfrak{D}_4}\sum_{j=1}^{6} I_j \mathfrak{D}_4^j , \tag{17d}$$

$$\frac{\Delta\overline{v_x v_r}}{\overline{v_x v_r}} \approx \frac{2b^2\Delta I}{\mathfrak{D}_5}\sum_{j=1}^{6} I_j \mathfrak{D}_5^j , \tag{17e}$$

$$\frac{\Delta\overline{v_r v_\theta}}{\overline{v_r v_\theta}} \approx \frac{2b^2\Delta I}{\mathfrak{D}_6}\sum_{j=1}^{6} I_j \mathfrak{D}_6^j , \tag{17f}$$

where, in general, $\mathfrak{D}_i^j$ is the fifth-order determinant formed from $\mathfrak{D}$ by eliminating the jth row and the ith column.

If the Reynolds stresses are known or assumed, the relative errors may be calculated from Eqs. (17). For illustration, we take from Laufer's air data [5], at a Reynolds number of 500,000 and three locations in the pipe, the values listed in Table 1.

The value for the error, ΔI, was taken[3] to be ±0.025, while the reciprocal sensitivity k has the value 2 ft $\sec^{-1}$ v^{-1} in air. In each case, the scale factor b was selected to yield maximum sensitivity without exceeding the maximum allowable root-mean-square voltage-meter reading of $I = 5$.

---

[3]Numerical values pertain to model 40-W, linear constant-temperature hot-film anemometer, Lintronic Laboratories.

Table 1. Values of Reynolds Stresses Taken from Laufer [5] at $N_{Re} = 500{,}000$

$\frac{r'}{r_0}$	Reynolds Stresses (ft^2/sec^2)					
	$\overline{v_x^2}$	$\overline{v_\theta^2}$	$\overline{v_r^2}$	$\overline{v_x v_r}$	$\overline{v_x v_\theta}$	$\overline{v_r v_\theta}$
0.05	7.36	5.50	3.55	−12.00	*a*	*a*
0.40	4.97	4.08	3.27	−7.56	*a*	*a*
0.90	2.87	2.66	2.63	−1.26	*a*	*a*

[a] Assumed to be zero.

Only four-wire orientations are required for this case, since $\overline{v_x v_\theta}$ and $\overline{v_r v_\theta}$ are assumed to be zero, leaving four unknowns. The orientation selections (tabulated below) were made for ease of calculation and do not represent an optimum condition or a convenient experimental situation:

	$j = 1$	$j = 2$	$j = 3$	$j = 4$
α, deg	0	0	45	90
β, deg	0	90	90	45

The relative errors calculated by using Eqs. (17) are listed in Table 2.

Table 2. Percentage Errors in the Determination of the Reynolds Stresses

$\frac{r'}{r_0}$	b	Errors (%)			
		$\frac{\Delta\overline{v_x^2}}{\overline{v_x^2}}$	$\frac{\Delta\overline{v_\theta^2}}{\overline{v_\theta^2}}$	$\frac{\Delta\overline{v_r^2}}{\overline{v_r^2}}$	$\frac{\Delta\overline{v_x v_r}}{\overline{v_x v_r}}$
0.05	0.4	±4.48	±6.00	±5.58	±3.80
0.40	0.4	±5.77	±7.02	±5.25	±5.07
0.95	0.2	±4.07	±4.39	±2.66	±11.51

ADDITIONAL NOTE FROM THE AUTHORS

With a slight modification, the proposed method may be used to determine the Reynolds stresses without employing a low-turbulence assumption. Square and time-average Eq. (3*b*) to yield

$$k^2 \overline{E^2(t)} + 2km\overline{E(t)} + m^2 = \overline{|\vec{V}_p(t)|^2} . \tag{18}$$

The quantity $\overline{|\vec{V}_p(t)|^2}$ can be related to the Reynolds stresses by combining Eqs. (5*a*), (9), and (10), from which

$$\overline{|V_p(t)|^2} = A(V_x^2 + \overline{v_x^2}) + B(V_y^2 + \overline{v_y^2}) + C(V_z^2 + \overline{v_z^2}) + D(V_x V_y + \overline{v_x v_y})$$
$$+ E(V_x V_z + \overline{v_x v_z}) + F(V_y V_z + \overline{v_y v_z}) . \tag{19}$$

For simplicity, define a reading, S, as follows:

$$S \equiv k^2 \overline{E^2(t)} + 2km\overline{E(t)} + m^2 - AV_x^2 - BV_y^2 - CV_z^2 - DV_x V_y$$
$$- EV_x V_z - FV_y V_z . \tag{20}$$

Combine Eqs. (18), (19), and (20) to yield the desired result:

$$S_i = A_i \overline{v_x^2} + B_i \overline{v_y^2} + C_i \overline{v_z^2} + D_i \overline{v_x v_y} + E_i \overline{v_x v_z} + F_i \overline{v_y v_z} . \tag{21}$$

As before, six readings, S_i, at six wire orientations (arranged so as to yield a nonzero determinant, $\mathfrak{D}$) fix the value of the six Reynolds stresses.

REFERENCES

[1] S. C. Traugott, *Influence of Solid-Body Rotation on Screen-Produced Turbulence*, NASA-TN-4135, p 42 (1958).

[2] J. O. Hinze, *Turbulence*, p 103, McGraw-Hill, New York, 1959.

[3] S. Ling, "Measurement of Flow Characteristics by the Hot-Film Technique," Ph.D. dissertation, University of Iowa, 1955.

[4] J. O. Hinze, *Turbulence*, pp 82–102, McGraw-Hill, New York, 1959.

[5] J. Laufer, *The Structure of Turbulence in Fully Developed Pipe Flow*, NASA report 1174 (1954).

ENERGY ANALYSIS OF PRESSURE SURGES IN CLOSED CONDUITS

D. J. Wood
Clemson College

and

T. E. Stelson
Carnegie Institute of Technology

The relationship between velocity and pressure changes for unsteady conduit flow is developed based on the conservation of energy. This approach has certain advantages over that based on the conservation of momentum, which is commonly used. These advantages stem primarily from the ease with which the effects of nonuniform conditions across a section of conduit may be considered. Use of the relationships based on the conservation of energy explains some disparities between experimental and analytical results which were not explained by momentum considerations.

Measurements of pressure rises due to sudden velocity extinction are presented and compared with analytical predictions based on the two approaches. Experimental values differ from the predictions based on momentum by as much as 20%. Much closer agreement is predicted by the more inclusive energy approach.

The observed reflection of a pressure surge at a completely closed valve was 10% below the value computed with the momentum analysis. This difference is easily predicted by energy considerations.

INTRODUCTION

The relationship between velocity and pressure in unsteady flow in a prismatic conduit has been established by several different methods. A fundamental method of analysis, employed by Allievi [*1*], Parmakian [*2*], Rich [*3*], and others, is the simultaneous solution of two partial differential equations. One differential equation is based on the conservation of momentum and the other on conservation of mass. The equations are:

$$\frac{\partial p}{\partial x} = \rho\left(\frac{\partial v}{\partial t} + v\,\frac{\partial v}{\partial x}\right) + f(v)\,, \tag{1}$$

$$\frac{\partial p}{\partial t} + v\,\frac{\partial p}{\partial x} = -\,c^2\rho\,\frac{\partial v}{\partial x}\,, \tag{2}$$

NOTATION

A	Cross-sectional area of conduit
c	Wave celerity in liquid
c_c	Wave celerity in conduit wall
D	Diameter of conduit
E_c	Elastic modulus of conduit
E_L	Elastic bulk modulus of liquid
L	Length of conduit
p, p_0, p_f, p_a	Pressure of liquid
t	Time variable
t'	Wall thickness
v, v_0, v_f, v_1	Mean velocity of liquid
x	Distance variable
ρ	Mass density of liquid
ϵ, ϵ_A	Strain, axial strain
$\sigma, \sigma_T, \sigma_A$	Stress, transverse stress, axial stress
μ	Poisson's ratio

where

c = celerity of wave,

p = pressure,

v = velocity,

t = time,

x = distance along conduit,

ρ = mass density,

$f(v)$ = pressure loss per unit length due to boundary shear, which is a function of velocity.

Even with simplifying assumptions, the simultaneous solution of these equations is difficult, usually requiring the use of sophisticated operational calculus techniques as described by Rich [4] and Wood [5].

An equally valid method of determining pressure-time and velocity-time relationships is to compute the magnitude of a pressure wave procreated at a point where the velocity undergoes a rapid change. The changes in velocity and pressure throughout the system are then determined by computing the effect of the pressure wave as it propagates throughout the system. The celerity, reflection, and transmission characteristics of the pressure wave must be known to the analyst. This method, sometimes referred to as a wave-plan solution, has an advantage over the more rigid mathematical treatment where solutions are limited to a number of relatively simple cases. The wave-plan solution when applied correctly can be used to analyze unsteady conditions in a highly complicated conduit network.

The relationship between pressure and velocity changes can be derived by considering either the conservation of mass, the conservation of momentum, or the conservation of energy. Consideration of any of these basic premises for the same limiting assumptions gives identical results.

Joukowsky [6] derived his classical expression by equating the volume of the net mass inflow into an elemental section of liquid and conduit (subjected to unsteady flow conditions) to the volume created by the compression of the liquid and the extension of the conduit walls. Joukowsky's expression is

$$\Delta p = -\rho c \Delta v \ . \tag{3}$$

The same expression was derived by Moody [7] using Newton's second law. The unbalanced forces (due to unsteady conditions) on an elemental volume of fluid were equated to the product of mass and acceleration of the fluid.

Several attempts have been made to obtain Eq. (3) by an analysis based on the conservation of energy [*8*, *9*, *10*]. These analyses were incorrect and agreed with Eq. (3) only for the special case of complete velocity extinction with negligible initial pressure. Incorrect reasoning, however, led these authors to conclude that there was general agreement between their results and Eq. (3).

ENERGY ANALYSIS

The correct relationship between pressure and velocity changes based on the conservation of energy is derived as follows: Neglecting thermal energy, the total energy associated with a certain volume of moving liquid consists of the kinetic energy of the liquid, the elastic energy stored in the liquid, and the elastic energy stored in the conduit which contains the liquid.

These energies may be expressed quantitatively if the following simplifying assumptions are made:

1. The velocity is constant throughout a cross section.
2. Changes in conduit area and liquid density (due to changes in pressure) can be neglected.
3. The conduit and liquid are elastic and Hookian.

Figure 1 shows the volume under consideration.

The kinetic energy of the liquid is given by

$$KE_L = \frac{1}{2} \rho A dx v^2 \ .$$

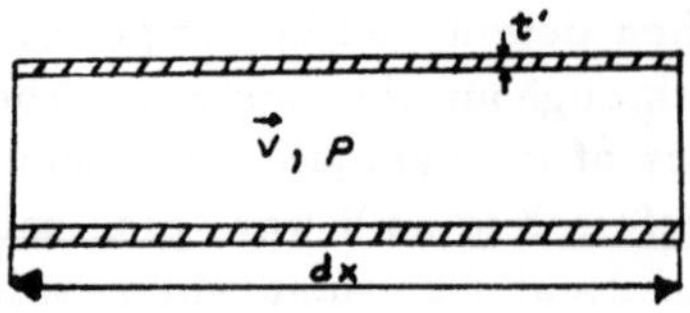

Fig. 1. Control Volume.

The energy stored in an elastic body is given by the area under the stress-strain curve (Fig. 2). The elastic energy due to the compression of the liquid is

$$PE_L = \frac{1}{2}\sigma\epsilon = \frac{1}{2}\frac{\sigma^2}{E_L} \text{ (per unit volume)},$$

where E_L = bulk modulus of liquid.

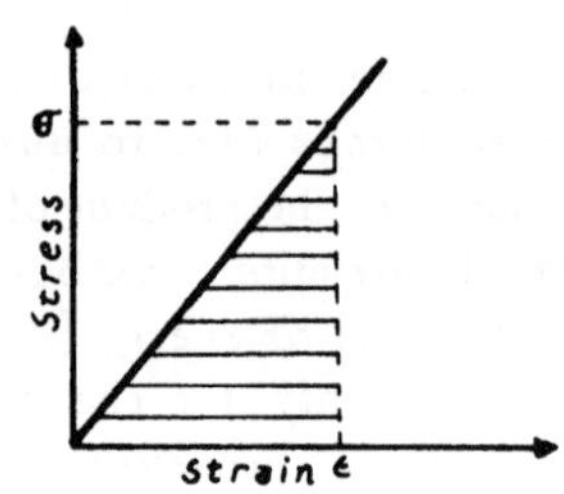

Fig. 2. Stress-Strain Curve for Elastic Hookian Material.

The stress in the liquid is the pressure under which it acts. Thus, the total energy stored by the liquid is

$$PE_L = \frac{1}{2}\frac{p^2}{E_L}(Adx) .$$

A similar expression for the elastic energy stored in the conduit walls is

$$PE_c = \frac{1}{2}\sigma\epsilon = \frac{1}{2}\frac{\sigma^2}{E_c} \text{ (per unit volume)},$$

where E_c = modulus of elasticity of conduit material restrained from lateral contraction. The stress is given by $\sigma = pD/2t'$, and the volume is approximately (for small wall thicknesses) equal to $\pi Dt'dx$. Thus, for the elemental length dx,

$$PE_c = \frac{1}{2}\left(\frac{pD}{2t'}\right)^2 \frac{\pi Dt'dx}{E_c} = \frac{p^2D}{E_c t'}\left(\frac{Adx}{2}\right) .$$

The total energy is given by

$$TE = KE_L + PE_L + PE_c$$

$$= \frac{1}{2} \rho A dx v^2 + \frac{1}{2} \frac{p^2 A dx}{E_L} + \frac{1}{2} \frac{p^2 D}{E_c t'} (A dx)$$

$$= \frac{A dx}{2} \left[\rho v^2 + p^2 \left(\frac{1}{E_L} + \frac{D}{E_c t'} \right) \right] .$$

The celerity of a wave in the conduit-liquid system is given by

$$c = \sqrt{\frac{E_L}{\rho} \Big/ \left(1 + \frac{E_L D}{E_c t'} \right)} .$$

Substituting this into the preceding equation gives

$$TE = \frac{A dx}{2} \left[\rho v^2 + \left(\frac{p}{c} \right)^2 \Big/ \rho \right] ,$$

or

$$TE = \frac{1}{2} (\text{vol}) \left[\rho v^2 + \left(\frac{p}{c} \right)^2 \Big/ \rho \right] . \tag{4}$$

With Eq. (4) it is possible to obtain the desired relationship between velocity and pressure changes for unsteady flow. Figure 3 depicts the effect of a

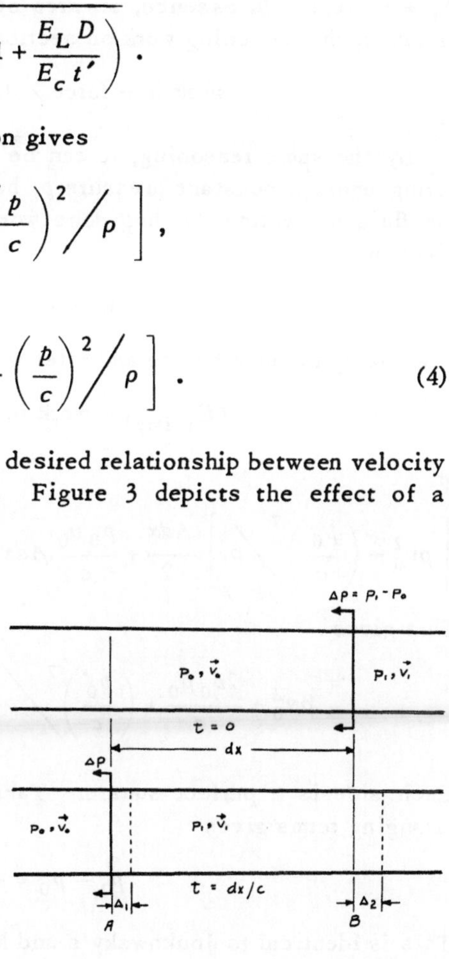

Fig. 3. Propagation of a Pressure Surge.

pressure wave as it moves from A to B in a closed conduit. The pressure wave traveling at a velocity equal to wave celerity (c) for the system changes the initial pressure and velocity (p_0, v_0) to new values (p_1, v_1).

The initial elastic and kinetic energy between A and B computed from Eq. (4) is

$$TE_{\text{initial}} = \left[\rho v_0^2 + \left(\frac{p_0}{c}\right)^2 \Big/ \rho \right] \frac{Adx}{2} .$$

The energy after wave passage is

$$TE_{\text{final}} = \left[\rho v_1^2 + \left(\frac{p_1}{c}\right)^2 \Big/ \rho \right] \frac{Adx}{2} .$$

It takes a finite time for the pressure wave to travel from A to B (time = dx/c). In this time the left boundary of the fluid has moved a distance $\Delta_1 = v_0\, dx/c$. In essence, an area of constant pressure p_0 has moved a finite distance, thereby doing work on section AB. This work is given by

$$\text{work in} = \text{force} \times \text{distance} = p_0 v_0\, dxA/c .$$

By the same reasoning, it can be seen that the right boundary of the fluid, acting under a constant pressure p, has moved a distance $\Delta_2 = v_1\, dx/c$. Thus, the fluid in section AB has done work on the fluid to the right. This work is given by

$$\text{work out} = p_1 v_1\, dxA/c .$$

The energy balance before and after wave passage is given by

$$TE_{\text{initial}} + \text{work in} = TE_{\text{final}} + \text{work out} ,$$

or

$$\left[\rho v_0^2 + \left(\frac{p_0}{c}\right)^2 \Big/ \rho \right] \frac{Adx}{2} + \frac{p_0 v_0}{c} Adx = \left[\rho v_1^2 + \left(\frac{p_1}{c}\right)^2 \Big/ \rho \right] \frac{Adx}{2} + \frac{p_1 v_1}{c} Adx .$$

This gives

$$\rho v_0^2 + \frac{2 v_0 p_0}{c} + \left(\frac{p_0}{c}\right)^2 \Big/ \rho = \rho v_1^2 + \frac{2 v_1 p_1}{c} + \left(\frac{p_1}{c}\right)^2 \Big/ \rho . \tag{5}$$

Each side is a perfect square. Taking the square root of both sides and rearranging terms gives

$$p_1 - p_0 = -\rho c (v_1 - v_0) .$$

This is identical to Joukowsky's and Moody's result.

Since it takes a finite time for the pressure wave to travel from A to B, a certain mass of liquid flows into the left end and out of the right end of section AB. The inflow and outflow do not possess the same energy, and, thus, a refinement of Eq. (5) could be made by adding the energy of the inflow and subtracting the energy of the outflow. It can be shown that these energies are in the ratio of v/c to the terms in Eq. (5). This is usually a very small number, and the refinement is unnecessary.

An advantage of the preceding analysis of unsteady conduit flow based on the conservation of energy over the more conventional analyses based on the conservation of mass and momentum is that the energy approach lends itself better to physical interpretation. Because of this, Eq. (5) can be modified to include effects which would be difficult to analyze by the more conventional methods.

EFFECTS OF VELOCITY DISTRIBUTION

The assumption that the velocity and the pressure are constant across a conduit cross section was employed when Eqs. (3) and (5) were obtained. However, for the flow of real fluids, especially at lower Reynolds numbers, this is not true. In real fluid flow, the velocity profile is never constant across a section of conduit. In addition, the flow is never strictly unidirectional. One of the quantities most severely affected by assuming ideal conditions is the kinetic energy of the flowing liquid.

The kinetic energy is usually computed from the mean velocity as

$$KE = \frac{1}{2}\rho\bar{v}^2 \text{ (per unit volume) },$$

where $\bar{v}$ = mean velocity of liquid. In order for this to be the true kinetic energy, the velocity must be constant across a section and the flow unidirectional. The more the actual flow conditions deviate from these assumed conditions, the greater is the kinetic energy. The kinetic energy computed from the mean velocity represents an absolute minimum for that discharge.

When a valve is suddenly closed, the initial kinetic energy in a certain section of conduit and the work done on the liquid in that section during wave passage are converted into elastic energy, resulting in a pressure rise in the conduit section. If none of this energy is converted into heat, the magnitude of the pressure wave generated by instantaneous stoppage of flow would depend on the initial kinetic energy of the flow. The value computed by Eqs. (3) and (5) (employing the mean velocity) would represent a minimum.

Equation (5) can be easily modified to study the effect of the difference between the actual kinetic energy and the mean-velocity kinetic energy. The kinetic-energy factor K is defined as the ratio of actual kinetic energy to the mean-velocity kinetic energy (per foot of conduit). Equation (5) becomes

$$K\rho^2c^2v_0^2 + \rho^2p_0v_0c + p_0^2 = p_f^2 , \tag{6}$$

where

p_0, v_0 = pressure, mean velocity before valve closure,

p_f = pressure after valve closure.

The ratio of initial pressure to final pressure is defined as A:

$$p_0/p_f = A .$$

Substituting this into Eq. (6) gives

$$(p_f)^2 (1 - A^2) - (2Ap_f)(\rho c v_0) - K(\rho c v_0)^2 = 0 \,. \tag{7}$$

The pressure rise for sudden stoppage of flow with a mean velocity v_0 computed from the Joukowsky equation is denoted as

$$\Delta p_0 = \rho c v_0 \,.$$

Substituting this into Eq. (7) and multiplying through by $1/\Delta p_0^2$ gives

$$\left(\frac{p_f}{\Delta p_0}\right)^2 (1 - A^2) - 2A\left(\frac{p_f}{\Delta p_0}\right) - K = 0 \,.$$

The quadratic formula gives

$$\frac{p_f}{\Delta p_0} = \frac{A \pm \sqrt{A^2 + K(1 - A^2)}}{1 - A^2} \,. \tag{8}$$

This result can be easily modified to give the ratio of the actual pressure rise (Δp_a) to the rise predicted by Joukowsky's equation. Since

$$\Delta p_a = p_f - p_0 \,,$$

dividing by Δp_0 gives

$$\frac{\Delta p_a}{\Delta p_0} = \frac{p_f}{\Delta p_0} - \frac{p_0}{\Delta p_0} \,,$$

$$p_0 = A p_f \,.$$

Thus,

$$\frac{\Delta p_a}{\Delta p_0} = \frac{p_f}{\Delta p_0}(1 - A) \,.$$

Substituting Eq. (8) for $p_f/\Delta p_0$ gives

$$\frac{\Delta p_a}{\Delta p_0} = \frac{A + \sqrt{A^2 + K(1 - A^2)}}{1 + A} \tag{9}$$

A look at the extremes for the initial pressure is didactic. When $p_0 = 0$, $A = 0$ and $\Delta p_a/\Delta p_0 = \sqrt{K}$ $(K > 1)$. This is when the effect of K is most severe. When $p_0/p_f = A \rightarrow 1$, or the pressure rise is small compared with the initial pressure,

$$\Delta p_a/\Delta p_0 = 1 \,.$$

In this case, the effect of the difference in the actual kinetic energy and the mean-velocity kinetic energy is greatly mitigated. The reason is that the work done by the initial pressure during compression is much greater than the initial kinetic energy of the liquid.

Figure 4 shows the effect of *K* on the ratio of the actual pressure rise to the computed mean-velocity pressure rise for the possible range of initial final pressure ratios. This chart makes it possible to analyze the experimental results for the instantaneous valve closure which have been presented in the literature. If the ratio of the calculated and observed pressure rises and the initial pressure to the final pressure can be found from the information in a publication, it is an easy matter to enter the chart and find what value of *K* would produce such a pressure rise if no energy were lost in the compression.

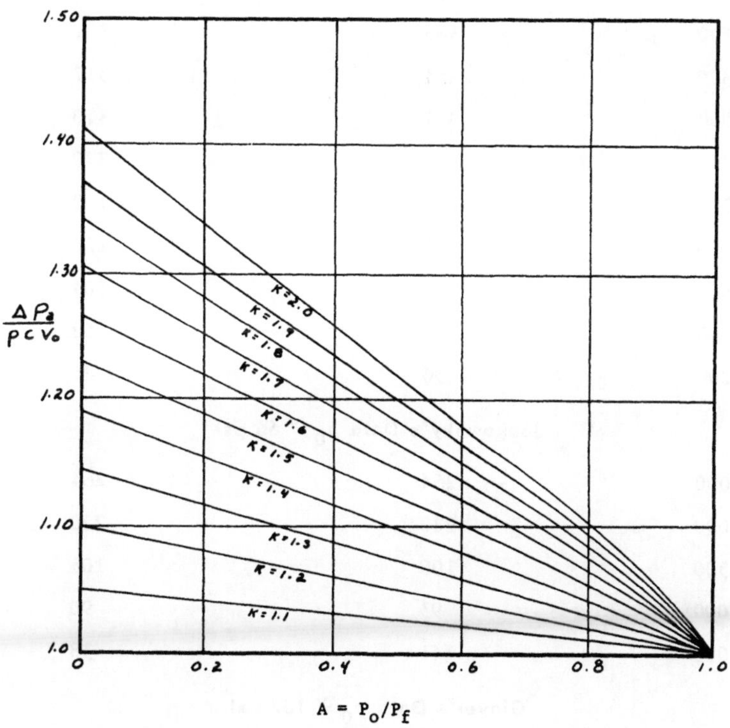

Fig. 4. Effect of Initial Pressure and Kinetic-Energy Factor on Pressure Rise for Instantaneous Valve Closure.

The authors measured pressure rises due to the rapid velocity extinction of flows of Reynolds numbers from 5000 to 40,000 [*11*]. In this series of tests, the initial pressures were negligible when compared with the pressure rises. Two other publications which give experimental results for sudden valve closures in the desired range of Reynolds numbers (less than 50,000) are the classical paper by Joukowsky [6] and a report by Glover [*12*] submitted to the Symposium on Water Hammer in 1933. These experimental results are summarized in Table 1.

Table 1. Summary of Experimental and Calculated Pressure-Rise Data

Reynolds Number	Observed Pressure Rise Δp_a (psi)	Calculated Pressure Rise $\Delta p_0 = \rho c \Delta v_0$ (psi)
	Authors' Data $p_0 \cong 0$ psi	
39,000	753	748
35,000	740	731
28,000	609	585
24,400	550	506
23,000	521	481
15,500	386	321
15,300	354	317
15,000	347	310
12,700	318	270
9,100	226	190
5,800	145	120
5,000	127	104
2,800	71	58
2,200	56	46
	Joukowsky's Data $p_0 \cong 66$ psi	
48,000	264	264
40,000	220	220
19,500	108	103
17,000	93	90
7,000	41	38
	Glover's Data $p_0 \cong 132$ psi	
5,700	33.5	31.7
5,700	34.7	33.0
3,900	36.5	33.0

For each of the experimental results, the observed pressure rises were somewhat higher than the value obtained by Joukowsky's equation. This disparity increased with decreasing Reynolds numbers. From Fig. 4 the values of the kinetic-energy factors (K) which would cause the observed pressure rises were obtained. The results are shown in Fig. 5.

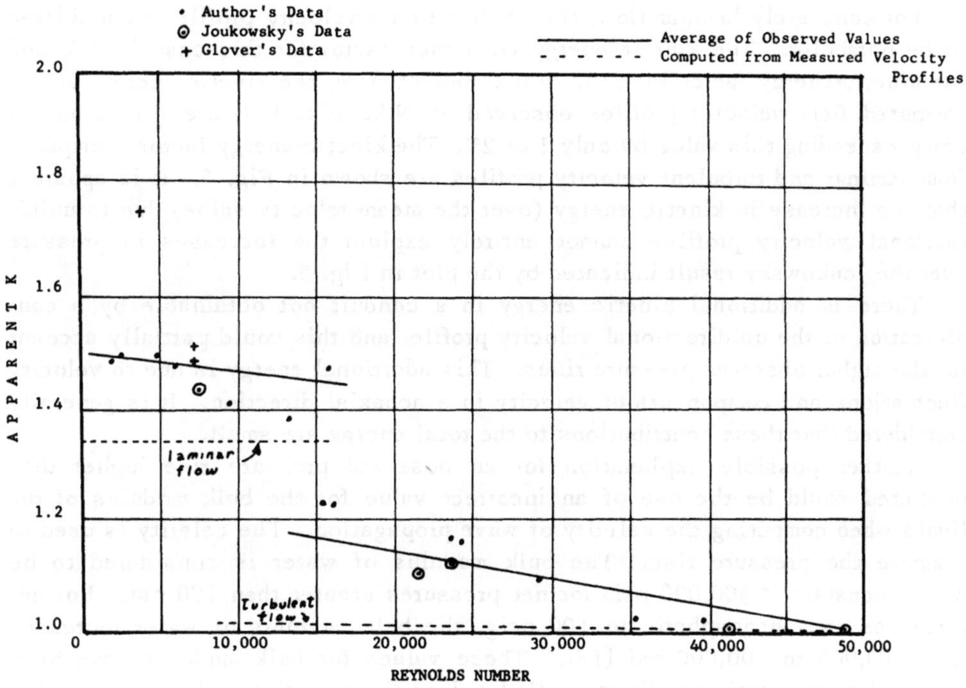

Fig. 5. Variation of Apparent Kinetic-Energy Factor with Reynolds Number.

Attributing the complete disparity between observed and predicted results to the differences between actual and mean-velocity, kinetic energy may be erroneous. However, Fig. 5 illustrates several interesting trends.

The tendency for the kinetic-energy factor to increase with decreasing Reynolds numbers is evident. In addition, there appears to be an abrupt change in the value of the kinetic-energy factor for flows of Reynolds numbers between 13,000 and 17,000. The latter might be interpreted as the transition between laminar and turbulent flow. The range of Reynolds numbers observed for the transition appears high, but the transition depends to a great extent on intangible conditions such as the initial quiescence of the fluid.

The kinetic-energy factors for various pipe flows can be computed from measured unidirectional velocity profiles. Quantitative agreement between these computed factors and the apparent kinetic-energy factors is not good.

The kinetic-energy factor used in this paper differs from the kinetic-energy correction factor defined in many standard texts on fluid mechanics. The kinetic-energy correction factor is the ratio of the actual kinetic energy to the mean-velocity kinetic energy per pound of transmitted fluid, and it is proportional to the integral of the third power of the velocity over the cross section of the conduit. The kinetic-energy factor defined in this paper is the ratio of the actual kinetic energy to the mean-velocity kinetic energy per foot of conduit, and it is proportional to the integral of the second power of the velocity over the cross section.

For completely laminar flow, the unidirectional velocity profile is considered to be parabolic. The kinetic-energy correction factor for this case is 2.0, and the kinetic-energy factor is 1.33. For turbulent flow, the kinetic-energy factors computed from velocity profiles observed by Nikuradse [*13*] are very close to unity, exceeding this value by only 1 or 2%. The kinetic-energy factors computed from laminar and turbulent velocity profiles are shown in Fig. 5. It is apparent that the increase in kinetic energy (over the mean-velocity value) due to unidirectional velocity profiles cannot entirely explain the increases in pressure over the Joukowsky result indicated by the plot in Fig. 5.

There is additional kinetic energy in a conduit not obtainable by a consideration of the unidirectional velocity profile, and this could partially account for the higher observed pressure rises. This additional energy is due to velocity fluctuations and components of velocity in a nonaxial direction. It is generally considered that these contributions to the total energy are small.

Another possible explanation for an observed pressure rise higher than predicted could be the use of an incorrect value for the bulk modulus of the liquid when computing the celerity of wave propagation. The celerity is used to compute the pressure rise. The bulk modulus of water is considered to be nearly constant (~300,000 psi) for net pressures greater than 100 psi. For net pressures from atmospheric to 100 psig, the bulk modulus of water increases from 200,000 to 300,000 psi [*14*]. These values for bulk modulus have been observed under static conditions. Under dynamic conditions, the bulk modulus may differ from the value measured under static conditions. This difference could be more significant for lower pressure rises.

The trend shown in Fig. 5 of the apparent kinetic-energy factors exceeding ones predicted from consideration of velocity profiles could be due to changes in the bulk modulus of a liquid subjected to dynamic conditions (if the liquid became stiffer). Furthermore, values reported by investigators for the bulk modulus often differ by 4 or 5% [*14*]. The use of an incorrect value introduces a systematic error into predicted values for pressure rises.

Additional experimental test data for rapid velocity extinction of flows of low Reynolds numbers and small ratios of initial pressure to pressure rises would help establish more definite conclusions. Additional experiments, however, would not alter the trends shown in Fig. 5 and predicted by an energy analysis of unsteady conduit flow.

CLOSED-END REFLECTION

The usefulness of the energy analysis is not limited to flows where the differences in the mean-velocity and actual kinetic energy are significant. The energy analysis can be fruitfully employed to explain other experimental observations which differ from previous analytical predictions.

Standard references on water hammer all indicate that a pressure surge undergoes a perfect positive reflection at a closed-end conduit [1, 2, 3]. Experimental observations, however, have indicated that these reflections are not perfect. An analysis of pressure-surge reflection at a closed-end conduit based on the conservation of energy offers both qualitative and quantitative insight into the reasons for imperfect reflections.

When a pressure wave is reflected from a closed end, part of the energy is used to stretch the pipe axially, and elastic energy is stored due to this stretching. (Only the energy stored in the pipe due to transverse strain was considered in the derivation of the water-hammer equation based on the conservation of energy.) Since part of the energy goes into axially stretching the pipe, the reflection cannot be perfect.

The exact amount of energy which goes into axially stretching the pipe is a function of the fixity of the pipe and Poisson's ratio for the conduit. An upper limit can easily be determined for the case where the end of the conduit is completely free to move. (Under these conditions, maximum energy would be expended in axially stretching the pipe.) Figure 6 shows this case before and after reflection from a dead end.

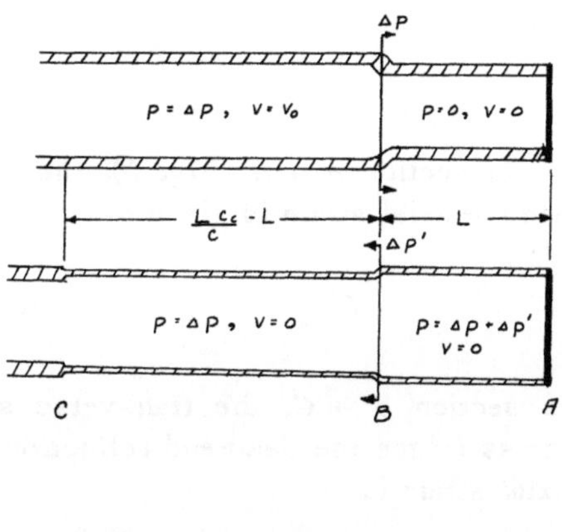

Fig. 6. Conditions Before and After Pressure-Surge Reflection from a Dead End.

The changes in cross-sectional area and wall thickness are practically negligible when the material is stressed within the elastic limit. These effects are emphasized in Fig. 6 to illustrate the position of the pressure wave in the conduit and the stress wave in the conduit wall. The energy equation (5) gives

$$\frac{AL}{2}\left[\rho v_0^2 + \frac{\Delta p^2}{\rho c^2} + \frac{2\Delta p v_0}{c}\right] = \frac{AL}{2}\left(\frac{\Delta p + \Delta p_1}{\rho c^2}\right)^2 + PE_{ca} ,$$

where PE_{ca} is the energy expended in axially stretching the conduit. Multiplying by ρc^2 and substituting $\Delta p = \rho c v_0$ gives

$$\frac{AL}{2}(4\Delta p^2) = \frac{AL}{2}(\Delta p + \Delta p_1)^2 + \rho c^2(PE_{ca}) . \tag{10}$$

The energy expended in axially stretching the conduit can be determined as follows: The stress wave in the conduit wall travels at the speed of wave propagation for the conduit material (c_c). In the time it takes for the pressure wave to travel a distance L in the conduit, the stress wave in the conduit wall has gone a distance Lc_c/c (see Fig. 6), and that length is acting under an axial stress given by

$$\sigma_a = \frac{pD}{4t'} = \frac{(\Delta p + \Delta p_1)D}{4t'} \; . \tag{11}$$

The energy stored in the conduit is given by

$$PE_{ca} = \frac{1}{2} \sigma_a \, \epsilon_a (\text{vol}) \; . \tag{12}$$

The axial strain (ϵ_a) is a function of Poisson's ratio (μ) and, hence, the transverse stress of the material:

$$\epsilon_a = \frac{1}{E_c} (\sigma_a - \mu \sigma_T) \; .$$

The transverse stress is given by

$$\sigma_T = \frac{pD}{2t} \; .$$

In section $A \to B$ (Fig. 6), the transverse stress is twice the axial stress, and the axial strain is given by

$$\epsilon_a(A \to B) = \frac{\sigma_a}{E_c} (1 - 2\mu) \; . \tag{13}$$

In section $B \to C$, the transverse stress is approximately equal to the axial stress (since the dead-end reflection is nearly perfect and $\Delta p \cong \Delta p_1$), and the axial strain is

$$\epsilon_a(B \to C) = \frac{\sigma_a}{E_c} (1 - \mu) \; . \tag{14}$$

The total energy stored in the conduit from Eqs. 12–14 is

$$PE_{ca} = \frac{\sigma_a^2}{2E_c} [(1 - 2\mu)(\text{vol } A \to \text{B}) + (1 - \mu)(\text{vol } B \to C)] \; . \tag{15}$$

The volume of the conduit material from A to B is

$$\text{vol } (A \to B) = \pi D t' L \; ,$$

and the volume from B to C is

$$\text{vol } (B \to C) = \pi D t' L \left(\frac{c_c}{c} - 1 \right) \; .$$

Substituting these into Eq. (15) and inserting the value of the axial stress from Eq. (11) gives

$$PE_{ca} = (\Delta p + \Delta p_1)^2 \left(\frac{\pi D^2 L}{8}\right) \left\{\frac{D}{4t' E_c}\left[\frac{c_c}{c} - \mu\left(\frac{1 + c_c}{c}\right)\right]\right\}.$$

Putting this result into Eq. (10) and canceling $AL/2$ gives

$$4\Delta p^2 = (\Delta p + \Delta p_1)^2 (1 + \alpha),$$

where α is a function of the physical properties of the liquid and pipeline and is given by

$$\alpha = \frac{D\rho c^2}{4t' E_c}\left[\frac{c_c}{c} - \mu\left(1 + \frac{c_c}{c}\right)\right]. \tag{16}$$

Combining terms gives $(\Delta p_1)^2(1 + \alpha) + 2\Delta p(1 + \alpha)\Delta p_1 - \Delta p^2(3 - \alpha) = 0$. The quadratic formula gives a result which reduces to

$$\Delta p_1 = \Delta p\left(\frac{2\sqrt{1 + \alpha}}{1 + \alpha} - 1\right). \tag{17}$$

A plot showing the loss in the reflected wave due to the energy expended in axially stretching the conduit is given in Fig. 7.

The authors, while conducting experiments on ½-in. copper tubing, observed a loss of 10.0% in the magnitude of a wave reflected from a dead end [11]. The coefficient α for the experimental pipeline was 0.10. The preceding analysis predicts a loss of 9.6%, which is in close agreement to the observed value.

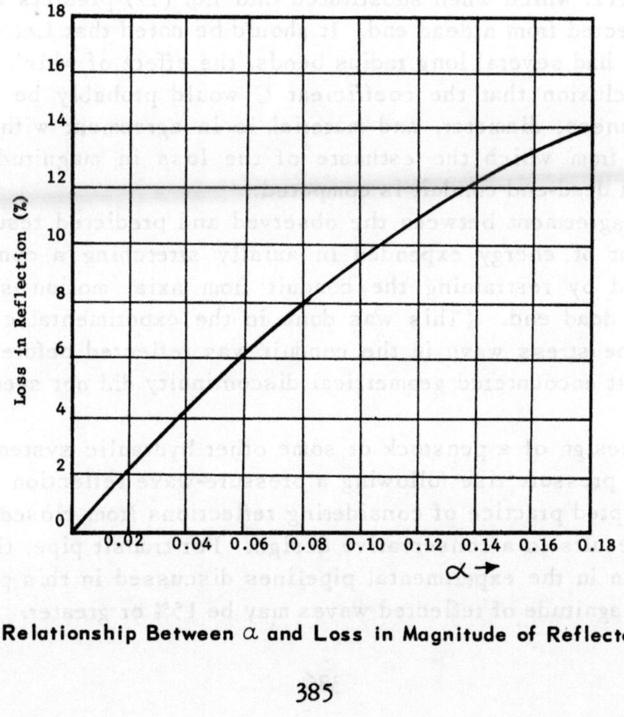

Fig. 7. Relationship Between α and Loss in Magnitude of Reflected Wave.

One other time in the literature there has been an indication that a loss in the magnitude of a pressure wave reflected from a dead-end conduit was observed. LeConte [*15*], while attempting to compute the time between resurges after sudden valve closure, observed the loss in reflection at a dead-end conduit and treated it in a different manner. The following passage is taken from LeConte's publication.

"The results of laboratory experiment make it clear that the velocity with which the water strikes the valve on any one of these surges is always greater than that with which it starts away on the next surge. This relation can be expressed as $V_0 = CV_{01}$, V_0 being the velocity of the return surge, and V_{01} that of the approach. The coefficient C, which seems to be of the nature of a 'restitution coefficient' in the mechanics of impact and collision, is nearly constant for the set of waves of one experiment and approximately so for different velocities in a single pipeline. Strictly, it should be a function of the velocity and pipe diameter and probably of the wall thickness and material of the pipe. It seems to diminish in numerical value as the velocity increases."

LeConte also emphasizes that the above effect is different from reductions in velocities due to friction in the pipeline. For his short experimental pipeline, the losses in velocities of adjacent surges due to friction are small. LeConte's experimental observations can be attributed to an imperfect reflection at the closed end of his experimental pipeline. If the reflection at the dead end were perfect, the surge would move away from the valve with the same velocity which was originally extinguished. Since the velocity has been reduced by the coefficient C, it follows from Joukowsky's equation that the wave causing the velocity change has been reduced by the same ratio.

The coefficient C observed by LeConte was approximately 0.87, or a 13.0% loss in magnitude of the wave was experienced. The coefficient α for LeConte's pipeline was 0.12, which when substituted into Eq. (17) predicts an 11.0% loss for a wave reflected from a dead end. It should be noted that LeConte's experimental pipeline had several long radius bends, the effect of which is not known. LeConte's conclusion that the coefficient C would probably be a function of pipe wall thickness, diameter, and material is in agreement with terms in the coefficient α, from which the estimate of the loss in magnitude of a wave reflected from a dead-end conduit is computed.

The close agreement between the observed and predicted results indicates that the amount of energy expended in axially stretching a conduit was not greatly affected by restraining the conduit from axial motion some distance away from the dead end. (This was done in the experimental cases.) Also, the fact that the stress wave in the conduit was reflected before the pressure wave at the first encountered geometrical discontinuity did not seem to alter the result.

Often the design of a penstock or some other hydraulic system is based on the anticipated pressure rise following a pressure-wave reflection from a closed end. The accepted practice of considering reflections from closed-end conduits to be perfect results in a conservative design. For transit pipe, the coefficient α is higher than in the experimental pipelines discussed in this paper, and the losses in the magnitude of reflected waves may be 15% or greater.

SUMMARY

Each of the preceding results could have been obtained by an analysis which utilized the conservation of momentum. For the first analysis dealing with flows with a nonlinear velocity profile, the derivation could have been made by applying Joukowsky's equation to a series of concentric, annular rings, each with a different fluid velocity. This would result in a conduit with zero fluid velocity and varying pressure across a section. The pressure must then be allowed to reach equilibrium with no loss in elastic energy.

LeConte's method of assigning a "restitution coefficient" to a pressure surge impinging on a closed end is an attempt to utilize a momentum approach. However, it involves the use of the experimentally observed restitution coefficient, and this approach is less acceptable than the one presented in this paper.

The energy analysis of elastic pressure waves provides a convenient and simple method of studying fundamental as well as refined aspects of the phenomenon.

REFERENCES

[1] L. Allievi, "Theory of Water Hammer," translated by E. E. Halmos, presented by Riccardo Garoni, Rome, Italy, 1925.

[2] J. Parmakian, *Water Hammer Analysis,* Prentice Hall, New York, 1955.

[3] G. R. Rich, *Hydraulic Transients,* 1st ed., Engineering Societies Monographs, McGraw-Hill, New York, 1951.

[4] G. R. Rich, "Water Hammer Analysis by the Laplace-Mellin Transformations," *Trans. ASME* **67**, 361–68 (1945).

[5] F. M. Wood, "The Application of Heavysides Operational Calculus to the Solution of Problems in Water Hammer," *Trans. ASME* **59**, 707–13 (1937).

[6] N. Joukowsky, "Water Hammers" (translated by O. Simin), *Proc. Am. Water Works Assoc.* **24**, 341–424 (1904).

[7] L. F. Moody, "Simplified Derivation of Water Hammer Formula," *Symposium on Water Hammer,* pp 25–28, ASME-ASCE, 1933.

[8] R. L. Daugherty, *Hydraulics*, pp 300–308, McGraw-Hill, New York, 1937.

[9] N. M. Sverdrup, "Pressure Surges in Hydraulic Circuits," *Prod. Eng.* **24**, 161–69 (1953).

[10] M. M. Warren, "Penstock and Surge Tank Problems," *Trans. Am. Soc. Civil Engrs.* **79**, 238–305 (1915).

[11] D. J. Wood, *Generation, Transmission and Reflection of Pressure Surges in Conduits,* Ph.D. Thesis, Carnegie Institute of Technology, September 1961.

[12] R. Glover, "Computation of Water Hammer Pressures in Compound Pipes," *Symposium on Water Hammer*, pp 64–69, ASME-ASCE, 1933.

[13] H. Schlichting, *Boundary Layer Theory*, McGraw-Hill, New York, 1955.

[14] S. L. Kerr, L. H. Kessler, M. B. Gamet, "New Method for Bulk Modulus Determinations," *Trans. ASME* **72**, 1143–57 (1950).

[15] J. N. LeConte, "Experiments and Calculations on the Resurge Phase of Water Hammer," *Trans. ASME* **59**, 691–94 (1937).

Experimental and Applied Mechanics

STRESSES IN ROTATING TAPERED DISKS WITH NONCENTRAL HOLES

James Hal Armstrong
Georgia Institute of Technology

An approximate solution for stresses in rotating tapered disks with noncentral holes is presented. The new feature of the solution is the "spoke effect." This "spoke effect" accounts for the additional strain experienced by the outer annulus of the disk due to the presence of the noncentral holes.

Frozen-stress photoelastic analyses of a tapered disk and of a uniform disk, both containing noncentral holes, are included in the paper. The agreement between the computed stresses and the actual stresses at critical points in both disks is well within 3%. At other locations in the disks the actual stresses are lower than the computed stresses. This apparent discrepancy is explained and discussed.

INTRODUCTION

In recent years considerable study has been given to the problems of rotating disks of arbitrary profile with or without central holes and of disks of uniform thickness with various arrangements of noncentral holes. Udoguchi [*1*] and Saito [*2*] have developed theoretical solutions for uniform rotating disks with noncentral holes. Prior to these solutions several investigators have made photoelastic studies of similar types of disks. Barnhart, Hale, and Meriam [*3*] have done one such investigation and have included in their report an approximate theoretical analysis which consisted of first determining the stresses existing in a solid disk and then applying appropriate stress concentration factors to take care of the effect of the holes. Their results, in general, check the photoelastic analysis fairly closely. However, in the outer annulus of the disk the actual stresses appear somewhat higher than those determined by their approximate theory.

In an attempt to apply their method to a similar disk, the author found the same type of discrepancy, which was even larger for the particular case then under investigation. It is proposed that this apparent discrepancy can be eliminated, or at least reduced, by utilizing the "spoke effect," which is described here and which was first proposed by the author in 1960 [*4*].

NOTATION

A_g	Gross circumferential area per hole at the radial distance of the noncentral holes
A_{sp}	Minimum cross-sectional area of a spoke, in.^{2}
E	Modulus of elasticity, tension, or compression, psi
f	Material fringe value, pounds shear per fringe inch
h	Thickness of disk at radius r, in.
L_{sp}	Length of spoke, in.
n	Fringe order in a photoelastic specimen
r	General radius, in.
r'	Average radius of outer annulus of disk, in.
R	Radius to where extended faces of a uniformly tapered disk intersect, in.
X	r/R
∂	Partial derivative symbol
δ	Additional extension of spoke, in.
ϵ	Unit strain, tension, or compression
γ	Unit strain, shear
ρ	Mass density of material, lb sec^2 in.$^{-4}$
σ_c	Tangential stress correction for outer annulus, psi
σ_r	Radial stress, psi
σ_{rh}	Radial stress in the disk without noncentral holes at the radial location for such holes, psi
σ_θ	Tangential stress, psi
ω	Angular speed of disk, radians/sec

APPROXIMATE THEORETICAL SOLUTION FOR A TAPERED DISK WITH NONCENTRAL HOLES

In this analysis the following assumptions concerning the disk and its material are made:

1. The material obeys Hooke's law; that is, the proportional limit is nowhere exceeded.
2. The material is homogeneous.
3. Except for the holes the thickness of the disk is a function of the radius only.

4. The disk revolves about its own axis and is essentially symmetrical with respect to that axis.

5. The disk is symmetrical with respect to its own plane.

6. The effect of gravity is neglected and there are no load components perpendicular to the plane of the disk.

7. The disk is relatively thin; the stresses parallel to the axis of rotation are considered negligible.

Consider any tapered disk having symmetrically spaced noncentral holes. One first obtains the solution for the "filled disk," that is, a disk similar to the given disk but having all noncentral holes filled in with the same material as that in the disk. Such solutions are available in the literature. For an arbitrary disk profile, where the exact solution does not exist, various approximate solutions are available; most of these consist in approximating the filled disk by means of a series of annular rings, each ring having a uniform, conical, parabolic, or hyperbolic profile.

After the solution for the filled disk is obtained, the "spoke effect" is next employed. The spokes, or regions between the holes, are under somewhat higher stress than the corresponding material in the filled disk. This means that such material will experience greater strain than that in the filled disk. This spoke strain allows the outer annulus to expand more than it would were there no holes present.

This additional expansion is allowed for by computing the effective strain in one of the spokes and from this strain computing the increase of the tangential stresses in the outer annulus on the basis of thin-ring theory, using the average radius of the outer annulus. As this correction is a relatively small one, thick-ring theory is thought to be an unnecessary refinement. Thin-ring theory neglects the effect on the radial stresses, so the spoke-effect correction is added to the tangential stresses only as a constant.

The final step is to apply appropriate stress concentration factors for the type of noncentral hole occurring in the disk. As the variation in thickness was accounted for in the initial calculation, these factors should be based on a plate of uniform thickness. The appropriate factors for both the tangential and the radial stresses must be applied. It should be noted that at the inner and outer boundaries of the disk the radial stresses have known values. The final calculation of the stresses should include the slight local adjustments necessary to bring such stresses to their known boundary values.

As a check for the method described, a disk of epoxy resin with a specific gravity of 1.10 was analyzed. The pertinent dimensions are given in Fig. 1, and the speed of rotation was 2670 rpm.

The stresses in the corresponding filled disk, that is, the disk without noncentral holes, were determined by Martin's method [5] and are shown in Fig. 2.

The spoke effect is next calculated. For this disk,

$$A_g = 1.85h \ ,$$

$$A_{sp} = 1.11h \ .$$

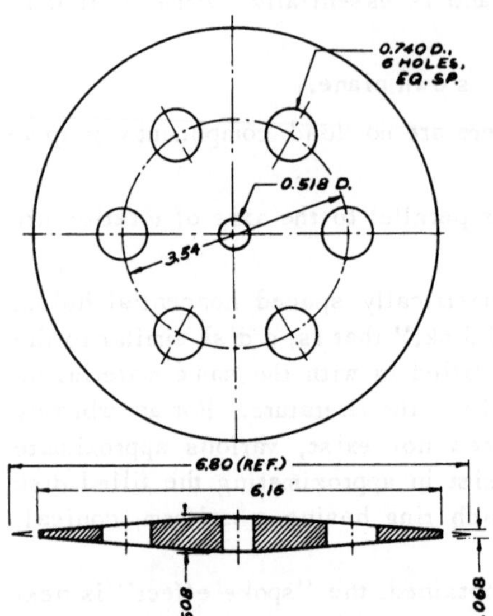

Fig. 1. Detail of Tapered Disk with Noncentral Holes.

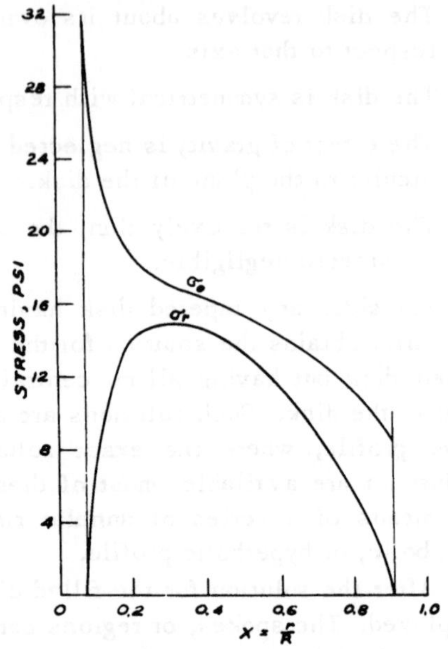

Fig. 2. Stresses in a Tapered Disk Containing Only a Central Hole.

The additional extension of each spoke, computed on the basis of the relative areas involved, is

$$\delta = \frac{\sigma_{rb}}{E}\left(\frac{A_g}{A_{sp}} - 1\right)L_{sp}$$

$$= \frac{12.9}{E}(1.67 - 1)(0.74)$$

$$= \frac{6.40}{E} \, . \tag{1}$$

The average radius r' of the outer annulus of the disk was 2.61 in., so the stress correction for the outer annulus is

$$\sigma_c = \frac{\delta(E)}{r'}$$

$$= \frac{6.40(E)}{E(2.61)}$$

$$= 2.44 \text{ psi} \, . \tag{2}$$

In the example here this was rounded to 2.4 and added to the tangential stresses outboard of the noncentral holes.

Appropriate stress concentration factors, in this case based on a circular hole in a large uniform plate, are then applied to the stresses thus far obtained. Barnhart *et al.* [3, p 48] give these relationships:

$$\sigma_\theta = S_1 + \frac{1}{2}(S_1 + S_2)\frac{a^2}{r^2} + \frac{3}{2}(S_1 - S_2)\frac{a^4}{r^4},$$

$$\sigma_r = S_2 - \frac{1}{2}(S_1 + S_2)\frac{a^2}{r^2} - \frac{3}{2}(S_1 - S_2)\frac{a^4}{r^4} + 2(S_1 - S_2)\frac{a^2}{r^2}, \tag{3}$$

where

a = radius of hole,

r = distance from center of hole,

S_1 = tangential stress in undrilled disk,

S_2 = radial stress in undrilled disk.

As previously mentioned, at the central hole and at the periphery, the radial stress is arbitrarily brought to zero to fit the known boundary values; this is a relatively small local adjustment.

The stresses resulting from these calculations are shown by the dashed curves in Fig. 3. A similar set of curves calculated for a uniform disk of the same material and having a similar hole plan is shown in Fig. 4.

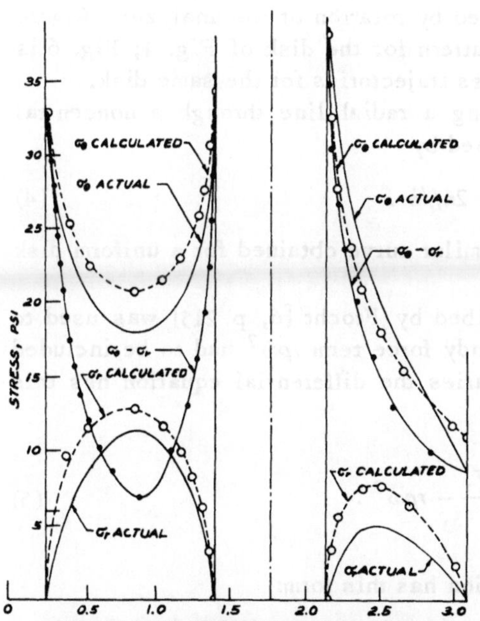

Fig. 3. Distribution of Stresses in a Tapered Disk with Holes.

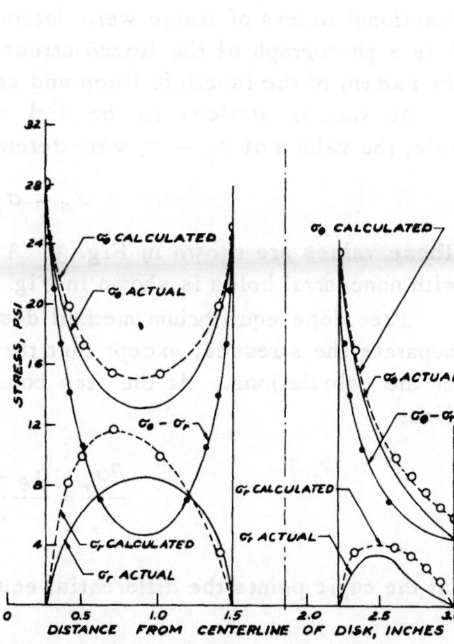

Fig. 4. Distribution of Stresses in a Uniform Disk with Holes.

A numerical integration of the calculated stresses over the radial area of the tapered disk yielded a total force of 16.78 lb, as compared with 16.80 lb from a similar integration for the filled disk. This close agreement is to be expected, since the stress concentration factors merely redistribute a given total force over a new area. One significant point which this method does not take into account is the reduction in the total force due to the loss of centrifugal force with removal of the hole material. This will be discussed in the latter portion of the following section.

PHOTOELASTIC ANALYSIS

Several disk models of Bakelite ERL 2795 epoxy resin cured with ZZ-L-0814 were made and rotated in a stress-freezing oven. These included a disk with the same dimensions as that analyzed in the previous section and shown in Fig. 1. The speed of rotation was also 2670 rpm. The resulting frozen stresses were then studied under monochromatic polarized light with techniques which are standard procedure in photoelastic analysis [6].

The automatically controlled stress-freezing cycle was as follows:

1. heating to 255°F, about 15 min;
2. soaking at this temperature, 120 min;
3. cooling to room temperature, 5–10°F per hour.

The calibration member was loaded in direct tension. For the particular disk under consideration the material fringe value was $f = 0.555$ lb per fringe inch. Fractional orders of fringe were determined by rotation of the analyzer. Figure 5 is a photograph of the frozen-stress pattern for the disk of Fig. 1; Fig. 6 is the pattern of the isoclinic lines and stress trajectories for the same disk.

At various stations in the disk along a radial line through a noncentral hole, the values of $\sigma_\theta - \sigma_r$ were determined by

$$\sigma_\theta - \sigma_r = 2nf/h \, . \tag{4}$$

These values are shown in Fig. 3. A similar curve obtained for a uniform disk with noncentral holes is shown in Fig. 4.

The slope equilibrium method described by Frocht [6, p 215] was used to separate the stresses, except that the body force term $r\rho\omega^2$ had to be included in the calculations. At the free boundaries the differential equation has this form:

$$\frac{\partial \sigma_r}{\partial_r} = \frac{\sigma_\theta - \sigma_r}{r} - r\rho\omega^2 \, . \tag{5}$$

At the cupic points the differential equation has this form:

$$\frac{\partial \sigma_r}{\partial_r} = - r\rho\omega^2 \, . \tag{6}$$

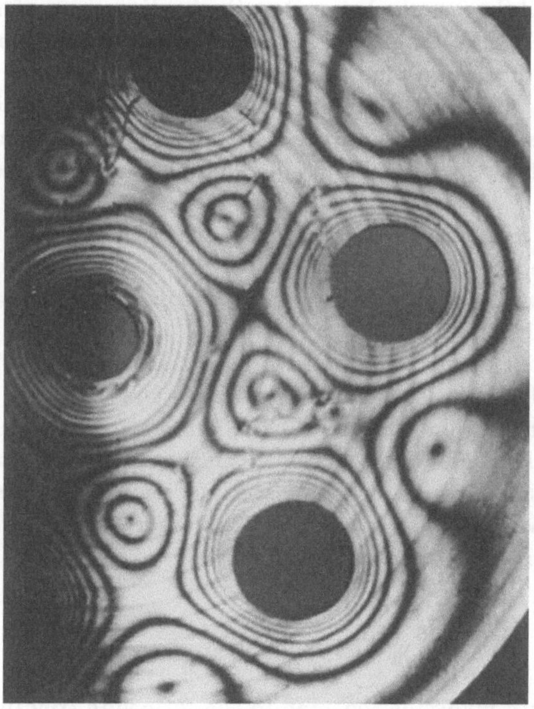

Fig. 5. Frozen-Stress Pattern for Tapered Disk with Holes.

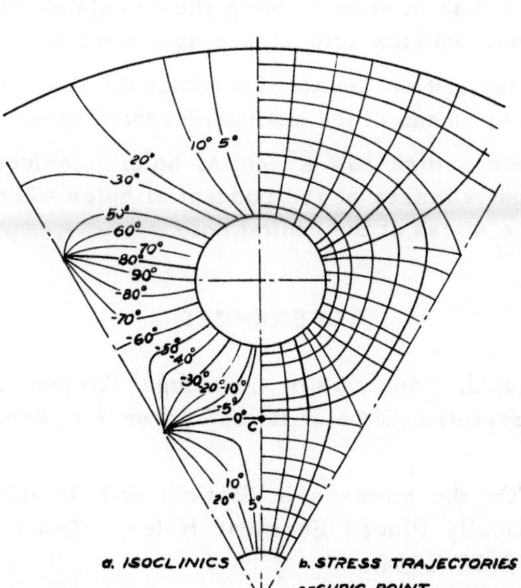

Fig. 6. Isoclinics and Stress Trajectories for the Disk of Fig. 5.

The slope equilibrium method requires final adjustment of the stresses by numerically integrating the tangential stresses over the radial area upon which they act. The resulting force plus the centrifugal force missing because of the drilled-out hole material should check the result of a similar integration over the radial area of the filled disk analyzed initially in the previous section. The relative height of the σ_θ curve is then adjusted until the integration results agree.

Figure 3 shows the curves of the stresses for the tapered disk analyzed here; the corresponding stresses calculated by the method of the previous section are also shown. Figure 4 shows the results of a similar analysis of a uniform disk with holes.

It can be seen that at the more critical points in each disk the agreement of the stresses is well within 3%. At other points, however, the actual stresses are lower than the theoretical stresses. The integration of this difference over the radial area involved gave 1.22 lb and exactly equals the value of the centrifugal force missing because of the material removed by the drilling of the holes. No attempt was made to correct for this discrepancy, as it occurs in the regions where the stresses are relatively low.

CONCLUSIONS

The results of this investigation lead to the following conclusions:

1. The maximum stress in a disk with holes occurs at the edge of a noncentral hole on a radial line through the center of the disk.
2. The "spoke effect" as described in this paper may be employed in the analysis of such a disk in order to bring the calculated value of the critical stress in the outer annulus into closer agreement with the actual value.
3. The stresses at the critical points throughout the disk can be reliably and easily calculated by means of the method presented here.
4. In the case where a disk has a central hole, introduction of noncentral holes creates local stresses at the noncentral holes which are of the order of magnitude of the stress at the central hole.

REFERENCES

[1] Teruyosbi Udoguchi, "Analysis of Centrifugal Stresses in a Rotating Disk Containing an Eccentric Circular Hole," *Japan Sci. Rev. Series 1* **1** 55–63 (1949).

[2] Hideo Saito, "On the Stresses in Statical and Rotating Circular Plates Having Symmetrically Placed Eccentric Holes," *Trans. Japan Soc. Mech. Engr.* **20**(95), 473–78 (1954).

[3] K. E. Barnhart, Jr., A. L. Hale, and J. L. Meriam, "Stresses in Rotating Disks Due to Noncentral Holes," *Proc. Soc. Exptl. Stress Anal.* **9**(1), 35–52 (1951).

[4] J. H. Armstrong, *Stresses in Rotating Tapered Disks with Noncentral Holes,* Ph.D. Thesis, Iowa State University of Science and Technology, 1960.

[5] H. M. Martin, "Stress Distribution in Rotating Disks of Conical Profile," *Engineering* **115**, 1–3, 115–16, 407, 630 (1923).

[6] M. M. Frocht, *Photoelasticity,* Vol 1, pp 215–51, 329–42, Wiley, New York, 1941.

PLASTIC BEHAVIOR OF CONICAL SHELLS[1]

Kurt H. Gerstle
University of Colorado
and
Richard Lance and E. T. Onat
Brown University

This paper reports the results of an investigation of the plastic behavior of circular conical shells of ductile materials under concentrated loads. The purpose of these tests was (1) to verify theoretical collapse loads for such shells, (2) to investigate the effects of large deflections on the carrying capacity of these shells, and (3) to establish velocity fields for shells steeper than those to which present theory applies. Procedure, results, and appropriate conclusions are presented.

INTRODUCTION

The behavior in the inelastic range of plane structures of perfectly plastic material has been fairly well understood for some time [*1*]; it has only been in recent years that attention has been directed to the investigation of nonplanar structures in the inelastic range. In particular, the inelastic response of shells of rotation seems to have been first discussed by Onat and Prager [*2*]; these authors established the yield condition for such structures based on Tresca's yield criterion; later, Hodge [*3*] established an approximate yield condition for similar structures. Based on this work, theoretical solutions for conical shells under concentrated loading have been derived independently by Onat [*4*] and by Hodge [*5*]. Onat applied the yield condition of Onat and Prager to simply supported conical shells with rigid center portion, to which the load was applied. Hodge used his yield condition to analyze the capacity of shells which were hinged to the center portion, to which the load was applied; it should be noted that according to the lower-bound theorem of limit analysis, this solution constitutes a lower-bound approximation to the capacity of shells in which the

[1] The research reported in this paper was performed under National Science Foundation Project NSF – G 8188. The first-named author performed his part of the project under a National Science Foundation science faculty fellowship.

NOTATION

P	Applied concentrated load on shell, lb
P_p	Concentrated collapse load on shell, lb
P_c	$2\pi M_0$; collapse load on equivalent flat plate, lb
σ_{yp}	Yield stress of perfectly plastic material, psi
M_0	Fully plastic moment per unit width of equivalent flat plate, in.-lb/in.
ϕ	Angle of inclination of shell generator, radians
s	Meridional coordinate, in.
l_0	Meridional radius of rigid center of shell, in.
l	Meridional radius of conical shell, in.
h	Shell thickness, in.
δ	Rise of shell, in.
δ_0	Initial rise of shell, in.
α	l/l_0; ratio of radius of rigid center to shell radius
β	$l \tan \phi/h$
v	Meridional component of displacement rate, in.
w	Normal component of displacement rate, in.
b	Constant; meridional radius to point of discontinuity in velocity field of Hodge's solution, in.

center portion is rigidly connected to the cone. Both of these solutions are valid only for shallow conical shells; for steeper shells, they become upper bounds to the correct collapse load.

It is one of the purposes of the present investigation to verify the theoretical results of Onat and Hodge. This seems particularly appropriate in view of the fact that these solutions vary considerably from each other. Furthermore, to extend these theoretical solutions to conical shells of steeper pitch, it becomes necessary to make suitable assumptions regarding the velocity field of the deformed shell. The measured deformations of plastically deformed shells may give a clue as to likely velocity fields to be assumed in setting up theoretical solutions.

The classical theory of plasticity is based on the assumption of deformations so small that the equilibrium of any portion of the structure may be calculated by using its original dimensions. While even in the elastic range this limitation may be unduly restrictive for some cases (such as very thin plates), in a solution considering plastic action with its accompanying larger deformations, a small deflection theory may give quite unrealistic answers. This question was considered by Onat and Haythornthwaite [6] as applied to circular plates under symmetric loading; in that investigation it was brought out that the geometry changes under load cause a considerable increase in carrying capacity beyond the limit load.

Shells of revolution under symmetrical load constitute particularly appropriate structures for investigation of the effects of geometry changes, since under deformations these bodies will continue to be shells of revolution. In particular, Onat's theoretical solution for the plastic deformations of conical shells of small rise, under concentrated central load, indicates that their generators will continue to remain straight lines. Thus, cones will deform into new cones of different rise whose carrying capacity can again be obtained theoretically. Depending on whether the applied load is such as to make the cone shallower or steeper, the effect of deformation is to weaken or strengthen the structure. In the first case, one can speak of instability, in the second of hardening of the structure under load.

This behavior can be shown graphically by means of a type of plot first used by Haythornthwaite and Ona [7], and used again in Fig. 1, in which the

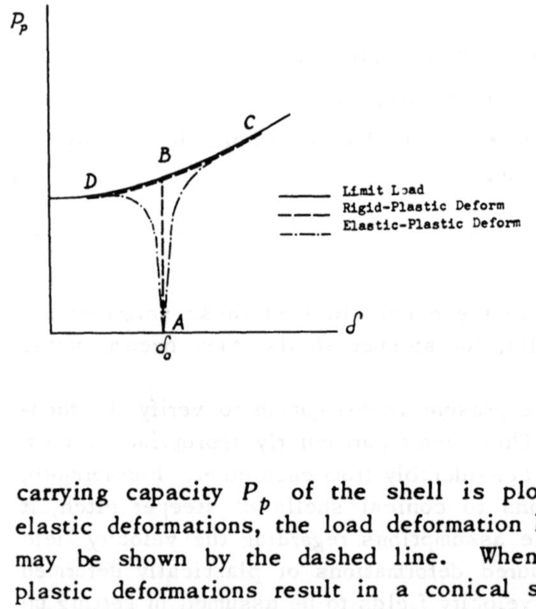

Fig. 1. Deformation vs Carrying Capacity of Conical Shells.

carrying capacity P_p of the shell is plotted vs shell rise δ. Neglecting elastic deformations, the load deformation history of a shell of initial rise δ_0 may be shown by the dashed line. When the capacity load P_p is reached, plastic deformations result in a conical shell of larger or smaller rise, depending on the direction of the applied load, with an appropriate change in carrying capacity; accordingly, for loads applied so as to cause further dishing of the shell, path ABC will be followed; for loads of such sense as to cause flattening, path ABD will result.

In the present investigation, conical shells were tested under both types of loading, and the resulting changes in carrying capacity were determined and compared with theoretical predictions.

THEORETICAL RESULTS

The collapse load of shallow conical shells under central concentrated load, as shown in Fig. 2*a*, based on Tresca's yield condition, has been obtained by Onat [4] as

$$P_p = 2\pi M_0 \cos^2\phi \left[\frac{1}{1-\alpha} + \frac{\beta^2}{3}(1+\alpha-5\alpha^2+3\alpha^3)\right],$$

where

$$\alpha = \frac{l_0}{l}, \quad \beta = \frac{l}{h}\tan\phi, \quad M_0 = \sigma_{yp}\frac{h^2}{4},$$

and l, h, l_0, and ϕ are defined in Fig. 2*a*. The corresponding plastic velocity field is given by

$$V = V_0, \quad W = W_0\,\frac{l-s}{l-l_0}.$$

The linear variation of W indicates conical deformation. These results constitute an exact solution only for values of

$$\beta^2 \leqq \frac{3}{2}\,\frac{\alpha}{1-\alpha}\cdot\frac{1}{[1-(\alpha/2)](1+\alpha)+\tfrac{3}{2}(\alpha^4-\alpha^3)}.$$

For deeper shells, they form an upper bound to the correct collapse load.

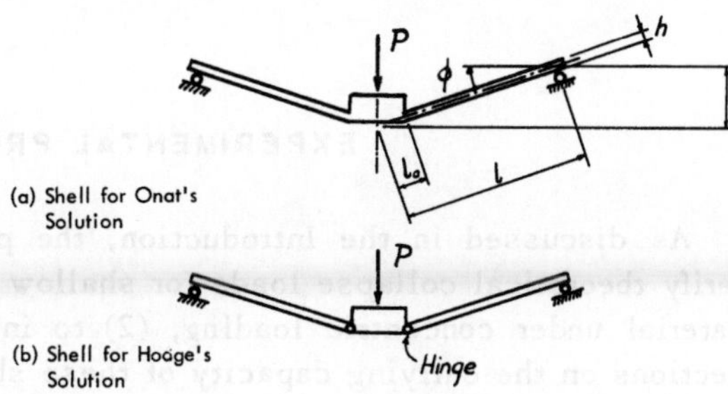

Fig. 2. Shell Geometry.

Hodge's solution [5] for the type of shell shown in Fig. 2*b* is obtained by simultaneous solution of the following equations:

$$P_p = 2\pi M_0 \cos^2\phi \left[1 + \beta\,\frac{1-b^2(1-2\log b)}{1-b}\right],$$

$$P_p = 2\pi M_0 \cos^2\phi \left[1 + 4\beta\,\frac{b(1-\log b)-(\alpha-1/e)}{\log b/\alpha}\right],$$

where the symbols are the same as above, and b is a constant to be eliminated. For relatively small values of α, the validity of this solution extends to shells of $\beta \leqq 0.316$. For deeper shells, this solution constitutes an upper bound. It can also be seen that in the case of rigid connection between shell and center portion, the solution is still statically admissible, and therefore forms a lower bound to the correct solution.

Figure 3 shows a comparison of Onat's and Hodge's results for shells of $\alpha = 0.10$ and $\beta = 10 \sin \phi / \cos^2 \phi$; these ratios are applicable to the specimens tested.

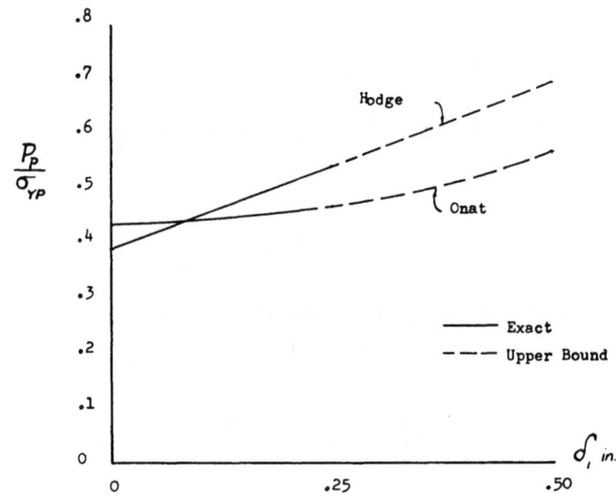

Fig. 3. Comparison of Onat's and Hodge's Results.

EXPERIMENTAL PROCEDURE

As discussed in the Introduction, the purpose of these tests was (1) to verify theoretical collapse loads for shallow conical shells of perfectly plastic material under concentric loading, (2) to investigate the effects of large deflections on the carrying capacity of these shells, and (3) to establish velocity fields for shells steeper than those to which present theory applies.

Test specimens were machined of hot-rolled mild-steel plates obtained commercially. In order to get information about the mechanical properties of the material, two sets of tension and compression specimens were milled from each plate with their axes both parallel and perpendicular to the direction of rolling. In this fashion it was established that effects of anisotropy, as well as differences in yield point in tension and compression, were negligible. Material properties did differ considerably from slab to slab. Table 1 lists the relevant characteristics of each specimen.

Dimensions of the shell specimens were so chosen that the load could be applied by means of a 200,000-lb universal testing machine. The diameter of the support ring was 10 in., with a slight overhang of the shell. The center of

Table 1. Test Specimens

Shell	Orientation	δ_0, Shell Rise (in.)	Yield Stress (ksi)
1	Vertex down	0.25	31.5
2	Vertex down	0.50	31.5
3	Vertex down	1.00	27.0
4	Vertex down	1.50	27.9
5	Vertex up	0.25	31.5
6	Vertex up	0.50	31.5
7	Vertex up	1.00	27.0
8	Vertex up	1.50	35.5

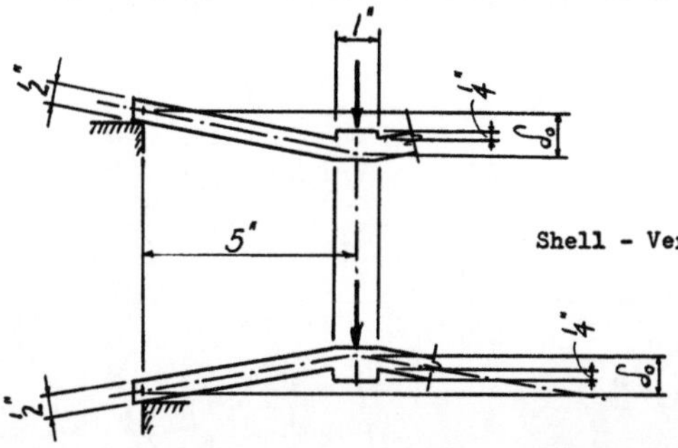

the shell consisted of a thickened solid boss of 1-in. diameter through which the load was applied by means of a 1-in.-diam hardened-steel punch attached to the movable cross head of the machine. In the theory, this boss is conceived as a perfectly rigid region. To avoid any danger of buckling under compressive stresses, the shell thickness was made $\frac{1}{2}$ in. Shells of four different degrees of steepness were investigated, namely, rises of $\frac{1}{4}$, $\frac{1}{2}$, 1, and $1\frac{1}{2}$ in. It should be noted that only the first of these four falls within the range of applicability of the presently available theoretical solution.

Table 1 shows dimensions and properties of the specimens tested. Two sets of shells were made, one to be tested vertex up, the other vertex down, as shown in the figures. The purpose of this arrangement was to obtain experimentally the curves *ABC* and *ABD* of Fig. 1, for various values of initial shell rise δ_0.

The shell support was designed to simulate the boundary conditions of the available theoretical solutions, namely, a perfectly frictionless concentric ring exerting a reaction normal to the shell surface. Accordingly, a brass ring, inserted in a heavy-steel ring to avoid distortion, was machined. Prior to placing the shell specimen on this ring, a graphite-base lubricant was applied to minimize friction. It was anticipated that under maximum shell reaction the brass support (machined to a radius of $\frac{1}{8}$ in.) would be flattened permanently. However, no permanent deformation of the brass ring could be noted visually.

The specimens were so instrumented that the data required for the three listed purposes of the tests could be obtained. To get load vs center-deflection curves, the applied load was read from the dial of the testing machine; the deflection of the vertex-up shells was measured by means of a dial gage with a least count of 0.001 in., resting against the deepened center portion of the shell immediately below the punch; for the vertex-down specimens, the travel of the cross head of the testing machine was obtained from the average of the readings of two dial gages arranged symmetrically. The velocity field of the deformed shell was determined by means of six additional gages arranged along a diameter of the shell, the readings of which enabled a profile of the shell deformations to be plotted.

The results of the tests are presented in the form of load-deflection curves and of profiles of the plastically deformed cross section of the shell. Figure 4,

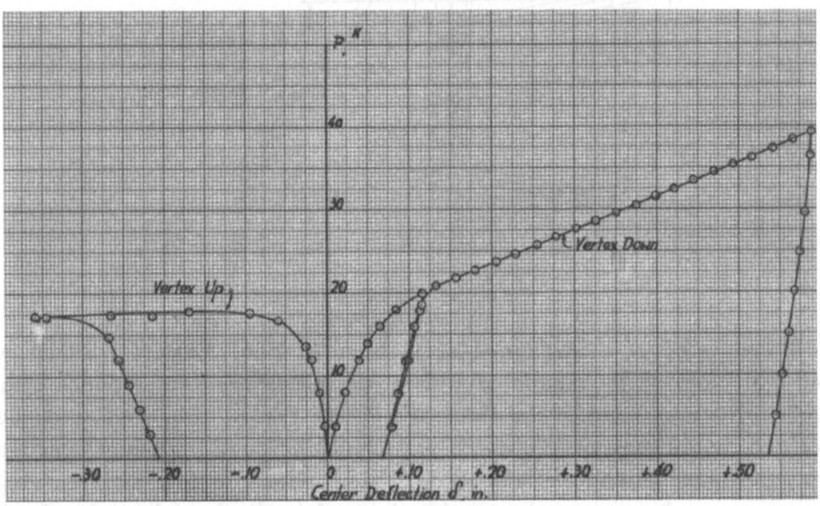

Fig. 4. Load-Deflection Curves for $\frac{1}{2}$-in. Shells.

showing a plot of the applied load vs the center deflection of the two shells with $\frac{1}{2}$-in. rise with vertex up and vertex down, is typical of the experimental results obtained. At least one unloading branch was obtained for each shell in order to obtain elastic properties. Tests were generally carried on until deformations approached the range of the dial gages.

To correlate these results with theory, it proved convenient to nondimensionalize the load-deflection curves in terms of the ratios P/P_c and δ/h, where P = applied load, P_c = collapse load of equivalent flat plate = $2\pi M_0$, and

δ and h are defined in Fig. 2*a*. In order to put all results on a common basis, the center deflections were referred to a base plane through the support. The resulting curves were superimposed upon theoretical curves of the collapse load vs shell rise based on Onat's and Hodge's theory, similarly nondimensionalized.

Figure 5 shows this comparison plot. All load-deflection curves are transcribed directly from the experimental results except the one corresponding to zero initial rise, which was taken from the tests performed by Foulkes and Onat [8] on flat plates of mild steel. The experimental curves indicate that the theoretical predictions of the first-order collapse load by Onat's theory are closely verified for shells up to a rise of two times the shell thickness. For steeper shells the upper-bound approximation of the theory becomes evident, while Hodge's results seem to underestimate the capacity of the shells.

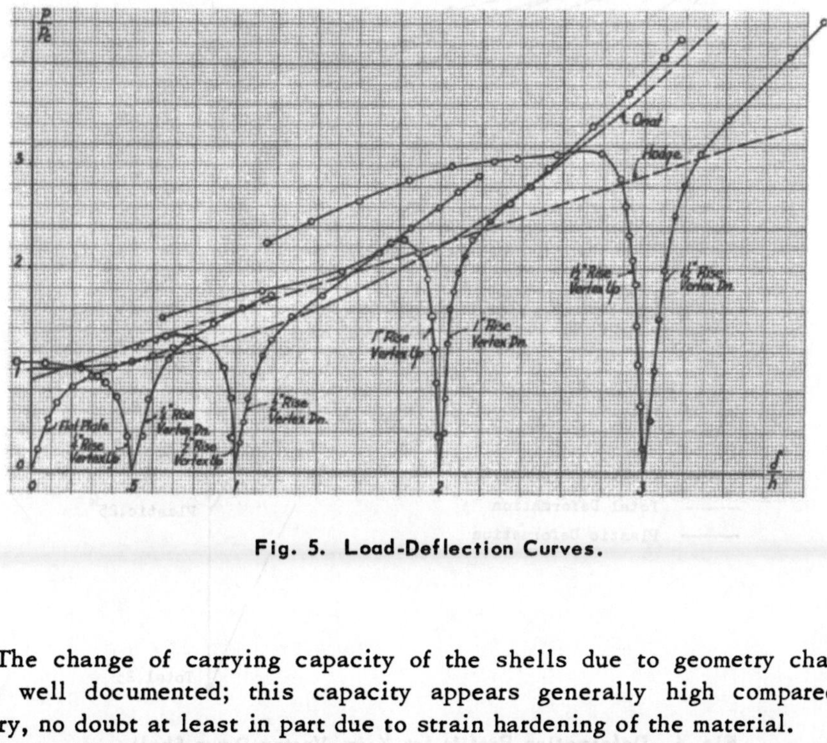

Fig. 5. Load-Deflection Curves.

The change of carrying capacity of the shells due to geometry change is also well documented; this capacity appears generally high compared with theory, no doubt at least in part due to strain hardening of the material.

It should be noted that the elastic curves due to unloading at various stages of deformation indicate the change in elastic stiffness due to geometry change. Thus, with increasing deformation, the vertex-up shells become more flexible, the vertex-down shells become stiffer.

It is also seen, by comparing the shapes of the curves in the vicinity of the first-order collapse load, that this load has a strict meaning only in cases in which geometry changes lead to instability. For those in which hardening takes place under deformation, it indicates the onset of a less-stiff regime, rather than exhaustion of carrying capacity.

A typical profile of the deformed cross section at various stages of loading is shown in Fig. 6. At each of three different load levels, the total deformation is plotted, as well as the elastic curve (obtained by prorating the profile changes during unloading), and the latter is subtracted from the former to give the plastic profile.

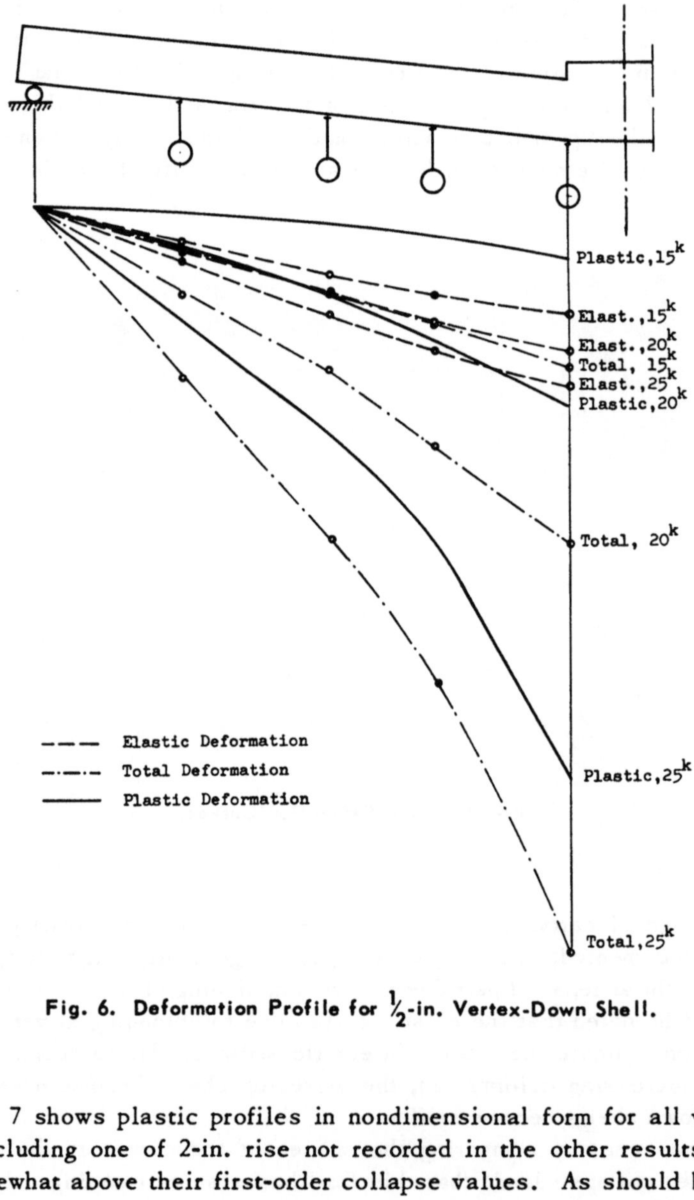

Fig. 6. Deformation Profile for ½-in. Vertex-Down Shell.

Figure 7 shows plastic profiles in nondimensional form for all vertex-down shells (including one of 2-in. rise not recorded in the other results), taken at loads somewhat above their first-order collapse values. As should be expected (because the membrane-flexural strength increases with steepness of the shell, while the local shear yield strength remains substantially constant), the effect of shear yielding becomes more pronounced for the steeper shells. It may be

Fig. 7. Plastic Profiles for Vertex-Down Shells.

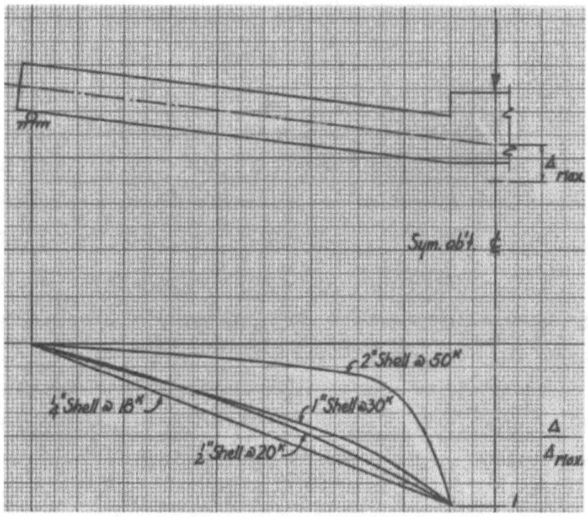

concluded that for shells of the considerable thickness used in these tests, the velocity fields are severely affected by shear deformations. To verify velocity fields predicted by a theory which neglects shear effects, much thinner shells should be used. Nevertheless, it is possible to recognize a definite outward directed concavity of the plastic profiles for the deeper shells.

CONCLUSIONS

From the experimental results it may be concluded that presently available theoretical solutions are able to predict fairly accurately the carrying capacity of shallow conical shells subject to central loading.

It is concluded that the first-order collapse load has physical meaning only in the case of shells in which large deformations lead to instability. In other cases, this load merely denotes the onset of a less-stiff regime; behavior of shells during this new regime may also be predicted by theory.

The plastic profiles appear to indicate that solutions for steep conical shells will yield velocity profiles of concave outward shape.

APPENDIX

Theoretical Basis of Onat's and Hodge's Solutions

In the theory of limit design of perfectly plastic structures, a correct solution must satisfy the following conditions:

1. Equilibrium conditions for every portion of the structure.
2. An appropriate yield condition for the perfectly plastic material; at no point in the structure may the stress exceed the value specified by this yield condition. In Cartesian stress space, the yield condition may be represented by a closed convex surface called the yield surface.

3. A displacement (or velocity) field which will allow the structure to deform into a kinematically admissible mechanism. This displacement field must be associated with the stresses by means of the flow rule, which demands that the vector representing the plastic deformation at any point of the structure is in the direction of the outward normal to the yield surface at the stress point under consideration. To draw this vector, the deformation coordinates are superimposed upon the corresponding stress coordinates.

An equilibrium solution which satisfies the yield condition (but not necessarily represents a kinematically admissible displacement field) is a lower bound on the correct collapse load.

An equilibrium solution which is based on a kinematically admissible displacement field (but does not necessarily satisfy the yield condition) leads to an upper bound on the correct collapse load.

The difference in the results of Onat and Hodge is due to the choice of yield condition. Onat assumes that yielding of the material at any point occurs when the maximum shear stress reaches a critical value, as first assumed by Tresca. This theory has been shown to reasonably well represent the behavior of ductile materials such as mild steel. For shells, this yield condition results in a nonlinear expression in terms of four stress resultants, tangential and meridional membrane forces, and moments. Applying the flow rule to the yield surface results in a conical displacement field for the plastic shell.

Hodge's theory is also based on Tresca's yield condition, but in the interest of linearizing the yield surface the walls of the shell are assumed to act as sandwich plates in which all material effective in resisting bending and axial forces is concentrated at the extreme fiber. The linearity of the yield condition is obtained at the expense of a large number of discontinuities. To reduce the number of these discontinuities (or edges between the planes of the convex yield surface), a further approximation is proposed by Hodge which neglects interaction between internal forces and moments [3]. If the flow rule for the conical shell is applied to the resulting approximate yield surface, the plastic deformation profile takes the shape of a logarithmic curve.

In view of the assumptions, it is concluded that within the range of validity of the solutions, Onat's results are exact for solid-walled shells of a Tresca material. It may be questioned whether mild steel, strictly speaking, follows Tresca's yield condition; if this yield condition is considered merely an approximation, then Hodge's solution may be considered on a par with Onat's.

REFERENCES

[1] P. G. Hodge, Jr., *Plastic Analysis of Structures*, McGraw-Hill, New York, 1959.

[2] E. T. Onat and W. Prager, "Limit Analysis of Shells of Revolution," *Koninkl. Ned. Akad. Wetenschap. Proc.* **B57**, 534 (1954).

[3] P. G. Hodge, Jr., "Yield Conditions for Rotationally Symmetric Shells Under Axisymmetric Loading," *Trans. ASME* **82**, 323 (1960).

[4] E. T. Onat, "Plastic Analysis of Circular Conical Shells," *Proc. A.S.C.E., J. Eng. Mech. Div.* (December 1960).

[5] P. G. Hodge, Jr., "Plastic Analysis of Circular Conical Shells," *J. Appl. Mechanics, Trans. ASME* **27**, 696 (1960).

[6] E. T. Onat and R. M. Haythornthwaite, "Load Carrying Capacity of Circular Plates at Large Deflection," *J. Appl. Mechanics* **23**, 49 (1956).

[7] R. M. Haythornthwaite and E. T. Onat, "The Load Carrying Capacity of Initially Flat Circular Steel Plates Under Reversed Loading," *J. Aeronaut. Sci.* **22**(12) (1955).

[8] J. Foulkes and E. T. Onat, "Tests of the Behavior of Circular Plates Under Transverse Load," Brown University Technical Report, OOR-3172/3 (May 1955).

FAILURE THEORIES FOR GENERALLY ORTHOTROPIC MATERIALS

John E. Griffith and William M. Baldwin
University of Florida

In this paper the distortion-energy theory of failure is extended to include generally orthotropic materials. Yielding, fracture, and fatigue behavior under a general state of stress are treated as separate failure conditions. For each type of failure, the criterion can be completely determined from uniaxial stress behavior, which need not be coincident with the principal axes of orthotropy. The comparison of theory with experimental results for a parallel laminated material shows good agreement.

INTRODUCTION

The increased use of orthotropic materials such as wood, some plastics, and laminates has placed a demand on engineers to consider the directional variance of mechanical properties. Orthotropic materials offer certain economical advantages if the orthotropic axes are favorably oriented with respect to the applied stresses. However, a serious reduction in strength or fatigue life results if either the applied stresses or the material axes shift in orientation. Therefore, it is necessary to consider the mechanical behavior to be dependent on orientation as well as magnitude of applied stresses.

A number of elasticity solutions have been established for orthotropic beams and plates under both static and dynamic load conditions [*1*]. These solutions frequently lead to calculations which are difficult except by computer methods and are subject to strong restrictions.

Macroscopic analyses on orthotropic-material behavior have been made by Hill [*2*], Fisher [*3*], and Hu and Marin [*4*] for yielding. In each case the analysis was limited to specially orthotropic materials; that is, principal axes of stress and orthotropy were coincident, and numerous tests were necessary to define the yield surface. On the other hand, microscopic studies on the mechanical properties of single crystals, which are generally anisotropic, have been made by metallurgists [*5*]. In these investigations plastic deformation was observed to occur by either a slip or twinning process on favorably oriented crystallographic planes. Although failure can be accurately predicted for single crystals, little success has resulted when the behavior of polycrystalline materials is considered to be derivable from single-crystal behavior.

ANALYSIS OF ELASTIC SOLIDS

The general equations of an anisotropic elastic solid can be written [6] as

$$\sigma_i = S_{ij}\,\epsilon_j \qquad (i, j = 1, 2, 3, 4, 5, 6)\,, \tag{1}$$

which is the general form of Hooke's law. The matrix S_{ij} represents 36 stiffness constants which from strain energy considerations must be symmetrical [7], or

$$S_{ij} = S_{ji}\,. \tag{2}$$

Thus, the number of independent elastic constants S_{ij} is reduced to 21. This is the usual representation of a generally anisotropic elastic material.

Orthotropic Solids. – If there are three mutually perpendicular planes of symmetry in the elastic solid, the body is described as generally orthotropic, and the elastic stress-strain relations are written in the form

$$\begin{aligned}
\sigma_1 &= S_{11}\epsilon_1 + S_{12}\epsilon_2 + S_{13}\epsilon_3\,,\\
\sigma_2 &= S_{12}\epsilon_1 + S_{22}\epsilon_2 + S_{23}\epsilon_3\,,\\
\sigma_3 &= S_{13}\epsilon_1 + S_{23}\epsilon_2 + S_{33}\epsilon_3\,,\\
\sigma_4 &= S_{44}\epsilon_4\,,\\
\sigma_5 &= S_{55}\epsilon_5\,,\\
\sigma_6 &= S_{66}\epsilon_6\,,
\end{aligned} \tag{3}$$

where the directions 1, 2, and 3 are the principal axes of orthotropy. Thus, the elastic constants S_{ij} are further reduced in number from 21 to 9.

Stress and Strain Transformations. – For a general state of applied stress σ_{mn}, the stress components σ_{ij} can be found from the stress transformation law, or

$$\sigma_{ij} = l_{im}\,l_{jn}\,\sigma_{mn}\,, \tag{4}$$

where $m, n, i, j = 1, 2, 3$ and l_{im} and l_{jn} are direction cosines relating the mn and ij coordinate systems.

In a similar manner ϵ_{ij} can be found from a general state of strain ϵ_{mn}, by the strain transformation law, or

$$\epsilon_{ij} = l_{im}\,l_{jn}\,\epsilon_{mn}\,. \tag{5}$$

Volumetric and Distortion Energy in Orthotropic Solids. – A system of equations equivalent to Eq. (3) can be written as

$$\begin{aligned}
\epsilon_1 &= C_{11}\sigma_1 + C_{12}\sigma_2 + C_{13}\sigma_3\,,\\
\epsilon_2 &= C_{12}\sigma_2 + C_{22}\sigma_2 + C_{23}\sigma_3\,,\\
\epsilon_3 &= C_{13}\sigma_3 + C_{23}\sigma_2 + C_{33}\sigma_3\,,
\end{aligned}$$

$$\epsilon_4 = C_{44}\sigma_4 ,$$
$$\epsilon_5 = C_{55}\sigma_5 ,$$
$$\epsilon_6 = C_{66}\sigma_6 , \tag{6}$$

where C_{ij} represents the nine compliance constants. The total strain energy per unit volume can be written as

$$U_T = \frac{\sigma_1 \epsilon_1}{2} + \frac{\sigma_2 \epsilon_2}{2} + \frac{\sigma_3 \epsilon_3}{2} + \sigma_4 \epsilon_4 + \sigma_5 \epsilon_5 + \sigma_6 \epsilon_6 , \tag{7}$$

or, in terms of compliance constants and stress components,

$$U_T = \sigma_1^2 \frac{C_{11}}{2} + \sigma_2^2 \frac{C_{22}}{2} + \sigma_3^2 \frac{C_{33}}{2} + C_{12}\sigma_1\sigma_2 + C_{13}\sigma_1\sigma_3 + C_{23}\sigma_2\sigma_3 + C_{44}\sigma_4^2 + C_{55}\sigma_5^2 + C_{66}\sigma_6^2 . \tag{8}$$

This total strain energy can be separated into two parts, energy of distortion and energy resulting from volume change. Let the axial stresses σ_1, σ_2, and σ_3 be decomposed into two states of stress, or

$$\sigma_1 = \sigma_1' + \sigma_1'' ,$$
$$\sigma_2 = \sigma_2' + \sigma_2'' , \tag{9}$$
$$\sigma_3 = \sigma_3' + \sigma_3'' ,$$

where the stresses denoted by the primes produce distortion only. Considering only these stresses, the resulting axial strains from Eq. (6) become

$$\epsilon_1' = C_{11}\sigma_1' + C_{12}\sigma_2' + C_{13}\sigma_3' ,$$
$$\epsilon_2' = C_{12}\sigma_1' + C_{22}\sigma_2' + C_{23}\sigma_3' , \tag{10}$$
$$\epsilon_3' = C_{13}\sigma_1' + C_{23}\sigma_2' + C_{33}\sigma_3' .$$

The sum of these strains must be zero if the stress system σ_1, σ_2, σ_3 produces no volume change. The remaining axial stress components, indicated by the double primes, produce a spherical state of strain ϵ_v. Thus, the volume change is

$$\epsilon_1 + \epsilon_2 + \epsilon_3 = \epsilon_1'' + \epsilon_2'' + \epsilon_3'' = 3\epsilon_v . \tag{11}$$

The spherical strain ϵ_v can be written in terms of the stress components which produce volume change only, or

$$\epsilon_v = \frac{\sigma_1''(C_{11} + C_{12} + C_{13}) + \sigma_2''(C_{22} + C_{12} + C_{23}) + \sigma_3''(C_{13} + C_{23} + C_{33})}{3} . \tag{12}$$

The numerator may be expressed in terms of the total axial stresses σ_1, σ_2, σ_3 by adding the axial strain components,

$$\epsilon_1 + \epsilon_2 + \epsilon_3 = \epsilon_1' + \epsilon_2' + \epsilon_3' + \epsilon_1'' + \epsilon_2'' + \epsilon_3'' .$$

In terms of stress this becomes

$$\begin{aligned}\epsilon_1 + \epsilon_2 + \epsilon_3 = {} & \sigma_1'(C_{11} + C_{12} + C_{13}) + \sigma_2'(C_{22} + C_{23} + C_{12}) \\ & + \sigma_3'(C_{13} + C_{23} + C_{33}) + \sigma_1''(C_{11} + C_{12} + C_{13}) \\ & + \sigma_2''(C_{12} + C_{22} + C_{23}) + \sigma_3''(C_{13} + C_{23} + C_{33}) . \end{aligned} \quad (13)$$

The sum of the first three terms on the right side of Eq. (13) is zero by Eq. (11). Combining Eqs. (12) and (13),

$$\epsilon_v = \frac{\sigma_1(C_{11} + C_{12} + C_{13}) + \sigma_2(C_{12} + C_{22} + C_{23}) + \sigma_3(C_{13} + C_{23} + C_{33})}{3} . \quad (14)$$

The strain energy due to volume change only is

$$U_v = \frac{(\sigma_1 + \sigma_2 + \sigma_3)}{2} \epsilon_v . \quad (15)$$

Substitution of Eq. (14) into Eq. (15) and subtracting the volumetric energy from the total strain energy of Eq. (8) yields the distortion energy. The result is

$$\begin{aligned} U_D = {} & \sigma_1^2\left(\frac{C_{11}}{3} - \frac{C_{12}}{6} - \frac{C_{13}}{6}\right) + \sigma_2^2\left(\frac{C_{22}}{3} - \frac{C_{12}}{6} - \frac{C_{23}}{6}\right) \\ & + \sigma_3^2\left(\frac{C_{33}}{3} - \frac{C_{13}}{6} - \frac{C_{23}}{6}\right) + \sigma_1\sigma_2\left(\frac{2C_{12}}{3} - \frac{C_{11}}{6} - \frac{C_{22}}{6} - \frac{C_{13}}{6} - \frac{C_{23}}{6}\right) \\ & + \sigma_2\sigma_3\left(\frac{2C_{23}}{3} - \frac{C_{12}}{6} - \frac{C_{22}}{6} - \frac{C_{13}}{6} - \frac{C_{33}}{6}\right) + \sigma_1\sigma_3\left(\frac{2C_{13}}{3} - \frac{C_{11}}{6} - \frac{C_{12}}{6} - \frac{C_{23}}{6} - \frac{C_{33}}{6}\right) \\ & + \sigma_4^2 C_{44} + \sigma_5^2 C_{55} + \sigma_6^2 C_{66} . \end{aligned} \quad (16)$$

If the constants are $C_{11} = C_{22} = C_{33} = 1/E$, $C_{44} = C_{55} = C_{66} = 1/2G$, $C_{12} = C_{13} = C_{23} = -\nu/E$, and $G = E/2(1 + \nu)$, Eq. (16) assumes the familiar isotropic form,

$$U_D = \frac{1}{12G}[(\sigma_1 - \sigma_2)^2 + (\sigma_2 - \sigma_3)^2 + (\sigma_3 - \sigma_1)^2 + 6(\sigma_4^2 + \sigma_5^2 + \sigma_6^2)] . \quad (17)$$

Plane Stress. – For plane stress in the 1–2 plane ($\sigma_3 = \sigma_5 = \sigma_6 = 0$) the distortion-energy equation reduces to

$$\begin{aligned} U_D = {} & \frac{\sigma_1^2}{3}\left(C_{11} - \frac{C_{12} + C_{13}}{2}\right) + \frac{\sigma_2^2}{3}\left(C_{22} - \frac{C_{12} + C_{23}}{2}\right) \\ & + \frac{\sigma_1\sigma_2}{3}\left(2C_{12} - \frac{C_{11} + C_{22} + C_{13} + C_{23}}{2}\right) + \sigma_4^2 C_{44} . \end{aligned} \quad (18)$$

For this special case six elastic constants must be known to express distortion energy.

If the applied stresses σ_{mn} do not coincide with the axes of orthotropy, the problem is one of a general orthotropic plate. In contrast, if the applied stresses σ_{mn} are coincident with axes 1 and 2, the plate is considered to be specially orthotropic [1]. The general case is of particular interest in connection with directional variation of material behavior. Let the x and y axes make an arbitrary angle α with the principal orthotropic axes 1 and 2, and consider the fol-following simple applied-stress cases.

Equal Triaxial Stress. – For equal triaxial stress, $\sigma_x = \sigma_y = \sigma_z$. For this state of stress,

$$U_D = 0 \, , \tag{19}$$

as in the isotropic case.

Pure Shear. – For pure shear, $\sigma_x = \sigma_y = 0$, $\sigma_{xy} \neq 0$. From the stress transformation of Eq. (4),

$$\begin{aligned} \sigma_1 &= -\sigma_{xy} \sin 2\alpha \, , \\ \sigma_2 &= \sigma_{xy} \sin 2\alpha \, , \\ \sigma_4 &= \sigma_{xy} \cos 2\alpha \, . \end{aligned} \tag{20}$$

For this stress system the distortion energy is

$$U_D = \sigma_{xy}^2 \left[\left(\frac{C_{11}}{2} + \frac{C_{22}}{2} + C_{12} \right) \sin^2 2\alpha + C_{44} \cos^2 2\alpha \right] . \tag{21}$$

A plot of Eq. (21) is shown in Fig. 1 for both constant U_D and constant σ_{xy}. The material constants for a glass-base silicone resin were used. The variation of U_D with the rotation α of the constant applied shear stress σ_{xy} is pronounced. Considerable variation in σ_{xy} with α for constant U_D is also apparent.

Pure Tension. – For pure tension, $\sigma_x \neq 0$, $\sigma_y = \sigma_{xy} = 0$. Transformation of σ_x to the orthotropic axes by Eq. (4) gives

$$\begin{aligned} \sigma_1 &= \sigma_x \cos^2 \alpha \, , \\ \sigma_2 &= \sigma_x \sin^2 \alpha \, , \\ \sigma_4 &= \sigma_x \sin \alpha \cos \alpha \, , \end{aligned} \tag{22}$$

and the distortion-energy variation with α becomes

$$U_D = \frac{\sigma_x^2}{3} \left[\left(C_{11} - \frac{C_{12} + C_{13}}{2} \right) \cos^4 \alpha + \left(C_{22} - \frac{C_{12} + C_{23}}{2} \right) \sin^4 \alpha \right.$$

$$\left. + \left(2C_{12} - \frac{C_{11} + C_{22} + C_{13} + C_{33}}{2} + 3C_{44} \right) \sin^4 \alpha \cos^4 \alpha \right] . \tag{23}$$

Both U_D vs α for constant σ_x, and σ_x vs α for constant U_D appear in Fig. 2 for the same glass-base silicone resin material [8].

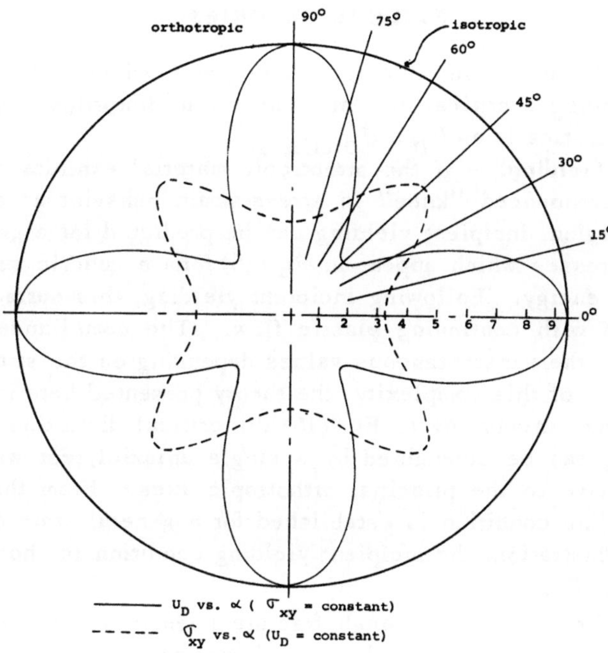

Fig. 1. Effect of Applied Shear Stress σ_{xy}.

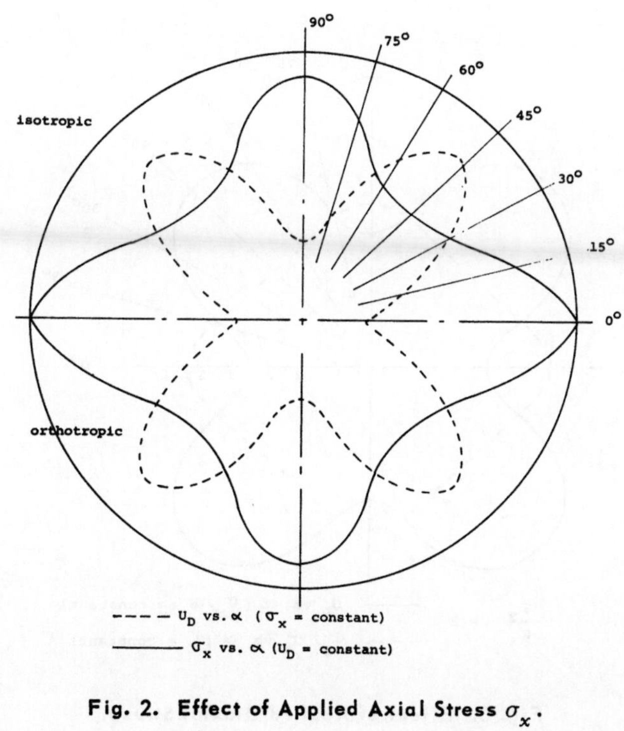

Fig. 2. Effect of Applied Axial Stress σ_x.

It is assumed that certain yielding, fracture, and fatigue failures are directly related to expending a critical amount of available distortion energy. This is expressed mathematically as $U_D = U_{D\ \text{critical}}$.

Failure by Yielding. – If the orthotropic material exhibits a well-defined yield point or pronounced "knee" in stress-strain behavior as an upper limit to the elastic region, incipient yielding can be predicted for a general state of stress. The stresses which appear in Eq.(16) form a quadric surface for constant distortion energy. Following incipient yielding, this surface changes in shape and size with continuing plastic flow. The compliance "constants" C_{ij} also change, their instantaneous values depending on the stress and strain history. Because of this complexity, the theory presented here is restricted to incipient yielding. According to Eq. (16) the critical distortion energy for incipient yielding can be determined by a single uniaxial test with any known orientation relative to the principal orthotropic axes. From this single test result, the yielding condition is established for a general state of stress σ_{mn}. For graphical illustration, the incipient yielding condition is shown for unequal biaxial stress in Fig. 3.

Failure by Fracture. – Although fracture phenomena are not well understood, the important role of prior plastic deformation in the initiation of cracks is evident [9]. Despite the appearance of a brittle or cleavage fracture, fracture generally occurs after a significant amount of plastic flow has taken place. Usually the plastic flow can be attributed to shear stresses. As previously

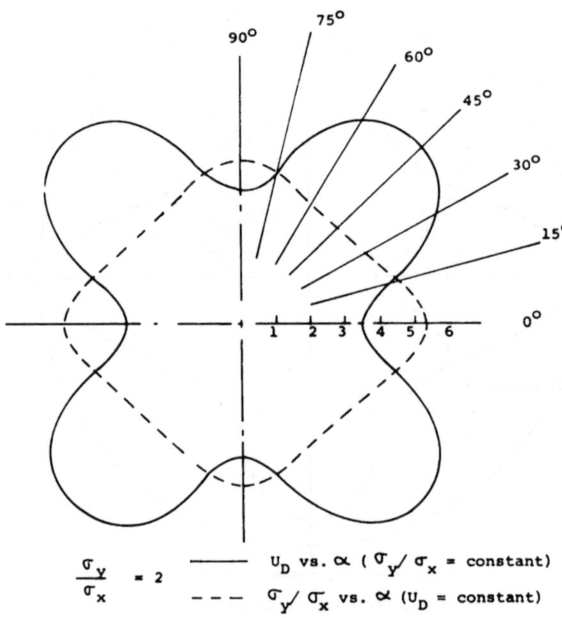

Fig. 3. Effect of Applied Biaxial Stress.

noted, the distortion energy is insensitive to hydrostatic stress and, therefore, is a measure of the energy produced by shear stresses. Thus, distortion energy, in the form of Eq. (16), appears to be a suitable criterion for fracture if the material tends to remain elastic up to fracture. This limitation is necessary to ensure that the compliances C_{ij} are essentially constant.

Stated mathematically, the criterion is simply $U_D = U_{D\ critical}$, where, again, $U_{D\ critical}$ is determined by a single uniaxial test to fracture, with arbitrary orientation with respect to the principal orthotropic axes. Figures 1, 2, and 3 serve as illustrations of the variation of the fracture condition with orientation of simple applied stresses.

Fatigue Failure. – During stress cycling a certain amount of energy is expended irreversibly for each cycle according to the representative uniaxial cycling path of Fig. 4. Here the inelastic flow is associated with distortion

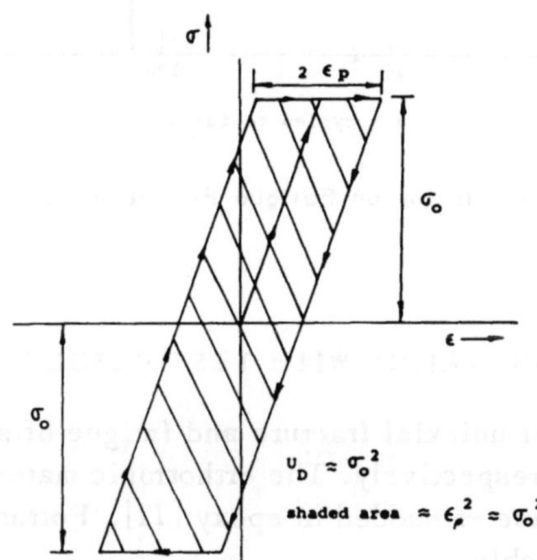

Fig. 4. Stress Path for Reversed Loading.

energy. With an increasing number of cycles, additional distortion energy is expended until a certain critical quantity is consumed resulting in fracture. A family of stress amplitude vs number of cycles to failure, σ vs N curves, is illustrated in Fig. 5. Each σ vs N curve corresponds to a particular orientation of the applied axial stress with respect to the principal orthotropic axes. Any vertical line (N = a constant) is considered to be a line of constant critical distortion energy. In fact, the fracture criterion of the previous section is simply the vertical line erected at $N = 1/4$. Presume that one of the σ vs N curves is determined experimentally for reversed axial stress. Various procedures for fitting an equation to this curve appear elsewhere [*10*]. From this single curve the fatigue failure for any combination of applied stresses with any spatial orientation can be determined. The axial stress case of Fig. 5 has been used as a simple illustration of the general criterion.

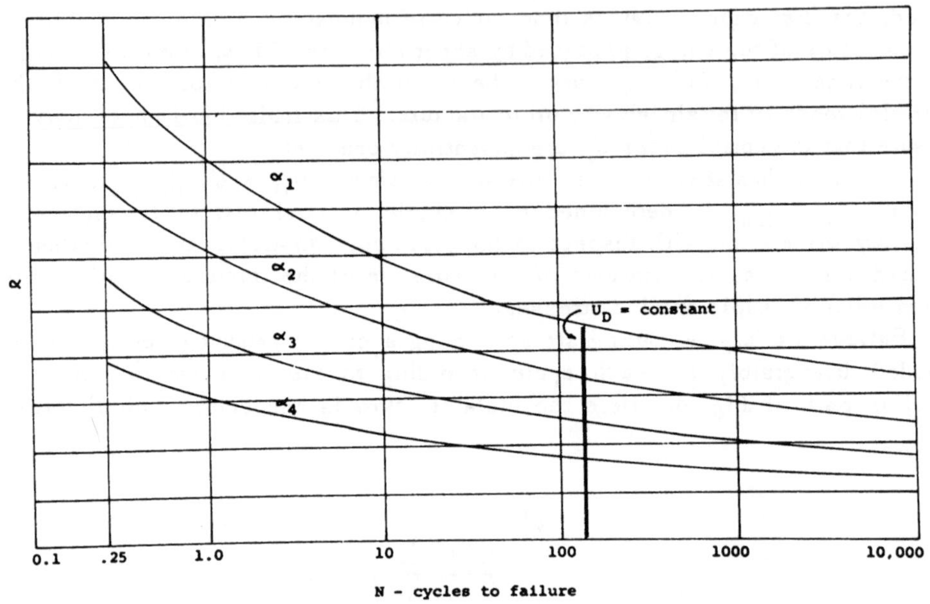

Fig. 5. Effect of Orientation on Fatigue Failure in Orthotropic Materials.

COMPARISON WITH TEST RESULTS

Experimental data for uniaxial fracture and fatigue of an orthotropic material appear in Figs. 6 and 7 respectively. The orthotropic material is a parallel laminated Fiberglas cloth base embedded in epoxy [*11*]. Fortunately, the compliance constants are also available.

In order to compare the distortion-energy theory of fracture with test data, suppose that a single axial test to fracture is conducted in the principal direction 1, and that the fracture stress for other orientations of uniaxial stress in the 1–2 plane is determined from this single test result. As indicated in Fig. 6, the predicted fracture stress at other orientations lies resonably close to test data.

Next, consider the test data of Fig. 7 for reversed axial fatigue at various orientations with the principal axes of the material. This time, presume that the σ vs N curve for the 15° orientation is selected as the test program for determination of fatigue behavior. From this single curve the fatigue behavior for other reversed axial stress orientations is found. Again, the agreement with test results is good.

CONCLUSIONS

Before accepting the extension of distortion-energy theories of failure to orthotropic materials, additional experiments are necessary, particularly tests with a more general state of stress. Also, the suitability of the theory to

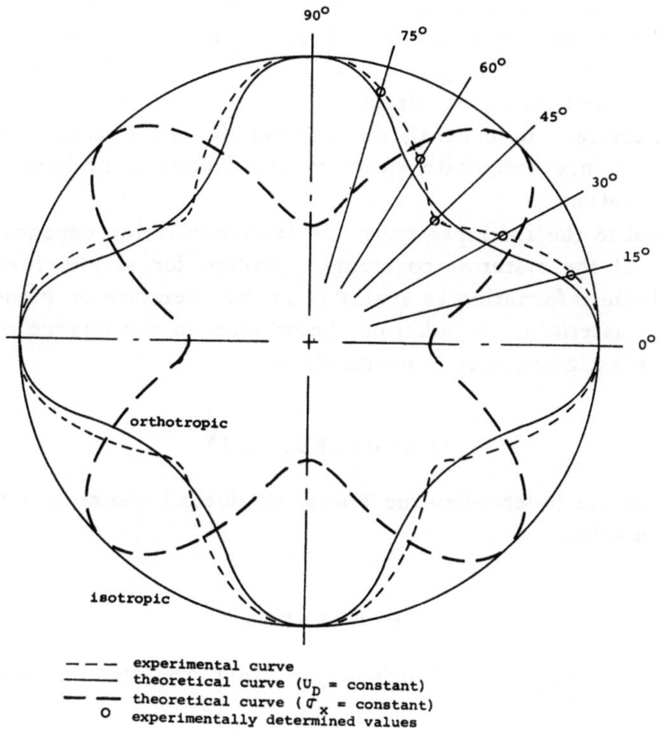

Fig. 6. U_D vs α (σ_x = a constant), σ_x vs α (U_D = a constant).

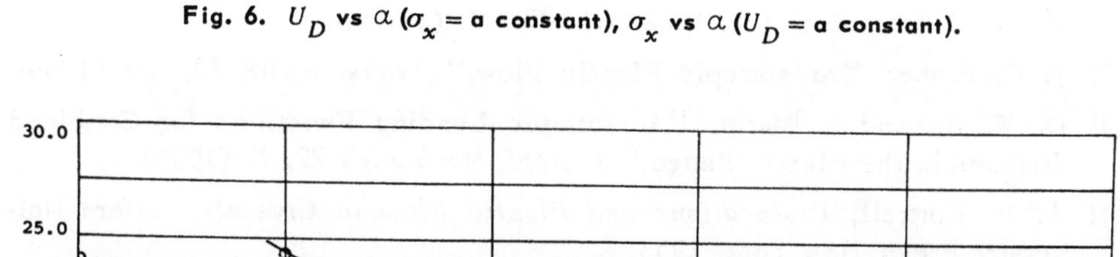

Fig. 7. Comparison of Reversed Axial Stress Fatigue Data with Theory.

various orthotropic materials should be explored. The test results used for comparison in this paper were obtained from a heterogeneous material, not homogeneous as implied by the single matrix of compliances. However, the fracture and fatigue failures could be directly attributed to the base material in the composite structure. In general, the critical distortion energy for each constituent material in a composite structure would have to be known, as well as details of fabrication.

As opposed to the isotropic case, the failure criterion depends on accurate determination of the material constants. Except for single-crystal analyses, surprisingly little information is available in the literature on elastic constants of orthotropic materials. In addition, knowledge on the change of compliance matrix with inelastic behavior is nonexistent.

ACKNOWLEDGMENT

The authors are indebted to the Forest Products Laboratories for furnishing experimental results.

REFERENCES

[1] R. F. S. Hearmon, *Applied Anisotropic Elasticity*, Oxford University Press, New York, 1961.

[2] R. Hill, "A Theory of the Yielding and Plastic Flow of Anisotropic Metals," *Proc. Roy. Soc. (London)* **193**, 281 (1948).

[3] J. C. Fisher, "Anisotropic Plastic Flow," *Trans. ASME* **71**, 349 (1949).

[4] L. W. Hu and J. Marin, "Anisotropic Loading Functions for Combined Stresses in the Plastic Range," *J. Appl. Mechanics* **22**, 77 (1955).

[5] A. H. Cottrell, *Dislocations and Plastic Flow in Crystals*, Oxford University Press, New York, 1953.

[6] I. S. Sokolnikoff, *Mathematical Theory of Elasticity*, Dover Publications, New York, 1944.

[7] A. E. H. Love, *A Treatise on the Mathematical Theory of Elasticity*, Dover Publications, New York, 1944.

[8] R. K. Witt, W. H. Hoppmann, and R. S. Buxbaum, *Bull. Am. Soc. Testing Materials* **194**, 53 (1953).

[9] A. H. Cottrell, *Fracture* (ed. by Averbach, Felbeck, Hahn, and Thomas), pp 20–44, Wiley, New York, 1959.

[10] F. R. Shanley, *Strength of Materials*, pp 693–98, McGraw-Hill, New York, 1957.

[11] Forest Products Laboratory, Report 1823-B.

COLLAPSE OF TUBES BY EXTERNAL PRESSURE[1]

C. R. Kennedy and J. T. Venard

Oak Ridge National Laboratory[2]

The problem of tube collapse by external pressure has been investigated. A graphical solution developed to simplify inelastic-collapse design problems was shown to agree with the test results. The von Karman reduced modulus was used in the graphical solution to correct for the stress redistribution caused by yielding. The effects of the geometric imperfections of ovality and wall-thickness variations on collapse pressure were shown to be related to the stress-strain behavior of the material. The concept of a "critical time" is discussed in regard to creep-buckling phenomenon.

INTRODUCTION

As a result of the interest in tube collapse by external pressure, dating from the first experiments of Fairbairn [*1*] in 1858, several theoretical and empirical solutions have been obtained. The code methods derived from these solutions, in general, contain safety factors of unknown magnitude which allow for geometric imperfections, material imperfections, and a confidence factor. The use of such code methods is overly restrictive under conditions where optimum design is required for efficient operation, particularly in reactor design at elevated temperatures, where neutron economy dictates a minimum mass in the flux field.

This investigation applies previous solutions for instantaneous collapse of perfect tubes to experimental results obtained at elevated temperatures. The effect of geometric imperfections in reducing the critical pressure for collapse was determined so that more appropriate safety factors could be used. The confidence factor is not discussed, since it must be determined independently for each design by balancing operating efficiency and initial cost against failure and replacement costs.

Time-dependent or creep-collapse experiments have not yet been performed, but the problem is considered in view of its importance in design at the elevated temperatures.

---

[1]Supported by the USAEC under contract No. W-7405-eng-26.

[2]Operated for USAEC by Union Carbide Corporation.

NOTATION

γ	Poisson's ratio (0.3)
P_{cr}	Critical pressure for collapse of perfect tube (psi)
a	Mean radius of tube (in.)
E_r	Reduced modulus of elasticity (psi)
h_0	Average or nominal wall thickness of tube (in.)
n	Number of lobes in collapsed tube
l	Tube length between supports (in.)
f	Ratio of ϕ for a given l/a and ϕ for a tube of infinite length
σ_θ	Stress in tangential direction (psi)
P	Pressure (psi)
E	Young's modulus (psi)
E_σ	Tangential modulus (psi)
σ_{max}	Maximum stress in tube wall (psi)
w_0	Maximum initial deflection of tube, $\frac{1}{4}(OD_{max} - OD_{min})$(in.)
v	Minimum wall-thickness variation $\left[\frac{(h_0 - h_{min})}{h_0}\right]$
ϵ_c	Creep strain (in. / in.)
t	Time (hr)
A, α, m	Material constants
ϵ_t	Total strain
$\epsilon_{c\theta}$	Tangential creep strain under tangential stress

EXPERIMENTAL PROCEDURE

Specimens

Tube specimens were fabricated from seamless type 304 stainless steel pipe and tubing to give ratios of radius to wall thickness ranging from 10 to 25. These ratios were obtained by honing the interior and grinding the exterior to closely held tolerances. The required range of ratios necessitated the use of several heats of material.

Deliberate wall-thickness variations of up to ±10% were produced by off-cente. grinding in a number of specimens. Specimens were also deformed in a press to obtain tubes of an oval shape. All specimens were annealed at 1900°F for 1 hr in hydrogen after machining and forming operations.

The finished and annealed specimens were provided with slip-fit end plugs welded by means of edge fusion weld, with a resulting unsupported length of 11.0 in. in most cases. End defects required that two 0.025-in.-wall specimens be shortened to a 9.0-in. free length, and two 0.015-in.-wall specimens to an 8.0-in. free length. Figure 1 is a schematic drawing of a typical specimen showing end plugs and the $\frac{1}{4}$-in. vent tube.

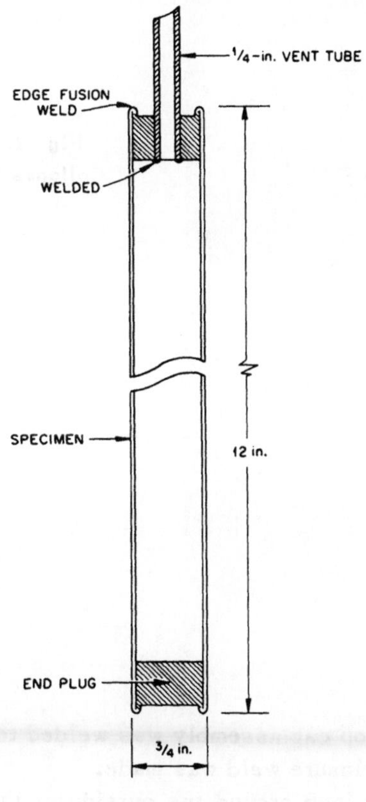

Fig. 1. Tube Collapse Specimen.

Tensile data were obtained by pulling 3.0-in. sections of typical specimens fitted with special end plugs for gripping.

Capsule

The experimental capsule consisted of a length of 2-in. sched-40 Inconel pipe with welded pipe-cap end closures; a cutaway view is shown in Fig. 2. The top cap was drilled to receive the specimen vent tube, three swaged thermocouples,

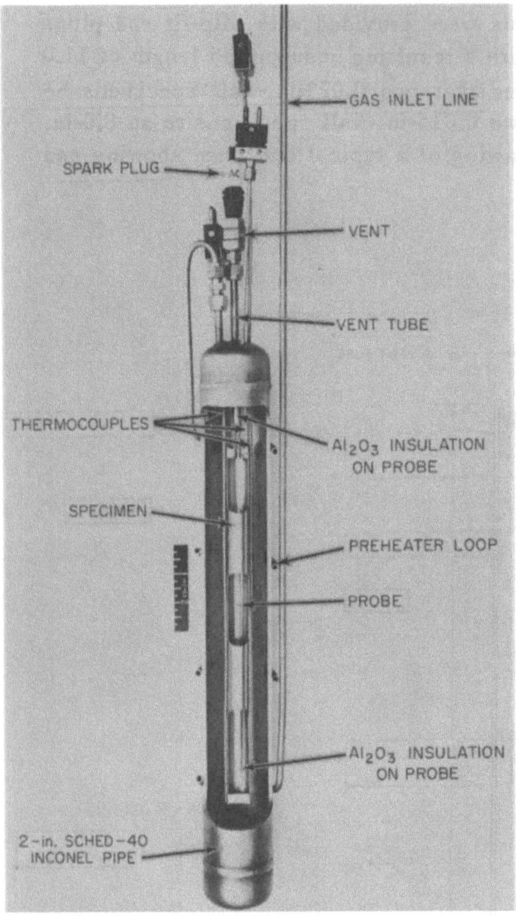

Fig. 2. Cutaway View of Tube-Collapse Capsule.

and a pressure line. The top cap assembly was welded together and inserted into the capsule, and then the closure weld was made.

An inlet gas preheater loop around the outside of the capsule prevented the specimen from cooling during pressurization. Test temperatures, as determined by the top, middle, and bottom thermocouples, were maintained within ±9°F over the specimen length during the test.

The capsule was placed in a furnace and brought to 1200°F, the test temperature. When equilibrium was reached, the capsule was pressurized at approximately 300 psig of argon per minute until collapse occurred. The capsule pressure was recorded on a strip-chart recorder reading the output of a strain-gage fluid-pressure cell. A relay system which was actuated by the probe shorting against the specimen wall and which simultaneously energized an event marker on the pressure time chart and a light on the control panel was initially used to indicate failure or specimen collapse. The probe system was abandoned, however, when it became evident that the discontinuity obtained on the pressure-vs-time recording plus the definite and quite audible snap of the collapse were sufficient indications of failure.

Time of test, maximum pressure, and loading rate for each test were obtained by examination of the pressure-vs-time chart.

EXPERIMENTAL RESULTS

The experimental results are presented in Table 1, which gives material identification, information on the specimen dimensions, collapsing pressure, and time of test. Stress-strain curves obtained for materials 23999X and 24555 are shown in Figs. 3 and 4 respectively. It is important to note the agreement between the two curves shown in Fig. 3 despite the strain rates differing by a factor of 30. Although not noted on Fig. 3, a third stress-strain curve was obtained at 0.00075 in. in.$^{-1}$ min^{-1} which corresponds almost precisely with curve *A*. Figure 5 presents some typical specimens after test. It was interesting to observe that specimens having wall-thickness variations of less than 5% exhibited a uniform two-lobe collapse, while those having 10 to 13% variations failed in a twisted and uneven manner along their length.

DISCUSSION AND CONCLUSIONS

The attention given to the problem of the collapse of long, thin, circular, cylindrical shells by external pressure during the past hundred years has resulted in a number of theoretical and empirical solutions which form the basis for the design of such vessels.

The solution for elastic collapse of perfect tubes can be accomplished by using the von Mises [2] relationship:

$$\phi = \frac{(1-\gamma^2)P_{cr}a}{E_r h_0} = \frac{1-\gamma^2}{(n^2-1)(1+n^2l^2/\pi^2a^2)^2} + \frac{h_0^2}{12a^2}\left(n^2 - 1 + \frac{2n^2-1-\gamma}{1+n^2l^2/\pi^2a^2}\right), \quad (1)$$

where

P_{cr} = critical pressure for collapse (psi),

a = mean radius of tube (in.),

h_0 = average or nominal wall thickness of tube (in.),

E_r = reduced modulus of elasticity (psi),

γ = Poisson's ratio (0.3),

l = tube length between supports (in.),

n = number of lobes in collapsed tube.

The results of calculations using Eq. (1) are shown in Fig. 6, where l/a is plotted vs ϕ. It is seen that a single curve for a given value of a/h_0 is formed by combining portions of curves for integral values of n. The critical pressure for any given tube may be obtained through Fig. 6; however, the tube geometry necessary to resist a given pressure cannot be obtained without the use of unwieldy

Table 1. Instantaneous Tube Collapse Data for Type 304 Stainless Steel at 1200° F
(All specimens annealed at 1900° F for 1 hr in hydrogen)

Test No.	Heat No.	Average Wall Thickness (in.)	Average Outside Diameter (in.)	Mean Radius / Average Wall Thickness	Wall-Thickness Variation (%)	Initial Deflection / Average Wall Thickness	Specimen Length (in.)	Collapse Pressure (psi)	Test Time (min)
638	23999X	0.01500	0.7520	24.57	3.33	0.02	9.0	320	1.0
639	23999X	0.01510	0.7520	24.40	4.64	0.02	9.0	310	1.0
663	23999X	0.01483	0.7518	24.85	5.39	0.03	12.0	360	1.5
664	23999X	0.01607	0.7518	22.89	13.50	0.01	12.0	330	1.0
613	McJunkin	0.02031	0.7402	17.72	4.97	<0.01	12.0	540	0.5
614	McJunkin	0.02056	0.7402	17.50	3.70	<0.01	12.0	560	3.5
630	23999X	0.02103	0.7528	17.40	1.78	0.01	12.0	660	2.0
659	23999X	0.02115	0.7529	17.30	1.65	0.01	12.0	660	2.0
615	24555	0.02510	0.7490	14.42	3.98	0.01	12.0	760	6.5
636	24555	0.02473	0.7490	14.64	3.23	0.01	10.0	780	3.0
637	24555	0.02505	0.7490	14.45	4.39	0.01	10.0	820	2.5
640	McJunkin	0.02509	0.7505	14.46	10.32	0.01	12.0	800	
655	24555	0.03100	0.7494	11.59	2.26	0.01	12.0	1040	4.0
656	24555	0.03100	0.7493	11.58	2.58	<0.01	12.0	1090	3.5
621	24555	0.03323	0.7492	10.77	0.83	<0.01	12.0	1200	4.0
622	24555	0.03305	0.7490	10.83	1.06	0.01	12.0	1150	5.0
631	24555	0.03325	1.0002	14.54	0.46	0.01	12.0	840	2.5
633	24555	0.03345	1.0001	14.45	0.45	<0.01	12.0	860	2.5
652	24555	0.02585	0.7521	14.05	1.35	0.89	12.0	160	0.5
654	24555	0.02570	0.7518	14.13	2.62	0.88	12.0	160	0.5
661	24555	0.02560	0.7520	14.19	2.34	0.26	12.0	360	1.0
662	24555	0.02580	0.7519	14.07	1.16	0.27	12.0	430	1.5
667	24555	0.02585	0.7518	14.04	1.74	0.14	12.0	360	1.5
668	24555	0.02600	0.7520	13.96	1.92	0.14	12.0	440	1.5

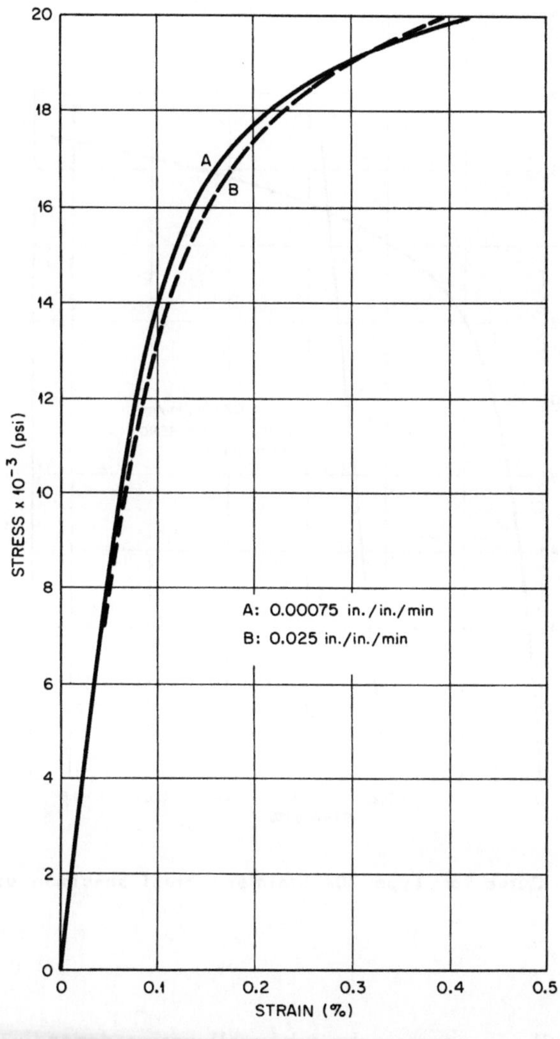

Fig. 3. Tensile Curves for Type 304 Stainless Steel Specimens at 1200°F – Heat 23999X.

trial-and-error procedures. As a means of eliminating the necessity for such procedures the following graphical solution was developed.

Equation (1) may be modified to

$$(P_{cr})_{l/a} = (f)\,\frac{E_r}{4(1-\gamma^2)}\left(\frac{h_0}{a}\right)^3 , \tag{2}$$

where (f) = the ratio of ϕ for a given l/a and ϕ for an infinite-length tube, obtained from Fig. 6. It is observed in Eq. (1) and more clearly shown in Eq. (2) that the critical pressure is a linear function of the reduced modulus for fixed values of l/a and a/h_0.

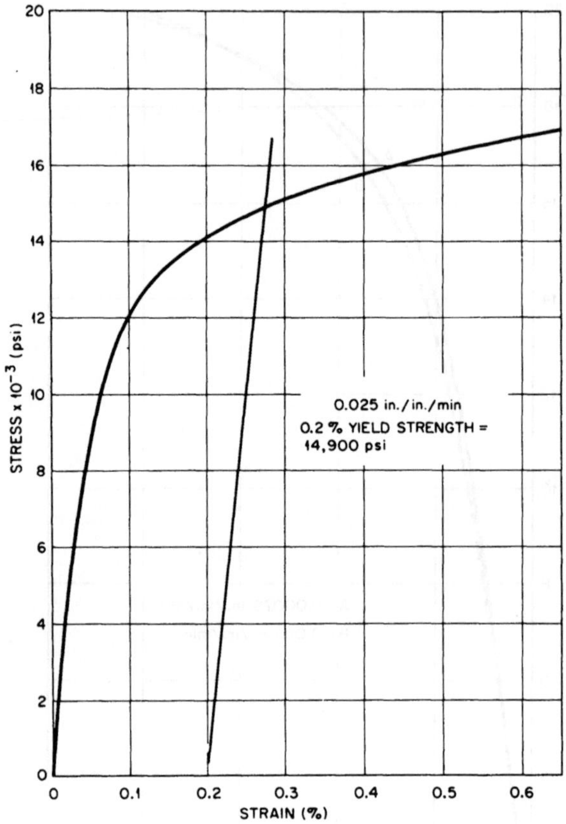

Fig. 4. Tensile Curve for Type 304 Stainless Steel Specimen at 1200°F – Heat 24555.

The stress distribution across the tube wall varies depending on whether the deformation is elastic or plastic. The average stress, however, is given by

$$\sigma_\theta = P\frac{a}{h_0}, \tag{3}$$

where σ_θ = stress in tangential direction (psi) and P = pressure (psi). By combining Eqs. (2) and (3),

$$\sigma_{\theta_{cr}} = KE_r, \tag{4}$$

where

$$K = \frac{(f)}{4(1-\gamma^2)}\left(\frac{h_0}{a}\right)^2 .$$

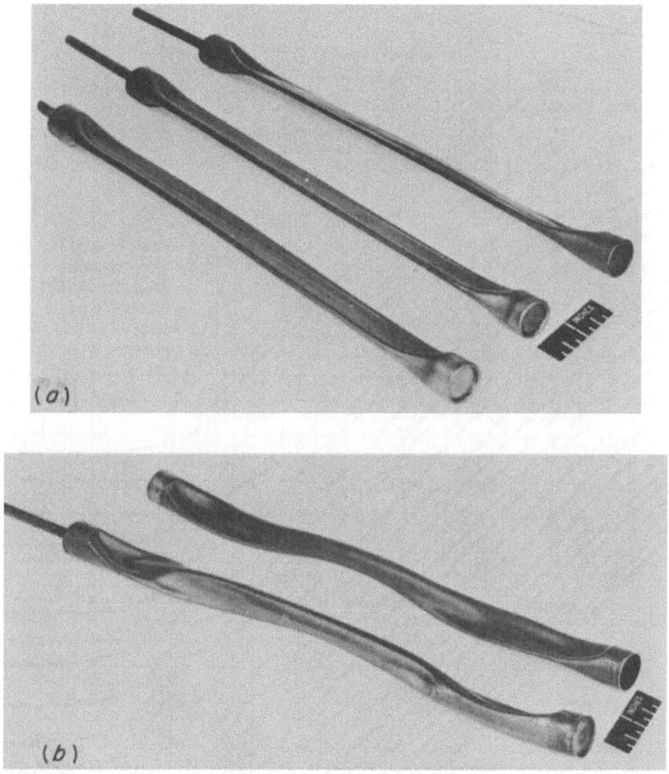

Fig. 5. Collapsed Tube Specimens for (*a*) Test Nos. 662, 656, and 663, with Wall-Thickness Variations of $\leqq 5\%$, and (*b*) Test Nos. 664 and 640, with Wall-Thickness Variations of 10 to 13%.

By substituting values of (f) and a/h_0 in Eq. (4), a series of straight parallel lines may be generated on a logarithmic plot of stress vs reduced modulus. Since the designer is primarily concerned with the tube geometry necessary to resist a particular pressure, Eq. (3) is used to construct straight parallel isobars superimposed on the previously obtained plot. Each isobar intersects the constant (f) and a/h_0 lines at a stress satisfying Eq. (3). A plot as described above is shown in Fig. 7 for infinite-length tubes $[(f) = 1]$.

The addition of a material line to Fig. 7 will now yield conditions of stability and instability for tubes under external pressure. This material line should ideally be obtained from compression tests; however, the use of tensile data to the strains of interest will not introduce significant error.

The reduced modulus, defined by von Karman [3], considers the stress redistribution due to yielding:

$$E_r = \frac{4EE_\sigma}{(E^{1/2} + E_\sigma^{1/2})^2}, \tag{5}$$

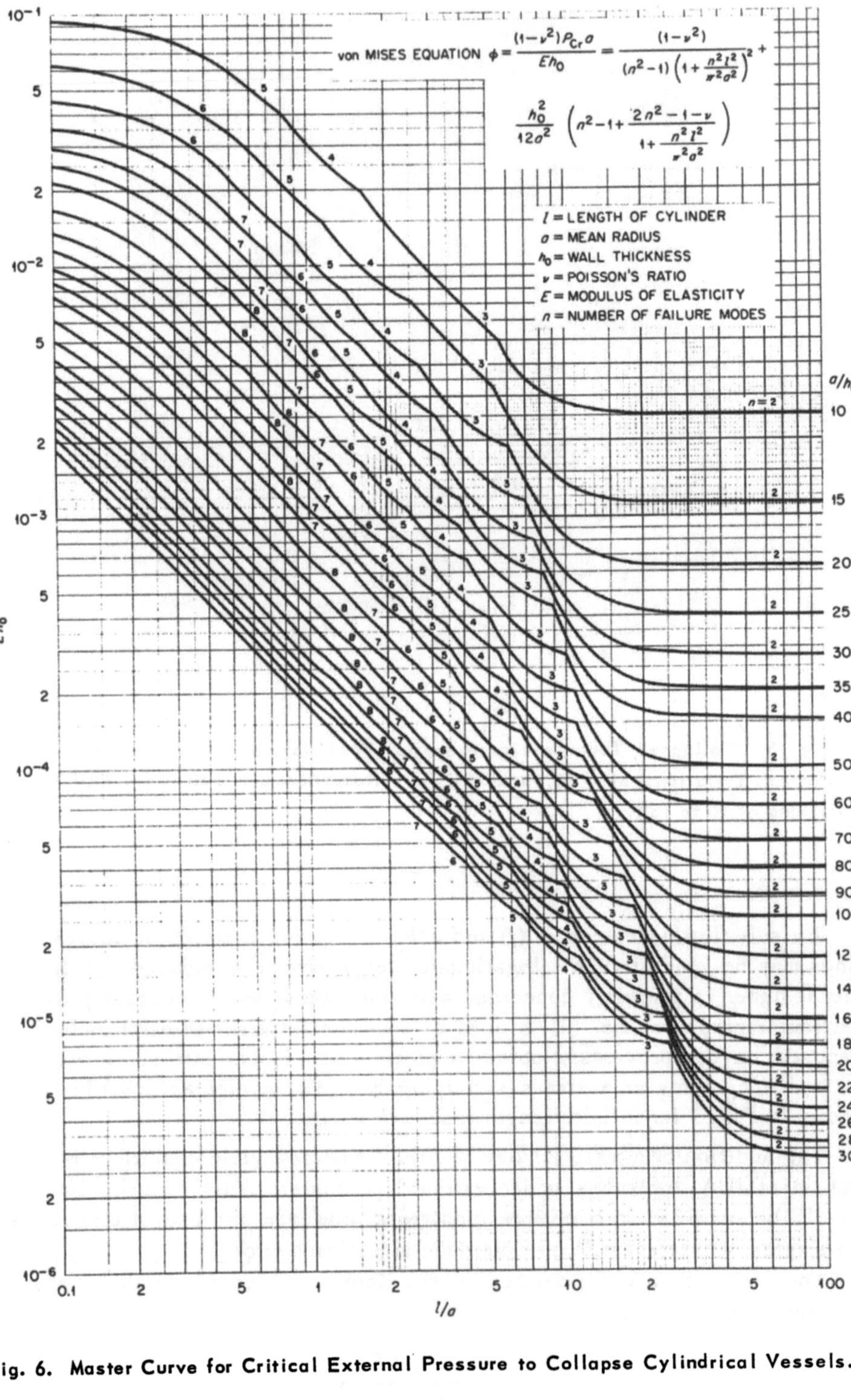

Fig. 6. Master Curve for Critical External Pressure to Collapse Cylindrical Vessels.

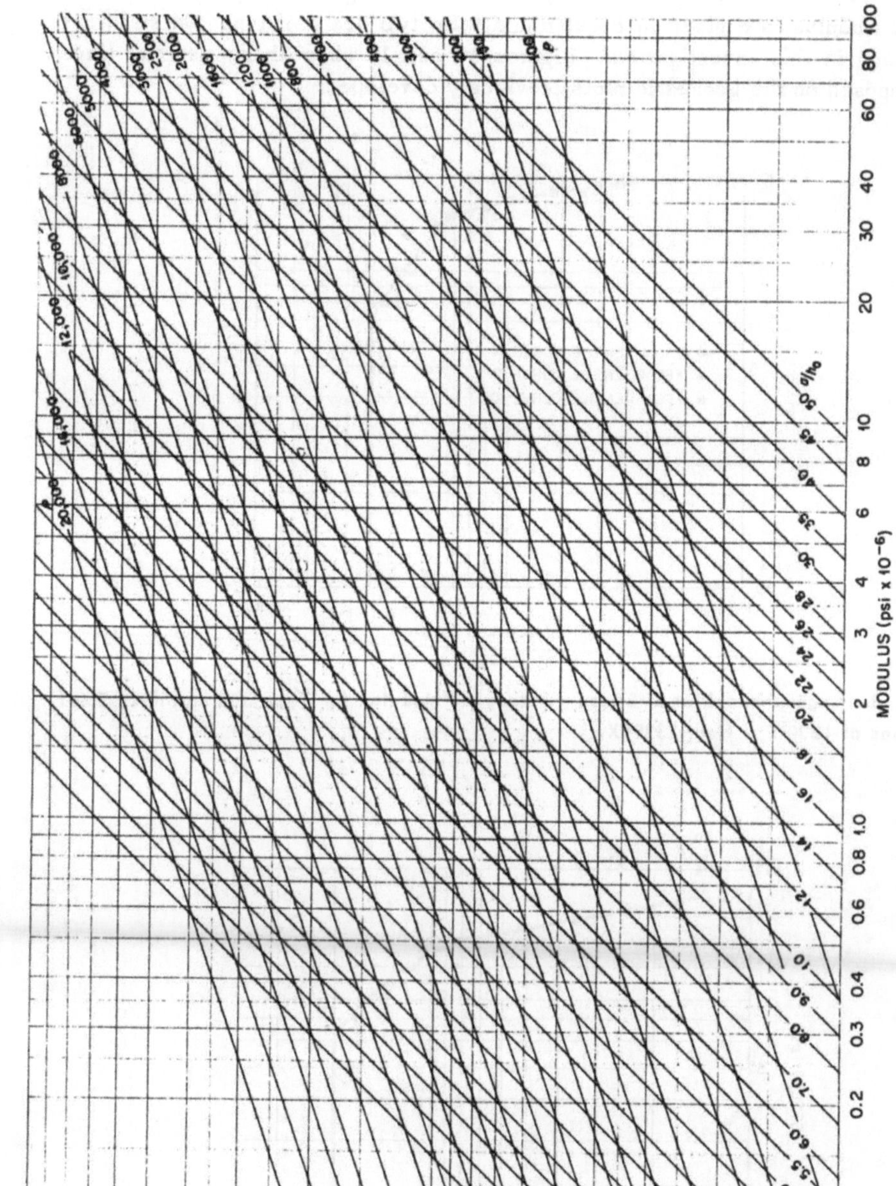

Fig. 7. Logarithmic Plot of Geometric Factors for Infinite-Length Tubes.

where E_r = reduced modulus (psi), E = Young's modulus (psi), and E_σ = tangent modulus (psi). The material lines, represented as a logarithmic plot of stress vs reduced modulus, are given in Figs. 8 and 9 for two heats of type 304 stainless steel used in this investigation. Figures 10 and 11 show these material lines superimposed on the geometric plots previously developed.

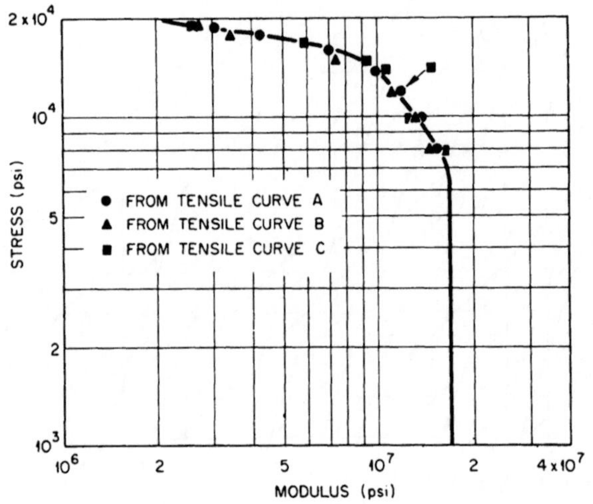

Fig. 8. Logarithmic Plot of Stress vs Reduced Modulus for Type 304 Stainless Steel Specimens at 1200°F – Heat 23999X.

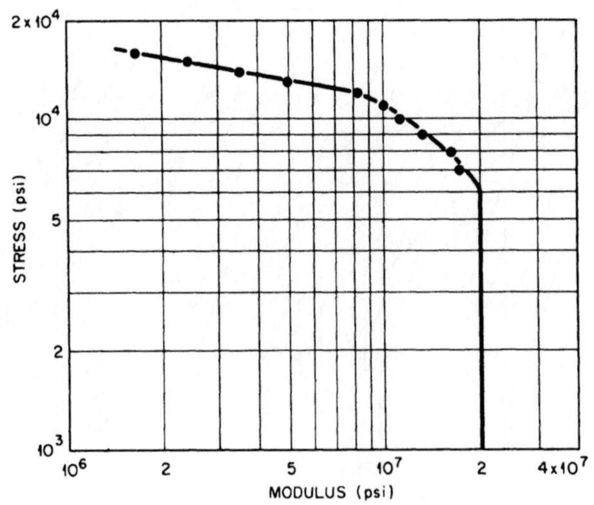

Fig. 9. Logarithmic Plot of Stress vs Reduced Modulus for Type 304 Stainless Steel Specimen at 1200°F – Heat 24555.

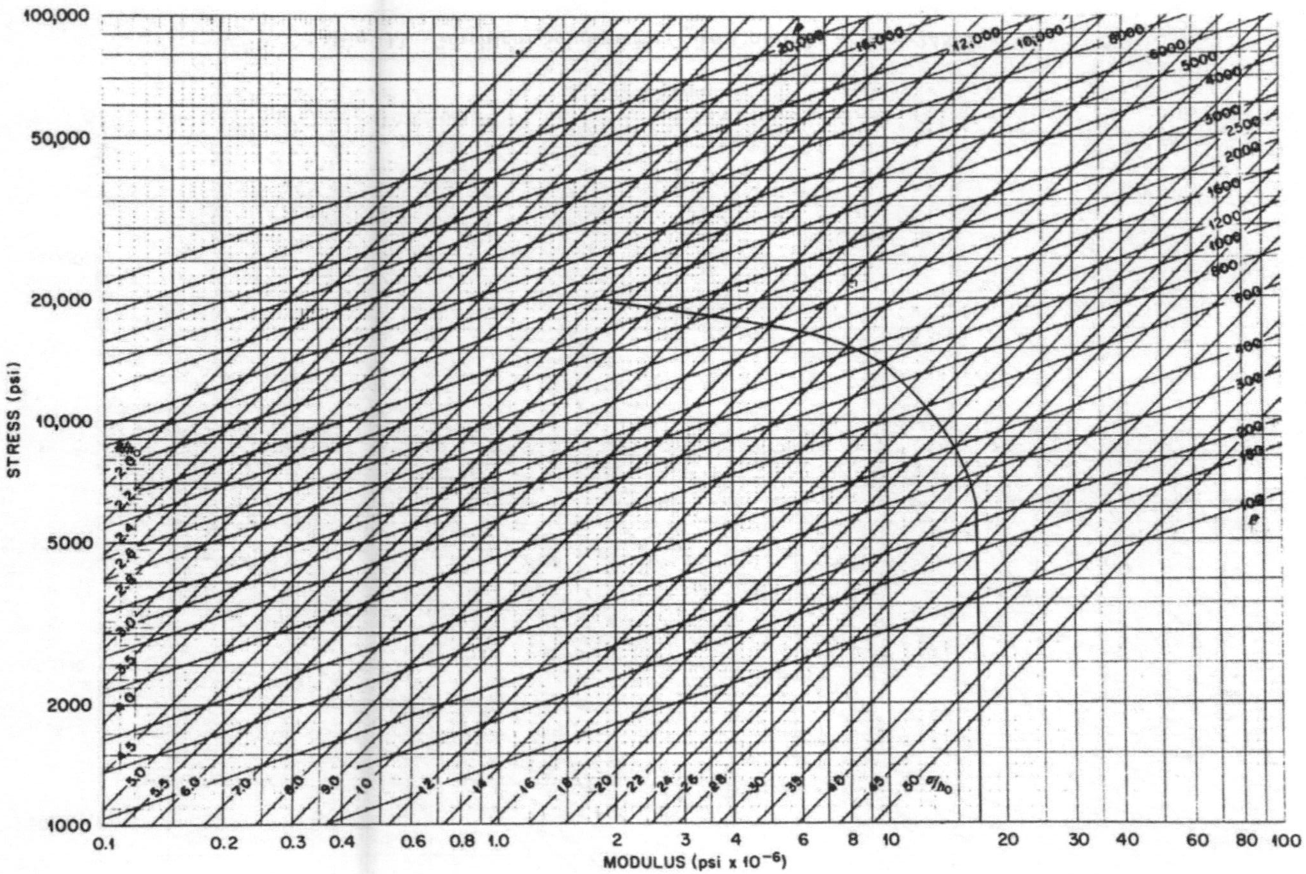

Fig. 10. Graphical Solution for Tube Collapse – Heat 23999X.

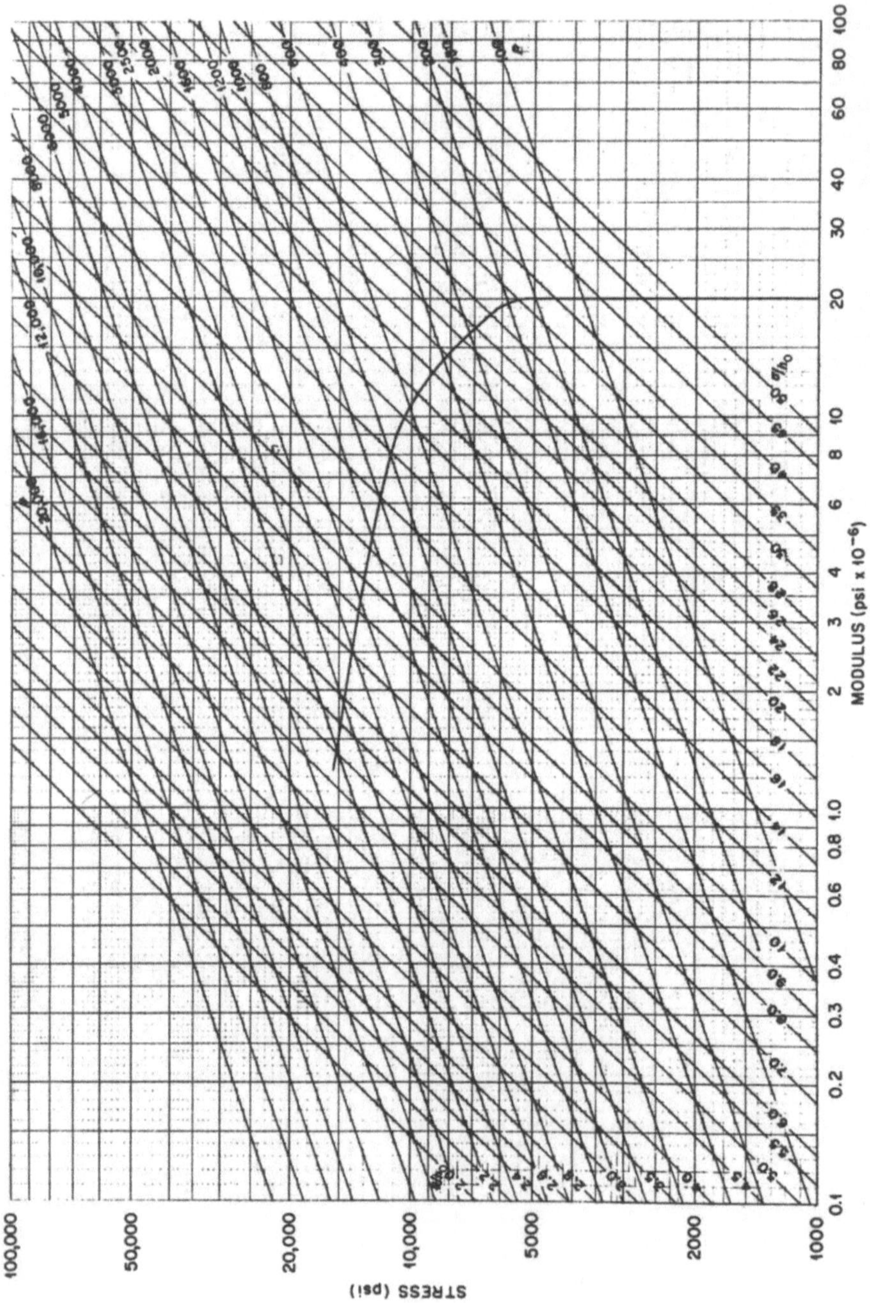

Fig. 11. Graphical Solution for Tube Collapse – Heat 24555.

Intersections of a/h_0 lines with isobars which occur on, above, or to the right of the material lines in Figs. 10 and 11 indicate situations which will result in instantaneous collapse. Combinations of a/h_0 and pressure lines intersecting below the material lines will not result in instantaneous collapse. Verification of these results with the experiments for the two heats of material is given in Fig. 12.

It should be recognized that the effect of increasing temperature is a lowering of the elastic limit of the material and that inelastic collapse becomes possible at lower pressures. The critical stress necessary for collapse in all tests shown in Fig. 12 was greater than the elastic limit of the material. Thus, without the graphical solution as shown in Figs. 10 and 11, it would be necessary to design to the elastic limit and accept the existence of an unknown safety factor, or to use a tedious trial-and-error solution to obtain exact answers.

The use of the graphic solution in Figs. 10 and 11 allows direct solution of the tube-collapse problem with any desired safety factor. The plots shown are applicable for infinite-length tubes only; for tubes with smaller l/a ratios, (f) must be determined from Fig. 6. The direct proportionality of (f) and critical pressure is no longer valid for inelastic collapse, since the reduced modulus varies with stress. The factor (f) for a given l/a ratio also varies slightly with the a/h_0 ratio, and therefore the same (f) cannot be used to shift all the a/h_0 lines in Figs. 10 and 11. A method which may be used to determine the effect of reducing the l/a ratio is to make an appropriate shift of the material line to the right for a given l/a ratio.

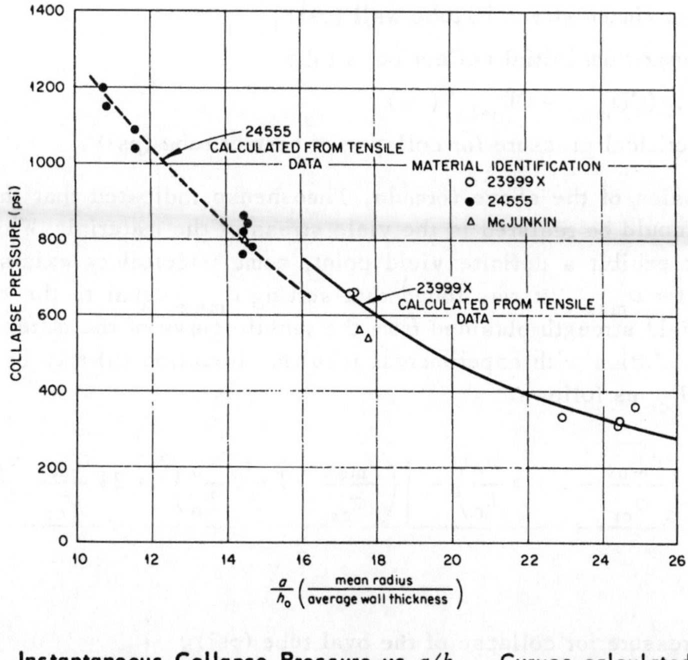

Fig. 12. Instantaneous Collapse Pressure vs a/h_0. Curves calculated from tensile data.

For design purposes the minimum expected material properties of type 304 stainless steel at 1200°F should be used. The graphic solution based on these minimum values is shown in Fig. 13 and includes the effect of lowering the l/a ratio. This was done by shifting the material line to the right through the use of Fig. 6.

This solution does not imply that tubes under a pressure less than the critical pressure for prolonged periods of time will not collapse, since creep is important at 1200°F. In fact, time-dependent behavior may be important even under rapid increases in pressure for certain materials and temperatures. For this particular case of type 304 stainless steel at 1200°F, the tensile data were unaffected by strain-rate variations in the testing range, and time dependency is therefore felt to be negligible in these results.

Selection of a suitable safety factor is dependent upon several criteria, for example, anticipated perfection of the tube, minimum expected strength of the material, and cost of fabrication and operation vs cost of failure and replacement. The major imperfections in tubing, ovality, and wall-thickness variation will now be evaluated.

The most severe reduction in collapse pressure results from tube ovality or out-of-roundness. Timoshenko [4] has shown that the maximum stress acting on the tube wall of an oval specimen is given by

$$\sigma_{max} = \frac{Pa}{h_0} + \frac{6Pa}{h_0^2} \frac{w_0}{1 - P/P_{cr}} , \tag{6}$$

where

σ_{max} = maximum stress in tube wall (psi) ,

w_0 = maximum initial deflection of tube

$= \frac{1}{4} (OD_{max} - OD_{min})$ (in.) ,

P_{cr} = critical pressure for collapse of perfect tube (psi) .

In his discussion of the above formula, Timoshenko indicated that the maximum stress which could be resisted is the yield stress of the material. With materials which do not exhibit a definite yield point, some uncertainty exists as to the proper value for σ_{max}. It was found that setting σ_{max} equal to the value of the 0.2%-offset yield strength obtained from the tensile curve of the material resulted in a good correlation with experimental results. Equation (6) may be solved for the ratio P_0/P_{cr} as follows:

$$\frac{P_0}{P_{cr}} = R = \frac{\left(\frac{\sigma_{max}}{\sigma_{cr}} + 1 + 6\frac{w_0}{h_0}\right) - \left[\left(\frac{\sigma_{max}}{\sigma_{cr}} - 1 - 6\frac{w_0}{h_0}\right)^2 + 24\frac{\sigma_{max}}{\sigma_{cr}}\frac{w_0}{h_0}\right]^{1/2}}{2} , \tag{7}$$

where

P_0 = pressure for collapse of the oval tube (psi) ,

P_{cr} = pressure for collapse of a perfect tube of same a/h_0 (psi) ,

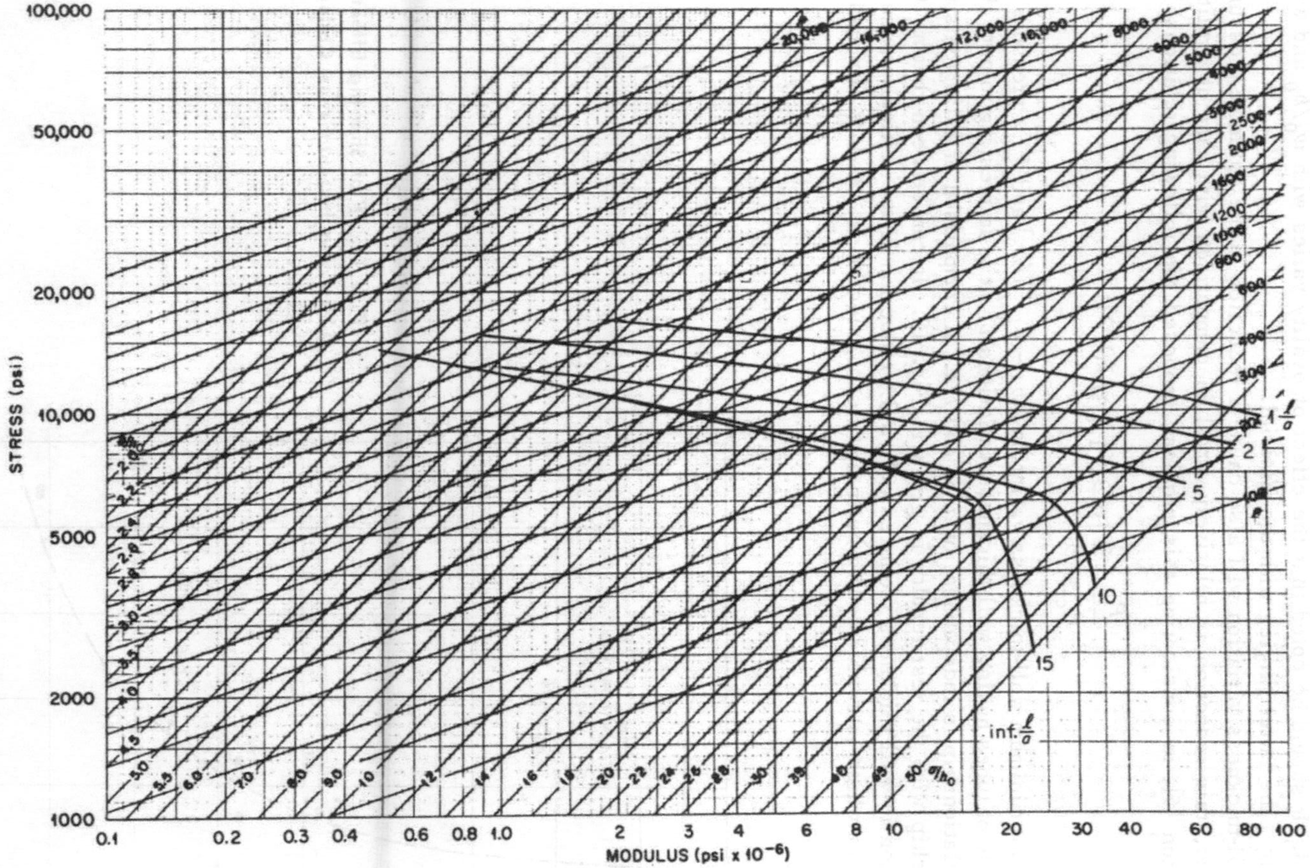

Fig. 13. Graphical Solution for Collapse Problem for Tubes of Less Than Infinite Length.

σ_{max} = 0.2%-offset yield strength of material (psi) ,

σ_{cr} = critical stress for collapse of a perfect tube of same a/h_0 (psi) .

The correlation of the foregoing equation and experimental values is shown in Fig. 14. It should be noted that the effect of ovality varies with w_0/h_0 and with σ_{max}/σ_{cr}, the latter varying with the material.

Wall-thickness variation will also cause a reduction in the collapse pressure; however, this reduction is relatively small in comparison with that of ovality. Ellington [5], in considering this problem, proposed the following relationship:

$$\frac{P_v}{P_{cr}} = \frac{1}{3}\left[9 - 6(1 + 3v^2)^{1/2}\right] , \tag{8}$$

where v = minimum wall-thickness variation $[(h_0 - h_{min})/h_0]$ and P_v = pressure for collapse of tubes having wall-thickness variations v. It is interesting to note that this relationship is independent of a/h_0 and l/a. A comparison of Ellington's relationship with the test results is given in Fig. 15. It can be seen that a 10% variation in wall thickness produces only a 3% reduction of the critical pressure, an effect which is considered to be insignificant, since most vendors can readily

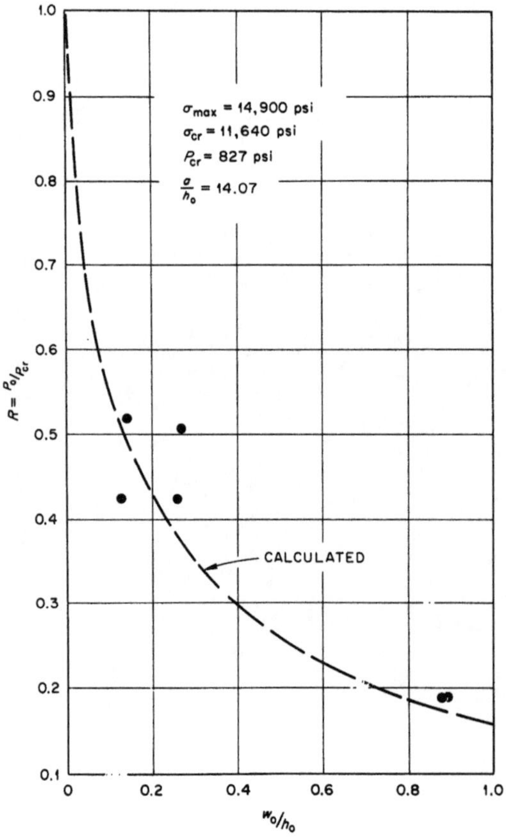

Fig. 14. Plot Showing Effect of Ovality on Instantaneous Collapse Pressure.

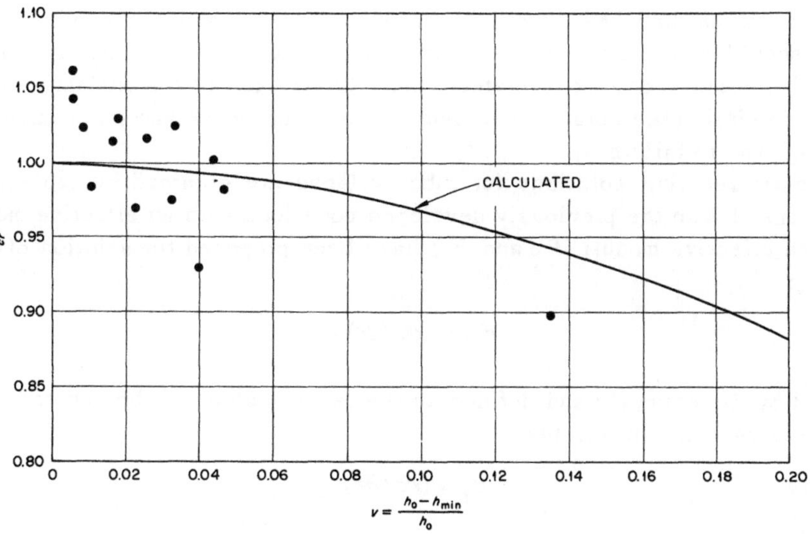

Fig. 15. Plot Showing Effect of Wall-Thickness Variation on Instantaneous Collapse Pressure.

produce tubing within such a tolerance and since the scatter in the experimental results exceeds 3%.

The foregoing discussion serves to demonstrate how the critical pressure for instantaneous collapse may be obtained and what the effects of tube imperfections are. Therefore, with the minimum expected strength of the material and the expected imperfections known, it is possible to design with any desired confidence factor tubes to resist collapse.

Time-Dependent Collapse

Although no experimental work has been performed on time-dependent collapse, it is felt that this phenomenon warrants discussion.

An exact solution of the creep-buckling problem is limited by the lack of proper and exact generalizations of the tension-creep stress-strain relations to multiaxial stress with changing principal stress directions. Several tentative solutions have been proposed [6, 7], which are approximate because of the many necessary simplifying assumptions made.

An approximate solution which may be more easily handled than those referred to above can be obtained by using an analysis of the "deformation-theory" type. Such an approach to inelastic buckling assumes that total strain at a given time is described by the equation resulting from principal strains being always proportional to principal stresses, with no rotation of the principal axis occurring. When the lateral deflection of the tube wall, or buckling, begins to occur because of increasing stresses, the magnitude of this deflection is computed as though the new increased stress had been applied throughout the entire loading period. This method should then provide a lower limit to the inelastic buckling load of members with limited geometric imperfections. Application of the inelastic-buckling solution to pin-jointed columns has demonstrated the conservatism of such a prediction of time to failure.

Inelastic-buckling solutions for tube collapse are obtained by replacing the reduced modulus in the previously developed equations with an effective modulus. Two such effective moduli ($E_{\dot{\epsilon}}$ and E_t) have been proposed for solution of column buckling:

$$E_{\dot{\epsilon}} = (\partial\sigma/\partial\epsilon)_{\dot{\epsilon}} , \tag{9}$$

proposed by Robotnov [8] and defined as the tangent slope of the constant strain-rate stress vs strain curve; and

$$E_t = (\partial\sigma/\partial\epsilon)_t , \tag{10}$$

proposed by Shanley [9] and defined as the tangent slope of the isochronous stress-strain curves.

Expressions for both moduli may be obtained from the creep equation

$$\epsilon_c = (\sigma/A)^{\alpha} t^m , \tag{11}$$

where ϵ_c = creep strain (in./in.), t = time (hr), and A, α, m = material constants, and from the total strain equation

$$\epsilon_t = \epsilon_c + \sigma/E , \tag{12}$$

where ϵ_t = total strain. The resulting expressions are then

$$E_{\dot{\epsilon}} = \left[\frac{\alpha\epsilon_c}{(1-m)\sigma} + \frac{1}{E}\right]^{-1} , \tag{13}$$

$$E_t = \left(\frac{\alpha\epsilon_c}{\sigma} + \frac{1}{E}\right)^{-1} . \tag{14}$$

It should be noted that if m were to become equal to 1, $E_{\dot{\epsilon}}$ would go to zero for all values of stress regardless of the material strength. Since it is not uncommon that $m = 1$ for many materials, it is felt that the Robotnov effective modulus is not suitable for use in tube-buckling equations. The stress-strain relationships to be considered in tube collapse are those occurring in the tangential direction, and it has been shown [*10*] that for capped end tubes,

$$\epsilon_{c_\theta} = \frac{3}{4}\epsilon_c , \tag{15}$$

where ϵ_{c_θ} = tangential creep strain under tangential stress σ_θ, and ϵ_c = uniaxial creep strain under a uniaxial stress equal to σ_θ.

Substituting in Eq. (14) yields

$$E_t = \left(\frac{\alpha \epsilon_{c\theta}}{\sigma_\theta} + \frac{1}{E}\right)^{-1} \tag{16}$$

or

$$E_t = \left(\frac{3}{4}\frac{\alpha \epsilon_c}{\sigma_\theta} + \frac{1}{E}\right)^{-1} . \tag{17}$$

The resulting tube collapse relationship from Eq. (2) is then

$$P_{cr} = (f)\,\frac{E_t}{4(1-\gamma^2)}\left(\frac{h_0}{a}\right)^3 \tag{18}$$

or

$$P = P_{cr}\frac{E_t}{E_r} . \tag{19}$$

As was previously pointed out, Eq. (19) does not yield an exact solution for the critical time of collapse, but yields, instead, the critical time for the loss of stability in the classical sense. After the critical time has elapsed, any small disturbance applied to the tube should result in deflections which increase with time. Collapse may then be expected at some time greater than the critical time for loss of stability calculated by Eq. (19).

A simple relationship or proportionality will possibly exist for critical time vs actual collapse time, but it must be determined experimentally. Until such experiments have been performed, the conservative prediction in Eq. (19) should be used.

An important result of the above analysis is that the previously developed graphical solution may again be utilized. The substitution of a material curve of E_t vs σ_θ corresponding to the design lifetime of the vessel for the material curve of E_r vs σ will produce the solution of Eq. (18).

CONCLUSIONS

The following conclusions on tube collapse at 1200°F for type 304 stainless steel have been reached:

1. The use of the von Karman reduced modulus in collapse equations adequately accounts for inelastic behavior of the material under rapidly applied loading.

2. The graphical method developed, which superimposes material behavior on tube geometry, allows the application of known safety factors in design to resist instantaneous collapse.

3. The effect of ovality, a serious imperfection which greatly reduces the pressure for instantaneous collapse, can be predicted.

4. The reduction of collapse pressure due to wall-thickness variations within normal manufacturing tolerances may be neglected.

5. A critical time, after which collapse may occur, exists for a tube under an external pressure less than the critical pressure for instantaneous collapse.

ACKNOWLEDGMENTS

The authors wish to thank F. L. Beeler and C. W. Walker for their assistance in the experimental program.

REFERENCES

[*1*] W. Fairbairn, *Phil. Trans. Roy. Soc. London* **148(A)**, 389–413 (1858).

[2] R. von Mises, *Z. Ver. deut. Ingr.* **58**, 750 (1914).

[3] Th. von Karman, *Forsch. Gebiete Ingenieurw., VDI–Forschungsheft* **81**, 20 (1910).

[4] S. Timoshenko, *Theory of Elastic Stability*, McGraw-Hill, New York, 1936.

[5] J. P. Ellington, *The Critical Pressure of a Tube with an Eccentric Bore*, DEG-43(R) (1960).

[6] N. J. Hoff, W. E. Jahsman, and W. Nachbar, *A Study of Creep Collapse of a Long Cylinder Under Uniform External Pressure*, Lockheed Aircraft Corporation Missile Systems Division Report LSMD-2360 (Mar. 28, 1958).

[7] J. P. Ellington, *Creep Collapse of Tubes Under External Pressure*, DEG-162(R) (1960).

[8] G. N. Robotnov and S. A. Shesterikov, *J. Mech. Phys. Solids* **6**, 27 (1957).

[9] R. F. Shanley, "Principles of Creep Buckling," chap. 19 in *Weight-Strength Analysis of Aircraft Structures*, McGraw-Hill, New York, 1952.

[*10*] C. R. Kennedy, W. O. Harms, and D. A. Douglas, *Trans. Am. Soc. Mech. Engrs.* **81(D)**, 599 (1959).

INFLUENCE OF THICKNESS-WISE TEMPERATURE GRADIENTS ON LARGE DEFLECTIONS AND STABILITY OF THIN ELASTIC SHELLS[1]

C. Libove and K. N. Tong
Syracuse University

On the basis of theoretical and experimental studies it is concluded that a thickness-wise temperature gradient which is constant over a thin elastic shell has negligible effect (if any) on the deflections and buckling strength of a clamped-edge shallow shell (spherical or nonspherical) under normal pressure and a complete spherical shell under uniform external pressure. In the case of the shallow shell with a clamped edge, it is also shown that a thickness-wise temperature gradient which is not constant over the shell can be replaced by an equivalent pressure distribution.

INTRODUCTION

The present paper describes three separate studies of the influence of thickness-wise temperature gradient on the behavior of thin elastic shells of constant thickness under normal pressure. The temperature gradients considered are of such a nature that they tend to produce curvature changes but no middle-surface areal dilatations of infinitesimal shell segments. (The conclusions obtained are also valid for the case in which the thermal strains tend to cause middle-surface areal dilatations, provided that such dilatations can take place without producing any resultant cross-sectional thrusts.)

The first section is concerned with shallow shells (spherical or nonspherical) clamped at the boundary. It is shown that a thickness-wise temperature gradient which is constant over the shell has no effect on the deflections and buckling strength of the shell under normal pressure and that a thickness-wise temperature gradient which is *not* constant over the shell can be replaced by an equivalent normal-pressure distribution. The analysis in this part is based on von Karman's large-deflection theory for flat plates, generalized to include small initial deviations from flatness and thermal deformations.

---

[1]Abridged version of Syracuse University Research Institute Report ME 790-612 T, February 1961, bearing the same title, on work performed for Oak Ridge National Laboratory and Union Carbide Nuclear Company.

NOTATION

ζ	Coordinate through thickness of shell
$\mu(\zeta)$	Coefficient of thermal expansion
h	Shell thickness
$T(\zeta)$	Temperature
E	Young's modulus, assumed constant
ν	Poisson's ratio, assumed constant
D	Plate flexural stiffness; $Eh^3/[12(1-\nu^2)]$
$\epsilon_T(\zeta)$	Thermal strain; μT
p	Pressure
χ	Spherical curvature change produced in an infinitesimal shell element by temperature alone; $(+12/h^3)\int_{-h/2}^{h/2} \epsilon_T \zeta \, d\zeta$
ϵ	Middle-surface strain that results from temperature alone in the same element; $(1/h)\int_{-h/2}^{h/2} \epsilon_T \, d\zeta$

The second section is concerned with complete spherical shells under a constant thickness-wise temperature gradient and uniform external pressure. It is shown that the temperature gradient has a negligible effect (if any) on the axisymmetric classical buckling pressure and on the postbuckling axisymmetric large deflections. The analysis is based on a study of Reissner's differential equations for large deflections of rotationally symmetric shells, generalized to include temperature strains.

The theoretical results suggest that a thickness-wise temperature gradient constant over a spherical shell will not alter its buckling strength under hydrostatic pressure regardless of whether the buckling is of the classical type or of the jump type postulated by Karman and Tsien, which depends on the existence of nonadjacent equilibrium states. In an effort to verify this prediction, a series of experiments was performed and is described in the third section. Because it is rather difficult to produce and control experimentally a thickness-wise temperature gradient, these experiments were performed with bimetallic shells at various uniform temperatures. The shells were hemispherical and were formed out of bimetallic sheets of brass and Invar, each constituting half the thickness. The differential thermal expansions produced bending moments equivalent to those that would be produced in homogeneous shells by temperature gradients. The test results generally supported the conclusion suggested by the theoretical analysis.

CLAMPED SHALLOW SHELLS

This section is concerned with the nonlinear deflections of thin, elastic shallow shells (spherical and other) of uniform thickness, clamped at the boundary and subjected simultaneously to normal loading and temperature gradients through the thickness. Its objective is to show (1) that a constant thickness-wise temperature gradient has no influence on the deflections and stability of a clamped shallow shell and (2) that a nonconstant-temperature gradient can be replaced by an equivalent pressure distribution.

Description of Shells

The shells are shallow in the sense that the middle surface deviates only slightly from a plane and its initial and final slopes relative to this plane are much smaller than unity. The usual assumption is made that material lines normal to the middle surface in the undeformed shell are also straight and normal to the middle surface in the deformed shell. The ratio of shell thickness to any of the curvature radii is assumed to be much smaller than 1.

The temperature distribution can be described with sufficient detail for incorporation into shallow-shell theory by means of two parameters, χ and ϵ, whose variations throughout the shell are assumed to be known. The parameter χ is the spherical curvature change that would be produced in an infinitesimal shell element by the temperature alone (i.e., if the element were removed from the rest of the shell by means of a closed cut normal to the middle surface), and ϵ is the middle surface strain that would result from temperature alone in the same element. The parameter χ depends mainly on the way the temperature varies through the thickness, and ϵ depends mainly on the mean temperature through the thickness. For the special case in which the temperature at a given location varies linearly from T_l at the lower surface to T_u at the upper surface and the thermal expansion coefficient μ is a constant, the χ and ϵ values at this location can be computed from the formulas

$$\chi = \frac{\mu(T_l - T_u)}{h}, \quad \epsilon = \mu\left(\frac{T_l + T_u}{2}\right), \tag{1}$$

where h is the shell thickness, and the temperatures are measured relative to some basic temperature distribution for which the unloaded shell would be stress-free. (Sign convention: χ positive implies that the center of curvature is on the T_u side of the shell, ϵ positive implies elongational strain.) In the more general case in which the temperature T and expansion coefficient μ vary in some arbitrary but known way through the thickness while Young's modulus and Poisson's ratio are constant, formulas given in the section on Notation may be used for evaluating χ and ϵ; in these formulas ζ is the coordinate locating a material point with respect to the middle surface, with $\zeta = +h/2$ corresponding to the lower surface and $\zeta = -h/2$ corresponding to the upper surface.

The middle surface of the unloaded and unheated shell can be described by means of its small initial normal deviations $w_0(x, y)$ from the XY plane of a right-hand. Cartesian coordinate system, with w_0 taken positive in the positive direction of the Z axis. Thus (x, y, w_0) represent the initial coordinates of a material point of the shell middle surface. The additional Z-wise deflection of the same material point resulting from load and temperature changes will be denoted by $w(x, y)$. Simultaneously, the point may have displacements $u(x, y)$ and $v(x, y)$ parallel to the X and Y axes respectively. The normal pressure will be designated by $p(x, y)$, positive if it acts in the direction of decreasing Z. The designation "upper surface" will be used for that surface of the shell on the larger Z side; "lower surface" will be used for the smaller Z side of the shell. In the case of shells with circular boundary, it may be convenient to employ an r, θ polar coordinate system lying in the XY plane. Then χ, ϵ, w_0, w, u, v, and p may equally well be regarded as functions of r and θ instead of x and y.

The condition of clamping at the boundary may be described in symbols as follows:

$$u = 0\,, \qquad v = 0\,, \qquad w = 0\,,$$
$$(\nabla w) \cdot \vec{n} = \frac{\partial w}{\partial x} \cos \alpha + \frac{\partial w}{\partial y} \sin \alpha = 0\,, \tag{2}$$

where $\vec{n}$ is the outward normal unit vector on the projection of the shell boundary on the XY plane, and α is the angle between the positive directions of the $\vec{n}$ vector and the X axis (measured by rotating the positive end of the X axis toward the positive end of the Y axis until it is parallel to, and pointing in the same direction as, the $\vec{n}$ vector).

The above conditions will be referred to as "full clamping." In addition, a "limited clamping" will be considered, in which those edge displacements that would result from the ϵ distribution alone are permitted and the conditions in Eq. (2) apply only to the additional displacements resulting from χ and p. The "limited-clamping" condition may be pertinent if, for example, the shallow shell under consideration is in reality a small portion of a large spherical shell in which ϵ is constant. Then the "base" of the shallow shell experiences a thermal expansion consistent with the constant ϵ distribution. Limited clamping, as defined above, may be considered only for those ϵ distributions which are associated with a physically possible set of displacements, for otherwise the phrase "those edge displacements that would result from the ϵ distribution alone" becomes meaningless. Therefore, limited clamping will be considered for ϵ distributions satisfying the compatibility condition

$$\nabla^2 \epsilon = 0\,,$$

where ∇^2 represents the two-dimensional Laplacian operator $\partial^2/\partial x^2 + \partial^2/\partial y^2$ in Cartesian coordinates, and

$$\frac{\partial^2}{\partial r^2} + \frac{1}{r}\frac{\partial}{\partial r} + \frac{1}{r^2}\frac{\partial^2}{\partial \theta^2}$$

in polar coordinates.

Results and Discussion

The following statements apply to the "fully clamped" shell with ϵ everywhere zero, or to the shell with "limited clamping" and $\nabla^2\epsilon$ everywhere zero:

1. If χ is constant through the shell, the deflections under load are unaffected by the temperature distribution.
2. If χ is not constant, the shell will deflect like an identical but unheated shell with a pressure distribution $\tilde{p}(x, y)$ given by

$$\tilde{p}(x, y) = p(x, y) - D(1 + \nu)\nabla^2\chi \,. \tag{3}$$

Statement (2) [which includes (1) as a special case] implies that a χ distribution can be replaced by an equivalent pressure distribution. To illustrate the usability of this result, let it be supposed that the critical uniform pressure for "snap-through" buckling must be determined for a shallow spherical shell with a circular fully clamped boundary having the following dimensions and properties:

$E = 30{,}000{,}000$ psi,

$\nu = 0.25$,

a = radius of boundary circle = 50 in.,

H = rise at center of shell = 5 in.,

R = radius of curvature of shell $\approx a^2/2H = 250$ in.,

$h = 0.5$ in.,

$\mu = 0.000006/^\circ$F.

The shell's inner (lower) surface is 200°F hotter than the outer (upper) surface at the center and 50°F hotter than the outer surface at the boundary. The temperature difference between the inner and the outer surface is axisymmetric and varies with radial distance r from the center of the shell in the following parabolic manner:

$$T_l - T_u = 200 - 150\frac{r^2}{a^2}\ {}^\circ\text{F} \,.$$

Suppose, furthermore, that T varies linearly through the thickness and that the middle-surface temperature remains constant at the base value (thus making ϵ zero everywhere).

On the basis of the foregoing information, the χ variation is computed from Eq. (1) as

$$\chi = \frac{0.000006}{0.5}\left[200 - 150\frac{r^2}{(50)^2}\right] \text{in.}^{-1} \,,$$

where r is measured in inches. Then (for axial symmetry)

$$\nabla^2\chi = \frac{\partial^2\chi}{\partial r^2} + \frac{1}{r}\frac{\partial\chi}{\partial r}$$

$$= \frac{0.000006}{0.5}\left[-\frac{300}{(50)^2} - \frac{300}{(50)^2}\right]$$

$$= -0.288 \times 10^{-5}\ \text{in.}^{-3} ,$$

$$-D(1+\nu)\nabla^2\chi = -\frac{Eh^3}{12(1-\nu)}\nabla^2\chi$$

$$= \frac{(30{,}000{,}000)(0.5)^3(+0.288)(10^{-5})}{12(0.75)}$$

$$= +1.20\ \text{psi} .$$

Thus the given temperature gradients are equivalent to a uniform pressure of 1.20 psi. In the absence of temperature effects the uniform snap-through buckling pressure for this shell (estimated conservatively from the experimental data in Fig. 19 of ref [1] is 45.6 psi. Hence, under simultaneous action of pressure and the given temperature distribution, the pressure required for snap-through buckling is 45.6 – 1.20 or 44.4 psi. In this case the temperature distribution has a small destabilizing effect on the shell.

In general, an axisymmetric χ distribution which varies from $\chi(0)$ at $r = 0$ to $\chi(a)$ at $r = a$ in the manner

$$\chi = \chi(0) + [\chi(a) - \chi(0)]\frac{r^2}{a^2} \tag{4}$$

may, in a shallow shell, be replaced by the constant equivalent pressure,

$$p = -\frac{4D(1+\nu)}{a^2}[\chi(a) - \chi(0)] . \tag{5}$$

Therefore, analytical or experimental results on clamped elastic shallow shells under constant pressure may be generalized to include χ variations in the form of Eq. (4) by merely replacing the pressure symbol p in those results by $p - \{4D(1+\nu)[\chi(a) - \chi(0)]/a^2\}$. The shells need not be spherical (e.g., this substitution may be made in the analytical results of ref [2] for not-quite-spherical shells); in fact, they need not be axisymmetric.

Analysis

The elastic analysis of thin shallow shells in which the slopes are small enough to permit the curvatures and the twist to be computed from expressions of the type $\partial^2 w/\partial x^2$, $\partial^2 w/\partial y^2$, and $\partial^2 w/\partial x\, \partial y$ can be based on von Karman's large deflection equations for flat plates [3, pp 415–18], generalized to include small initial deviations from flatness [4] and the deformations χ and ϵ due to temperature.

Needed for the generalization are the following "stress-strain-temperature" relations for the middle-surface strains:

$$\epsilon_x = \frac{1}{hE}(N_x - \nu N_y) + \epsilon \,,$$

$$\epsilon_y = \frac{1}{hE}(N_y - \nu N_x) + \epsilon \,, \tag{6}$$

$$\gamma_{xy} = \frac{2(1+\nu)}{hE} N_{xy} \,,$$

and for the curvatures and twist:

$$\frac{\partial^2 w}{\partial x^2} = \frac{12}{Eh^3}(M_x - \nu M_y) + \chi \,,$$

$$\frac{\partial^2 w}{\partial y^2} = \frac{12}{Eh^3}(M_y - \nu M_x) + \chi \,, \tag{7}$$

$$\frac{\partial^2 w}{\partial x\, \partial y} = \frac{12(1+\nu)}{Eh^3} M_{xy} \,.$$

(The undefined notation is that of ref [3, pp xiii and xiv].) Also needed are the generalized equilibrium equations,

$$\frac{\partial N_x}{\partial x} + \frac{\partial N_{xy}}{\partial y} = 0 \,,$$

$$\frac{\partial N_y}{\partial y} + \frac{\partial N_{xy}}{\partial x} = 0 \,, \tag{8}$$

$$\frac{\partial M_x}{\partial x} + \frac{\partial M_{xy}}{\partial y} = Q_x \,,$$

$$\frac{\partial M_y}{\partial y} + \frac{\partial M_{xy}}{\partial x} = Q_y \,, \tag{9}$$

$$\frac{\partial Q_x}{\partial x} + \frac{\partial Q_y}{\partial x} + p = N_x \frac{\partial^2 (w_0 + w)}{\partial x^2} + N_y \frac{\partial^2 (w_0 + w)}{\partial y^2} + 2N_{xy} \frac{\partial^2 (w_0 + w)}{\partial x\, \partial y} \,,$$

and the strain-displacement relations

$$\epsilon_x = \frac{\partial u}{\partial x} + \frac{1}{2}\left[\frac{\partial(w_0 + w)}{\partial x}\right]^2 - \frac{1}{2}\left(\frac{\partial w_0}{\partial x}\right)^2,$$

$$\epsilon_y = \frac{\partial v}{\partial y} + \frac{1}{2}\left[\frac{\partial(w_0 + w)}{\partial y}\right]^2 - \frac{1}{2}\left(\frac{\partial w_0}{\partial y}\right)^2, \tag{10}$$

$$\gamma_{xy} = \frac{\partial u}{\partial y} + \frac{\partial v}{\partial x} + \left[\frac{\partial(w_0 + w)}{\partial x}\right]\left[\frac{\partial(w_0 + w)}{\partial y}\right] - \left(\frac{\partial w_0}{\partial x}\right)\left(\frac{\partial w_0}{\partial y}\right).$$

Equations (8) imply the existence of a stress function F such that

$$N_x = h\,\frac{\partial^2 F}{\partial y^2}, \qquad N_y = h\,\frac{\partial^2 F}{\partial x^2}, \qquad N_{xy} = -h\,\frac{\partial^2 F}{\partial x\,\partial y}. \tag{11}$$

Using the first two of Eqs. (9) to eliminate Q_x and Q_y in the third, gives

$$\frac{\partial^2 M_x}{\partial x^2} + 2\,\frac{\partial^2 M_{xy}}{\partial x\,\partial y} + \frac{\partial^2 M_y}{\partial y^2} + p = h\left[\frac{\partial^2 F}{\partial y^2}\,\frac{\partial^2(w_0 + w)}{\partial x^2} + \frac{\partial^2 F}{\partial x^2}\,\frac{\partial^2(w_0 + w)}{\partial y^2} - 2\,\frac{\partial^2 F}{\partial x\,\partial y}\,\frac{\partial^2(w_0 + w)}{\partial x\,\partial y}\right]. \tag{12}$$

Then, elimination of M_x, M_y, M_{xy} via Eqs. (7) gives

$$D\nabla^4 w - D(1 + \nu)\nabla^2\chi + p = h\left[\frac{\partial^2 F}{\partial y^2}\,\frac{\partial^2(w_0 + w)}{\partial x^2} + \frac{\partial^2 F}{\partial x^2}\,\frac{\partial^2(w_0 + w)}{\partial y^2} - 2\,\frac{\partial^2 F}{\partial x\,\partial y}\,\frac{\partial^2(w_0 + w)}{\partial x\,\partial y}\right], \tag{13}$$

where ∇^4 represents the biharmonic operator $\frac{\partial^4}{\partial x^4} + 2\frac{\partial^4}{\partial x^2\partial y^2} + \frac{\partial^4}{\partial y^4}$ in Cartesian coordinates, and $\left(\frac{\partial^2}{\partial r^2} + \frac{1}{r}\frac{\partial}{\partial r} + \frac{1}{r^2}\frac{\partial^2}{\partial \theta^2}\right)\left(\frac{\partial^2}{\partial r^2} + \frac{1}{r}\frac{\partial}{\partial r} + \frac{1}{r^2}\frac{\partial^2}{\partial \theta^2}\right)$ in polar coordinates.

By differentiation of Eqs. (10) it is seen that the three middle-surface strains and the normal displacement satisfy the compatibility equation,

$$\frac{\partial^2 \epsilon_x}{\partial y^2} + \frac{\partial^2 \epsilon_y}{\partial x^2} - \frac{\partial^2 \gamma_{xy}}{\partial x\,\partial y} = \left\{\left[\frac{\partial^2(w_0 + w)}{\partial x\,\partial y}\right]^2 - \left[\frac{\partial^2(w_0 + w)}{\partial x^2}\right]\left[\frac{\partial^2(w_0 + w)}{\partial y^2}\right]\right\} - \left[\left(\frac{\partial^2 w_0}{\partial x\,\partial y}\right) - \left(\frac{\partial^2 w_0}{\partial x^2}\right)\left(\frac{\partial^2 w_0}{\partial y^2}\right)\right]. \tag{14}$$

Writing ϵ_x, ϵ_y, γ_{xy} in terms of N_x, N_y, N_{xy} through Eqs. (6) and then in terms of F through Eqs. (11), transforms the last result into

$$\nabla^4 F + E\nabla^2 \epsilon = E\left\{\left[\frac{\partial^2(w_0 + w)}{\partial x\, \partial y}\right]^2 - \left[\frac{\partial^2(w_0 + w)}{\partial x^2}\right]\left[\frac{\partial^2(w_0 + w)}{\partial y^2}\right] - \left(\frac{\partial^2 u_0}{\partial x\, \partial y}\right)^2 + \left(\frac{\partial^2 w_0}{\partial x^2}\right)\left(\frac{\partial^2 w_0}{\partial y^2}\right)\right\} . \quad (15)$$

Equations (13) and (15) represent two equations in two unknowns, w and F. With the temperature terms left out, they become identical with the equations obtained in a different manner by Marguerre [5].

Inasmuch as w and F are the only unknowns appearing in Eqs. (13) and (15), it is desirable to have boundary conditions in which w and F are also the only unknowns. Two of the four boundary conditions for full clamping [Eqs. (2)] are already in terms of w. It remains to express the other two conditions ($u = 0$, $v = 0$) in terms of w and F. This will be done first for a shell whose boundary has a convex projection on the XY plane, that is, such that any straight line connecting two points of the projected boundary lies entirely within the area enclosed by the projected boundary (see Fig. 1).

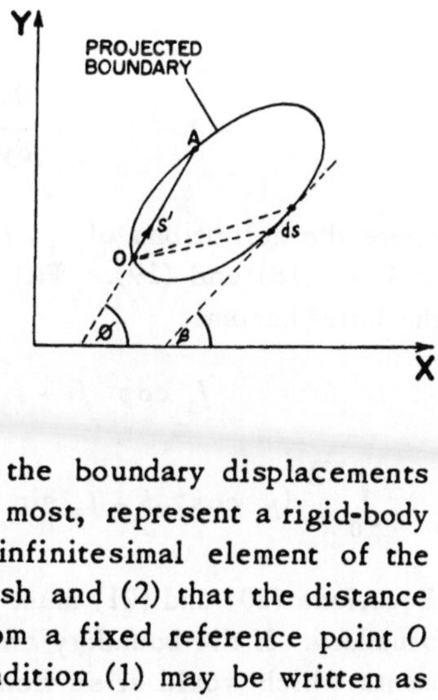

Fig. 1. Boundary of Shell Projected on XY Plane.

The necessary and sufficient conditions for the boundary displacements u, v parallel to the XY plane to vanish (or, at most, represent a rigid-body displacement) are (1) that the strain of every infinitesimal element of the projected boundary such as ds (see sketch) vanish and (2) that the distance of every projected boundary point, such as A, from a fixed reference point O of the projected boundary remain unchanged. Condition (1) may be written as [ref 6, p 18]

$$\frac{\partial u}{\partial x}\cos^2\beta + \left(\frac{\partial u}{\partial y} + \frac{\partial v}{\partial x}\right)\sin\beta\cos\beta + \frac{\partial v}{\partial y}\sin^2\beta = 0 \quad (16)$$

at all boundary points, and condition (2) as

$$\int_0^{\overline{OA}}\left[\frac{\partial u}{\partial x}\cos^2\phi + \left(\frac{\partial u}{\partial y} + \frac{\partial v}{\partial x}\right)\sin\phi\cos\phi + \frac{\partial v}{\partial y}\sin^2\phi\right] ds' = 0 \quad (17)$$

for all possible A, where the integrand is evaluated at those interior points of the shell middle surface whose projections were originally along line OA. The quantities $\partial u/\partial x$, $\partial u/\partial y + \partial v/\partial x$, $\partial v/\partial y$ appearing in the last two equations can be expressed in terms of w and F through Eqs. (10), (6), and (11). The resulting expressions are

$$\frac{\partial u}{\partial x} = \frac{1}{2}\left(\frac{\partial w_0}{\partial x}\right)^2 - \frac{1}{2}\left[\frac{\partial (w_0 + w)}{\partial x}\right]^2 + \epsilon + \frac{1}{E}\left(\frac{\partial^2 F}{\partial y^2} - \nu \frac{\partial^2 F}{\partial x^2}\right),$$

$$\frac{\partial v}{\partial y} = \frac{1}{2}\left(\frac{\partial w_0}{\partial y}\right)^2 - \frac{1}{2}\left[\frac{\partial (w_0 + w)}{\partial y}\right]^2 + \epsilon + \frac{1}{E}\left(\frac{\partial^2 F}{\partial x^2} - \nu \frac{\partial^2 F}{\partial y^2}\right), \tag{18}$$

$$\frac{\partial u}{\partial y} + \frac{\partial v}{\partial x} = \left(\frac{\partial w_0}{\partial x}\right)\left(\frac{\partial w_0}{\partial y}\right) - \left[\frac{\partial (w_0 + w)}{\partial x}\right]\left[\frac{\partial (w_0 + w)}{\partial y}\right] - \frac{2(1+\nu)}{E}\frac{\partial^2 F}{\partial x\, \partial y} .$$

These will be written for later use in the short-hand form

$$\frac{\partial u}{\partial x} = f_1(w, F, w_0) + \epsilon ,$$

$$\frac{\partial v}{\partial y} = f_2(w, F, w_0) + \epsilon , \tag{19}$$

$$\frac{\partial u}{\partial y} + \frac{\partial v}{\partial x} = f_3(w, F, w_0) ,$$

where the definitions of f_1, f_2, and f_3 are readily apparent through comparison of Eqs. (18) and (19). With the substitution of Eqs. (19) into (16) and (17), the latter become

$$f_1 \cos^2 \beta + f_3 \sin \beta \cos \beta + f_2 \sin^2 \beta + \epsilon = 0 , \tag{20}$$

$$\int_0^{\overline{OA}} (f_1 \cos^2 \phi + f_3 \sin \phi \cos \phi + f_2 \sin^2 \phi)\, ds' + \int_0^{\overline{OA}} \epsilon\, ds' = 0 . \tag{21}$$

Equations (20) and (21) must be satisfied for all points A along a fully clamped boundary. If the boundary has "limited clamping" – that is, if those displacements which would arise from temperature alone are permitted – then the right sides of Eqs. (20) and (21) become ϵ and $\int_0^{\overline{OA}} \epsilon\, ds'$, respectively, in place of zero. Now the boundary conditions (20) and (21) are replaced by

$$f_1 \cos^2 \beta + f_3 \sin \beta \cos \beta + f_2 \sin^2 \beta = 0 , \tag{20a}$$

$$\int_0^{\overline{OA}} (f_1 \cos^2 \phi + f_3 \sin \phi \cos \phi + f_2 \sin^2 \phi)\, ds' = 0 . \tag{21a}$$

Equations (20) and (21) [or (20*a*) and (21*a*)], which have w and F as their only unknowns, take the place of conditions $u = 0$, $v = 0$ of Eqs. (2). The remaining conditions of Eqs. (2), namely,

$$w = 0 , \quad \frac{\partial w}{\partial x} \cos \alpha + \frac{\partial w}{\partial y} \sin \alpha = 0 , \tag{22}$$

must be added to Eqs. (20) and (21) [or (20*a*) and (21*a*)] to obtain a complete statement of the boundary conditions.

The replacement of the boundary conditions $u = 0$, $v = 0$ by Eqs. (20) and (21) was predicated on the assumption that all points of the boundary could be joined to a single point O by a fanlike arrangement of radiating straight lines lying entirely inside the boundary curve. If the boundary is not convex but the enclosed area can be covered by several adjacent fanlike arrays of straight lines radiating from several points, then it may be necessary to write equations like (21) for each of the several segments of the boundary. (Figure 2 illustrates a covering with three arrays of radiating lines.) In such a case w and F will still be the only unknowns appearing in the boundary conditions.

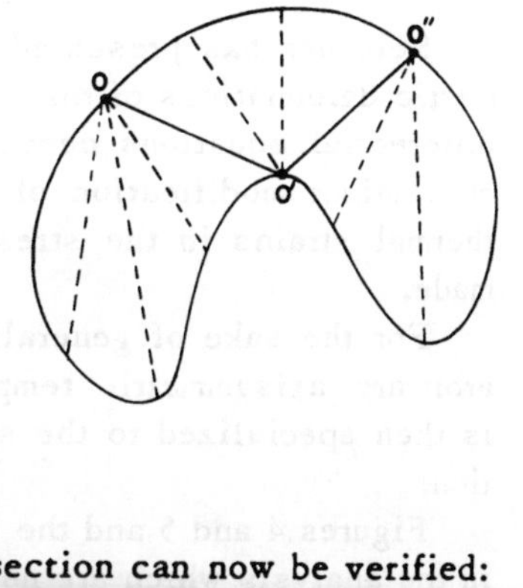

Fig. 2. Nonconvex Projected Boundary.

The results (1) and (2) stated in the previous section can now be verified: If ϵ is everywhere zero, or if the boundary has limited clamping, ϵ does not appear in the boundary conditions [(20), (21), and (22) in the former case; (20*a*), (21*a*), and (22) in the latter]. If, further, $\nabla^2 \epsilon = 0$, then ϵ also vanishes from the differential equation (15). Now the entire system of differential equations and boundary conditions for F and w become those of the unheated shell with the substitute pressure distribution $\tilde{p}(x,y)$ equal to the actual pressure distribution $p(x,y)$ minus $D(1 + \nu)\nabla^2 \chi$.

COMPLETE SPHERICAL SHELLS

Description of Shells and Results

This section contains an investigation of an elastic thin-walled complete spherical shell with a temperature which varies radially (i.e., through the thickness) but is constant circumferentially. Two phenomena (see Fig. 3) are

Fig. 3. Phenomena Considered in the Second Section: (*a*) Axisymmetric Classical Buckling; (*b*) Postbuckling Axisymmetric Large Deflections.

considered in connection with such a shell: (1) the axisymmetric classical buckling under uniform external pressure and (2) the axisymmetric postbuckling large deflections. Young's modulus and Poisson's ratio are assumed constant.

The purpose of the investigation is to determine what influence, if any, the temperature gradient has on these phenomena. By examination of the differential equations, it is concluded that the temperature gradient has a negligible influence; that is, the buckling pressure in case (1) and the deflections in case (2) are, for all practical purposes, unaffected by the temperature gradient. The analysis on which these results are based follows.

Special Notation and Basic Equations

Reissner has presented [7, 8] a large-displacement theory for the axisymmetric deformations of thin elastic shells of revolution with small strains. The differential equations needed for the present study can be obtained by means of a minor modification of Reissner's development, namely, the inclusion of thermal strains in the stress-strain relations. This modification will now be made.

For the sake of generality an arbitrary axisymmetric shell subjected to an arbitrary axisymmetric temperature variation is analyzed first; this analysis is then specialized to the spherical shell with purely radial temperature variation.

Figures 4 and 5 and the following definitions explain those symbols required in the analysis which are not already defined in the Notation.

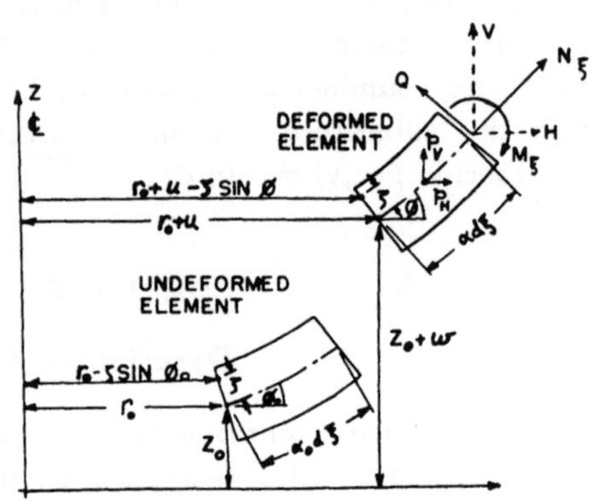

Fig. 4. Side View of Element of Shell in Undeformed and in Deformed State.

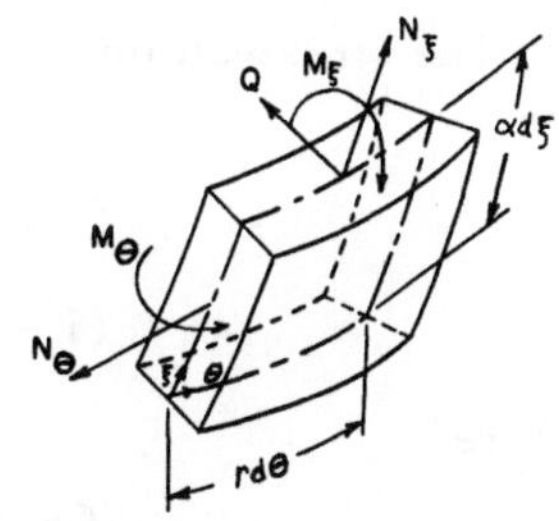

Fig. 5. Element of Shell Showing Stress Resultants and Couples.

C	Stretching stiffness, Eh
ξ	Parameter associated with material points on meridional profile of shell middle surface
$r_0'(\xi)$, $z_0(\xi)$	Coordinates of middle surface of undeformed shell
$u(\xi)$, $u(\xi)$	Radial and axial displacement components of a point on the middle surface
$r(\xi)$, $z(\xi)$	Coordinates of middle surface of deformed shell ($r = r_0 + u$, $z = z_0 + w$)
$\alpha_0\, d\xi$	Length of an infinitesimal portion of the meridional profile of the middle surface of the undeformed shell $[\alpha_0 = \alpha_0(\xi)]$
$\alpha\, d\xi$	Similarly defined for deformed shell $[\alpha = \alpha(\xi)]$
$\phi_0(\xi)$	Angle between radial direction and a line tangent to the meridional profile of the middle surface of the undeformed shell
$\phi(\xi)$	Similarly defined for deformed shell
$\beta(\xi)$	$\phi_0 - \phi$
$p_H(\xi)$, $p_V(\xi)$	Radial and axial distributed loading per unit area of middle surface of deformed shell
$H(\xi)$, $V(\xi)$	Radial and axial cross-sectional stress resultants per unit circumferential width of deformed-shell middle surface
$N_\xi(\xi)$, $Q(\xi)$	Normal and shearing cross-sectional stress resultants per unit circumferential width of deformed-shell middle surface
$M_\xi(\xi)$	Meridional bending moment per unit circumferential width of deformed-shell middle surface
$M_\theta(\xi)$	Circumferential bending moment per unit meridional length of deformed-shell middle surface
ϵ_ξ, ϵ_θ	Meridional and circumferential normal strains at a point
σ_ξ, σ_θ	Meridional and circumferential normal stresses at a point
M	Value of M_ξ and M_θ that would be needed to suppress thermally induced curvature change, $-D(1+\nu)\chi$

The stress-strain relations [7, Eqs. (33)], revised to include thermal strains, are,

$$(1-\nu^2)\sigma_\xi = E[(\epsilon_\xi - \epsilon_T) + \nu(\epsilon_\theta - \epsilon_T)] \,,$$
$$(1-\nu^2)\sigma_\theta = E[(\epsilon_\theta - \epsilon_T) + \nu(\epsilon_\xi - \epsilon_T)] \,, \tag{23}$$

where

$$\epsilon_\xi = \epsilon_{\xi M} + \zeta\chi_\xi \,, \qquad \epsilon_\theta = \epsilon_{\theta M} + \zeta\chi_\theta \,, \tag{24}$$

$$\epsilon_{\xi M} = \frac{\alpha - \alpha_0}{\alpha_0} = \frac{\cos\phi_0}{\cos\phi}\left(1 + \frac{u'}{r_0'}\right) - 1 \,,$$
$$\epsilon_{\theta M} = \frac{u}{r_0} \,, \tag{25}$$

$$\chi_\xi = -\frac{\phi' - \phi_0'}{\alpha_0} = \frac{\beta'}{\alpha_0} \,,$$
$$\chi_\theta = -\frac{\sin\phi - \sin\phi_0}{r_0} \,. \tag{26}$$

The primes denote differentiation with respect to ξ. In writing Eqs. (24), terms of the order ζ/R_ξ and ζ/R_θ have been neglected by comparison with unity.

With Eqs. (24) substituted into Eqs. (23), the latter can be multiplied by $d\zeta$ and integrated between the limits $-h/2$ and $+h/2$ to give

$$(1-\nu^2)N_\xi = Eh[(\epsilon_{\xi M} - \epsilon) + \nu(\epsilon_{\theta M} - \epsilon)] \,,$$
$$(1-\nu^2)N_\theta = Eh[(\epsilon_{\theta M} - \epsilon) + \nu(\epsilon_{\xi M} - \epsilon)] \,, \tag{27}$$

where

$$N_\xi = \int_{-h/2}^{+h/2} \sigma_\xi \, d\zeta \,, \quad N_\theta = \int_{-h/2}^{+h/2} \sigma_\theta \, d\zeta \,, \quad \epsilon = \frac{1}{h}\int_{-h/2}^{+h/2} \epsilon_T \, d\zeta \,, \tag{28}$$

$$N_\xi = H\cos\phi + V\sin\phi \,. \tag{29}$$

Multiplying Eqs. (23) by $\zeta\, d\zeta$ and integrating gives

$$(1-\nu^2)M_\xi = \frac{Eh^3}{12}[(\chi_\xi - \chi) + \nu(\chi_\theta - \chi)] \,,$$
$$(1-\nu^2)M_\theta = \frac{Eh^3}{12}[(\chi_\theta - \chi) + \nu(\chi_\xi - \chi)] \,, \tag{30}$$

where

$$M_\xi = \int_{-h/2}^{+h/2} \sigma_\xi \zeta \, d\zeta \,, \quad M_\theta = \int_{-h/2}^{+h/2} \sigma_\theta \zeta \, d\zeta \,, \quad \chi = +\frac{12}{h^3}\int_{-h/2}^{+h/2} \epsilon_T \zeta \, d\zeta \,. \tag{31}$$

The equilibrium equations remain, as in ref [7],

$$(rV)' + rap_V = 0 ,$$
$$(rH)' - \alpha N_\theta + r\alpha p_H = 0 , \tag{32}$$
$$(rM_\xi)' - (\alpha \cos\phi)M_\theta + r\alpha(H \sin\phi - V \cos\phi) = 0 ,$$

where

$$r = r_0 + u = r_0\left(1 + \frac{u}{r_0}\right) = r_0(1 + \epsilon_{\theta M}) ,$$
$$\alpha = \alpha_0(1 + \epsilon_{\xi M}) . \tag{33}$$

If small strains (but not necessarily small rotations) are assumed, then, according to Eqs. (33), r and α in the equilibrium equations may be replaced by r_0 and α_0, respectively, and the following modified form of Eqs. (32) is obtained [8]:

$$(r_0 V)' + r_0 \alpha_0 p_V = 0 ,$$
$$(r_0 H)' - \alpha_0 N_\theta + r_0 \alpha_0 p_H = 0 , \tag{32a}$$
$$(r_0 M_\xi)' - (\alpha_0 \cos\phi)M_\theta + r_0 \alpha_0(H \sin\phi - V \cos\phi) = 0 .$$

The compatibility condition, obtained by eliminating u between the strain-displacement relations (25), also remains unaffected by temperature and is

$$(\cos\phi_0)(r_0 \epsilon_{\theta M})' - (\cos\phi)(r_0' \epsilon_{\xi M}) = r_0'(\cos\phi - \cos\phi_0) . \tag{34}$$

Reduction to Two Simultaneous Equations

Equations (27), (30), and (32*a*) [with $\epsilon_{\xi M}$, $\epsilon_{\theta M}$, χ_ξ, χ_θ, and N_ξ eliminated via Eqs. (25), (26), and (29)] constitute seven equations in the seven unknowns u, ϕ, M_ξ, M_θ, V, H, and N_θ. This system, without the temperature terms ϵ and χ, was reduced by Reissner to two simultaneous equations for ϕ and ψ, where

$$\psi \equiv r_0 H . \tag{35}$$

The presence of the temperature terms offers no difficulty to this reduction. Repeating Reissner's steps, one first expresses some of the remaining quantities in the terms of ϕ and ψ:

From Eqs. (32*a*),

$$r_0 V = -\int r_0 p_V \alpha_0 \, d\xi . \tag{36}$$

From Eq. (29),

$$r_0 N_\xi = \psi \cos\phi + (r_0 V) \sin\phi . \tag{37}$$

From Eqs. (32*a*),

$$\alpha_0 N_\theta = \psi' + \alpha_0 r_0 p_H \,.$$

From Eqs. (30) and (26),

$$M_\xi = M - D\left(\frac{\phi' - \phi_0'}{\alpha_0} + \nu\,\frac{\sin\phi - \sin\phi_0}{r_0}\right),$$
$$M_\theta = M - D\left(\frac{\sin\phi - \sin\phi_0}{r_0} + \nu\,\frac{\phi' - \phi_0'}{\alpha_0}\right). \tag{38}$$

From Eqs. (25) and (27),

$$u = r_0\,\epsilon + \frac{1}{C}\left\{\frac{r_0}{\alpha_0}\psi' + (r_0^2 p_H) - \nu[\psi\cos\phi + (r_0 V)\sin\phi]\right\}. \tag{39}$$

From Eqs. (27) and (37),

$$\epsilon_{\xi M} = \epsilon + \frac{1}{C}\left\{\frac{1}{r_0}[\psi\cos\phi + (r_0 V)\sin\phi] - \frac{\nu}{\alpha_0}(\psi' + \alpha_0 r_0 p_H)\right\},$$
$$\epsilon_{\theta M} = \epsilon + \frac{1}{C}\left\{\frac{1}{\alpha_0}(\psi' + \alpha_0 r_0 p_H) - \frac{\nu}{r_0}[\psi\cos\phi + (r_0 V)\sin\phi]\right\}. \tag{40}$$

The two simultaneous equations are then obtained by (1) substituting Eqs. (38) into the third of the equilibrium equations (32*a*) and (2) substituting Eqs. (40) into the compatibility equations (34). The results, with the temperature terms underlined, are:

$$(\phi'' - \phi_0'') + \frac{(r_0 D/\alpha_0)'}{(r_0 D/\alpha_0)}(\phi' - \phi_0') - \left(\frac{\alpha_0}{r_0}\right)^2(\cos\phi)(\sin\phi - \sin\phi_0)$$
$$+ \nu\,\frac{\alpha_0}{r_0}\left[(\cos\phi - \cos\phi_0)\phi_0' + \frac{(D\nu)'}{D\nu}(\sin\phi - \sin\phi_0)\right]$$
$$= \frac{\alpha_0^2}{r_0 D}\left[\psi\sin\phi - (r_0 V)\cos\phi + \underline{M(\cos\phi - \cos\phi_0) - \frac{r_0}{\alpha_0}M'}\right], \tag{41}$$

$$\psi'' + \frac{(r_0/C\alpha_0)'}{(r_0/C\alpha_0)}\psi' - \left[\left(\frac{\alpha_0}{r_0}\right)^2\cos^2\phi - \frac{\nu\alpha_0}{r_0}\phi'\sin\phi + \frac{C\alpha_0}{r_0}\left(\frac{\nu}{C}\right)'\cos\phi\right]\psi$$
$$= \frac{\alpha_0^2 C}{r_0}\left[\underline{(1+\epsilon)(\cos\phi - \cos\phi_0) - \frac{r_0}{\alpha_0}\epsilon'}\right] + \frac{\alpha_0}{r_0}(\nu r_0 V)'\sin\phi$$

$$+ (r_0 V)\left[\left(\frac{\alpha_0}{r_0}\right)^2 \cos\phi \sin\phi_0 + \nu \frac{\alpha_0}{r_0}\left(\phi' \cos\phi - \frac{C'}{C}\sin\phi\right)\right]$$

$$- \frac{\alpha_0}{r_0}(r_0^2 p_H)' - \left[\nu \left(\frac{\alpha_0}{r_0}\right)^2 \cos\phi - \frac{\alpha_0}{r_0}\frac{C'}{C}\right] r_0^2 p_H \, . \quad (42)$$

Reduction to the Case of Spherical Shells of Uniform Thickness

For a spherical shell of radius a, ξ may be taken as the angle measured from the symmetry axis (see Fig. 6). Then

$$\begin{aligned} r_0 &= a \sin\xi \, , \\ z_0 &= -a \cos\xi \, , \\ \alpha_0 &= a \, , \\ \phi_0 &= \xi \, . \end{aligned} \quad (43)$$

With these expressions substituted, and with D, ν, C taken as constants, the following special forms of Eqs. (41) and (42) are obtained:

$$\phi'' + (\phi' - 1)\cot\xi - \cot^2\xi \frac{\cos\phi}{\cos\xi}\frac{\sin\phi - \sin\xi}{\cos\xi} + \nu\frac{\cos\phi - \cos\xi}{\sin\xi}$$

$$= \frac{a}{D}(\sec\xi)\,[\psi \sin\phi - (r_0 V - \underline{M})\underline{(\cos\phi - \cos\xi)}$$

$$- r_0 V \cos\xi - \underline{M' \sin\xi}] \, , \quad (44)$$

$$\psi'' + \psi' \cot\xi - \psi\left(\cot^2\xi \frac{\cos^2\phi}{\cos^2\xi} - \nu\phi' \frac{\sin\phi}{\sin\xi}\right)$$

$$= (r_0 V)\cot\xi\left(\frac{\cos\phi \sin\phi}{\cos\xi \sin\xi} + \nu\phi' \frac{\cos\phi}{\cos\xi}\right) + (r_0 V)' \nu \frac{\sin\phi}{\sin\xi}$$

$$- \nu(r_0^2 p_H)\frac{\cot\xi \cos\phi}{\sin\xi \cos\xi} - \frac{(r_0^2 p_H)}{\sin\xi} + aC\left[\underline{(1+\epsilon)\left(\frac{\cos\phi - \cos\xi}{\sin\xi}\right) - \epsilon'}\right] . \quad (45)$$

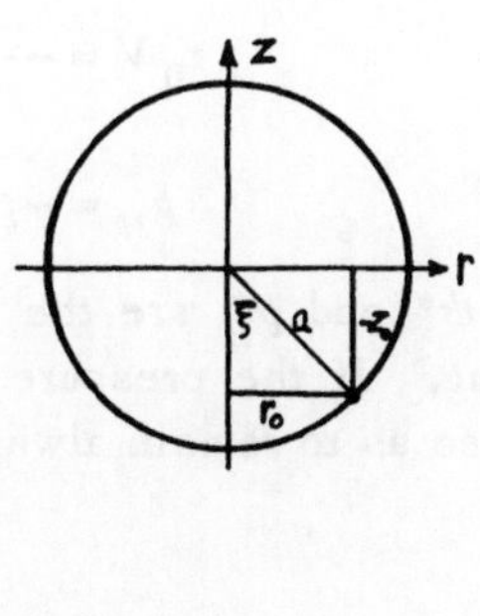

Fig. 6. Coordinates for Spherical Shell.

Substituting Eqs. (43) into Eq. (39) gives the following radial-displacement formula:

$$u = \epsilon a \sin \xi + \frac{1}{C}\{\psi' \sin \xi + p_H a^2 \sin^2 \xi - \nu[\psi \cos \phi + (r_0 V) \sin \phi]\} . \quad (46)$$

Differential Equations for Classical Buckling of Spherical Shells

When M and ϵ are constant, the spherical shape remains a possible equilibrium configuration under the action of a uniform external pressure p (which is thought of as acting on the middle surface). The state of stress, deformation, and loading is described by the following equations:

$$\begin{aligned}
&\phi = \phi_0 = \xi ,\\
&N_\xi = -\frac{1}{2} pa ,\\
&Q = 0 ,\\
&\psi = r_0 H = r_0(N_\xi \cos \phi - Q \sin \phi) = -\frac{1}{2} pa^2 \sin \xi \cos \xi ,\\
&r_0 V = -\frac{1}{2} pa^2 \sin^2 \xi ,\\
&p_H = -p \sin \phi = -p \sin \xi .
\end{aligned} \quad (47)$$

By substitution it may be verified that these conditions satisfy the differential Eqs. (44) and (45).

If an equilibrium configuration is now sought which is infinitesimally close to the above under the same external pressure, one may write

$$\begin{aligned}
&\phi_0 - \phi = \beta ,\\
&\cos \phi \approx \cos \phi_0 + \beta \sin \phi_0 = \cos \xi + \beta \sin \xi ,\\
&\sin \phi \approx \sin \phi_0 - \beta \cos \phi_0 = \sin \xi - \beta \cos \xi ,\\
&\psi = -\frac{1}{2} pa^2 \sin \xi \cos \xi + \psi^* ,\\
&r_0 V = -\frac{1}{2} pa^2 \sin^2 \xi ,\\
&p_H = -p \sin \xi + p_H^* ,
\end{aligned} \quad (48)$$

where ψ^* and p_H^* are the infinitesimal changes in ψ and p_H occurring during buckling. If the pressure vectors arise from a hydrostatic loading, they will rotate so as to remain always normal to the middle surface; then

$$p_H = -p \sin \phi = -p \sin \xi - p(\sin \phi - \sin \xi) = -p \sin \xi + \beta p \cos \xi \ ,$$

and p_H^* can be identified as $\beta p \cos \xi$. (If the pressure vectors did not rotate, one would have $p_H = -p \sin \phi_0$, and p_H^* would be zero.)

Using Eqs. (48) to eliminate ϕ, ψ, $r_0 V$, and p_H in Eqs. (44) and (45), and retaining only terms up to the first degree in the infinitesimal quantities β, ψ^*, and p^* gives the following differential equations suitable for the analysis of axisymmetric classical buckling:

$$\beta'' + \beta' \cot \xi - \beta(\nu + \cot^2 \xi) = -\frac{a}{D}\left[\psi^* + \underline{\beta\left(M + p\frac{a^2}{2}\right)}\right], \tag{49}$$

$$\psi^{*\prime\prime} + \psi^{*\prime} \cot \xi + \psi^*(\nu - \cot^2 \xi) = -\beta\left[(1-\nu)p\frac{a^2}{2} - (1+\epsilon)aC\right]$$

$$- \nu r_0^2 p_H^* \frac{\cot \xi}{\sin \xi} - \frac{(r_0^2 p_H^*)'}{\sin \xi} . \tag{50}$$

Substituting Eqs. (48) into Eq. (46) and retaining again infinitesimal terms only up to the first degree, one obtains the following formula for the radial displacement of the infinitesimally buckled shells:

$$u = \epsilon a \sin \xi - \frac{pa^2(1-\nu)}{2C} \sin \xi + \frac{1}{C}(\psi^{*\prime} \sin \xi$$

$$+ p_H^* a^2 \sin^2 \xi - \nu \psi^* \cos \xi) \ , \tag{51}$$

in which the first two terms on the right side represent the prebuckling displacement.

Boundary Conditions

The classical buckling problem represented by Fig. 3*a* and Eqs. (49) and (50) has the following boundary conditions:

$$\xi = 0 \text{ and } \pi \ , \qquad \beta = 0 \ , \qquad u = 0 \ . \tag{52}$$

The boundary condition on u, when rewritten in terms of ψ^* via Eq. (51), becomes

$$\xi \to 0 \text{ and } \pi \ , \qquad \psi^{*\prime} \sin \xi - \nu \psi^* \cos \xi \to 0 \ . \tag{53}$$

The postbuckling large-deflection problem, represented by Fig. 3*b* and Eqs. (44) and (45), also has the boundary conditions (52). Using Eq. (46) to transform the u boundary condition, one obtains (for ϵ regular at $\xi = 0$)

$$\xi \to 0 \text{ and } \pi \ , \qquad \psi' \sin \xi - \nu \psi \cos \phi \to 0 \ . \tag{54}$$

Assessing the Importance of the Temperature Terms

Examination of Eqs. (52) to (54) shows that the temperature terms ϵ and M do not appear explicitly in the boundary conditions. To prove the result stated at the outset, it only remains to show that, for ϵ and M constant, the temperature terms are also insignificant in the differential equations.

Considering first the equations for classical buckling, one notes that ϵ appears only in Eq. (50) and only in the combination $1 + \epsilon$. Since ϵ has the order of magnitude of a strain, it may be clearly omitted by comparison with 1. The other temperature term, M, appears only in the combination $M + \frac{1}{2}pa^2$ in Eq. (49). In order to assess its importance, let it be assumed that the thermal strain ϵ_T varies linearly with ζ from the value ϵ_{Ti} at the inner surface ($\zeta = +h/2$) to the value $\epsilon_{Ti} + \Delta\epsilon_T$ at the outer surface ($\zeta = -h/2$). Then $\dot{\chi} = -\Delta\epsilon_T/h$ and $M = Eh^2\Delta\epsilon_T/[12(1-\nu)]$. Thus, when p is equal to the classical buckling pressure without temperature effects, namely [9]

$$\left[2/\sqrt{3(1-\nu^2)}\right](Eh^2/a^2) ,$$

we have

$$M : \frac{1}{2}pa^2 = \Delta\epsilon_T : 4\sqrt{\frac{3(1-\nu)}{1+\nu}} . \tag{55}$$

This indicates that M is minute compared with $\frac{1}{2}pa^2$ and may, therefore, be dropped from the combination $M + \frac{1}{2}pa^2$, or that if it is left in, any relative change in classical buckling pressure due to its presence would be of the order $\Delta\epsilon_T$, that is, negligible for engineering calculations.

The negligibility of M can be deduced in still another way: In Eq. (50) substitute for p_H^* its value $\beta p \cos\xi$ and rewrite the equation so that p appears only in the combination $M + \frac{1}{2}pa^2$. Then Eq. (50) takes the form

$$\begin{aligned}\psi^{*\prime\prime} + \psi^{*\prime}\cot\xi + \psi^*(\nu - \cot^2\xi) = {} & -\beta\left(M + \frac{1}{2}pa^2\right)[2 + (1+\nu)\cos 2\xi] \\ & + \beta\{(1+\epsilon)aC + \underline{M[2 + (1+\nu)\cos 2\xi]}\} \\ & + \left[2\beta\left(M + \frac{1}{2}pa^2\right)\sin\xi\cos\xi\right] - \underline{(2\beta M \sin\xi\cos\xi)'} . \end{aligned} \tag{50a}$$

But aC is equal to Eha, and M is of the order $Eh^2\Delta\epsilon_T/12(1-\nu)$. Therefore,

$$aC : M = 1 : \frac{\Delta\epsilon_T}{12(1-\nu)}\,\frac{h}{a} . \tag{56}$$

Hence the underlined M terms in Eq. (50a) are negligible by comparison with the aC terms and may be omitted. With the underlined M terms out of Eq. (50a), and with ϵ dropped as negligible in the combination $1 + \epsilon$, Eqs. (49) and (50a) become those for an unheated shell with the parameter $\frac{1}{2}pa^2$ replaced everywhere by $M + \frac{1}{2}pa^2$. Therefore classical buckling of the heated shell should occur when

$$M + \frac{1}{2}pa^2 = \frac{1}{2}p_{cr}a^2 , \tag{57}$$

where p_{cr} is the buckling pressure $\left[2/\sqrt{3(1-\nu^2)}\right](Eh^2/a^2)$ of the unheated shell. But in view of condition (55), M on the left side of Eq. (57) is negligibly small compared with the right side, and one is left with the result that the buckling pressure of the heated shell is, for all practical purposes, the same as that of the unheated shell.

Turning now to the problem of postbuckling large deflections (Fig. 3b), consider Eqs. (44) and (45) with M' and ϵ' set equal to zero. Once again ϵ occurs only in the combination $1 + \epsilon$ and it may, therefore, be dropped from Eq. (45). The term M appears in the combination $r_0 V - M$ in Eq. (44). To assess the importance of M, some value must be assumed for p that is known to be structurally significant, and the resulting value of $r_0 V$ must be compared with M. Assuming that p is still of the order of the classical buckling pressure, say, for conservatism, one-sixth of it, one gets from Eqs. (48)

$$r_0 V = -\frac{1}{2}\left[\frac{2E}{6\sqrt{3(1-\nu^2)}}\,\frac{h^2}{a^2}\right] a^2 \sin^2\xi \approx -0.1\,Eh^2 \sin^2\xi .$$

But, again assuming a linear ϵ_T variation through the thickness,

$$M \approx 0.1\Delta\epsilon_T Eh^2 .$$

Thus

$$r_0 V : M = \sin^2\xi : \Delta\epsilon_T .$$

Therefore, M is negligible by comparison with $r_0 V$ everywhere except in the neighborhood of the apexes, where $\sin^2\xi$ can be as small as, or smaller than, $\Delta\epsilon_T$. However, in those regions the entire term $(r_0 V - M)(\cos\phi - \cos\xi)$ becomes negligible by comparison with the next term $r_0 V \cos\xi$, because $\cos\phi$ and $\cos\xi$ are both close to unity. To see this more clearly, consider a place where $\sin^2\xi$ is of the order $\Delta\epsilon_T$. Then $r_0 V - M$ is of the order $\Delta\epsilon_T$. To make $\cos\phi - \cos\xi$ as large as is reasonably possible, assume that this place corresponds to the kink at the edge of a small dimple of reversed curvature (see Fig. 7). Then $\phi = 0$ there, $\cos\phi - \cos\xi \approx \frac{1}{2}\xi_e^2$, where ξ_e is the angle between the axis of the shell and the radius drawn to the kink, and

$$(r_0 V - M)(\cos\phi - \cos\xi) = 0(\Delta\epsilon_T \cdot \xi_e^2) = 0[(\Delta\epsilon_T)^2] .$$

On the other hand,

$$r_0 V \cos\xi = 0(\Delta\epsilon_T) .$$

Hence, even in the neighborhood of the apexes the M term may be dropped from the equation.

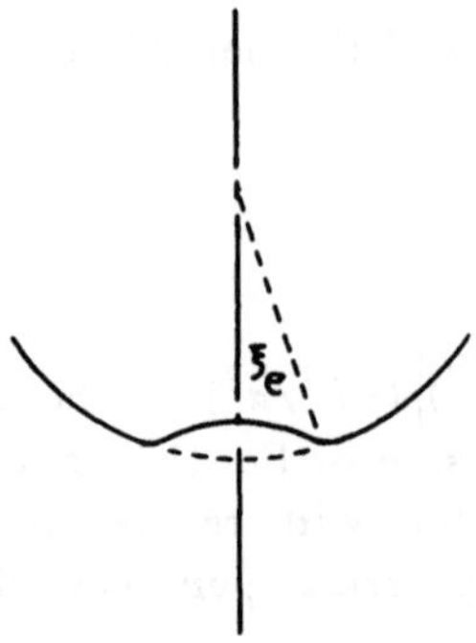

Fig. 7. Dimple in Spherical Shell.

EXPERIMENTS WITH BIMETALLIC SHELLS

The analytical results obtained above have important bearing on the design of spherical vessels subjected to an external pressure and a thickness-wise temperature gradient. The theory contends that a uniform temperature gradient over the surface of a hemispherical shell produces a thermal-stress distribution which has a negligible effect upon its buckling strength and can thus be ignored in the design procedure. An experimental verification of this contention is highly desirable. However, a uniform-temperature gradient is difficult both to produce and to maintain. Inasmuch as even a qualitative verification of the absence of a pronounced effect would be of enough value, it was decided to simulate the thermal stress by that in a bimetallic shell at different homogeneous temperatures.

The experiments performed can be generally described as follows: A series of bimetallic hemispheres formed out of sheet stock, whose thickness was half Invar and half brass, were heated or cooled to different temperatures ranging from 40 to 250°F. The buckling pressures at these temperatures were obtained to determine if the thermal stress induced by the differential expansion of the two materials had an important effect on the buckling strength. The result was essentially negative.

Specimens, Apparatus, and Procedure

The specimens used in these experiments were all hemispheres having a nominal diameter of 8.4 in. and a thickness of 0.021 in. The hemispheres terminated at the equator into a short cylindrical section of about $\frac{1}{2}$ in. These specimens were formed at Oak Ridge National Laboratory by a process commonly known as "Hydroform" from bimetallic sheet stocks consisting of brass and Invar. Two different series were used, one with brass on the outside and the other with Invar on the outside. The forming process was sufficiently accurate so that in the finished shells the deviation from sphericity was within 0.008 in. and the uniformity of wall thickness was held to 0.001 in. Samples of the sheet stocks were milled to determine the division between the two materials by observing the change in color from that of Invar to that of brass.

Measurement by micrometer of the thickness remaining after one material was milled away revealed that the division was within 0.001 in. of being at the middle.

The tests were performed in a heated pressure vessel which consisted of a heavy base plate and a dome, as schematically shown in Fig. 8. As can be

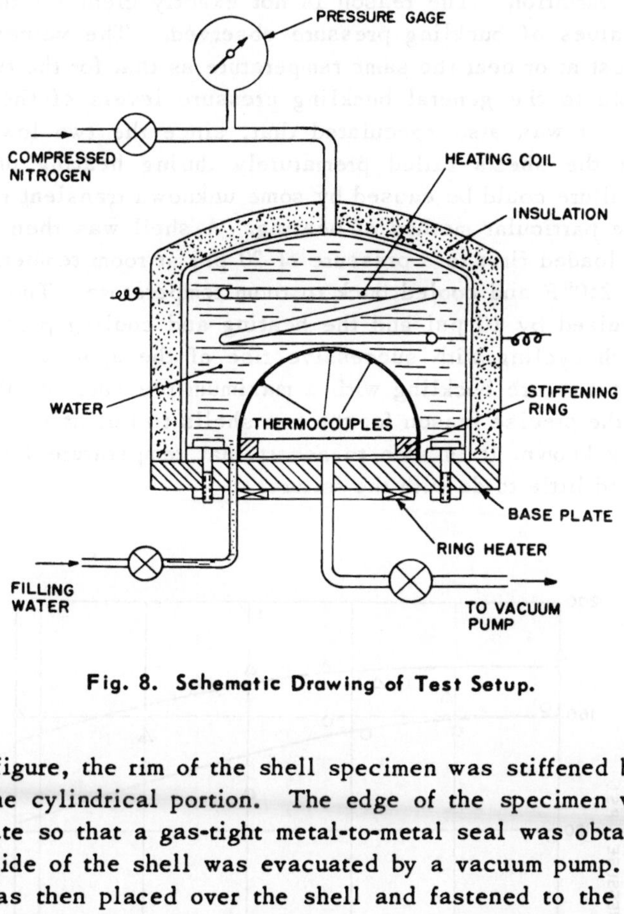

Fig. 8. Schematic Drawing of Test Setup.

seen in this figure, the rim of the shell specimen was stiffened by a split ring inserted in the cylindrical portion. The edge of the specimen was lapped on the based plate so that a gas-tight metal-to-metal seal was obtained when the air at the inside of the shell was evacuated by a vacuum pump. The dome of the vessel was then placed over the shell and fastened to the plate, and its interior was completely filled with water. The assembled rig was then connected to a nitrogen gas bottle through a pressure regulation valve. Afterward the content was slowly heated to reach the desired temperature. Before the desired temperature was reached the vessel was periodically bled to maintain a net pressure on the shell of not more than 20 psi. Three thermocouples were attached to different parts of the shell specimen to monitor the temperature. For those tests conducted below room temperature, the cooling was effected by placing the rig outdoors on a wintery day.

When the desired temperature was reached and stabilized, the vessel was pressurized by compressed nitrogen gas at the approximate rate of 30 psi/min until buckling occurred. The buckling pressure as indicated by a precision Bourdon gage (Heise 0–300 psi range) was recorded.

Results and Discussion

Eighteen shells were tested – nine with Invar on the outside and nine with brass on the outside. The buckling pressure vs temperature relation is shown in Fig. 9. The collapsed shells are shown in Fig. 10.

It is seen from the results that, with exception of a single shell in each series (shells No. 10 and No. 14), the buckling pressure was not very sensitive to temperature variation. The reason is not exactly clear for the two exceptionally low values of buckling pressure observed. The values obtained in repeating the test at or near the same temperature as that for the two low points were comparable to the general buckling pressure levels of the other shells in the series. It was also speculated that, since the two low values were obtained when the shells failed prematurely during heating under a slight pressure, the failure could be caused by some unknown transient thermal stress produced in the particular method of heating. A shell was then set up in the usual way and loaded first to a pressure of 20 psi at room temperature. It was then heated to 250°F and cooled back to room temperature. The pressure was subsequently raised by 10 psi and the heating and cooling process repeated. Successive such cycling with successive rise of the applied pressure failed to induce any premature buckling with a maximum pressure of 80 psi. Therefore, although the precise reason for the two shells to fail at so low a pressure was not clearly known, it is safe to assert that temperature level or thermal stress per se had little to do with the failure.

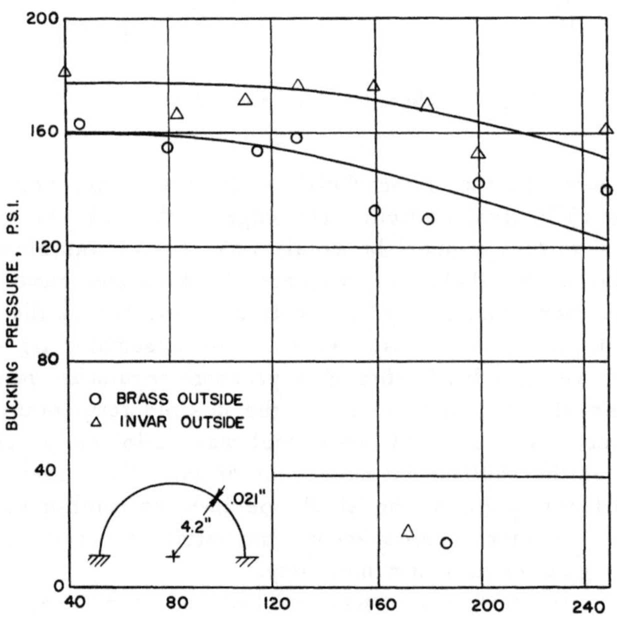

Fig. 9. Buckling Pressure vs Temperature of an Invar-Brass Bimetallic Hemispherical Shell.

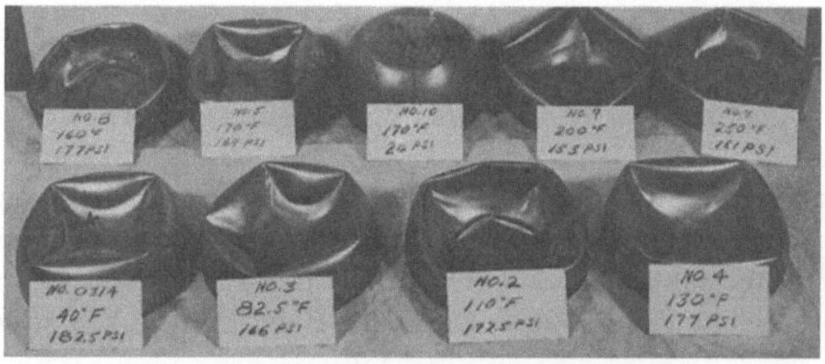

Invar-Exterior Shells

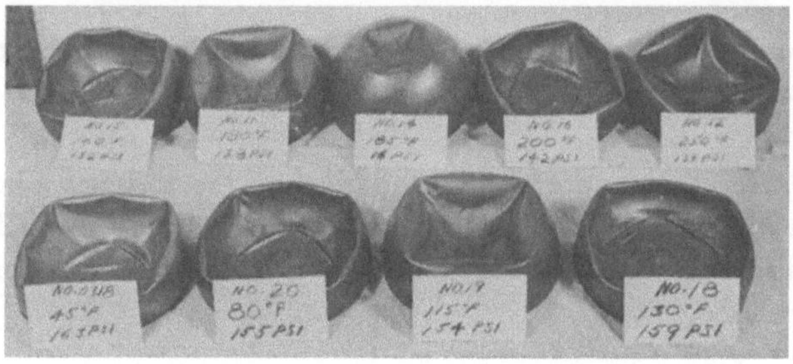

Fig. 10. Collapsed Shells.

A slight decrease of the buckling pressure with temperature seemed to be the general trend, possibly due to a stress- and temperature-induced lowering of the tangent modulus of brass.[2] The variation, however, was only a little more than the expected scatter of tests at a given temperature. Taking the average coefficients of expansion of brass and Invar to be $11 \times 10^{-6}/^\circ F$ and $1 \times 10^{-6}/^\circ F$, respectively,[3] the moduli of elasticity of the materials to be 16×10^6 and 21×10^6 psi, respectively, and Poisson's ratio to be 0.3, one finds the thermal stress at a temperature 100°F above the stress-free value to be of the order of 13,000 psi and the internal bending moment to be 1.3 in.-lb/in. As a comparison, this internal-bending moment is equivalent to that produced in a homogeneous brass shell by a temperature difference of 155°F across a 0.02-in. thickness. This represents a heat flux of the order of 1700 Btu ft^{-2} sec^{-1}.

It is also interesting to note that the buckling pressures for shells with brass exterior were consistently lower than those with Invar exterior. This can be qualitatively explained[2] by the fact that the dimple mode of buckling superimposes, on the prebuckling stress, predominantly compressive stress in the outer portion of the shell thickness. The prebuckling stress is always compressive for the brass, due to temperature and pressure individually or in

---

[2]Such an explanation was suggested to the writers by Professor N. J. Hoff of Stanford University.

[3]The actual value of the coefficient of expansion of Invar was perhaps much lower because of prior cold work. This uncertainty is, however, unimportant.

combination. Therefore, when the brass is on the outside its tangent modulus will be brought into play, and at the temperatures and pressures achieved in the tests the tangent modulus of brass can be significantly lower than its elastic modulus.

Because of the prior history of the sheet material and the subsequent forming process, the residual-stress pattern and the physical-property distribution in the finished specimens were exceedingly complex. The purpose of these experiments, however, was qualitative in nature. The two important findings were:

1. the insensitivity of the buckling pressure toward temperature variations from 40 to 250°F,
2. the same trend of lower buckling pressure at higher temperatures regardless of whether the exterior material has a higher or lower coefficient of thermal expansion.

It can, therefore, be safely concluded that for a hemispherical shell the existence of a thermal stress distribution which is self-equilibrating across the thickness and uniform over the spherical surface has negligible effect upon its buckling pressure.

REFERENCES

[1] A. Kaplan and Y. C. Fung, "A Nonlinear Theory of Bending and Buckling of Thin Elastic Shallow Spherical Shells," *NACA* TN 3212 (August 1954).

[2] Bernard Budiansky, *Buckling of Clamped Shallow Spherical Shells*, Harvard University, Cambridge, Mass. (1959) (ONR Contract Nonr 1866(02) Technical Report No. 5); also, pp 64–94 in *The Theory of Thin Elastic Shells*, North-Holland Publishing Co., Amsterdam, 1960.

[3] S. Timoshenko and S. Woinowsky-Krieger, *Theory of Plates and Shells*, 2d ed., McGraw-Hill, New York, 1959.

[4] P. C. Hu, E. E. Lundquist, and S. B. Batdorf, "Effect of Small Deviations from Flatness on Effective Width and Buckling of Plates in Compression," *NACA* TN 1124 (September 1946).

[5] K. Marguerre, "Zur Theorie der gekrümmten Platte grosser Formänderung," *Proceedings of the Fifth International Congress for Applied Mechanics*, pp 93–101, Wiley, New York, 1939.

[6] S. Timoshenko and J. N. Goodier, *Theory of Elasticity*, 2d ed., McGraw-Hill, New York, 1951.

[7] Eric Reissner, "On the Theory of Thin Elastic Shells," *Reissner Anniversary Volume, Contributions to Applied Mechanics*, J. W. Edwards, Ann Arbor, Mich., 1949.

[8] Eric Reissner, "On Axisymmetrical Deformation of Thin Shells of Revolution," *Proceedings of Symposia in Applied Mechanics, Vol. III, Elasticity*, McGraw-Hill, New York, 1950.

[9] S. Timoshenko, *Theory of Elastic Stability*, p 496, McGraw-Hill, New York, 1936.

BEHAVIOR OF BEAM-COLUMNS IN THE INELASTIC RANGE[1]

B. B. Muvdi
Michigan College of Mining and Technology

and

O. M. Sidebottom
University of Illinois

Two theories are presented for constructing either moment-load or load-deflection relations as well as for determining the collapse loads for beam-columns. In each case, trial and error solutions are required which use constant depth-of-yielding interaction curves. A stepwise theory is presented which gives results as accurate as desired; however, the theory is not practical because of the excessive time required. An approximate theory is presented which gives results in close agreement with the stepwise theory and with experimental data. This theory requires the elastic solution for maximum elastic conditions.

INTRODUCTION

This investigation was undertaken to examine the behavior of a certain class of menbers known as beam-columns, when subjected to loads higher than those necessary to initiate yielding in the member. The term beam-column is applied to long, slender members which are subjected to transverse as well as axial loads, that is, members that simultaneously perform the functions of beams and columns.

While the problem of column behavior beyond the elastic range has been examined extensively in recent years [*1–13*], that of beam-columns has received very little attention. Aside from the well-known solutions for the behavior of beam-columns in the elastic range [*14*], only two published papers were found that treat the subject of beam-columns in the inelastic range. The first paper was published by Osgood [*15*] in 1947. He proposed an approximate method

---

[1]Sponsored by the Materials Laboratory of the Wright Air Development Division, Dayton, Ohio (contract No. AF 33 (616)-7600 with R. F. Klinger acting as project engineer). This investigation was conducted as a part of the work of the Engineering Experiment Station of the University of Illinois, of which Professor R. J. Martin is the Director, in the Department of Theoretical and Applied Mechanics, of which Professor T. J. Dolan is the Head.

NOTATION

A	Cross-sectional area of beam-column
I	Second moment of area
h	Depth of beam-column
c_1	Distance from centroid to most-stressed compression fibers
c_2	Distance from centroid to most-stressed tension fibers
x	Coordinate measuring distance along beam-column from one end
y	Coordinate measuring deflection of centroidal axis
l	Length of beam-column
l/r	Slenderness ratio
P	Compression load on beam-column
η_1, η_2	Eccentricity of P at each end of beam-column
Q	Transverse load on beam-column
$l-d$	Location of Q from origin
M	Moment
M_Q	Moment due to Q acting alone
σ_{e_1}	Yield stress in compression
σ_{e_2}	Yield stress in tension
E	Modulus of elasticity
ϵ_{e_1}	σ_{e_1}/E
ϵ_{e_2}	σ_{e_2}/E
P_e	$\sigma_{e_1} A$
M_e	Smaller of $\sigma_{e_1} I/c_1$ and $\sigma_{e_2} I/c_2$
γ	M_e/P_e
β	$\sqrt{P/EI}$
ψ	y'', curvature of beam-column
x_c	Section of maximum moment when yielding is impending
ψ_c	Curvature at x_c
y_c	Deflection at x_c
C	ψ_c/y_c, assumed to be independent of depth of yielding
k	Ratio of elastic depth to h at x_c

which essentially consisted of replacing the given beam-column by an equivalent eccentrically loaded column. Available methods were then applied to the resulting eccentrically loaded columns. Theoretical solutions were presented for several cases, but no test data were given. The second paper was published in 1955 by Ketter [*16*], who presented a solution based on the principle of virtual displacements. Ketter's method is essentially based on the premise that instability in the beam-column will set in when the rate of increase of the internal resisting moment (M_{int}) with respect to the curvature (ψ) becomes equal to the rate of increase of externally applied moment (M_{ext}) with respect to curvature. Symbolically, this instability condition may be expressed as follows:

$$\frac{\Delta M_{int}}{\Delta \psi} = \frac{\Delta M_{ext}}{\Delta \psi} \,.$$

By assuming ideal plasticity (i.e., nonstrain hardening material) and a given configuration for the deflection curve, Ketter determined the combination of axial load and bending moment that would cause instability in members subjected to constant axial load with increasing bending moment. However, all cases considered involved members that were symmetrically loaded.

In the present paper, a new method is proposed that would make it possible to analyze beam-columns made of materials with any strain hardening characteristics. The approach is sufficiently general to consider symmetrically and unsymmetrically loaded members, as well as concentrically and eccentrically applied axial loads. Furthermore, the shape of cross section of the member does not influence the ease with which the solution can be performed.

ANALYSIS

Two methods will be presented for constructing either moment-load or load-deflection curves for beam-columns. The collapse load is defined as the maximum load obtained from either the moment-load or load-deflection curve.

The theories are based on the usual assumptions that plane sections remain plane and that the stress-strain relation for each fiber in the beam-column is given by either the tension or compression stress-strain diagram of the material. In addition, the theories assume that none of the fibers will unload elastically. The latter assumption will be satisfied if the lateral load is first applied to the beam-column and the axial load is then increased until collapse. If the axial load is applied first, followed by the lateral load, it is possible that some fibers may be unloaded. It is the opinion of the authors that a theory including elastic unloading would be almost impossible to develop and use. Furthermore, elastic unloading would increase the strength of the beam-column, so that the theories presented will be conservative in these cases.

In order to construct the moment-load or load-deflection curves, two independent relations between the variables axial load, moment, and curvature are developed. The first load-moment-curvature relation is obtained by considering the equilibrium of the internal forces and moments, and is represented graphically by dimensionless interaction curves together with a moment-curvature equation. In this investigation, interaction curves for constant depth of yielding are used. The second axial load-moment-curvature relation is established by equating the internal and external moments, and by considering the deformation characteristics of the beam-columns.

By use of the interaction curves and a numerical procedure described by Chwalla [4], the configuration of the inelastically deformed member can be derived as accurately as desired. This stepwise method is described in the Appendix to this paper; however, the method is considered impractical because of the time required to obtain a solution for even the simplest problem, unless the method is programmed into a computer.

A theory based on certain approximations regarding the configuration of the inelastically deformed beam-column was found to compare favorably with both the exact theory and test results. This approximate theory requires that the elastic solution to the beam-column be known, so that the magnitude of the axial load, maximum moment, deflection, and curvature at the critical section[2] can be determined for maximum elastic conditions.

Interaction Curves

A relationship between axial load and moment may be developed in the form of dimensionless interaction curves as shown schematically in Fig. 1. In this figure, P_e is the axial load required to initiate yielding when acting alone, and M_e is the moment required to initiate yielding when acting alone. Such relations for rectangular and T-sections were previously developed by Sidebottom and Clark [*1*] and will be used in this paper without derivation.

A study of Fig. 1 indicates that, with the exception of the straight line defining initiation of yielding, all curves consist of two parts, a straight line and a curve. The combinations of P/P_e and M/M_e represented by the straight-line portion are those necessary to cause yielding on one side of the member only, while the other side is still elastically strained. The point of intersection between the two portions of the interaction curve defines the combination of P/P_e and M/M_e that produces the specified depth of yielding on one side of the member and incipient yielding on the other side. The curved portion, however, defines the various combinations of P/P_e and M/M_e that would cause inelastic action on both sides, such that the total depth of inelastic penetration is equal to the specified depth of yielding.

Once the interaction curves for constant depth of yielding are known, they may be used to obtain a relation between axial load, moment, and curvature. It has been shown by Sidebottom and Clark [*1*] that the curvature of the member for any point on any one of the interaction curves shown in Fig. 1 is given by the relation

$$\psi = \frac{\epsilon_{e_1} + \epsilon_{e_2}}{kh} \, \frac{M/M_e}{M_u/M_e}, \tag{1}$$

---

[2]The term "critical section" is used in its ordinary sense to signify the section of the member at which the bending moment has its maximum value.

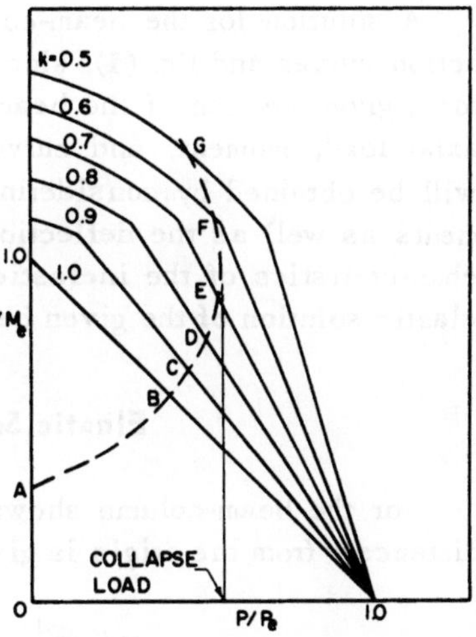

Fig. 1. Schematic Representation of Interaction Curves for Constant Depth of Yielding.

where ϵ_{e_1} and ϵ_{e_2} define the strains corresponding to the compressive and tensile yield stresses, respectively, as shown in Fig. 2. The quantity kh is the depth of the section remaining elastic in which k is a factor that varies from unity for purely elastic conditions to zero for fully plastic conditions. The term

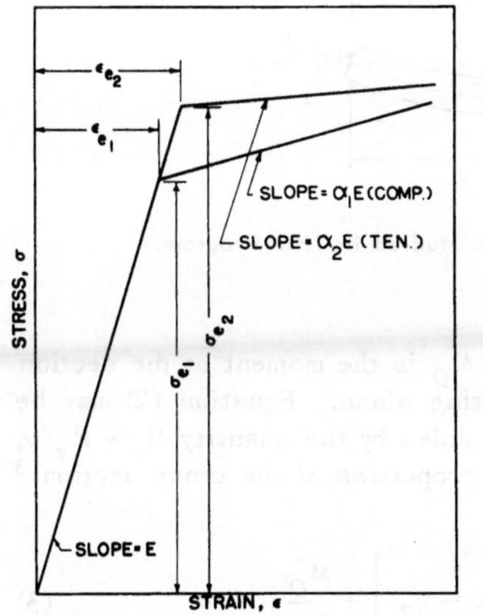

Fig. 2. Schematic Representation of Stress-Strain Relations in Tension and Compression.

M/M_e is the dimensionless moment that exists at the given section of the member, and M_u/M_e is the dimensionless moment at the upper end of the straight-line portion of the interaction curve. The ratio M/M_u assumes a constant value of unity in the curved portion of the interaction curve.

A solution for the beam-column cannot be obtained by using only the interaction curves and Eq. (1), since the point on the interaction curve which is valid for a given section of the beam-column is not known. Another relation between axial load, moment, and curvature for the inelastically deformed beam-column will be obtained by considering the equality between internal and external moments as well as the deflection characteristics of the member. The deflection characteristics of the inelastically deformed member are approximated from the elastic solution of the given beam-column.

Elastic Solution for the Beam-Column

For the beam-column shown in Fig. 3 the bending moment at any section a distance x from the origin is given by the equation

$$M = P\left[y+\eta_1-\frac{x}{l}(\eta_1-\eta_2)\right]+M_Q, \qquad (2)$$

which applies to sections on either side of the transverse load Q. The symbols η_1 and η_2 are the eccentricities of the load P at the left and right ends respectively, y is the deflection of the centroidal axis of the member at the point

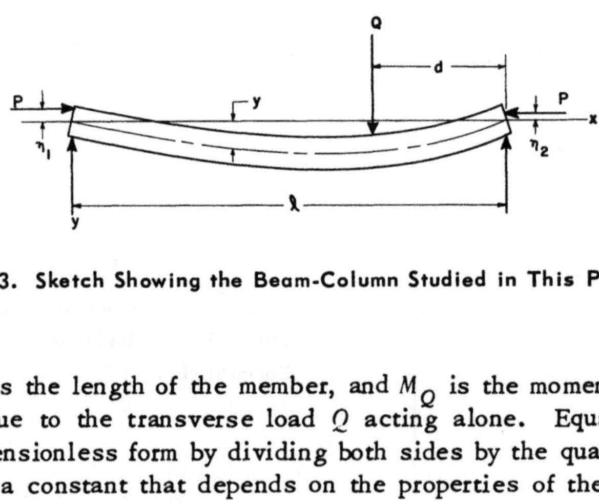

Fig. 3. Sketch Showing the Beam-Column Studied in This Program.

considered, l is the length of the member, and M_Q is the moment at the section in question due to the transverse load Q acting alone. Equation (2) may be written in dimensionless form by dividing both sides by the quantity $M_e = P_e/\gamma$, in which γ is a constant that depends on the properties of the cross section.[3] Thus

$$\frac{M}{M_e} = \gamma\left(\frac{P}{P_e}\right)\left[y+\eta_1-\frac{x}{l}(\eta_1-\eta_2)\right]+\frac{M_Q}{M_e}. \qquad (3)$$

It should be noted that Eqs. (2) and (3) are valid for either elastic or inelastic conditions.

---

[3]In the case of a rectangular section, for example, $M_e = \sigma_{e_1} I/c_1 = P_e h/6$, so that γ is equal to $6/h$.

The elastic deflection of the beam-column shown in Fig. 3 may be obtained by superposing certain solutions as given by Timoshenko [14]. The resulting equations may be expressed as follows:

$$y = \eta_1\left[\frac{\sin\beta(l-x)}{\sin\beta l} - \frac{l-x}{l}\right] + \eta_2\left[\frac{\sin\beta x}{\sin\beta l} - \frac{x}{l}\right] + \frac{Q\sin\beta d}{P\beta\sin\beta l}\sin\beta x - \frac{Qd}{Pl}x\,,\quad x \leqq (l-d)\,, \tag{4a}$$

$$y = \eta_1\left[\frac{\sin\beta(l-x)}{\sin\beta l} - \frac{l-x}{l}\right] + \eta_2\left[\frac{\sin\beta x}{\sin\beta l} - \frac{x}{l}\right] + \frac{Q\sin\beta(l-d)}{P\beta\sin\beta l}\sin\beta(l-x) - \frac{Q(l-d)(l-x)}{Pl}\,,\quad x \geqq (l-d)\,, \tag{4b}$$

where $\beta = \sqrt{P/EI}$, E is the modulus of elasticity, and I is the moment of inertia of the section.

The elastic curvatures may be obtained by differentiating Eqs. (4) twice; thus

$$\psi = y'' = -\eta_1\beta^2\frac{\sin\beta(l-x)}{\sin\beta l} - \eta_2\beta^2\frac{\sin\beta x}{\sin\beta l} - \frac{Q\beta\sin\beta d}{P\sin\beta l}\sin\beta x\,,\quad x \leqq (l-d)\,, \tag{5a}$$

$$\psi = y'' = -\eta_1\beta^2\frac{\sin\beta(l-x)}{\sin\beta l} - \eta_2\beta^2\frac{\sin\beta x}{\sin\beta l} - \frac{Q\beta\sin\beta(l-d)}{P\sin\beta l}\sin\beta(l-x)\,,\quad x \geqq (l-d)\,. \tag{5b}$$

The bending moment at any section along the span of the member is then given by the equation

$$M = -EI\psi\,. \tag{6}$$

For a given member, the deflection y_c and curvature ψ_c at the critical section may be determined for any value of P from Eqs. (4) and (5) respectively. Examination of the ratio ψ_c/y_c revealed that, in general, this ratio did not vary greatly as the axial load P increased from zero to its maximum elastic value. This behavior led to the assumption that the ratio ψ_c/y_c remains constant as the beam-column is strained inelastically. Essentially, then, a relation between curvature and deflection at the critical section in the inelastic range is assumed to be of the form

$$\psi_c = Cy_c\,, \tag{7}$$

where the constant C is determined as the ratio of curvature to deflection at the critical section when yielding is impending. Obviously, the assumption expressed by Eq. (7) is only an approximation which will lead to some error, particularly if the depth of yielding at collapse is appreciable. Fortunately, the depth of yielding at collapse is small for most beam-columns, being much less than one-half the section depth in most cases (i.e., $k > 0.5$). Furthermore, it should be noted that in cases where the moment due to the eccentric load P opposes that due to the transverse load Q, the assumption expressed by Eq. (7) may not be sufficiently adequate. Such cases were not examined in this program, but if they should arise, it is recommended that the more accurate stepwise method presented in the Appendix be used.

Eliminating the deflection y_c between Eqs. (3) and (7) and solving for P/P_e, the required relation between axial load, moment, and curvature is obtained, namely,

$$\frac{P}{P_e} = \frac{M/M_e - (M_Q/M_e)_c}{\gamma[\psi_c/C + \eta_1 - x_c)\eta_1 - \eta_2)/l]}, \tag{8}$$

where the subscript c designates conditions at the critical section.

Method of Solution

The purpose of this section is to show how the various relations established in the preceding section may be used to determine the moment-load curve or the load-deflection curve as well as the load P that causes complete collapse of a beam-column when it is subjected to a constant value of the transverse load Q.

The value of the axial load P that would initiate yielding in the member may be determined, by trial and error, using Eqs. (5) and (6). The procedure consists in first assuming a value for P and determining the quantity $\beta = \sqrt{P/EI}$. The value of x_c locating the critical section is now determined by maximizing Eq. (5). The maximum moment is then given by Eq. (6), and the maximum compressive stress at the critical section by the equation

$$(\sigma_{\max})_{\text{comp}} = -\frac{(M_{\max})c_1}{I} - \frac{P}{A}, \tag{9}$$

where A is the area of the section, and c_1 is the distance from the neutral axis to the outermost fibers stressed in compression. This process is repeated until the stress determined by Eq. (9) becomes equal to the compressive yield stress of the material.[4]

With the value of P that initiates yielding known, the constant C may be easily found from Eq. (7), in which the values of y_c and ψ_c are determined by the use of Eqs. (4) and (5) respectively. Furthermore, this value of P may be used to locate the intersection of the moment-load curve and the interaction curve defining the beginning of yielding ($k = 1$); see point B in Fig. 1.

---

[4]For all conditions examined in this investigation, the maximum stress at initiation of yielding was compressive. It is conceivable, however, that in other cases the maximum stress may be tensile. In these cases, Eq. (9) becomes $(\sigma_{\max})_{\text{tens}} = (M_{\max})c_2/I - P/A$.

The solution is then carried into the inelastic range by means of the interaction curves along with Eqs. (1) and (8).

In applying Eq. (8), the assumption is made that the critical section does not move as the member deforms inelastically. This is equivalent to saying that x_c and $(M_Q/M_e)_c$ retain the same values they assumed at initiation of yielding. With members which are symmetrically loaded (i.e., $d = l/2$ and $\eta_1 = \eta_2$), x_c does retain the same value, namely $l/2$, in the inelastic as well as in the elastic range. However, with members which are not symmetrically loaded (i.e., $d \neq l/2$ and/or $\eta_1 \neq \eta_2$), x_c does not, in general, retain the same value throughout the loading process. Test data indicated that the above assumption did not introduce a serious error into the theory.

Now consider the problem of determining the value of P/P_e that would produce a given depth of inelastic penetration (i.e., $k = k_1$ = a constant); see Fig. 1. The condition that has to be satisfied is that M/M_e be of such magnitude that the value of P/P_e given by Eq. (8) be equal to that given by the relation $P/P_e = f(M/M_e)$ as expressed by the interaction curves. To this end, some value of M/M_e is assumed and the curvature ψ_c is determined by Eq. (1). Equation (8) is then used to obtain a value of P/P_e, which is compared with that given by the interaction curve corresponding to $k = k_1$. This process is repeated until the two values of P/P_e are the same or very nearly so. This usually requires not more than three trials. Thus the problem is reduced to the solution of two simultaneous equations in the unknowns M/M_e and P/P_e. The deflection y_c, if needed, may now be obtained from Eq. (7).

Once the correct value of P/P_e is established for $k = k_1$, a point is located on the appropriate interaction curve. Other values of k are then considered and the same process repeated to establish a set of points such as C, D, E, F, and G in the inelastic domain (see Fig. 1). The location of point A on the M/M_e axis is dictated by the value of the transverse load Q. A smooth curve is then constructed through the points $A, B, \ldots, G$ to give the moment-load curve. The collapse load for the beam-column may be obtained from the moment-load curve, and in dimensionless form is the maximum value of P/P_e.

If additional points are needed in the elastic range (i.e., between points A and B in Fig. 1), they may be easily determined from the elastic solution by using Eq. (4) and the equilibrium condition expressed by Eq. (3).

MATERIALS AND APPARATUS

In order to examine the validity of the analysis discussed in the second section, tests were performed on beam-columns made of three materials, namely, 2024-T4 aluminum alloy, SAE 1020 steel, and 17-7PH stainless steel. The 2024-T4 aluminum samples were tested in the as-received condition. The SAE 1020 steel was annealed, prior to machining, by soaking at 1600°F for 3 hr and furnace cooling. The 17-7PH stainless steel specimens were precipitation-hardened[5] after machining. The SAE 1020 steel samples were machined from a

[5]This heat treatment consisted in soaking at 1400°F for 90 min, cooling to 60°F in 60 min, soaking at 60°F for 30 min, soaking at 1050°F for 90 min, and finally air cooling.

$3\frac{1}{2}$-in.-diam bar, while the 2024-T4 aluminum and the 17-7PH stainless steel specimens were machined from $\frac{1}{2}$-in. plates. The various beam-columns tested in this program and their physical characteristics are shown in Table 1. Either one or two beam-columns per condition were tested in this study.

Table 1. Materials and Physical Characteristics of Specimens Tested in This Program (l/r = slenderness ratio)*

Material	Rectangular Sections 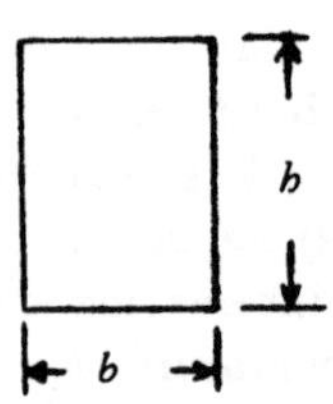			T-Sections 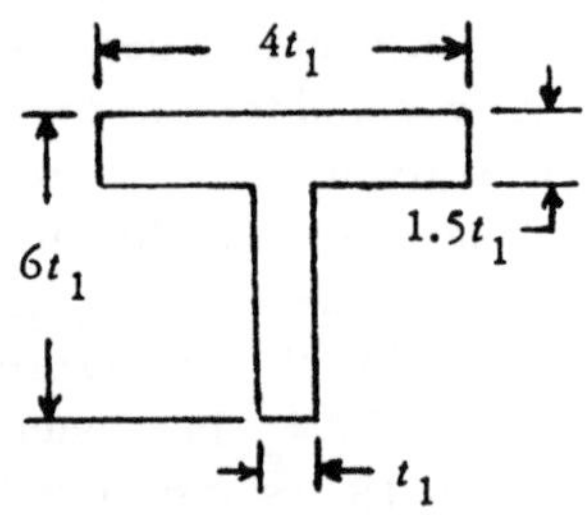
	$l/r = 50$	$l/r = 75$	$l/r = 100$	$l/r = 60$
2024-T4 aluminum alloy	$b = 0.500$	$b = 0.500$	$b = 0.500$	
	$h = 0.625$	$h = 0.625$	$h = 0.500$	$t_1 = 0.100$
SAE 1020 steel	$b = 0.500$	$b = 0.500$	$b = 0.500$	
	$h = 0.625$	$h = 0.625$	$h = 0.500$	$t_1 = 0.100$
17-7PH stainless steel	$b = 0.350$	$b = 0.400$		
	$h = 0.500$	$h = 0.420$		

*All units are in inches.

Standard tensile and compressive tests were performed on the aluminum alloy to determine the various properties needed for the analysis. However, only compressive tests were performed on the mild and stainless steels. Results of previous tests by Sidebottom and associates [*17*] on the last two materials have indicated that the difference between tensile and compressive properties was negligible for all practical purposes. The various properties used in the present analysis are shown, for the three materials, in Table 2. These values represent an average of at least six tests. By use of these properties, interaction curves were developed for both rectangular and T-sections.

The testing apparatus used in this program is shown in Fig. 4. A Riehle testing machine having a capacity of 120,000 lb was used to apply the axial

Table 2. Properties of the Three Materials Used in This Investigation

	E (psi)	σ_{e_1} (psi)	σ_{e_2} (psi)	ϵ_{e_1} (in./in.)	ϵ_{e_2} (in./in.)	α_1	α_2
	$\times 10^6$	$\times 10^6$	$\times 10^6$	$\times 10^{-4}$	$\times 10^{-4}$		
2024-T4 aluminum alloy	10.8	37.5	51.5	34.7	47.7	0.189	0
SAE 1020 steel	30.0	30.3	30.3	10.1	10.1	0	0
17-7PH stainless steel	28.0	170.0	170.0	60.7	60.7	0.111	0.111

Fig. 4. Test Apparatus to Apply Transverse and Axial Loads to Beam-Columns.

load through fixtures[6] provided with adjustable knife edges. The transverse load was applied by means of dead weights through the pulley system shown, and the deflection of the specimen was measured by means of a 1/1000-in. dial indicator (see Fig. 4).

RESULTS AND CONCLUSIONS

The experimental part of this investigation included tests on 46 beam-columns made of three different materials, as discussed in the third section of this paper. The results of the various tests are compared with theoretical values in Tables 3 and 4 for symmetrically and unsymmetrically loaded beam-columns respectively. In order to see if the test data were reproducible, two specimens were subjected to identical loading for many loading conditions. The results presented indicate that the two collapse loads agreed within 2% in each case.

A graphical comparison is also made between the results of the proposed approximate theory and those of the experiments. The comparison is made on the basis of moment-load curves (i.e., M/M_e vs P/P_e). This method of presentation was selected over the more conventional load-deflection plots because it conveniently indicates the value of M_Q/M_e to which the beam-column was initially subjected before the axial load was applied. In view of space limitations, however, only a few representative graphs are shown. A more complete presentation of results may be found in ref [*19*].

In order to examine the degree of accuracy attainable by the approximate method, a comparison was made between moment-load curves obtained by this method and those developed by the stepwise theory (see the Appendix). Since the latter method involves rather lengthy computations, particularly with unsymmetrically loaded members as indicated in the Appendix, the comparison was limited to two cases only, namely, one 2024-T4 aluminum member and one SAE 1020 steel member. Both members were symmetrically loaded with zero end eccentricities, had a slenderness ratio of 75, and were subjected to transverse loads at midspan such that M_Q/M_e was equal to 0.50. The results of both methods are illustrated in Fig. 5, together with the values exhibited by test. In the case of 2024-T4 aluminum, the two theoretical moment-load curves were nearly identical and were found to compare very well with the test values. In the case of SAE 1020 steel, however, a slight difference of about 5% is observed between the collapse loads as predicted by the two methods. While this difference was in favor of the stepwise theory, it is too small to justify the added labor required by this method. Furthermore, the approximate theory was conservative.

Figure 6 shows the results of tests on two symmetrically loaded 2024-T4 aluminum beam-columns, one with and one without end eccentricities. Figure 7 presents the results of tests on two symmetrically loaded 17-7PH stainless steel

---

[6] A complete description of these fixtures may be found in a paper by Sidebottom *et al.* [*18*]. The two end fixtures add a total of 1.20 in. to the machined length of the specimen.

Table 3. Values of P_{ca}, P_{ca}/P_{ct}, and P_{ca}/P_{ya} for the Various Symmetrically Loaded Beam-Columns Studied in This Program. P_{ca} = actual collapse load; P_{ct} = theoretical collapse load; P_{ya} = actual yield load

Material	Shape	l/r	M_Q/M_e	$\frac{\eta_1}{b}$	$\frac{\eta_2}{b}$	P_{ca} (lb)	P_{ca}/P_{ct}	P_{ca}/P_{ya}
2024-T4 aluminum alloy	Rect.	100	0.25	0	0	2130	1.04	1.05
		75	0.25	0	0	4240	1.00	1.09
		50	0.25	0	0	7750	1.03	1.21
		50	0.25	0	0	7620	1.01	1.18
		100	0.50	0	0	1680	1.04	1.37
		100	0.50	0	0	1680	1.04	1.37
		75	0.50	0	0	3580	1.05	1.45
		75	0.50	0	0	3510	1.03	1.43
		50	0.50	0	0	6360	1.06	1.70
		50	0.50	0	0	6330	1.05	1.69
		100	0.25	0.15	0.15	1690	1.03	1.17
		75	0.25	0.15	0.15	3250	1.06	1.28
		50	0.25	0.15	0.15	5130	1.05	1.50
	T-Section	60	0.50	0	0	1850	1.01	1.27
		60	0.25	0.15	0.15	1550	1.03	1.21
SAE 1020 steel	Rect.	100	0.50	0	0	3410	1.07	1.44
		100	0.50	0	0	3400	1.07	1.44
		75	0.50	0	0	5370	1.05	1.49
		75	0.50	0	0	5400	1.06	1.50
		50	0.50	0	0	6400	1.02	1.52
		50	0.50	0	0	6450	1.03	1.54
		100	0.75	0	0	2460	1.06	2.24
		75	0.75	0	0	3980	1.04	2.36
		50	0.75	0	0	5260	1.07	2.59
		50	0.75	0	0	5100	1.04	2.54
	T-Section	60	0.50	0	0	2450	1.02	1.17
		60	0.50	0.15	0.15	1640	1.00	1.30
		60	0.50	0.15	0.15	1680	1.03	1.33
17-7PH stainless steel	Rect.	75	0.50	0	0	4770	1.00	1.17
		50	0.25	0	0	13200	1.01	1.10
		75	0.50	0.15	0.15	3890	1.00	1.30
		50	0.25	0.15	0.15	9100	1.04	1.27

Table 4. Values of P_{ca}, P_{ca}/P_{ct}, and P_{ca}/P_{ya} for the Various Unsymmetrically Loaded Beam-Columns Studied in This Program. P_{ca} = actual collapse load; P_{ct} = theoretical collapse load; P_{ya} = actual yield load

Material	Shape	l/r	M_Q/M_e	$\frac{\eta_1}{h}$	$\frac{\eta_2}{h}$	P_{ca} (lb)	P_{ca}/P_{ct}	P_{ca}/P_{ya}
2024-T4 aluminum alloy	Rect.	100	0.50	0	0	1740	1.00	1.22
		100	0.50	0	0	1710	0.98	1.20
		75	0.50	0	0	3610	0.96	1.32
		50	0.50	0	0	6480	0.96	1.58
		50	0.50	0	0	6490	0.96	1.58
	T-Section	60	0.50	0	0	1840	0.95	1.15
		60	0.50	0	0	1890	0.97	1.20
SAE 1020 steel	Rect.	100	0.50	0	0	3660	1.05	1.41
		75	0.50	0	0	5680	1.04	1.47
		50	0.50	0	0	6530	1.01	1.52
		75	0.50	0.15	0.15	3590	1.00	1.69
		75	0.50	0.15	0	4400	0.96	1.47
	T-Section	60	0.75	0	0	2210	1.01	1.28
17-7PH stainless steel	Rect.	75	0.25	0	0	6480	1.00	1.02

beam-columns with zero end eccentricities. Figure 8 exhibits the results of tests on two SAE 1020 steel unsymmetrically loaded beam-columns, with equal and unequal end eccentricities. In all instances, close agreement is observed between theory and experiment.

The results presented in this paper may be summarized as follows:

1. The results obtained by the proposed approximate method compared well with those obtained by the stepwise theory. The differences observed between the results of the two methods are too small to justify the added labor required by the exact method.

2. Very good agreement was observed, in all instances, between the moment-load curves predicted by the approximate method and those obtained by test, particularly up to the collapse load. While close agreement was generally encountered beyond the collapse load, there were a few instances in which the discrepancy between the predicted and actual curves became pronounced.

3. The difference between the predicted and actual values of collapse loads was, in general, much less than 7%. Thus the proposed approximate method was found to be sufficiently adequate for all practical purposes.

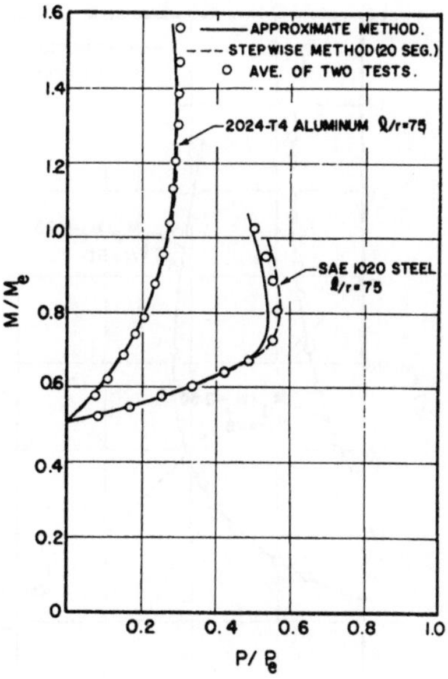

Fig. 5. Comparison of Moment-Load Curves for Symmetrically Loaded 2024-T4 Aluminum and SAE 1020 Steel Beam-Columns Determined by Two Theoretical Methods and by Test ($\eta_1 = \eta_2 = 0$).

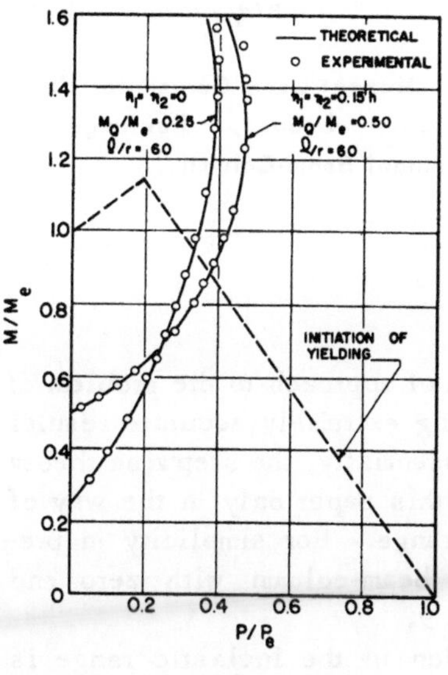

Fig. 6. Moment-Load Curves for Symmetrically Loaded T-Section Beam-Columns of 2024-T4 Aluminum.

4. In general, inelastic action was found to induce a considerable increase in the load-carrying capacity beyond that associated with initiation of yielding. The increase in the load-carrying capacity was observed to range from a maximum of 159% for SAE 1020 steel to a minimum of about 2% for 17-7PH stainless steel.

5. The increase in the load-carrying capacity was found to vary directly with M_Q/M_e, and inversely with l/r and ϵ_e.

6. Members with end eccentricities of 15% of the section depth attained a much higher increase in load-carrying capacity than members with zero end eccentricities.

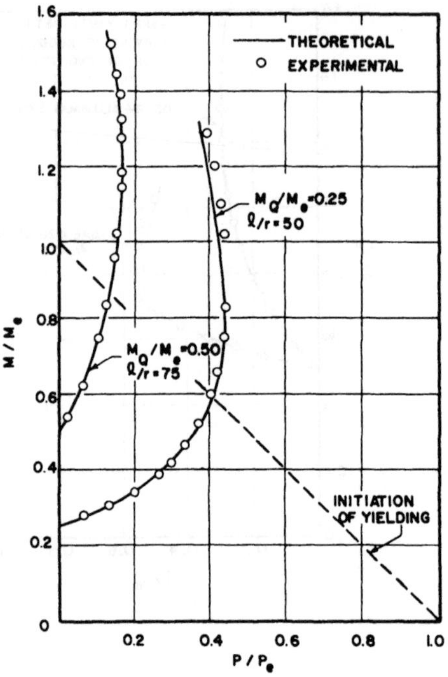

Fig. 7. Moment-Load Curves for Symmetrically Loaded, Rectangular 17-7PH Stainless Steel Beam-Columns ($\eta_1 = \eta_2 = 0$).

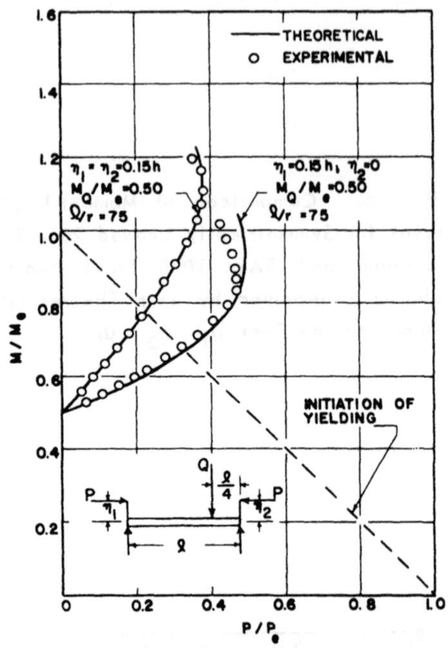

Fig. 8. Moment-Load Curves for Unsymmetrically Loaded, Rectangular SAE 1020 Steel Beam-Columns.

APPENDIX

A brief outline is given here of another method of approach to the problem of beam-columns. This method is capable of yielding extremely accurate results and is referred to here as the stepwise theory. Essentially, the stepwise theory differs from the approximate method described in this paper only in the way of relating curvature to deflection in the inelastic range. For simplicity in presenting this approach, a symmetrically loaded beam-column with zero end eccentricities will be considered, as shown in Fig. 9.

The method of relating curvature to deflection in the inelastic range is similar in principle to that used by Chwalla [4] in his solutions of column problems. It consists of subdividing the member into a number of equal segments λ as shown in Fig. 9. The assumption is then made that within each segment, the curvature remains constant (i.e., within each segment, the deflection curve is a circular arc). The deflection y_N at any section of the member may then be found by the relation

$$y_N = y_0 - \sum_{=1}^{N} \left(\frac{2N - 2i + 1}{2} \right) \lambda^2 \psi_{i-1}, \tag{10}$$

where y_0 is the deflection at midspan.

Fig. 9. Symmetrically Loaded Beam-Column.

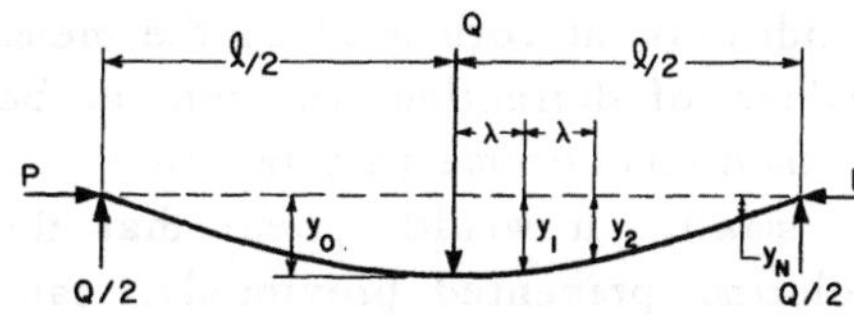

For the beam-column shown in Fig. 9, the dimensionless moment at any section of the beam is given by the relation [see Eq. (3)]

$$\frac{M}{M_e} = \gamma\left(\frac{P}{P_e}\right)y + \mu\left(\frac{M_Q}{M_e}\right), \tag{11}$$

in which $\gamma = P_e/M_e$, and μ is a factor varying linearly from unity at the center section to zero at the end of the member. Equations (10) and (11) thus provide one P-M-ψ relation. The second P-M-ψ relation is obtained by using Eq. (1) and interaction curves of the type shown in Fig. 1.

The solution of the problem is obtained by means of the above relations in the following manner. As in the previous solution, point A (see Fig. 1) is determined by the magnitude of the transverse load Q, and point B is determined from the elastic solution. For a given depth of inelastic penetration (k=k_1), some value of P/P_e is assumed and the corresponding value of M/M_e determined from the appropriate interaction curve. With $\mu = 1$, the value of y_0 is then determined from Eq. (11). Equation (1) is now used to determine the curvature ψ_0 at the center of the member. The deflection y_1 may now be determined by Eq. (10). By use of the value of y_0 and the appropriate value for μ in Eq. (11), the dimensionless moment M/M_e corresponding to the section of the member at y_1 may be established.[7] Then the curvature ψ_1 is found by Eq. (1). The values of M_u/M_e and k for use in Eq. (1) are found from the interaction curves by interpolation. This process is continued until the deflection at the end of the member is determined. The assumed value of P/P_e is correct if the computed value of deflection at the end of the beam-column is zero or very nearly so. Thus, to determine one point on the moment-load curve, this numerical procedure may have to be repeated three or four times until the boundary condition is satisfied.

Theoretically, very accurate results may be obtained by this method of solution if the segments λ are made extremely small. This, however, would entail very lengthy if not prohibitive computational work, particularly if conditions other than those shown in Fig. (9) are to be met. For example, when the transverse load Q is not applied at midspan, one more unknown is introduced into the problem, namely, the slope of the deflection curve at the point of application of Q. Therefore, in applying the method outlined above, it becomes necessary to assume a value for this slope in addition to the value of P/P_e. The procedure is then applied to both portions of the beam-column until the

---

[7] In determining the value of M/M_e from Eq. (11) for any given section, the procedure consisted in adding the maximum value due to P to the average value due to Q. Thus, for example, $(M/M_e)_1 = \gamma(P/P_e)y_0 + \frac{1}{2}[(M_Q/M_e)_0 + (M_Q/M_e)_1]$.

conditions at both ends of the member are satisfied (i.e., until the determined values of deflection are zero at both ends of the member). Obviously, this method can become very laborious, even when the number of segments considered is small. It would appear that this method, compared with the approximate solution presented previously, can become economically feasible only if a technique is developed to program the calculations on a computing machine.

REFERENCES

[1] O. M. Sidebottom and M. E. Clark, *Theoretical and Experimental Analysis of Members Loaded Eccentrically and Inelastically,* University of Illinois Engineering Experiment Station, Bulletin No. 447, 1958.

[2] T. von Karman, "Untersuchungen uber Knickfestigkeit," *Forsch. Gebiete Ingenieurw., V.D.I. Forschungsheft,* Heft 81, 1910.

[3] E. Chwalla, "Die Stabilitat zentrisch und exzentrisch gedruckter Stabe aus Baustahl," *Anz. Akad. Wiss. Wien, Math.-naturw. Kl., Sitzungsberichte Abt. IIa,* **137**, 469–512 (1928).

[4] E. Chwalla, "Theorie des aussermittig gedruckten Stabes aus Baustahl," *Stahlbau Suppl. Bautechnik* **7**, 161–65, 173–76, 180–84 (1934).

[5] W. R. Osgood, "The Double-Modulus Theory of Column Action," *Civil Eng.* **5**(3), 173–75 (1935).

[6] F. R. Shanley, "Inelastic Column Theory," *J. Aeronaut. Sci.* **14**(5), 261–67 (1947).

[7] C. T. Wang, "Inelastic Column Theories and an Analysis of Experimental Observations," *J. Aeronaut. Sci.* **15**(5), 283–92 (1948).

[8] P. Cicala, "Column Buckling in the Elastoplastic Range," *J. Aeronaut. Sci.* **17**(8), 508–12 (1950).

[9] W. D. Jordan, *Inelastic Behavior of Eccentrically Loaded Columns,* Ph.D. Thesis, University of Illinois, 1952.

[10] P. P. Bijlaard, "Buckling of Columns with Equal and Unequal End Eccentricities and Equal and Unequal Rotational End Strains," Proceedings *2nd U.S. National Congress Applied Mechanics,* 1954, pp 555–62.

[11] T. V. Galambos and R. L. Ketter, "Columns Under Combined Bending and Thrust," *Proc. Am. Soc. Civil Engrs.* **85**, 1990 (1959).

[12] G. A. Costello, *The Creep Buckling of Columns Made of Canvas Laminate,* Ph.D. Thesis, University of Illinois, 1959.

[13] O. M. Sidebottom, E. Pocs, and G. A. Costello, *The Effect of End Conditions on the Collapse Loads of Columns,* Dept. of Theo. and Appl. Mech. Report No. 585, University of Illinois, 1960.

[14] S. Timoshenko, *Theory of Elastic Stability,* pp 1–63, McGraw-Hill, New York, 1936.

[15] W. R. Osgood, "Beam-Columns," *J. Aeronaut. Sci.* **14**(3), 167–70 (1947).

[16] R. L. Ketter, "Stability of Beam-Columns Above the Elastic Limit," *Proc. Am. Soc. Civil Engrs.* **81**(692) 1–27 (1955).

[17] O. M. Sidebottom *et al.*, *The Effects of Inelastic Action on the Resistance to Various Types of Loads of Ductile Members Made from Various Classes of Metals, WADC Tech.* Rept. No. 56/330, Pt. XII (1959).

[18] O. M. Sidebottom, M. E. Clark, and S. Dharmarajan, *The Effects of Inelastic Action on the Resistance to Various Types of Loads of Ductile Members Made from Various Classes of Metals, WADC Tech.* Rept. No. 56/330, Pt. VIII (1958).

[19] B. B. Muvdi and O. M. Sidebottom, *Inelastic Design of Load Carrying Members. Part IV. The Behavior of Beam-Columns in the Inelastic Range, WADD Tech.* Rept. No. 60/580 (1961).

A MATRIX METHOD FOR ANALYZING BEAM GRILLAGES[1]

F. J. Stanek

Oak Ridge National Laboratory[2]

This paper demonstrates the application of matrix theory to the solution of beam-grillage problems. In this application it is assumed that only a concentrated interaction force is transmitted from one beam to the other at their common intersection. The evaluation of the interaction forces is obtained by solving a set of simultaneous equations which are derived by equating the deflection of two beams at their common intersection. The tedious notation and bookkeeping problem in obtaining the set of simultaneous equations is overcome with the aid of matrix theory.

INTRODUCTION

Almost any beam-grillage arrangement for which the elementary beam-deflection theory holds may be analyzed by the method presented. The length and the bending rigidity of the individual beams which make up the beam grillage need not be the same, and the ends of any beam may be simply supported or fixed. The external loading may be made up of both concentrated forces and uniformly distributed loads; however, only the latter is considered in the presentation.

The practical limit of the method is primarily in the number of interaction forces which have to be determined. This number is dependent upon the number of beams and their arrangement in the grillage network. Fortunately, most applications, such as in certain types of nuclear reactors, require an arrangement which is symmetrical about the two center lines. This greatly reduces the number of interaction forces which have to be determined.

The presentation of the procedure is confined to beam grillages with symmetrical arrangements, and the relatively small number of required general beam formulas are tabulated for symmetrically loaded beams only. The shear deflections are not considered, as they do not affect the basic fundamentals of the method presented. They may be taken into account by appropriately augmenting the beam-deflection formulas tabulated.

---

[1]Supported by the USAEC under contract No. W-7405-eng-26.

[2]Operated for USAEC by Union Carbide Corporation.

NOTATION

L Length of individual beam (in.)

z' Coordinate along length of individual beam (in.)

z z'/L, dimensionless coordinate of individual beam

c Distance to concentrated load from end of beam (in.)

k c/L, dimensionless position of concentrated load

q Uniform external load on individual beam (lb/in.)

Q qL, total load on beam (lb)

w Lateral deflection of beam, positive in direction q is acting (normally downward) (in.)

M Bending moment in beam, positive when compressive stress is at top surface (in.-lb)

E Modulus of elasticity (lb/in.^{2})

I Second moment of area of cross section of individual beam about bending axis (in.^{4})

D $L^3/24\,EI$, bending rigidity of individual beam (in./lb)

i,j Position point at intersection of the ith row (x beam) and the jth column (y beam)

Δ_{ij} Deflection w at intersection i,j (in.)

Δ_v Matrix vector of the deflections

P_v Matrix vector of the interaction forces

Q_v Matrix vector of the external forces

P_{ij} Interaction force at intersection i,j (lb)

P Major interaction matrix. Also used to represent concentrated forces on the beam in the list of general formulas

X_i Interaction matrix of the ith beam parallel to the x axis

Y_j Interaction matrix of the jth beam parallel to the y axis

X, Y Total interaction matrix for all beams parallel to the x and the y axis respectively

The rules for the multiplication of matrices, together with matrix representation of a set of simultaneous equations, are the essential phases of the matrix theory for an understanding of the methods presented. The procedure is explained by carrying through, in detail, two illustrative examples: first, a beam grillage symmetrical about the two center lines only; then one which also possesses symmetry about the two diagonals. The procedures for the two cases are slightly different. A summary listing the steps required for analyzing a beam grillage of each type is given at the end of the paper.

BEAM FORMULAS (BENDING ONLY)

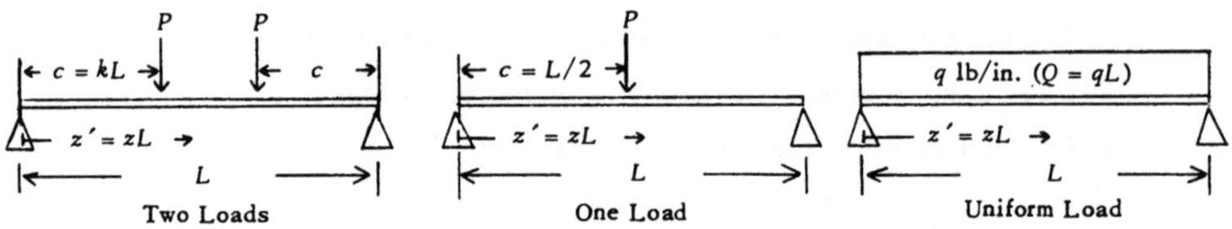

Type Load	Range of z	Lateral Deflection	Bending Moment
		Simply Supported	
Two loads	$0 \leqq z \leqq k$	$w = PD4z[3k(1-k) - z^2]$	$M = PLz$
($k < 0.5$)	$k \leqq z \leqq 0.5$	$w = PD4k[3z(1-z) - k^2]$	$M = PLk$ = a constant
One load ($k = 0.5$)	$0 \leqq z \leqq 0.5$	$w = PD2z(0.75 - z^2)$	$M = (PL/4)\, z$
Uniform load	$0 \leqq z \leqq 1$	$w = QDz(z^3 + 1 - 2z^2)$	$M = (QL/2)z(1-z)$
		Fixed Ends	
Two loads	$0 \leqq z \leqq k$	$w = PD4z^2[3k(1-k) - z]$	$M = PL[z - k(1-k)]$
($k < 0.5$)	$k \leqq z \leqq 0.5$	$w = PD4k^2[3z(1-z) - k]$	$M = PLk^2$ = a constant
One load ($k = 0.5$)	$0 \leqq z \leqq 0.5$	$w = PD2z^2(0.75 - z)$	$M = PL\ /2)\,(z - 0.25)$
Uniform load	$0 \leqq z \leqq 1$	$w = QDz^2(1-z)^2$	$M = (QL/12)[6z(1-z) - 1]$

PROCEDURE FOR GRILLAGE WITH SYMMETRY ABOUT THE TWO CENTER LINES

The procedure for the derivation of the necessary matrix equations will now be illustrated for a grillage with symmetry about the two center lines. The beam grillage used for this example is shown in Fig. 1. Special note should be made of the *ij* notation for the intersections and the designation of the individual beams. The x and y axes are used merely as references.

Since each set of four beams in this grillage is symmetrical about its corresponding center line, only two beams of each set must be considered. These are shown with their corresponding interaction loadings in Figs. 2 and 3 respectively.

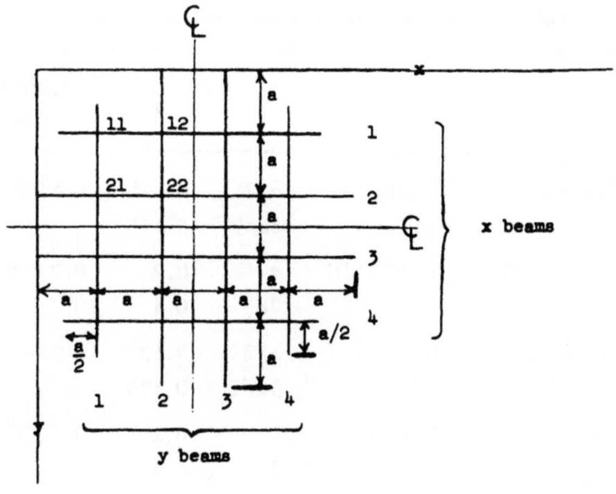

Fig. 1. Beam-Grillage Arrangement.

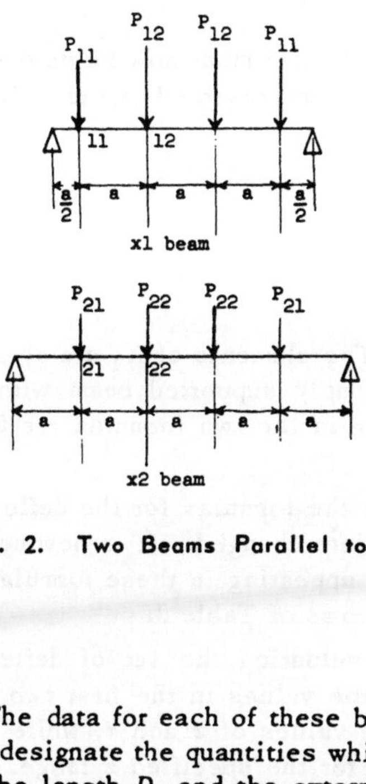

Fig. 2. Two Beams Parallel to the *x* Axis.

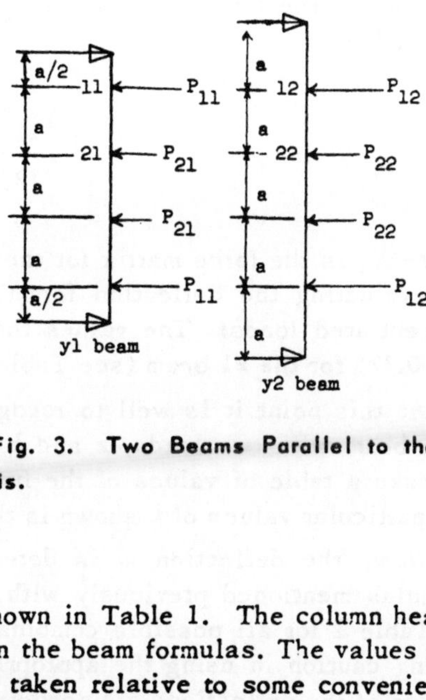

Fig. 3. Two Beams Parallel to the *y* Axis.

The data for each of these beams are shown in Table 1. The column headings designate the quantities which appear in the beam formulas. The values for EI, the length D, and the external load are taken relative to some convenient, but arbitrary, reference beam. The data for the $y1$ beam were selected as the reference in this example. The need for the various quantities shown will become apparent as they are used during the presentation which follows.

The derivation of the matrix deflection equation for the beams parallel to the x axis is illustrated for the two beams shown in Fig. 2. The corresponding equation for the beams shown in Fig. 3 will then follow accordingly in a more concise manner.

Table 1. Beam Data for the First Example

Beam No.	EI	Length	D	c	k	External Load	End Support
$x1$	$2EI$	$4a = L$	$0.5D$	$a/2$	0.125	$0.5QD$	Simple
				$3a/2$	0.375		
$x2$	$2.5EI$	$5a = 1.25L$	$0.78125D$*	a	0.2	$0.97656QD$*	Simple
				$2a$	0.4		
$y1$	EI	$4a = L$	D	$a/2$	0.125	QD	Simple
				$3a/2$	0.375		
$y2$	$1.5EI$	$5a = 1.25L$	$1.3021D$	a	0.2	$1.6276QD$	Simple
				$2a$	0.4		

*Note: $D_{x2} = (1.25^3/2.5)D = 0.78125D$; $(QD)_{x2} = (1.25Q)(0.78125D) = 0.97656QD$.

The deflections at points 11 and 12 in the $x1$ beam must now be determined in terms of the forces P_{11} and P_{12}, which will be represented in the following matrix form:

$$\begin{bmatrix} \Delta_{11} \\ \Delta_{12} \end{bmatrix} = X_1 \begin{bmatrix} P_{11} \\ P_{12} \end{bmatrix},$$

where X_1 is the force matrix for the x_1 beam. The elements of X_1 are obtained by evaluating the deflection formula for the simply supported beam with two concentrated loads. The values for k appearing in the two formulas are 0.125 and 0.375 for the $x1$ beam (see Table 1).

At this point it is well to recognize that in the formulas for the deflection of a beam with two loads, z and k are merely interchanged. It is now helpful to make a table of values of the functions of k appearing in these formulas for the particular values of k shown in the first two rows of Table 2.

Now, the deflection w is determined by evaluating the set of deflection formulas mentioned previously with the aid of the values in the first two rows of Table 2 for all possible combinations of the values of z and k, while exercising caution in using the appropriate formula for the specified z range. The values of the deflections are now arranged as shown in Table 3.

Each value in Table 3 is the dimensionless deflection w/PD of the beam at zL due to the concentrated load at kL. For example, the value of 0.34375 in the first row is the deflection at $0.125L$ due to the load at $0.375L$, whereas this value in the second row is the deflection at $0.375L$ due to the load at $0.125L$. The physical significance of each value in the table should be clearly understood.

Table 2. Values of $3k(1-k)$, $4k$, and k^2

k	$3k(1-k)$	$4k$	k^2
0.125	0.328125	0.500	0.015625
0.375	0.703125	1.500	0.140625
0.200	0.4800	0.800	0.0400
0.400	0.7200	1.600	0.1600

Table 3. Initial Interaction Matrix for the $x1$ Beam

z	k	
	0.125	0.375
0.125	0.15625	0.34375
0.375	0.34375	0.84375

The matrix deflection equation for the $x1$ beam can now be written in the following form:

$$\begin{bmatrix} \Delta_{11} \\ \Delta_{12} \end{bmatrix} = D_{x1} \begin{bmatrix} 0.15625 & 0.34375 \\ 0.34375 & 0.84375 \end{bmatrix} \begin{bmatrix} P_{11} \\ P_{12} \end{bmatrix} .$$

Since $D_{x1} = D/2$ (see Table 1), the final form of the matrix deflection equation for the $x1$ beam is

$$\begin{bmatrix} \Delta_{11} \\ \Delta_{12} \end{bmatrix} = D \begin{bmatrix} 0.078125 & 0.171875 \\ 0.171875 & 0.421875 \end{bmatrix} \begin{bmatrix} P_{11} \\ P_{12} \end{bmatrix} . \tag{1}$$

Because z and k will have identical values in beam-grillage problems, as shown in the above example, the force (interaction) matrix will be symmetrical provided k and z follow the same sequence and k is not equal to 0.5. In matrix notation, Eq. (1) is written as

$$\Delta_{vx1} = X_1 P_{vx1} , \tag{1a}$$

where the subscript v designates a single columnar matrix.

For the $x2$ beam, the values for k and z are 0.2 and 0.4, for which the required functions of k have been tabulated in the last two rows of Table 2. With this, the elements of the initial interaction matrix are calculated and tabulated as shown in Table 3*a* (corresponding to Table 3 for the $x1$ beam).

Table 3*a*. Initial Interaction Matrix for *x*2 Beam

z	k	
	0.2	0.4
0.2	0.352	0.544
0.4	0.544	0.896

Since $D_{x2} = 0.78125D$, the matrix deflection equation for the $x2$ beam is

$$\begin{bmatrix} \Delta_{21} \\ \Delta_{22} \end{bmatrix} = D \begin{bmatrix} 0.275 & 0.425 \\ 0.425 & 0.700 \end{bmatrix} \begin{bmatrix} P_{21} \\ P_{22} \end{bmatrix}, \tag{2}$$

which is written in matrix notation in the following manner:

$$\Delta_{vx2} = X_2 P_{vx2} \, . \tag{2a}$$

The matrix deflection equations for the beams parallel to the x axis may now be written in one matrix equation by "stacking" the columnar matrices, keeping the main diagonals of the individual interaction matrices "in line," and placing zeros in all other elements, as shown below:

$$\begin{bmatrix} \Delta_{vx1} \\ \Delta_{vx2} \end{bmatrix} = \begin{bmatrix} X_1 & 0 \\ 0 & X_2 \end{bmatrix} \begin{bmatrix} P_{vx1} \\ P_{vx2} \end{bmatrix} .$$

Specifically, Eqs. (1) and (2) are combined into the following single matrix equation:

$$\begin{bmatrix} \Delta_{11} \\ \Delta_{12} \\ \Delta_{21} \\ \Delta_{22} \end{bmatrix}_x = D \begin{bmatrix} 0.078125 & 0.171875 & 0 & 0 \\ 0.171875 & 0.421875 & 0 & 0 \\ 0 & 0 & 0.27500 & 0.42500 \\ 0 & 0 & 0.42500 & 0.70000 \end{bmatrix} \begin{bmatrix} P_{11} \\ P_{12} \\ P_{21} \\ P_{22} \end{bmatrix}, \tag{3}$$

where the subscript x has now been placed on the outside of the Δ vector only. Equation (3) is written in the following concise form:

$$\Delta_{vx} = XP_{vx} \, . \tag{3a}$$

In general, the matrix deflection equation for all beams parallel to the x axis is obtained by arranging in the following manner:

$$\begin{bmatrix} \Delta_{v1} \\ \Delta_{v2} \\ \Delta_{v3} \\ \cdot \\ \cdot \\ \cdot \\ \Delta_{vn} \end{bmatrix}_x = \begin{bmatrix} X_1 & 0 & 0 & \dots & 0 \\ 0 & X_2 & 0 & \dots & 0 \\ 0 & 0 & X_3 & \dots & 0 \\ \cdot & \cdot & \cdot & \cdot & \cdot \\ \cdot & \cdot & \cdot & \cdot & \cdot \\ \cdot & \cdot & \cdot & \cdot & \cdot \\ 0 & 0 & 0 & \cdots & X_n \end{bmatrix} \begin{bmatrix} P_{v1} \\ P_{v2} \\ P_{v3} \\ \cdot \\ \cdot \\ \cdot \\ P_{vn} \end{bmatrix} .$$

The matrix deflection equation for the beams parallel to the y axis is now derived in a similar manner.

The initial interaction matrices for the $y1$ and the $y2$ beams are identical to those for the $x1$ and $x2$ beams respectively. Hence the deflection equation for these two beams is the following matrix equation:

$$\begin{bmatrix} \Delta_{11} \\ \Delta_{21} \\ \Delta_{12} \\ \Delta_{22} \end{bmatrix}_y = D \begin{bmatrix} 0.15625 & 0.34375 & 0 & 0 \\ 0.34375 & 0.84375 & 0 & 0 \\ 0 & 0 & 0.45833 & 0.70833 \\ 0 & 0 & 0.70833 & 1.1667 \end{bmatrix} \begin{bmatrix} P_{11} \\ P_{21} \\ P_{12} \\ P_{22} \end{bmatrix} , \tag{4}$$

which can be written in the following concise form:

$$\Delta_{vy} = Y P_{vy} . \tag{4a}$$

The basic matrix necessary to determine the interaction forces at the designated beam intersection points can now be developed for the first example. First, however, each of Eqs. (3) and (4) must be augmented to include the deflections due to the external loads at the designated points (intersections). Since each beam carries a uniform load of q lb/in., an evaluation of the beam formula for a simply supported beam with a uniform load is required at all necessary z values. This is done in Table 4.

Table 4. The $w/(QD)_i$ Deflections

Beam No.	z	$w/(QD)_i$
$x1$	0.125	0.12134
$y1$	0.375	0.28931
$x2$	0.200	0.18560
$y2$	0.400	0.29760

The relative value of QD for each beam is shown in Table 1. Hence, the deflections due to the external load are obtained and arranged in the following manner:

$$\begin{bmatrix}\Delta_{11}\\\Delta_{12}\\\Delta_{21}\\\Delta_{22}\end{bmatrix}_x = QD\begin{bmatrix}(0.5)\begin{pmatrix}0.12134\\0.28931\end{pmatrix}\\(0.97656)\begin{pmatrix}0.18560\\0.29760\end{pmatrix}\end{bmatrix} = QD\begin{bmatrix}0.06067\\0.14465\\0.18125\\0.29062\end{bmatrix},$$

$$\begin{bmatrix}\Delta_{11}\\\Delta_{21}\\\Delta_{12}\\\Delta_{22}\end{bmatrix}_y = QD\begin{bmatrix}(1.0)\begin{pmatrix}0.12134\\0.28931\end{pmatrix}\\(1.6276)\begin{pmatrix}0.18560\\0.29760\end{pmatrix}\end{bmatrix} = QD\begin{bmatrix}0.12134\\0.28931\\0.30208\\0.48437\end{bmatrix}.$$

These two equations, respectively, are represented in the following simplified form:

$$\Delta_{vx} = Q_{vx},$$

$$\Delta_{vy} = Q_{vy}.$$

Adding the above to Eqs. 3*a* and 4*a*, respectively, the complete deflection equations for the beams parallel to the x and y axes are written in the following form:

$$\Delta_{vx} = XP_{vx} + Q_{vx}, \tag{5}$$

$$\Delta_{vy} = -YP_{vy} + Q_{vy}. \tag{6}$$

In this analysis it is assumed that the Y beams are above the x beams; hence, matrix Y is preceded by a negative sign (multiplied by -1) in Eq. (6) because of the action and reaction concept in which it is assumed that the interaction forces act *downward* on the x beams and *upward* on the y beams.

In the usual analysis of this type of statically indeterminate problem, each equation in the required set of simultaneous equations is obtained by equating the corresponding Δ's in Δ_{vx} and Δ_{vy}. In this matrix method, the two Δ columnar matrices are equated; however, before this can be done the Δ's in Δ_{vy} and the P's in P_{vy} must be in the same sequence as in Δ_{vx} and P_{vx} respectively. Hence, the matrices in Eq. (6) must be rearranged so that the Δ_{vy} and the P_{vy} columnar matrices have the same sequence as the corresponding columnar matrices in Eq. (5). The procedure for obtaining this rearrangement is given later. Assuming, for the moment, that this has been done and using primes to designate the matrices, Eq. (6) is rewritten in the following form:

$$\Delta'_{vy} = -Y'P'_{vy} + Q'_{vy}, \tag{6a}$$

and then, since $\Delta_{vx} = \Delta'_{vy}$, the following equation is written:

$$XP_{vx} + Q_{vx} = -Y'P'_{vy} + Q'_{vy}. \tag{7}$$

Because $P_{vx} \equiv P'_{vy} = P_v$, Eq. (7) is simplified to the following form:

$$(X + Y')P_v = Q'_{vy} - Q_{vx} \quad \text{or simply} \quad PP_v = Q_v , \tag{7a}$$

where $P = X + Y'$ is the required interaction matrix, and $Q_v = Q'_{vy} - Q_{vx}$.

Equation (6) is rearranged to the appropriate form by the following matrix rules:

1. Any two rows in Δ_{vy} may be interchanged provided that the corresponding rows in Y and Q_{vy} are simultaneously interchanged.
2. Any two rows in P_{vy} may be interchanged provided the corresponding columns in Y are simultaneously interchanged.

These rules may be verified by writing Eq. (6) in algebraic form, rearranging the equations and the terms in each equation to the desired sequence, and then writing the rearranged equations back into matrix form. The following is the desired form of Eq. (6):

$$\begin{bmatrix} \Delta_{11} \\ \Delta_{12} \\ \Delta_{21} \\ \Delta_{22} \end{bmatrix}_y = -D \begin{bmatrix} 0.15625 & 0 & 0.34375 & 0 \\ 0 & 0.45833 & 0 & 0.70833 \\ 0.34375 & 0 & 0.84375 & 0 \\ 0 & 0.70833 & 0 & 1.1667 \end{bmatrix} \begin{bmatrix} P_{11} \\ P_{12} \\ P_{21} \\ P_{22} \end{bmatrix} + QD \begin{bmatrix} 0.12134 \\ 0.30208 \\ 0.28931 \\ 0.48437 \end{bmatrix} . \tag{6b}$$

Performing the operations in Eq. (7*a*) yields the following set of simultaneous equations required to determine the interaction forces P_{ij}:

$$\begin{bmatrix} 0.23438 & 0.17188 & 0.34375 & 0 \\ 0.17188 & 0.88020 & 0 & 0.70833 \\ 0.34375 & 0 & 1.1188 & 0.42500 \\ 0 & 0.70833 & 0.42500 & 1.8667 \end{bmatrix} \begin{bmatrix} P_{11} \\ P_{12} \\ P_{21} \\ P_{22} \end{bmatrix} = Q \begin{bmatrix} 0.06067 \\ 0.15743 \\ 0.10806 \\ 0.19374 \end{bmatrix} . \tag{8}$$

The solution of Eq. (8) is

$$P_{11} = 0.091587Q , \qquad P_{12} = 0.12508Q ,$$

$$P_{21} = 0.051504Q , \qquad P_{22} = 0.044598Q .$$

In order to determine the deflection along a particular beam, for example, the deflection $w/(QD)_{x2}$ along the $x2$ beam, the initial interaction matrix for that beam (such as in Table 3*a*) must be extended to include the desired values of z, and a column must be added for the deflection due to the external loads, as is done in the first four columns of Table 5. The total deflection is then given in the last column. The bending moment along the beam is obtained in a similar manner by using the appropriate formulas.

Table 5. Deflection Along the $x2$ Beam

z	$\frac{w}{P_{21}D_{x2}}$ ($k = 0.2$)	$\frac{w}{P_{22}D_{x2}}$ ($k = 0.4$)	$\frac{w}{(QD)_{x2}}$ (external load)	$\frac{(0.0412)w^*}{P_{21}D_{x2}} = \frac{w}{(QD)_{x2}}$	$\frac{(0.0357)w^*}{P_{22}D_{x2}} = \frac{w}{(QD)_{x2}}$	Total $\frac{w}{(QD)_{x2}}$
0.1	0.188	0.284	0.0981	0.0077	0.0101	0.1159
0.2	0.352	0.544	0.1856	0.0145	0.0194	0.2195
0.3	0.472	0.756	0.2541	0.0194	0.0270	0.3005
0.4	0.544	0.896	0.2976	0.0224	0.0320	0.3520
0.5	0.568	0.944	0.3125	0.0234	0.0337	0.3696

*Note: $P_{21} = 0.0515Q = 0.0412Q_{x2}$, and $P_{22} = 0.0446Q = 0.0357Q_{x2}$, see Table 1.

PROCEDURE FOR GRILLAGE WITH COMPLETE SYMMETRY

A beam grillage is considered to have complete symmetry when there are the same number of beams n in each direction and the deflections along the xi and yi beams are identically equal for $i = 1, 2, 3, \ldots, n$. The procedure for this case is illustrated for a square beam grillage with six beams in each direction

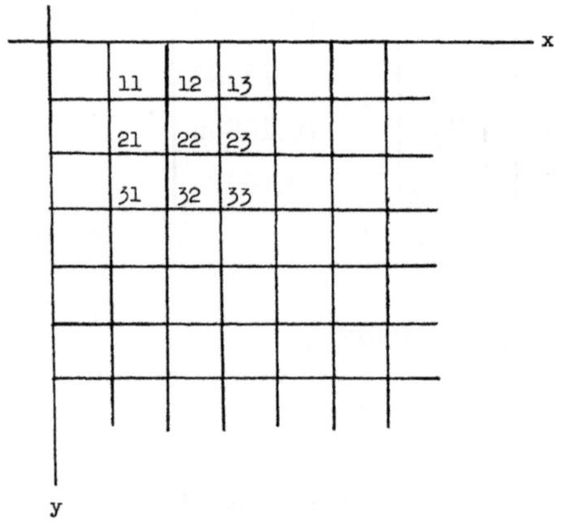

Fig. 4. A 6 × 6 Beam Grillage.

(Fig. 4). All beams are simply supported and have the same rigidity and the same spacing (i.e., $k = 1/7$, $2/7$, and $3/7$). Hence, the beam data table is not shown. The interaction matrix for each beam is the same and is

$$X_i = Y_i = \begin{bmatrix} 0.19825 & 0.27988 & 0.40816 \\ 0.27988 & 0.48980 & 0.74635 \\ 0.40816 & 0.74635 & 0.87463 \end{bmatrix}. \tag{9}$$

With the identification of the intersection points as shown in Fig. 4, the matrix deflection equation for beams parallel to the x axis is shown below. Since D is the same for all beams, it may be absorbed in the Δ's, as in the discussion which follows.

$$\begin{bmatrix} \Delta_{11} \\ \Delta_{12} \\ \Delta_{13} \end{bmatrix} = [X_i] \begin{bmatrix} P_{11} \\ P_{12} \\ P_{13} \end{bmatrix} + Q \begin{bmatrix} 0.13744 \\ 0.24573 \\ 0.30487 \end{bmatrix}, \tag{10a}$$

$$\begin{bmatrix} \Delta_{21} \\ \Delta_{22} \\ \Delta_{23} \end{bmatrix} = [X_i] \begin{bmatrix} P_{21} \\ P_{22} \\ P_{23} \end{bmatrix} + Q \begin{bmatrix} 0.13744 \\ 0.24573 \\ 0.30487 \end{bmatrix}, \tag{10b}$$

$$\begin{bmatrix} \Delta_{31} \\ \Delta_{32} \\ \Delta_{33} \end{bmatrix} = [X_i] \begin{bmatrix} P_{31} \\ P_{32} \\ P_{33} \end{bmatrix} + Q \begin{bmatrix} 0.13744 \\ 0.24573 \\ 0.30487 \end{bmatrix}. \tag{10c}$$

For complete symmetry, $P_{ii} \equiv 0$ and hence the deflections Δ_{ii} are not required. Therefore Eqs. (10) can be immediately simplified. The required deflections are readily obtained from Eqs. (9) and (10) by omitting the row and column corresponding to the position of Δ_{ii} and P_{ii} respectively. This should be noted while verifying each of Eqs. (11):

$$\begin{bmatrix} \Delta_{12} \\ \Delta_{13} \end{bmatrix} = \begin{bmatrix} 0.48980 & 0.74635 \\ 0.74635 & 0.87463 \end{bmatrix} \begin{bmatrix} P_{12} \\ P_{13} \end{bmatrix} + Q \begin{bmatrix} 0.24573 \\ 0.30487 \end{bmatrix}, \tag{11a}$$

$$\begin{bmatrix} \Delta_{21} \\ \Delta_{23} \end{bmatrix} = \begin{bmatrix} 0.19825 & 0.40816 \\ 0.40816 & 0.87463 \end{bmatrix} \begin{bmatrix} P_{21} \\ P_{23} \end{bmatrix} + Q \begin{bmatrix} 0.13744 \\ 0.30487 \end{bmatrix}, \tag{11b}$$

$$\begin{bmatrix} \Delta_{31} \\ \Delta_{32} \end{bmatrix} = \begin{bmatrix} 0.19825 & 0.27988 \\ 0.27988 & 0.48980 \end{bmatrix} \begin{bmatrix} P_{31} \\ P_{32} \end{bmatrix} + Q \begin{bmatrix} 0.13744 \\ 0.24573 \end{bmatrix}. \tag{11c}$$

These equations are now written in one matrix equation by the "stacking" method illustrated in the previous example:

$$\begin{bmatrix} \Delta_{12} \\ \Delta_{13} \\ \Delta_{21} \\ \Delta_{23} \\ \Delta_{31} \\ \Delta_{32} \end{bmatrix}_x = \begin{bmatrix} 0.48980 & 0.74635 & 0 & 0 & 0 & 0 \\ 0.74635 & 0.87463 & 0 & 0 & 0 & 0 \\ 0 & 0 & 0.19825 & 0.40816 & 0 & 0 \\ 0 & 0 & 0.40816 & 0.87463 & 0 & 0 \\ 0 & 0 & 0 & 0 & 0.19825 & 0.27988 \\ 0 & 0 & 0 & 0 & 0.27988 & 0.48980 \end{bmatrix} \begin{bmatrix} P_{12} \\ P_{13} \\ P_{21} \\ P_{23} \\ P_{31} \\ P_{32} \end{bmatrix} + Q \begin{bmatrix} 0.24573 \\ 0.30487 \\ 0.13744 \\ 0.30487 \\ 0.13744 \\ 0.24573 \end{bmatrix}, \tag{12a}$$

or in concise matrix notation,

$$\Delta_{vx} = XP_{vx} + Q_v \,. \tag{12b}$$

Because of the symmetry conditions in this example, it is not necessary to write a matrix deflection equation for the beams parallel to the y axis corresponding to Eq. (12*a*). The six unknowns P_{ij} appearing in Eq. (12*a*) can be be reduced to three unknowns by the condition

$$P_{ij} = -P_{ji} \,. \tag{13a}$$

Equation (12*a*) also provides the three necessary equations for their solution by the condition

$$\Delta_{ij} = \Delta_{ji} \,. \tag{13b}$$

The procedure now is to multiply through Eq. (12*b*) by a matrix F so that $F\Delta_{vx} = 0$, and to make a substitution for P_{vx} of the form GP'_{vx}, where G is a matrix to be determined, and P'_{vx} is a columnar matrix containing the least possible number of unknowns. Thus, Eq. (12*b*) takes the form

$$F\Delta_{vx} = FXGP'_{vx} + FQ_v \,. \tag{12c}$$

Matrices F and G are established by using the principles of matrix multiplication and the conditions stated in Eqs. (13*b*) and (13*a*) respectively. For example, in determining matrix F, the zero columnar matrix (the objective) is first written at the right, as shown below, then preceded by a columnar matrix of $\Delta_{ij} - \Delta_{ji}$ (its equivalence because of condition 13*b*). This, in turn, is preceded by the columnar matrix from which it must originate, namely, Δ_{vx}. Now with the principles of matrix multiplication, matrix F is established so that when it premultiplies Δ_{vx}, the $\Delta_{ij} - \Delta_{ji}$ columnar matrix will result. All elements of each row in matrix F will be zero except in the two columns which correspond to the position of the Δ's to be combined:

$$F\Delta_{vx} = \begin{bmatrix} 1 & 0 & -1 & 0 & 0 & 0 \\ 0 & 1 & 0 & 0 & -1 & 0 \\ 0 & 0 & 0 & 1 & 0 & -1 \end{bmatrix} \begin{bmatrix} \Delta_{12} \\ \Delta_{13} \\ \Delta_{21} \\ \Delta_{23} \\ \Delta_{31} \\ \Delta_{32} \end{bmatrix} = \begin{bmatrix} \Delta_{12} - \Delta_{21} \\ \Delta_{13} - \Delta_{31} \\ \Delta_{23} - \Delta_{32} \end{bmatrix} = \begin{bmatrix} 0 \\ 0 \\ 0 \end{bmatrix} .$$

Matrix F operates on a given columnar matrix (Δ_{vx}) to yield a certain objective, namely, a zero columnar matrix. Matrix G, on the other hand, operates on the objective (a columnar matrix of the least number of unknown P_{ij}'s) so as to yield a given columnar matrix, namely, P_{vx}. This is accomplished with the aid of condition (13a). Otherwise the procedures are essentially the same, which should become evident when the operations in the following are verified.

All the elements in each row of matrix G will be zero except in the column which corresponds to the position of the P_{ij} to be selected.

$$GP'_{vx} = \begin{bmatrix} 1 & 0 & 0 \\ 0 & 1 & 0 \\ -1 & 0 & 0 \\ 0 & 0 & 1 \\ 0 & -1 & 0 \\ 0 & 0 & -1 \end{bmatrix} \begin{bmatrix} P_{12} \\ P_{13} \\ P_{23} \end{bmatrix} = \begin{bmatrix} P_{12} \\ P_{13} \\ -P_{12} \\ P_{23} \\ -P_{13} \\ -P_{23} \end{bmatrix} = \begin{bmatrix} P_{12} \\ P_{13} \\ P_{21} \\ P_{23} \\ P_{31} \\ P_{32} \end{bmatrix}.$$

Equation (12c) is now of the following form:

$$0 = \begin{bmatrix} 1 & 0 & -1 & 0 & 0 & 0 \\ 0 & 1 & 0 & 0 & -1 & 0 \\ 0 & 0 & 0 & 1 & 0 & -1 \end{bmatrix} [X] \begin{bmatrix} 1 & 0 & 0 \\ 0 & 1 & 0 \\ -1 & 0 & 0 \\ 0 & 0 & 1 \\ 0 & -1 & 0 \\ 0 & 0 & -1 \end{bmatrix} \begin{bmatrix} P_{12} \\ P_{13} \\ P_{23} \end{bmatrix} + \begin{bmatrix} 1 & 0 & -1 & 0 & 0 & 0 \\ 0 & 1 & 0 & 0 & -1 & 0 \\ 0 & 0 & 0 & 1 & 0 & -1 \end{bmatrix} [Q_v]. \qquad (12d)$$

While premultiplying matrix X by F, note that the first row of the product is the first row of X minus the third row of X; the second row of the product is the second row of X minus the fifth row of X; and the third row is the fourth row of X minus the sixth row of X. Therefore the following may be immediately written:

$$FX = \begin{bmatrix} 0.48980 & 0.74635 & -0.19825 & -0.40816 & 0 & 0 \\ 0.74635 & 0.87463 & 0 & 0 & -0.19825 & -0.27988 \\ 0 & 0 & 0.40816 & 0.87463 & -0.27988 & -0.48980 \end{bmatrix}.$$

Then, in a similar manner,

$$FQ_{vx} = \begin{bmatrix} 0.10829 \\ 0.16743 \\ 0.05914 \end{bmatrix} Q .$$

In postmultiplying the matrix FX by G, the product is obtained by writing in columns rather than in rows, as was done above; that is, the first column of the product $(FX)G$ is the first column of FX minus the third column of FX; the second column of the product is the second column of FX minus the fifth column of FX; and the third column is the fourth column of FX minus the sixth column of FX:

$$(FX)G = \begin{bmatrix} 0.68805 & 0.74635 & -0.40816 \\ 0.74635 & 1.07288 & 0.27988 \\ -0.40816 & 0.27988 & 1.36443 \end{bmatrix}.$$

Substituting each of the above into Eq. (12c) and transposing FQ_{vx} to the other side of the equal sign, immediately yields the following set of simultaneous equations necessary to determine P_{12}, P_{13}, and P_{23}:

$$\begin{bmatrix} 0.68805 & 0.74635 & -0.40816 \\ 0.74635 & 1.07288 & 0.27988 \\ -0.40816 & 0.27988 & 1.36443 \end{bmatrix} \begin{bmatrix} P_{12} \\ P_{13} \\ P_{23} \end{bmatrix} = Q \begin{bmatrix} -0.10829 \\ -0.16743 \\ -0.05914 \end{bmatrix} . \quad (14)$$

The solution to Eq. (14), and hence all six interaction forces, is

$$P_{12} = (-P_{21}) = 0.0804Q , \quad P_{13} = (-P_{31}) = -0.2054Q ,$$
$$P_{23} = (-P_{32}) = -0.02526Q ,$$

where the negative sign indicates that the force is acting upward on the beam parallel to the x axis and downward on the beam parallel to the y axis.

Each of the above procedures is summarized into nine basic steps in the section which follows.

SUMMARY

Steps for the Solution of a Beam Grillage with Symmetry About the Two Center Lines

1. Draw sketch of beam-grillage arrangement, using the i, j notation for the identification of the intersection points.
2. Make beam data table, summarizing the necessary data for each beam.
3. Make table of values of $3k(1 - k)$, $4k$, and k^2 for all values of k when the beams have simply supported ends, and $3k(1 - k)$ and $4k^2$ for the beams with fixed ends.
4. Evaluate the appropriate deflection formula for all possible combinations of z and k values for each necessary beam and arrange in the initial interaction matrix form. Include a column of values for the deflections due to the external loading.
5. Set up the matrix deflection equation for all necessary beams parallel to the x axis, taking into account the ratio (if different from 1) of the bending rigidity D and the external load Q. The equation should be of the following form:

$$\Delta_{vx} = XP_{vx} + Q_{vx} .$$

6. Repeat step 5 for all beams parallel to the y axis. The equation should be of the following form:

$$\Delta_{vy} = YP_{vy} + Q_{vy} .$$

7. Rearrange the rows in Δ_{vy} so as to have the sequence in Δ_{vx}, while simultaneously rearranging the corresponding rows in Y and Q_{vy}, thereby obtaining Y'' and Q'_v.

8. Rearrange the rows in P_{vy} so as to have the sequence in P_{vx}, while simultaneously rearranging the corresponding columns in Y'', thereby obtaining Y'.

9. Determine $P = X + Y'$ and $Q_v = Q'_{vy} - Q_{vx}$, and write the necessary set of simultaneous equations in the following form:

$$PP_v = Q_v \, .$$

Steps for the Solution of a Beam Grillage with Complete Symmetry

1. Draw a sketch of the beam-grillage arrangement, using the i, j notation for the intersection points.

2. Make a beam data table summarizing the necessary data for each beam parallel to the x axis if all beams are not identical.

3. Make table of values of $3k(1 - k)$, $4k$, and k^2 for all values of k for the beams with simply supported ends, and $3k(1 - k)$ and $4k^2$ for the beams with fixed ends.

4. Evaluate the appropriate deflection formula for all possible combinations of z and k values for each necessary beam parallel to the x axis and arrange in the initial interaction matrix form. Include a column of values for the deflections due to the external loading.

5. Set up the matrix deflection equation for all necessary beams parallel to the x axis, taking into account the ratio (if different from 1) of the bending rigidity D and the external load Q. Omit Δ_{ii} since $P_{ii} = 0$. The equation should be of the form

$$\Delta_{vx} = XP_{vx} + Q_{vx} \, .$$

6. Determine matrix F so that $F\Delta_{vx} = 0$ by using the condition that

$$\Delta_{ij} = \Delta_{ji} \, .$$

7. Determine matrix G so that $GP_v = P_{vx}$, where P_v contains the least number of unknown interaction forces, by using the condition that

$$P_{ij} = -P_{ji} \, .$$

8. Determine FX and $FQ_{vx} = Q_v$ by combining the rows indicated by F.

9. Determine $(FX)G = P$ by combining the columns indicated by G.

10. Set up necessary set of simultaneous equations in the following form:

$$PP_v = -Q_v \, .$$

SEISMIC VIBRATIONS OF NUCLEAR REACTOR TYPE STRUCTURES[1]

F. J. Witt
Oak Ridge National Laboratory[2]

D. R. Carver
Louisiana State University

and

R. L. Maxwell
University of Tennessee

This paper describes the analysis of the dynamic behavior of a composite system. On the assumption of small oscillations and three degrees of freedom, the equations of motion are derived and solved, with special emphasis on determining the resulting stresses. Specifically, the method of analysis is applied to the composite structure (core, pressure vessel, and supporting skirt), which typifies the configuration used in nuclear reactors. General analytical solutions to the equations of motion are derived, and for a specific structure both analytical and analog solutions are computed and compared.

INTRODUCTION

Seismic disturbances may cause failure or malfunctioning of important components of a nuclear reactor, and, in areas where significant disturbances occur, the dynamic behavior of reactor components must be examined. Of particular interest is the reaction of the core region to a disturbance, since such a region is often constructed from structurally weak material (graphite, for instance). A typical design is shown in Fig. 1 by the dashed lines, and this is the structure which is analyzed.

The analysis is made considering only horizontal and angular displacements of the structure, with vertical displacements, which are usually smaller than horizontal displacements, being neglected. This is feasible since relatively slender vertical structures such as the core columns described below are much more prone to damage from lateral than from axial vibrations. Nozzles on the

---

[1]Supported by the USAEC under contract No. W-7405-eng-26.

[2]Operated for USAEC by Union Carbide Corporation.

Fig. 1. Model of a Reactor Illustrating the Three Degrees of Freedom of Motion.

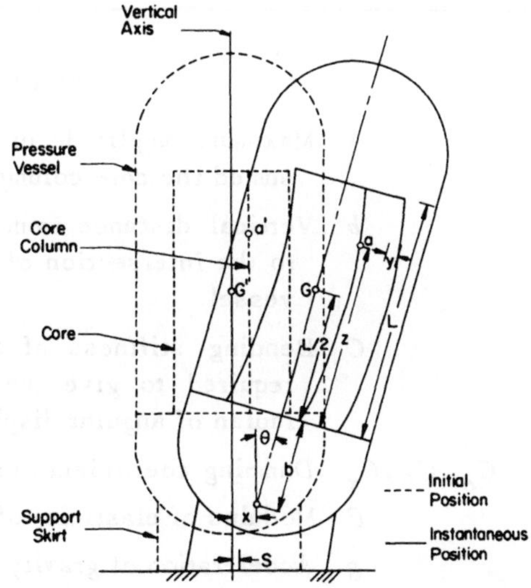

pressure vessel and piping are not specifically taken into account. The core is assumed to consist of a series of separate simply supported beams (or columns) under lateral loading and with identical geometries. Any structures enclosing the pressure vessel are assumed to follow the motion of the earth, with very little or no buildup of ground acceleration.

EQUATIONS OF MOTION WITHOUT DAMPING

The simplified model, which includes only the support skirt, pressure vessel, and core, is shown in Fig. 1. The pressure vessel is assumed to have horizontal and rotational motions, with the top and bottom of the core rigidly fixed to the pressure vessel. As noted above, it is assumed that the columns of which the core consists deflect together as beams simply supported at the ends; hence it is assumed they deflect into a sine curve of the form

$$y = A \sin \pi z/L \ , \tag{1}$$

where A is a function of time. This freedom of movement is illustrated in Fig. 1 by a', which is the location of a point in the core before disturbance, and by a, which is the instantaneous location of this point after disturbance. Also, G' and G illustrate the instantaneous displacement of the center of gravity of the pressure vessel. The absolute horizontal motions for a and G are $x + (b + z)\theta + y$ and $x + (b + L/2)\theta$ respectively. Figure 2 is a horizontal-force moment diagram for the reactor showing external and inertia forces. Deflections are considered small. The equations of motion are developed first without considering damping, and the damping effects are then included.

NOTATION

A Maximum amplitude of the sine wave into which it is assumed the core columns bend (function of time)

b Vertical distance from the lower end of the core columns to the intersection of the supporting skirt with the reactor vessel

C Bending stiffness of the supporting skirt – the moment required to give the top rim of the supporting skirt 1 radian of angular displacement about any diameter

C_A, C_θ, C_x Damping coefficients for the A, θ, and x degrees of freedom

E Modulus of elasticity of core material

g Acceleration of gravity

G Location of the average center of gravity of the entire reactor; it is assumed to be at the center of gravity of the pressure vessel

h Distance between the plane of the top rim of support skirt and the point G

I Moment of inertia of a cross section of a core column with respect to the neutral axis of the cross section

I_c Mass moment of inertia of the core about a horizontal axis through G

I_G Mass moment of inertia of the pressure vessel and core about a horizontal axis through G

I_p Mass moment of inertia of the pressure vessel about a horizontal axis through G

k_1 Shear stiffness of the supporting skirt – the force required to displace the top rim of the supporting skirt a unit distance horizontally with respect to the bottom rim

L Length of the core columns

M Total mass of the reactor

m_1 Mass of the reactor excluding the core

m_2 Mass of the reactor core

n Number of core columns

S Absolute horizontal displacement of the earth during an earthquake

x Absolute horizontal displacement of the intersection of the supporting skirt with the reactor vessel, positive to the right (function of time)

y Displacement of any point on a core column measured perpendicular to a straight line through the upper and lower ends of the column, positive to the right

z Axis along the length of the undeflected core column measured from the lower end of the columns, positive upward

ρ Mass of the reactor core per unit length (m_2/L)

θ Absolute angular rotation of the axis of the reactor pressure vessel (function of time)

δ_x Damping factor for the x degree of freedom, that is, the ratio of the damping present to the critical damping in the x coordinate

δ_A Damping factor for the A degree of freedom, that is, the ratio of the damping present to the critical damping in the A coordinate

δ_θ Damping factor for the θ degree of freedom, that is, the ratio of the damping present to the critical damping in the θ coordinate

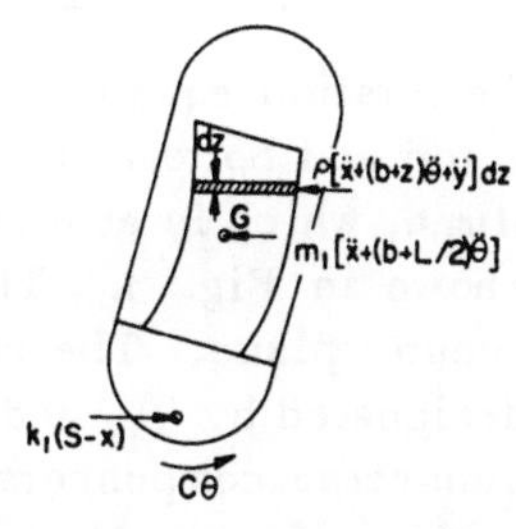

Fig. 2. Horizontal Force-Moment Diagram for the Pressure Vessel and Core.

Summing the horizontal forces in Fig. 2 yields

$$k_1(S - x) = m_1 \left[\ddot{x} + \left(b + \frac{L}{2} \right) \ddot{\theta} \right] + \int_0^L \rho[\ddot{x} + (b + z)\ddot{\theta} + \ddot{y}]\, dz \,, \tag{2}$$

where the dots denote derivatives with respect to time. Since $y = A \sin \pi z/L$,

$$\rho \int_0^L [\ddot{x} + (b + z)\ddot{\theta} + \ddot{y}]\, dz = \rho L \left[\ddot{x} + \left(b + \frac{L}{2} \right) \ddot{\theta} + \frac{2}{\pi} \ddot{A} \right] , \tag{3}$$

and because $\rho L = m_2$, Eq. (2) becomes

$$M\ddot{x} + Mh\ddot{\theta} + \frac{2m_2}{\pi} \ddot{A} + k_1 x = k_1 S \,, \tag{4}$$

where $h = b + L/2$.

Each column is assumed to be simply supported at the top and bottom and, in dynamic equilibrium, should satisfy the vibrating-beam equation. Thus,

$$EI\frac{\partial^4 y}{\partial z^4} = -\frac{\rho}{n}[\ddot{x} + (b+z)\ddot{\theta} + \ddot{y}]\ , \tag{5}$$

in which n is the number of columns. When the expression for y is substituted into this equation, it becomes

$$\frac{EI\pi^4}{L^4} A \sin\frac{\pi z}{L} = -\frac{\rho}{n}\left[\ddot{x} + (b+z)\ddot{\theta} + \ddot{A}\sin\frac{\pi z}{L}\right]\ . \tag{6}$$

No attempt is made to satisfy the equation at each point z along the axis of the columns. Instead, the total forces along the beams are made to be in equilibrium; that is, both sides of Eq. (6) are integrated with respect to z from 0 to L. The resulting equation is then solved exactly. Thus,[3]

$$2EI\frac{\pi^3}{L^3} A + \frac{m_2}{n}\left[\ddot{x} + \left(b + \frac{L}{2}\right)\ddot{\theta} + \frac{2\ddot{A}}{\pi}\right] = 0\ , \tag{7}$$

or

$$\frac{m_2}{2n}\ddot{x} + \frac{m_2 b}{2n}\ddot{\theta} + \frac{m_2}{n\pi}\ddot{A} + EI\frac{\pi^3}{L^3} A = 0\ . \tag{8}$$

Before the rotation equation is derived, the equations for the dynamic shear forces at the top and bottom of the columns will be found. A free-body diagram of the ith column, which is at a distance q_i from the center plane of the pressure vessel, is shown in Fig. 3. The coordinate q_i is considered positive to the right of the center plane. The shearing forces at the top and bottom of the ith column are designated by H_{1i} and H_{2i} respectively. Since the displacements are small, the transverse components of the horizontal forces are considered equal to the forces themselves. Also, the vertical components of displacement of an element of a column are taken to equal $q\theta$ rather than $(q + y)\theta$, since the product $y\theta$ is of the second order. In the remaining derivation all nonlinear terms are neglected. However, the effects of the nonlinearities were carefully examined and found to be negligible.

The overturning moment due to the shift in the location of the center of gravity of the core is negligible. Also, changes in the moment of inertia of the core due to deflection of the core columns are neglected. A summation of the horizontal forces in Fig. 3 yields

$$H_{1i} + H_{2i} = \frac{\rho}{n}\int_0^L [\ddot{x} + (b+z)\ddot{\theta} + \ddot{y}]\, dz\ . \tag{9}$$

[3]The variational procedure with the same assumptions yields a slightly different Eq. (8). The coefficients of the coupling terms, $m_2/2n$ and $m_2 b/2n$, are $4m_2/\pi^2 n$ and $4m_2 b/\pi^2 n$, respectively, by the variational procedure. Neither equation is exactly correct.

Fig. 3. Free-Body Diagram of a Core Column.

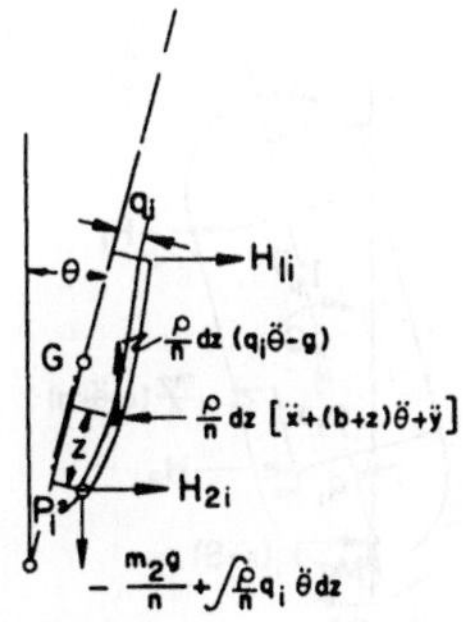

As before, $\ddot{y} = \ddot{A} \sin \pi z/L$ and $\rho L = m_2$, so that Eq. (9) becomes

$$H_{1i} + H_{2i} = \frac{m_2}{n}\left[\ddot{x} + \left(b + \frac{L}{2}\right)\ddot{\theta} + \frac{2\ddot{A}}{\pi}\right] . \tag{10}$$

The summation of moments about point P_i, neglecting secondary effects due to the column bending, yields

$$H_{1i} L = \frac{\rho}{n}\int_0^L [\ddot{x} + (b + z)\ddot{\theta} + \ddot{y}]z \, dz + \frac{\rho}{n}\int_0^L (q_i \ddot{\theta} - g)y \, dz . \tag{11}$$

Hence,

$$H_{1i} = \frac{m_2}{n}\left[\frac{\ddot{x}}{2} + \left(\frac{b}{2} + \frac{L}{3}\right)\ddot{\theta} + \frac{\ddot{A}}{\pi} + \frac{2(q_i\ddot{\theta} - g)A}{\pi L}\right] . \tag{12}$$

Equations (10) and (12) may be combined to give

$$H_{2i} = \frac{m_2}{n}\left[\frac{\ddot{x}}{2} + \left(\frac{b}{2} + \frac{L}{6}\right)\ddot{\theta} + \frac{\ddot{A}}{\pi} - \frac{2(q_i\ddot{\theta} - g)A}{\pi L}\right] . \tag{13}$$

If H_1 is the total shear force at the top of the core and H_2 at the bottom, then

$$H_1 = \sum_i H_{1i} , \tag{14}$$

$$H_2 = \sum_i H_{2i} . \tag{15}$$

Thus, because of symmetry, these shear forces are

$$H_1 = m_2\left[\frac{\ddot{x}}{2} + \left(\frac{b}{2} + \frac{L}{3}\right)\ddot{\theta} + \frac{\ddot{A}}{\pi} - \frac{2gA}{\pi L}\right] , \tag{16}$$

$$H_2 = m_2\left[\frac{\ddot{x}}{2} + \left(\frac{b}{2} + \frac{L}{6}\right)\ddot{\theta} + \frac{\ddot{A}}{\pi} + \frac{2gA}{\pi L}\right] . \tag{17}$$

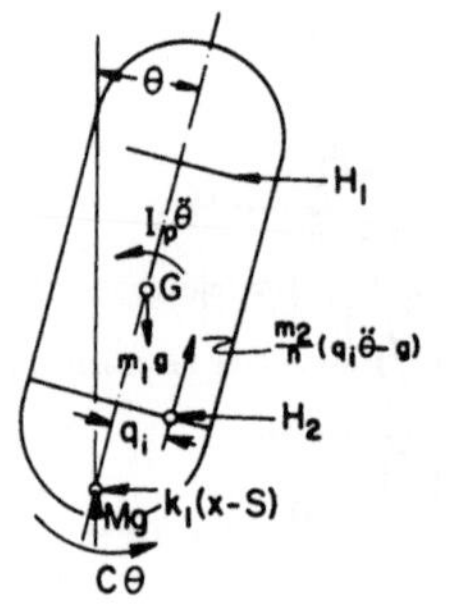

Fig. 4. Free-Body Diagram of the Pressure Vessel.

By use of the above expressions for H_1 and H_2, the equation of rotational motion will now be derived. From the free-body diagram in Fig. 4 the moments about the center of gravity G of the pressure vessel are summed. This gives

$$I_p\ddot{\theta} + C\theta + H_1\frac{L}{2} - H_2\frac{L}{2} - k_1(x - S)\left(b + \frac{L}{2}\right) - Mg\theta\left(b + \frac{L}{2}\right)$$
$$+ \frac{m_2}{n}\left[\sum_i (q_i\ddot{\theta} - g)q_i\right] = 0\ . \quad (18)$$

By substituting for H_1 and H_2, Eq. (18) becomes

$$I_p\ddot{\theta} + m_2\left(\frac{L^2}{12} + \frac{1}{n}\sum_i q_i^2\right)\ddot{\theta} + (C - Mgb)\theta - k_1 bx - \frac{2m_2 gA}{\pi} = -k_1 bS\ . \quad (19)$$

In Eq. (19),

$$m_2\left[\frac{L^2}{12} + \frac{1}{n}\sum_i q_i^2\right]\ddot{\theta} = \left\{\frac{m_2}{n}\left[\sum_i\left(\frac{L^2}{12} + q_i^2\right)\right]\right\}\ddot{\theta}\ . \quad (20)$$

Considering the graphite column as a thin rod, $(m_2/n)(L^2/12 + q_k^2)$ is the moment of inertia of the kth column about the center of gravity of the core. Thus, Eq. (19) becomes

$$(I_p + I_c)\ddot{\theta} + (C - Mgb)\theta - k_1 bx - \frac{2m_2 gA}{\pi} = -k_1 bS\ , \quad (21)$$

or

$$I_G\ddot{\theta} + (C - Mgb)\theta - k_1 bx - \frac{2m_2 gA}{\pi} = -k_1 bS\ . \quad (22)$$

EQUATIONS OF MOTION WITH DAMPING

In the previous section an undamped system was considered and the corresponding equations were derived. It will now be assumed that the system

Fig. 5. Damping Diagram for the Pressure Vessel and Core.

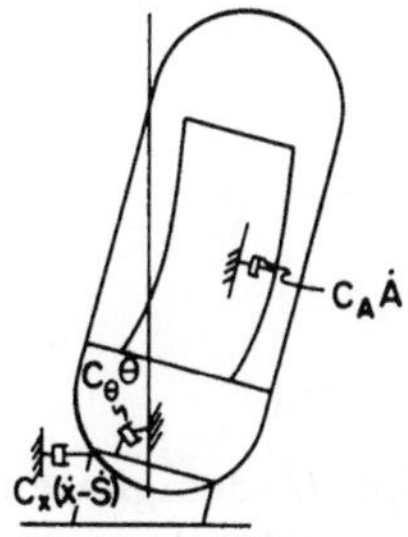

is subjected to viscous damping as indicated in Fig. 5. For each equation of motion the damping coefficient is of the form $2\delta\sqrt{KM}$, where M is the mass of the system described by the equation, K is the spring constant, and δ is the ratio of the actual damping to the critical damping. Thus, in order to include the effects of viscous damping on the equations of motion [Eqs. (4), (8), and (22)], the x coordinate is coupled to the S motion by a retarding force $C_x(\dot{x} - \dot{S})$, the bending vibration is damped by a force $C_A \dot{A}$, and rotation is damped by a force $C_\theta \dot{\theta}$. The coefficients C_x, C_A, and C_θ are given by

$$C_x = 2\delta_x \sqrt{k_1 M}\,, \tag{23}$$

$$C_A = 2\delta_A \sqrt{\frac{EI\pi^3}{L^3} \cdot \frac{m_2}{n\pi}}\,, \tag{24}$$

$$C_\theta = 2\delta_\theta \sqrt{(C - Mgh)I_G}\,, \tag{25}$$

where δ_x, δ_A, and δ_θ are the damping factors and are defined as the ratio of the damping present to the critical damping for the x, A, and θ degrees of freedom respectively.

Thus, the equations of motion become

$$M\ddot{x} + C_x \dot{x} + k_1 x + Mh\ddot{\theta} + \frac{2m_2}{\pi} \ddot{A} = k_1 S + C_x \dot{S}\,, \tag{26}$$

$$\frac{m_2}{n\pi} \ddot{A} + C_A \dot{A} + \frac{EI\pi^3}{L^3} A + \frac{m_2}{2n} \ddot{x} + \frac{m_2}{2n} h\ddot{\theta} = 0\,, \tag{27}$$

$$I_G \ddot{\theta} + C_\theta \dot{\theta} + (C - Mgh)\theta - k_1 hx - \frac{2m_2 gA}{\pi} = -k_1 hS\,. \tag{28}$$

SOLUTION OF THE SYSTEM OF EQUATIONS

The system of equations may be solved by the so-called classical method [1]. By this method the natural frequencies are determined, and the transient and steady-state portions of the solution may be separated. The remainder of this section is devoted to a discussion of the solution of the system of equations.

The homogeneous part of the solution is considered first. It is assumed that

$$x_H = D' e^{\alpha t} , \tag{29}$$

$$A_H = F' e^{\alpha t} , \tag{30}$$

$$\theta_H = G' e^{\alpha t} , \tag{31}$$

where subscript H designates the homogeneous portion, and D', F', G', and α are constants to be determined. When these expressions are substituted into Eqs. (26)–(28), the following system of linear simultaneous equations in D', F', and G' is obtained:

$$\begin{bmatrix} M\alpha^2 + C_x \alpha + k_1 & \dfrac{2m_2}{\pi}\alpha^2 & Mb\alpha^2 \\ \dfrac{m_2}{2n}\alpha^2 & \dfrac{m_2}{n\pi}\alpha^2 + C_A \alpha + \dfrac{EI\pi^3}{L^3} & \dfrac{m_2 b\alpha^2}{2n} \\ -k_1 b & -\dfrac{2m_2 g}{\pi} & I_G \alpha^2 + C_\theta \alpha + (C - Mgb) \end{bmatrix} \begin{bmatrix} D' \\ F' \\ G' \end{bmatrix} = \begin{bmatrix} 0 \\ 0 \\ 0 \end{bmatrix} . \tag{32}$$

For a nontrivial solution to exist, the determinant of the matrix must be zero. This yields a sixth-degree polynomial in α and is called the characteristic equation. Thus,

$$P_1 \alpha^6 + P_2 \alpha^5 + P_3 \alpha^4 + P_4 \alpha^3 + P_5 \alpha^2 + P_6 \alpha + P_7 = 0 , \tag{33}$$

where

$$P_1 = (Mm_2 - m_2^2) \frac{I_G}{n\pi} , \tag{34}$$

$$P_2 = \frac{m_2 I_G}{n\pi} C_x + MI_G C_A + [Mm_2 - m_2^2] \frac{C_\theta}{n\pi} , \tag{35}$$

$$P_3 = \left(\frac{MEI\pi^3}{L^3} + \frac{k_1 m_2}{n\pi} \right) I_G + (Mn_2 - m_2^2) \frac{(C - Mgb) + k_1 b^2}{n\pi}$$

$$+ I_G C_x C_A + MC_A C_\theta + \frac{m_2}{n\pi} C_x C_\theta , \tag{36}$$

$$P_4 = \left[\frac{EI\pi^3}{L^3} I_G + \frac{m_2}{n\pi} (C - Mgb + m_2 gb) \right] C_x + [k_1 I_G + M(C - Mgb) + k_1 Mb^2] C_A$$

$$+ \left(\frac{MEI\pi^3}{L^3} + k_1 \frac{m_2}{n\pi} \right) C_\theta + C_x C_A C_\theta , \tag{37}$$

$$P_5 = [k_1 I_G + M(C - Mgh)] \frac{EI\pi^3}{L^3} + k_1 \frac{m_2}{n\pi} (C - Mgh + m_2 gh) + k_1 \frac{Mh^2 EI\pi^3}{L^3}$$

$$+ \frac{EI\pi^3}{L^3} C_x C_\theta + k_1 C_A C_\theta + (C - Mgh) C_x C_A \, , \quad (38)$$

$$P_6 = \frac{EI\pi^3}{L^3} (C - Mgh) C_x + k_1 (C - Mgh) C_A + k_1 \frac{EI\pi^3}{L^3} C_\theta \, , \quad (39)$$

$$P_7 = k_1 \frac{EI\pi^3}{L^3} (C - Mgh) \, . \quad (40)$$

Each root of the polynomial guarantees the existence of a solution. Hence, the complete homogeneous solution becomes

$$x_H = \sum_{j=1}^{6} D_j e^{\alpha_j t} \, , \quad (41)$$

$$A_H = \sum_{j=1}^{6} F_j e^{\alpha_j t} \, , \quad (42)$$

$$\theta_H = \sum_{j=1}^{6} G_j e^{\alpha_j t} \, . \quad (43)$$

The 18 constants of integration are reduced to six by solving two of the equations in the system for two of the unknowns, D_j and F_j, for example, in terms of the third, G_j. This gives

$$D_j = \frac{\dfrac{m_2^2 gh\alpha_j^2}{n\pi} + \left(\dfrac{m_2 \alpha_j^2}{n\pi} + C_A \alpha_j + \dfrac{EI\pi^3}{L^3} \right) (I_G \alpha_j^2 + C_\theta \alpha_j + C - Mgh)}{k_1 h \left(\dfrac{m_2 \alpha_j^2}{n\pi} + C_A \alpha_j + \dfrac{EI\pi^3}{L^3} \right) - \dfrac{m_2^2 g\alpha_j^2}{n\pi}} G_j \, , \quad (44)$$

$$F_j = - \frac{m_2 \alpha_j^2 (k_1 h^2 + I_G \alpha_j^2 + C_\theta \alpha_j + C - Mgh)}{2n \left[k_1 h \left(\dfrac{m_2}{n\pi} \alpha_j^2 + C_A \alpha_j + \dfrac{EI\pi^3}{L^3} \right) - \dfrac{m_2 g\alpha_j^2}{n\pi} \right]} G_j \, . \quad (45)$$

The roots of Eq. (33) occur in complex conjugate pairs of the form

$$\alpha_j = \tau_j \pm i\sigma_j \, . \quad (46)$$

When the expressions given in Eqs. (44)–(46) are used in Eqs. (41)–(43), the following set of equations is obtained:

$$x_H = \sum_{j=1}^{3} e^{\tau_j t} (H_j \cos \sigma_j t - I_j \sin \sigma_j t) , \tag{47}$$

$$A_H = \sum_{j=1}^{3} e^{\tau_j t} (J_j \cos \sigma_j t - K_j \sin \sigma_j t) , \tag{48}$$

$$\theta_H = \sum_{j=1}^{3} e^{\tau_j t} (L_j \cos \sigma_j t - M_j \sin \sigma_j t) , \tag{49}$$

where

$$\begin{aligned} H_j &= (2D_j)_R , & J_j &= (2F_j)_R , & L_j &= (2G_j)_R , \\ I_j &= (2D_j)_I , & K_j &= (2F_j)_I , & M_j &= (2G_j)_I . \end{aligned} \tag{50}$$

The subscripts R and I signify the real and imaginary parts, respectively, which are obtained by using the plus sign in Eq. (46).[4] It should be noted that the real parts of the roots of the characteristic equation are the damping exponents and that the imaginary parts are the resonant frequencies.

The forcing function S is assumed to be of the form

$$S = \sum_k S_k \cos (w_k t + \phi_k) . \tag{51}$$

Since the system of equations is linear, particular solutions are additive. Thus, only one term of Eq. (51) needs to be considered. Subscript p will denote particular solutions. If

$$S = S_0 \cos (w_0 t + \phi_0) = (S_0 \cos \phi_0) \cos w_0 t - (S_0 \sin \phi_0) \sin w_0 t , \tag{52}$$

then the particular solutions are of the form

$$x_p = N_0 \cos w_0 t + O_0 \sin w_0 t , \tag{53}$$

$$A_p = P_0 \cos w_0 t + Q_0 \sin w_0 t , \tag{54}$$

$$\theta_p = R_0 \cos w_0 t + T_0 \sin w_0 t . \tag{55}$$

By substituting Eqs. (53)–(55) into the system of Eqs. (26)–(28), a linear simultaneous system results, and it is given in the matrix form

$$\rho\beta = \gamma , \tag{56}$$

[4]A real forcing function will be used; hence, the solutions are real. The above results follow.

where

$$\rho = \begin{bmatrix} -Mw_0^2 + k_1 & C_x w_0 & -\frac{2m_2}{\pi} w_0^2 & 0 & -Mbw_0^2 & 0 \\ C_x w_0 & Mw_0^2 - k_1 & 0 & \frac{2m_2}{\pi} w_0^2 & 0 & Mbw_0^2 \\ -\frac{m_2 w_0^2}{2n} & 0 & -\frac{m_2 w_0^2}{n\pi} + \frac{EI\pi^3}{L^3} & C_A w_0 & -\frac{m_2}{2n} bw_0^2 & 0 \\ 0 & -\frac{m_2}{2n} w_0 & -C_A w_0 & -\frac{m_2}{n\pi} w_0^2 + \frac{EI\pi^3}{L^3} & 0 & -\frac{m_2 bw_0^2}{2n} \\ -k_1 b & 0 & -\frac{2m_2 g}{\pi} & 0 & -I_G w_0^2 + C - Mgb & C_\theta w_0 \\ 0 & -k_1 b & 0 & -\frac{2m_2 g}{\pi} & -C_\theta w_0 & -I_G w_0^2 + (C - Mgb) \end{bmatrix}, \tag{57}$$

$$\beta = \begin{bmatrix} N_0 \\ O_0 \\ P_0 \\ Q_0 \\ R_0 \\ T_0 \end{bmatrix}, \quad \gamma = \begin{bmatrix} S_0(k_1 \cos\phi_0 - C_x w_0 \sin\phi_0) \\ S_0(k_1 \sin\phi_0 + C_x w_0 \cos\phi_0) \\ 0 \\ 0 \\ -k_1 bS_0 \cos\phi_0 \\ k_1 bS_0 \sin\phi_0 \end{bmatrix}. \tag{58,59}$$

The solution of this system yields the particular solution. Thus, particular solutions are of the form

$$x_p = \sum_k (N_k \cos w_k t + O_k \sin w_k t), \tag{60}$$

$$A_p = \sum_k (P_k \cos w_k t + Q_k \sin w_k t), \tag{61}$$

$$\theta_p = \sum_k (R_k \cos w_k t + T_k \sin w_k t). \tag{62}$$

The complete general solution is then

$$x = \sum_{j=1}^{3} e^{\tau_j t}(H_j \cos\sigma_j t - I_j \sin\sigma_j t) + \sum_k (N_k \cos w_k t + O_k \sin w_k t), \tag{63}$$

$$A = \sum_{j=1}^{3} e^{\tau_j t}(J_j \cos\sigma_j t - K_j \sin\sigma_j t) + \sum_k (P_k \cos w_k t + Q_k \sin w_k t), \tag{64}$$

$$\theta = \sum_{j=1}^{3} e^{\tau_j t}(L_j \cos \sigma_j t - M_j \sin \sigma_j t) + \sum_k (R_k \cos w_k t + T_k \sin w_k t) . \quad (65)$$

The usual initial conditions on the general solution are

$$x = A = \theta = \dot{x} = \dot{A} = \dot{\theta} = 0 . \quad (66)$$

AN EXAMPLE – THE EXPERIMENTAL GAS-COOLED REACTOR

While the results of the previous section may be applied to a considerable variety of structures, the analysis was made specifically for the Experimental Gas-Cooled Reactor being built at Oak Ridge, Tennessee. For this structure Eqs. (26)–(28) [2] become

$$\ddot{x} + 19.083\ddot{\theta} + 0.1946\ddot{A} + 410.82\delta_x \dot{x} + 42{,}193x = 410.82\delta_x \dot{S} + 42{,}193S , \quad (67)$$

$$\ddot{x} + 19.083\ddot{\theta} + 0.6366\ddot{A} + 86.321\delta_A \dot{A} + 3043A = 0 , \quad (68)$$

$$\ddot{\theta} + 323.32\delta_\theta \dot{\theta} + 26{,}134\theta - 6546x - 0.04A = -6546S . \quad (69)$$

The damping factors are generally not known and may be considered as parameters. A reasonable range for these factors is from 0.0 to 0.4, and for most structures they are less than 0.2 [3]. Various combinations of these factors were taken, and corresponding damping exponents and damped natural frequencies were found. The results of this procedure are shown in Table 1. Theoretical analyses have shown that considerable damping is required to alter

Table 1. Natural Frequencies and Damping Exponents for Certain Damping Factors

Damping Factors (δ_x, δ_A, δ_θ respectively)	Damping Exponent, Natural Frequency (radians)		
0, 0, 0	0, 58.53	0, 102.1	0, 452.3
0.03, 0.03, 0.03	−1.318, 58.52	−3.45, 101.98	−11.89, 452.2
0.05, 0.05, 0.05	−2.196, 58.49	−5.749, 101.8	−19.81, 452.1
0.1, 0.1, 0.1	−4.388, 58.38	−11.49, 101.22	−39.64, 451.3
0.4, 0.4, 0.4	−16.72, 59.88	−45.91, 81.37	−159.4, 437.0
0.01, 0.15, 0.01	−4.896, 58.98	−9.172, 100.5	−5.155, 452.2
0.05, 0.1, 0.05	−3.772, 58.54	−8.631, 101.4	−20.24, 451.99
0.05, 0.15, 0.05	−5.367, 58.71	−11.50, 100.7	−20.66, 451.9
0.1, 0.2, 0.1	−7.567, 58.60	−17.24, 99.52	−40.47, 450.9
0.15, 0.2, 0.1	−7.828, 58.44	−18.27, 99.54	−53.98, 449.6

noticeably the natural frequencies of structures of this type. This is borne out in Table 1. Using accelerograms from recorded quakes, Biot [4] determined the response of a single-degree-of-freedom ideal oscillator with varying natural periods and found that the maximum acceleration occurred when the period of the oscillator was between 0.1 and 0.5 sec. Particular attention was given to this range of period in the analysis. The horizontal ground acceleration in the Oak Ridge area was set at 5% of the acceleration of gravity.[5]

While Eq. (51), considered as a Fourier series, could be made to describe any random motion, such a consideration may not be necessary. Should the impressed frequency coincide with a natural frequency, amplitudes of the forced vibration should approach maximum values. The ground accelerations associated with an earthquake may be larger than those predicted for a given region, and the most important feature of the structure is that there must not be large magnifications of the displacement amplitudes when resonances occur. Hence, a forcing function with a frequency near that of the natural frequency for the structure was used in this investigation.

In the solution that follows, the damping factors were 5, 15, and 5% of critical for the x, A, and θ coordinates respectively.[6] For this case the natural frequencies are 9.3, 16.0, and 71.9 cps (58.71, 100.7, and 451.9 radians/sec). For the forcing function (Eq. 51), S_0, w_0, and ϕ_0 were chosen so that resonance existed at a natural frequency, and the maximum acceleration was $0.05g$.

Solutions were obtained by the analytical method previously described and also on a Donner 3100 analog computer.[7] For the forcing function

$$S = -4.670 \times 10^{-4} \cos 58.71t \; ; \qquad (70)$$

the displacements which were given by each method of solution are plotted for each of the coordinates in Figs. 6–8. For the analytical solution the transient, steady-state, and general solutions are given. These results compare very favorably, and differences may be attributed to the inherent inaccuracy of the analog computer. In Fig. 9 the analog solution given in Fig. 6 is shown for an extended time interval. Curves for accelerations and shearing stresses at the top and bottom of the core were also obtained.

Solutions were found for several other forcing functions. Comparisons of the ratios of maximum steady-state displacements and accelerations to the ground displacements and accelerations for frequencies of 58.71 and 100.7 radians/sec are shown in Table 2 for both methods of solution.

The magnitudes of the stresses and displacements for this example were small despite the selection of a frequency very near the natural frequencies of the system.

---

[5]This was the value suggested by the U.S. Coast and Geodetic Survey.

[6]These values seemed reasonable after considering ref [5].

[7]The authors wish to acknowledge the contribution of B. Y. Cotton, who performed most of the numerical calculations, and M. W. Milligan, who set up the problem on the analog computer.

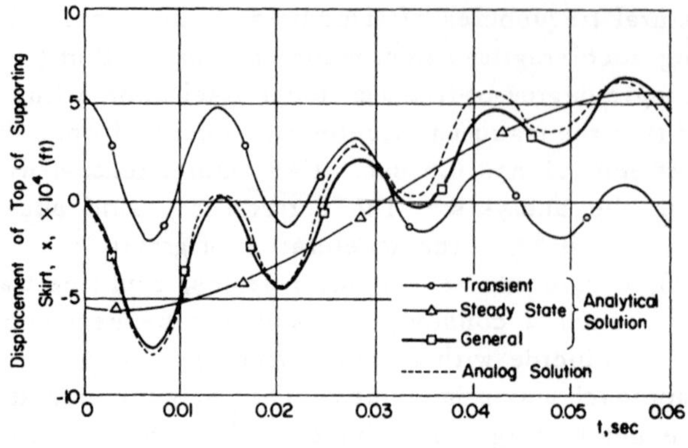

Fig. 6. Time-Dependent Displacement of the Top of the Supporting Skirt.

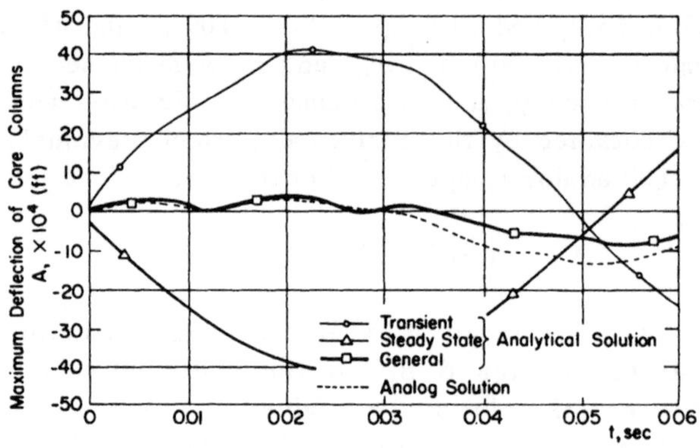

Fig. 7. Time-Dependent Maximum Deflection in Core Columns.

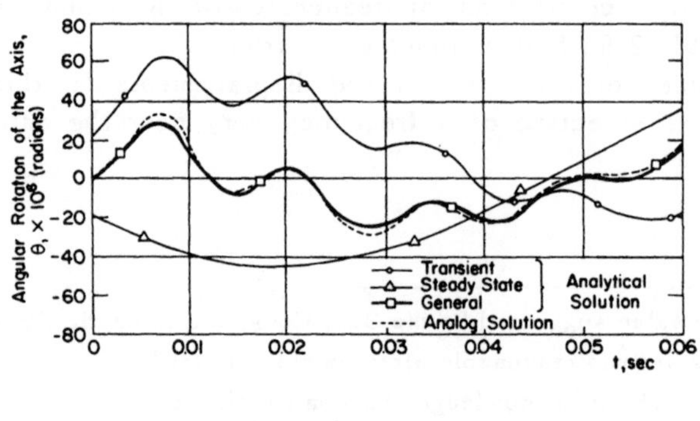

Fig. 8. Time-Dependent Angular Rotation of the Axis of the Pressure Vessel.

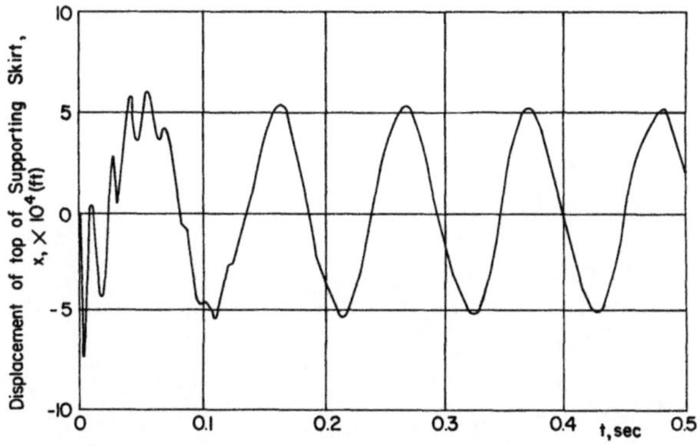

Fig. 9. Time-Dependent Displacement of the Top of the Supporting Skirt (Analog Solution over Extended Time Interval).

Table 2. Comparison of the Ratios of the Maximum Steady-State Displacements and Accelerations for Frequencies of 58.71 and 100.7 radians/sec

Frequency (radians/sec)	$\frac{\max x}{\max S}$ and	$\frac{\max \ddot{x}}{\max \ddot{S}}$	$\frac{\max A}{\max S}$ and	$\frac{\max \ddot{A}}{\max \ddot{S}}$	$\frac{\max \theta \text{ (radians)}}{\max S}$ and	$\frac{\max \ddot{\theta} \text{ (radians/sec}^2)}{\max \ddot{S}}$
	Analog Solution	Exact Solution	Analog Solution	Exact Solution	Analog Solution	Exact Solution
58.71[a]	1.10	1.19	9.05	8.4	0.120	0.097
100.7[b]	0.90	0.93	6.07	5.05	0.103	0.095

[a]The analog results are for 58.43 radians/sec.
[b]The analog results are for 100.5 radians/sec.

REFERENCES

[1] T. von Karman and M. A. Biot, *Mathematical Methods*, pp 37–42, McGraw-Hill, New York, 1940.

[2] *Experimental Gas-Cooled Reactor Preliminary Hazards Summary Report*, ORO-196 (May 1959). In this report each graphite column consisted of five blocks of equal length which were fastened together by threaded graphite pins. However, monolithic columns are actually being used.

[3] F. B. Seely and N. E. Ensign, *Analytical Mechanics for Engineers*, 4th ed., pp 368–79, Wiley, New York, 1952.

[4] M. A. Biot, "Analytical and Experimental Methods in Engineering Seismology," *Trans. Am. Soc. Civil Engrs.* **108**, 365 (1945).

[5] E. Y. W. Tsui, "Aseismic Design of Structures by Rigidity Criterion," *J. Structural Division, Proc. ASCE* **85**, 81–106 (1959).

MIX
Papier aus verantwortungsvollen Quellen
Paper from responsible sources
FSC® C105338
www.fsc.org

If you have any concerns about our products,
you can contact us on
ProductSafety@springernature.com

In case Publisher is established outside the EU,
the EU authorized representative is:
Springer Nature Customer Service Center GmbH
Europaplatz 3, 69115 Heidelberg, Germany

Printed by Libri Plureos GmbH
in Hamburg, Germany